Optimization of Stochastic Systems

Topics in Discrete-Time Dynamics
Second Edition

This is a volume in
ECONOMIC THEORY, ECONOMETRICS,
AND MATHEMATICAL ECONOMICS

Consulting Editor: Karl Shell, *Cornell University*

A list of recent titles in this series appears at the end of this volume.

Optimization of Stochastic Systems

Topics in Discrete-Time Dynamics, Second Edition

Masanao Aoki

Computer Science Department
School of Engineering and Applied Science
University of California, Los Angeles
Los Angeles, California

ACADEMIC PRESS, INC.
Harcourt Brace Jovanovich, Publishers
Boston San Diego New York
Berkeley London Sydney
Tokyo Toronto

ACADEMIC PRESS, INC.
1250 Sixth Avenue, San Diego, CA 92101

United Kingdom Edition published by
ACADEMIC PRESS INC. (LONDON) LTD.
24–28 Oval Road, London NW1 7XD

Library of Congress Cataloging-in-Publication Data

Aoki, Masanao.
Optimization of stochastic systems.

(Economic theory, econometrics, and mathematical economics)
Bibliography: p.
Includes index.
1. Control theory. 2. Stochastic processes.
3. Mathematical optimization. I. Title.
II. Series.
QA402.3.A58 1989 629.8′312 88-7668
ISBN 0-12-058851-X

89 90 91 92 9 8 7 6 5 4 3 2 1
Printed in the United States of America

Table of Contents

Preface to the Second Edition

The first edition of this book was written mainly for audiences with physical science and engineering backgrounds. Nevertheless, it reached some readers with economic and management science training. Analytical training of graduate students in economics and management sciences had progressed much in the last 20 years, and many new research results and optimization algorithms have also become available. My own interest in the meantime has shifted to the analysis of dynamics and optimization problems of economic and management science origin. With these developments and changes, I decided to rewrite much of the first edition to make it more accessible to graduate students and professionals in social sciences. I have also incorporated some new analytic tools that I deem useful in analyzing the dynamic and stochastic problems which confront these readers. I hope that my efforts successfully bring intertemporal optimization problems closer to economics professionals.

New topics introduced into this second edition appear mostly in Chapters 2, 4, 5, 6, and 8. Martingales and martingale differences are introduced early in Chapter 2. Some limit theorems and asymptotic properties of linear state space models driven by martingale differences are presented. Because many excellent books are available on martingales and their limit theorems, derivations and proofs are mostly sketchy, and readers are referred to these sources.

The results in Chapter 2 are applied in Chapters 5, 6, and 8, among other places. The notion of dynamic aggregation and its relation to cointegration and error-correction models are developed in Chapter 4. Some recursive parameter estimation schemes and their statistical properties are included in Chapters 5 and 6. Here again, books devoted entirely to these topics are available in the literature, and much had to be omitted to keep the second edition to a manageable size. In an appendix to Chapter 7, a potentially very powerful tool in proving convergence of adaptive schemes is outlined. Rational expectations models and their solution methods are developed in Chapter 8 because of their wide-spread interest to economists. A very important class of problems in sequential decision problems revolves around questions of approximating nonlinear dynamics or more generally complex situations with a sequence of less complex ones. Chapter 9 does not begin to do justice to this class of problems but is included as being suggestive of works to be done.

When I first started contemplating the revision of the first edition, I benefited from a list of excellent suggestions from Rick van der Ploeg, though I did not necessarily incorporate all of his suggestions. Conversations with Thomas Sargent and Victor Solo were useful in organizing the material into the form of the second edition. I also benefited from discussions with Hashem Pesaran and correspondences with L. Broze in finalizing Chapter 8.

Some material in this book was used as lecture notes in a graduate course in the Department of Economics, University of California, Los Angeles, the winter quarter of 1987. I thank the participants in the course for many useful comments.

Los Angeles, California
November, 1988

CHAPTER 1

Deterministic Models and Their Control Problems

This book is mostly about sequential decision problems under uncertainty or imperfect information, how to analyze them, and how to construct optimal decision rules for them and so on. As such, the book mostly deals with stochastic dynamic models. This chapter is an exception, being preparatory in nature, and contains some basic facts about deterministic models and their control problems to make the material of this book more self-contained and more easily accessible to economists.

Decision problems faced by economic agents are certainly sequential (multistage) in time and must usually be solved under risk or uncertainty. The usual textbook social planner's intertemporal optimization problem quickly becomes intractable analytically when we deviate from the representative agent framework. Even a slight attempt at some realism leads us into optimization problems that defy exact analytical solutions. There are many reasons for this difficulty. Laws of dynamics could be significantly nonlinear, the functional form of the objective functions could be nonquadratic or nonlogarithmic, and so on. We later examine some aspects of analyzing stochastic sequential decision problems under uncertainty, including issues of approximate solutions and of convergence of suboptimal decision rules or learning rules to optimal ones. To set the stage, however, we start with discussions of simple deterministic decision problems and of deterministic models. Another reason for studying deterministic models is that estimation and control aspects of stochastic control can be separated under certain conditions. This separation means that controlling processes with estimated state vectors is a deterministic problem, and it can be solved separately from that of estimating unknown parameters or state variables. Decision rules can be constructed by treating the estimated as given. Even when such separation is not optimal, we may adopt this for approximate analysis. For this reason as well, we will study deterministic control problems first.

A typical deterministic infinite horizon social planner's problem is an example of the control problems we discuss later. The problem is

$$\max \sum_{t=0}^{\infty} \beta_t u(c_t)$$

subject to

$$k_{t+1} = (1 - \delta)k_t + i_t$$
$$c_t + i_t = f(k_t)$$

where β_t is a (time varying) discount factor, k_t is capital stock, i_t is investment, c_t is consumption, and $f(k_t)$ is output. Note that k_t is governed by a first-order difference equation. This is an example of state equations we will introduce shortly. Here c_t is the decision variable.

A slightly different version of this is the optimal saving problem for a representative consumer

$$\max \sum_{t=0}^{\infty} \beta_t u(c_t)$$

subject to

$$w_{t+1} + c_t = (1 + r_t)w_t,$$

where w_t is the wealth and r_t is the interest rate. The variable w_t is the state variable for this problem. A stochastic version of these and other problems is discussed in Chapters 2 and 8. Before we return to those, we first present basic ideas behind dynamic programming which is used to tackle these intertemporal optimization problems.

1.1 Dynamic Programming and State Space (Markovian) Representation

A sequential decision process or problem requires the decision maker to make decisions sequentially in time; i.e., to choose $u_0, u_1, \ldots$ in that order.* A T-horizon problem is any sequential decision problem with

* Although deterministic sequential problems with complete information can equivalently be restated as static problems, this approach does not generalize to stochastic or incomplete information problems.

T decisions, $u_0, u_1, \ldots u_{T-1}$. Infinite horizon problems require sequential choices of $u_0, u_1, \ldots$ extending to infinity. State vectors are associated with sequential decision problems. Markovian properties of dynamic models facilitate formulations of optimization problems and discussions of solution procedures. We begin by treating deterministic processes in which the Markovian structure is most easily seen. We use superscript t to denote the collection of the variable, e.g., $y^t = \{y_0, y_1, \ldots, y_t\}$.

To state sequential decision problems we need dynamic models to represent the intertemporal aspect of constraints imposed on the problems. Often a set of variables serves to characterize the dynamic aspects completely and acts as a state vector in the dynamic programming formulation we shortly discuss. In the social planner's problem, k_t is the state variable, while w_t is the state variable in the optimal saving problem. Dynamic models can be stated in several ways. The two most common ones are time-domain and transform-domain representations. The first represents the time evolution of the problems by a set of difference equations and algebraic identities for state variable x_t and decision variable u_t. In the second, the same information is captured by a set of algebraic relations for the variables in transform such as lag- or z-transforms. In dynamic programming formulation, time-domain description is more natural and dynamic constraints are set as state equations. Knowledge of the initial conditions represented by x_0 and a sequence of current and future "input" variables including all relevant exogenous ones uniquely determine the time transition of x_t to x_{t+1}, $t = 1, 2, \ldots$.

In state space form, models are specified by

$$x_{t+1} = \mathrm{f}(x_t, u_t, t), \qquad t = 0, 1, \ldots, \tag{1}$$

where x_0 is given and u_t is a set of input or decision variables. A suitably defined (sometimes function-valued) x_t can always put dynamics as the first-order difference equation. This representation is useful because it is readily extended to the usual Markovian processes in stochastic context. The next state vector x_{t+1} can be calculated from x_t and u_t (and the knowledge of time) without the need for knowing $x_{t-1}, x_{t-2}, \ldots$ or $u_{t-1}, u_{t-2}, \ldots$. In this sense x_t and u_t are sufficient for x^t, u^t. Model (1) is called Markovian for this reason. (Optimal) decision or control at time t is a function of x_t in Markovian models. In a more general model, a separate observation equation is specified in addition to the dynamic transition equation,

$$y_t = \mathrm{g}(x_t, u_t, t). \tag{2}$$

In deterministic models, however, we generally assume that x_t itself is observed directly because if x_t can't be expressed uniquely from y^t and u^{t-1}, the observation Eq. (2) is inadequate in revealing x_t which is needed to use (1) to update or propagate x_t to the next time period, x_{t+1}. We only deal with models in which this "observability" condition, to be discussed shortly, is satisfied and x_t or equivalently x_0 can be exactly known or deductible from the data. This is the full information situation in deterministic decision problems. State space equations are vector ARMA (1, 1) or as a special case VAR (1). A stochastic linear state space model then is of the form

$$x_t = Ax_{t-1} + Bu_{t-1} + \xi_t$$

$$y_t = Cx_t + \eta_t$$

where ξ_t and η_t are exogenous vectors. In special cases the term η_t may be absent. The important point to note is that there is only one time lag in the model.

The optimal value of u_t will generally depend on the whole past decisions, u^{t-1}, and on past data, y^t. A variable s_t is said to be sufficient if the optimal u_t can be calculated as the function of s_t and s_{t+1} can be calculated from s_t, y_t, and u_t without knowing u_{t-1}. In effect s_t summarizes all past decision and observation history. In deterministic Markovian models, state vectors are sufficient.

As policy π is a sequence of u_t, $t = 0, 1, \ldots$, where $u_t = \psi(x^t, u^{t-1})$ for some ψ. When ψ is independent of t, then the policy is stationary. In sequential decision problems, the "cost-to-go" from time t is the cost to be incurred under policy π conditional on the current state vector x, denoted by $\gamma_{\pi,t}(x)$. It is generated by the backward recursion $\gamma_{\pi,t}(x) = \phi[x, x_+, u, t, \gamma_{\pi,t+1}(x)]$ where, from Eq. (1), $x_+ = f(x, u, t)$ is the next state to which x is transferred by the transition dynamics $f(.,.,.)$. A special case of importance is when ϕ is linear and increasing in the last argument $\gamma_{\pi,t}(x) = w(x, x_+, u, t) + \beta_t \gamma_{\pi,t+1}(x_+)$ where $\beta_t > 0$ may also depend on x, x_+, and u_t. A finite-horizon decision problem may have the terminal cost $\psi(x_T)$ at the end of the decision horizon. The minimum cost at the beginning of time t is a function of t and the state vector at time t. Denote the minimal cost by *

$$\gamma_t^*(x) = \inf_{\pi} \gamma_{\pi,t}(x).$$

* Sometimes a more elaborate notation $\gamma_{t,T}^*(x)$ is used to indicate that $T - t$ is the time-to-go where t is the current time and T denotes the end of the planning horizon.

The Bellman principle of optimality yields

$$\gamma_t^*(x) = \inf_u [W(x, x_+, u, t) + \beta_0 \gamma_{t+1}^*(x_+)], \qquad t = 0, 1, \ldots, T-1,$$

$$\gamma_T^*(x) = \psi(x).$$

This functional equation embodies the fact that if $u_t \ldots u_{T-1}$ is optimal from t on, then the subsequent $u_{t+1}, \ldots u_{T-1}$ must be optimal from $t+1$ onward. This is also called the Bellman's functional equation.

Some problems do not have definite ending points but may terminate at a random time when conditions for termination are fulfilled. Lifetime savings and consumption plans, gambling and guidance of missiles are examples of such situations. The plan terminates when the economic agent dies and gambling ceases either when the gambler or the bank breaks. In these problems, the termination time is a random variable. In a more simple infinite-horizon problem, the horizon is indeed infinite. In such problems, one must take care not to deal with infinite cost. One way to bound the cost to be finite is to work with discounted streams of costs, another is to work with some "average" cost per period. With the infinite-horizon problem, the cost-to-go satisfies a simpler Bellman equation since a subscript is not needed from the minimal cost expression. The Bellman equation becomes

$$\gamma(x) = \inf_u [W(x, x_+, u, t) + \beta\gamma(x_+)].$$

In infinite horizon problems, we need conditions to ensure that the principle of optimality applies and that return or value functions are well defined. For example, in quadratic cost minimization problems, lower bounds of cost functions are zero. A technical condition, called reachability to be discussed in Section 4, ensures that the costs are not unbounded for a (sub) set of feasible decisions. In maximization problems, some conditions to limit the growth of value functions are needed. See Appendix at the end of this chapter. In the next example, consumption grows geometrically, but the utility function grows only linearly in t and the value function turns out to be bounded.

Example 1: Saving and Consumption. The next example is a version of the problem discussed in Whittle (1982, p. 26) using the logarithmic utility function of consumption. Suppose that x_t is the amount of asset at time t. Assume that it grows with time according to the dynamics

$$x_{t+1} = \alpha(x_t - c_t), \qquad \alpha > 1,$$

where c_t is the amount of asset consumed during the current period. The optimality equation for maximizing $\Sigma_0^T \beta^t g(c_t)$, where $g(c_t)$ is the utility of consuming c_t, is

$$J_t(x) = \max_{0 \leqslant c \leqslant x} [g(c) + \beta J_{t+1}\{\alpha(x - c)\}],$$

where $x_t = x$ and where $J_t(x) = \max \Sigma_{s=t}^T \beta^s g(u_s)$ starting from $x_t = x$. It is convenient to reexpress this equation in terms of the "time-to-go" notion and define

$$F_t(x) = J_{T-t}(x). \tag{3}$$

Then the Bellman's principle of optimality yields

$$F_t(x) = \max_{0 \leqslant c \leqslant x} \{g(c) + \beta F_{t-1}[\alpha(x - c)]\}.$$

In the infinite-horizon situation, $F_t(\cdot)$ is independent of t and the functional equation

$$F(x) = \max_{0 \leqslant c \leqslant x} \{g(c) + \beta F[\alpha(x - c)]\} \tag{4}$$

defines the discounted sum of utility streams over the infinite horizon. Assuming an internal solution, i.e., $0 < c < x$, the first-order condition for optimality is*

$$g'(c) = \alpha\beta F'[\alpha(x - c)].$$

Now adopt the logarithmic utility of consumption $g(c) = \ln c$. The infinite-horizon problem is easily solved by positing $F(x) = a \ln x + b$ and determining the constants a and b. The first condition for maximization yields $c = x/(1 + \beta a)$. On substituting this into (4), the constants must satisfy

$$a = 1 + \beta a$$

or

$$a = 1/(1 - \beta),$$

and

$$b = \beta b + \ln(1 - \beta) + \beta(1 - \beta)^{-1} \ln(\alpha\beta)$$

since $1 + \beta a = (1 - \beta)^{-1}$ and $a\beta = \beta(1 - \beta)^{-1}$ or

$$b = (1 - \beta)^{-1}[\ln(1 - \beta) + \beta(1 - \beta)^{-2} \ln(\alpha\beta)].$$

* Value functions need not be differentiable everywhere. For example, in a time optimal control problem, there are at least two points at which the contour of constant cost is not differentiable. Benveniste and Schenikman (1979) gives conditions for differentiability once.

The optimal rule is a stationary rule

$$c = (1 - \beta)x$$

under which the capital stock still grows if $\alpha\beta > 1$.

Now, returning to the finite-horizon problem, we verify by induction that

$$F_t(x) = a_t \ln x + b_t$$

and

$$c_t(x) = x/(1 + \beta a_t),$$

starting from the assumption that $F_0(x) = a_0 x$, where

$$a_t = 1 + \beta a_{t-1}$$

and

$$b_t = \beta b_t - (1 + \beta a_{t-1}) \ln(1 + \beta a_{t-1}) + \beta a_{t-1} \ln(\alpha\beta a_{t-1}), \qquad t \geqslant 1.$$

As $t \to \infty$, $a_t \to (1 - \beta)^{-1}$ and b_t approaches b.

Example 2: Optimal Growth. In a deterministic version of Brock and Mirman (1972) model productivity shock enters the production function multiplicatively. Then dynamic programming formulation of the deterministic social planning problem is

(5) $$V(k_{t-1}) = \max_{c_t} [u(c_t) + \beta V(k_t)]$$

where

$$k_t = (1 - \delta)k_{t-1} + i_t$$

and

$$c_t + i_t = f(k_{t-1}).$$

The first-order condition is obtained by differentiating the right-hand side with respect to c_t:

(6) $$u'(c_t) = \beta V'(k_{t-1}^{+}),$$

where

$$k_{t-1}^{+} = (1 - \delta)k_{t-1} + f(k_{t-1}) - c_t.$$

Equation (6) is solved for c_t as $c_t = \phi(k_{t-1})$. Substitute this into (5) to obtain

(7) $$V(k) = u(\phi(k)) + \beta V[(1 - \delta)k + f(k) - \phi(k)]$$

where the subscript is dropped.

Differentiate (7) with respect to k to obtain

$$V'(k) = u'(\phi(k))\phi'(k) + \beta V'(k^+)[(1 - \delta) + f'(k) - \phi'(k)], \tag{8}$$

where $k^+ = (1 - \delta)k + f(k) - \phi(k)$.

From (6) and (8),

$$V'(k) = u'(c_t)[1 - \delta + f'(k)].$$

We see that this and (6) lead us to the expression

$$u'(c_t) = \beta u'(c_{t+1})[(1 - \delta) + f'(k_t)].$$

(This corresponds to (2.2) of Brock and Mirman [1972]).

Exercise 1

Rework Example 1 with $g(c) = \dfrac{1}{1-\sigma} c^{1-\sigma}$, $\sigma > 0$, $\sigma \neq 1$ and $T = \infty$.

Exercise 2

Use a linear technology, $f(k) = Ak$, and logarithmic utility $u(c) = \ln c$ in Example 2 to show that capital stock grows as $k_{t+1} = \alpha k_t$ with $\alpha = (1 - \delta + A)\beta$ and the optimal consumption is $c_t = \gamma k_{t-1}$ with $\gamma = (1 - \delta + A)(1 - \beta)$.

Exercise 3

Redo Exercise 2 with $\ln c = \dfrac{1}{1-\sigma} c^{1-\sigma}$.

1.2 State Space Models

In a simple model, stacking lagged endogenous variables is all that is required to put a model in state space form. A simple example is a model

$$z_{t+2} + 2z_{t+1} + 3z_t = u_{t+1},$$

which is restated as

$$\mathbf{x}_{t+1} = \begin{pmatrix} -2 & -3 \\ 1 & 0 \end{pmatrix} \mathbf{x}_t + \begin{pmatrix} 1 \\ 0 \end{pmatrix} u_t$$

by defining

$$\mathbf{x}_t = \begin{pmatrix} z_t \\ z_{t-1} \end{pmatrix} \quad \text{and} \quad z_t = (1 \quad 0)\mathbf{x}_t.$$

An essential feature of state space representation is that state vectors appear with only a unit lag and that exogenous variables at a single time point, such as at t, $t-1$ or $t+1$, appear in the dynamic equations.

Economic models are often specified by a set of (vector) autoregressive moving-average models with exogenous variables, called (V)ARMAX models for short. These models can be converted into state space models in several equivalent ways. A standard procedure is illustrated by the next univariate example. For further detail, see Aoki (1976, Chapter 2; 1987, Chapter 4) when y_t is a vector.

Example 3: ARMA to State Space Representation. Consider an ARMA (3, 2) model

$$y_t = \phi_1 y_{t-1} + \phi_2 y_{t-2} + \phi_3 y_{t-3} + b_0 x_t + b_1 x_{t-1} + b_1 x_{t-1} + b_2 x_{t-2}.$$

Define

$$z_{1t} = y_t - b_0 x_t,$$

and

$$z_{2t-1} = z_{1t} - \phi_1 y_{t-1} - b_1 x_{t-1}.$$

Substituting y_t out from the first equation of this example, this relation becomes

$$z_{1t} = \phi_1 z_{1t-1} + z_{2t-1} + \gamma_1 x_{t-1}$$

where

$$\gamma_1 = b_1 + \phi_1 b_0.$$

Next define

$$\begin{aligned} z_{3t-1} &= z_{2t} - \phi_2 y_{t-1} - b_2 x_{t-1} \\ &= z_{2t} - \phi_2 z_{1t} - \gamma_2 x_{t-1} \end{aligned}$$

where

$$\gamma_2 = b_2 + \phi_2 b_0.$$

Then z_{3t} becomes

$$\begin{aligned} z_{3t} &= z_{2t+1} - \phi_2 y_t - b_2 x_t \\ &= z_{1t+2} - \phi_1 y_{t+1} - b_1 x_{t+1} - \phi_2 y_t - b_2 x_t \\ &= y_{t+2} - \phi_1 y_{t+1} - \phi_2 y_t - b_0 x_{t+2} - b_1 x_{t+1} - b_2 x_t \\ &= \phi_3 y_{t-1} \\ &= \phi_3 (z_{1t} + b_0 x_t). \end{aligned}$$

The state space model is obtained by collecting the preceding as

$$\begin{bmatrix} z_1 \\ z_2 \\ z_3 \end{bmatrix}_t = \begin{bmatrix} \phi_1 & 1 & 0 \\ \phi_2 & 0 & 1 \\ \phi_3 & 0 & 0 \end{bmatrix} \begin{bmatrix} z_1 \\ z_2 \\ z_3 \end{bmatrix}_{t-1} + \begin{bmatrix} \gamma_1 \\ \gamma_2 \\ \gamma_3 \end{bmatrix} x_t,$$

where $\gamma_3 = \phi_3 b_0$, and relating y_t to the state variables by

$$y_t = (1 \quad 0 \quad 0)\begin{pmatrix} z_1 \\ z_2 \\ z_3 \end{pmatrix}_t + b_0 x_t.$$

The state vector is $(z_1 \quad z_2 \quad z_3)'_t$.

1.3 Dynamic Modes

Each pair of eigenvalues and eigenvectors of a matrix **A** in a linear model

$$\mathbf{x}_{t+1} = \mathbf{A}\mathbf{x}_t$$

may be thought of as representing a mode of dynamic responses or behavior of the model, a particular mode of response being "excited" by an initial condition vector $\mathbf{x}_0$. We refer to matrix **A** as a dynamic matrix to distinguish it from other matrices in dynamic models we introduce later. We can state the notion of modes more concretely when the dynamic matrix **A** has n linearly independent eigenvectors by expanding **A** into its polar decomposition form

$$\mathbf{A} = \sum_i \lambda_i \mathbf{u}_i \mathbf{v}'_i.$$

Here u_i is the normalized eigenvector of the eigenvalue λ_i, and v'_i is the ith row of the matrix $[u_1 \ldots u_n]^{-1}$. Then the initial condition vector $\mathbf{x}_0$ such that $\mathbf{v}'_i \mathbf{x}_0 = 0$ for all i except k causes $\mathbf{x}_t$ to be proportional to $\mathbf{u}_k$

$$\mathbf{x}_t = \lambda_k^t \mathbf{u}_k (\mathbf{v}'_k \mathbf{x}_0), \qquad t = 1, 2, \ldots$$

This is the kth modal response of the dynamics. The kth mode is asymptotically stable if the kth eigenvalue is less than one in magnitude. More generally, when n linearly independent vectors do not exist, we use generalized eigenvectors and the Jordan canonical form or the Schur decomposition forms instead. More will be said about the magnitudes of $\mathbf{x}_t$ or of some functions of state vectors later in Chapter 2.

In AR or ARMA models for a univariate series $\{y_t\}$, the roots of the AR parts determine the dynamic modes because they are the eigenvalues of the associated dynamic matrix. This relation is easily seen by an example. Suppose $\phi(L)y_t = u_t$, where $\phi(L) = 1 - \phi_1 L - \phi_2 L^2 - \phi_3 L^3$. We can convert this model to state space form by defining the stacked vector $(y_t, y_{t-1}, y_{t-2})'$ as a state vector:

$$\begin{bmatrix} y_t \\ y_{t-1} \\ y_{t-2} \end{bmatrix} = \begin{bmatrix} \phi_1 & \phi_2 & \phi_3 \\ 1 & 0 & 0 \\ 0 & 1 & 0 \end{bmatrix} \begin{bmatrix} y_{t-1} \\ y_{t-1} \\ y_{t-3} \end{bmatrix} + \begin{bmatrix} u_t \\ 0 \\ 0 \end{bmatrix}. \tag{9}$$

Now the preceding dynamic matrix can easily be verified to have the characteristic polynomial $\phi(\lambda) = 1 - \phi_1\lambda - \phi_2\lambda^2 - \phi_3\lambda^3$. When $\phi(L)$ is factored as $\phi(L) = \Pi_{i=1}^3 (1 - \alpha_i L)$, then $1/\alpha_i$, $i = 1, 2, 3$, are the eigenvalues of the dynamic matrix of the state space representation.

In this representation $y_t = (1 \quad 0 \quad 0)(y_t y_{t-1} y_{t-2})'$ and it is inconvenient to apply to the Kalman filter because there is no disturbance term on the observation equation. We convert this into another state space form using procedure introduced earlier. This seemingly roundabout way of introducing state variables leads to a state representation that is amenable directly to prediction by Kalman filter. Let $x_{1t} = y_t - u_t$. Advance t by one and define x_{2t} by

$$x_{1t+1} = \phi_1 y_t + x_{2t}$$

i.e., $x_{2t} = \phi_2 y_{t-1} + \phi_3 y_{t-2}$. Advance t by one and define x_{3t} by $x_{2t+1} = \phi_3 y_t + x_{3t}$, i.e., $x_{3t} = \phi_3 y_{t-1}$. Collecting the components

$$\begin{pmatrix} x_1 \\ x_2 \\ x_3 \end{pmatrix}_{t+1} = \begin{pmatrix} \phi_1 & 1 & 0 \\ \phi_2 & 0 & 1 \\ \phi_3 & 0 & 0 \end{pmatrix} \begin{pmatrix} x_1 \\ x_2 \\ x_3 \end{pmatrix}_t + \begin{pmatrix} \phi_1 \\ \phi_2 \\ \phi_3 \end{pmatrix} u_t$$

$$y_t = (1 \quad 0 \quad 0)\mathbf{x}_t + u_t.$$

This dynamic matrix also has $\phi(\lambda)$ as its characteristic polynomial. It sometimes simplifies further analysis if AR or ARMA models are converted into state space forms which display explicitly the eigenvalues, i.e., the roots of $\phi(L)$ of the AR part. Suppose the AR model is given by

$$\prod_{i-1}^{p} (1 - \alpha_i L) y_t = u_t.$$

Then, instead of using $(y_t, y_{t-1} \ldots y_{t-p+1})'$ as the state vector, put it into the partial fraction expansion form

$$y_t = \frac{1}{\prod_{i=1}^{p} (1 - \alpha_i L)} u_t = \sum_{i=1}^{p} \frac{r_i}{1 - \alpha_i L} u_t, \tag{10}$$

where r_i is the residual, and define x_{it} by

$$(1 - \alpha_i L) x_{it} = r_i u_t, \qquad i = 1 \ldots p. \tag{11}$$

For simpler exposition, assume that the roots are all distinct. Let $z_{it} = x_{it} - r_i u_t$ in (11). Then $z_{it+1} = \alpha_i x_{it} = \alpha_i (z_{it} + r_i u_t)$ or

$$z_{it+1} = \alpha_i z_{it} + \alpha_i r_i u_t,$$

and

$$x_{it} = z_{it} + r_i u_t$$

and from (10)

$$y_t = \Sigma x_{it} = \Sigma z_{it} + r u_t$$

where

$$r = \sum_{i=1} r_i.$$

Written jointly, the AR model in state space representation is given by

$$z_{t+1} = D z_t + b u_t$$

and

$$y_t = e' z_t + r u_t,$$

where

$$D = \mathrm{diag}\,(\alpha_1 \ldots \alpha_p), \qquad b' = (\alpha_1 r_1, \ldots, \alpha_p r_p).$$

Each term z_{it} is the state variable associated with the ith dynamic mode of the model, $\phi(L) y_t = u_t$. This procedure can be extended to multivariate series, and it is known as Gilbert's method. See Aoki (1987a, p. 57) for further particulars of the method. Decomposition of stochastic dynamics by partial fraction expansion turns out to be useful. See Appendix B at the end of the chapter.

1.4 Some System Properties

We touched on stability, which is one of the important properties of dynamic models. There are two more properties—observability and reachability—which play important roles as technical conditions, i.e., to insure finiteness of cost in decision problems.

Observability

If the initial state $\mathbf{x}_0$ of the dynamic system

(12) $$\mathbf{x}_{k+1} = \mathbf{A}\mathbf{x}_k, \qquad k = 0, \ldots$$

can be determined uniquely from the data $\mathbf{y}_0, \mathbf{y}_1, \ldots, \mathbf{y}_N$ where

$$\mathbf{y}_k = \mathbf{C}\mathbf{x}_k,$$

then the system is called observable. This notion is the same as the existence of a unique solution to the algebraic equation

$$\mathbf{y}_0^+ = \mathbf{O}_N \mathbf{x}_0,$$

where

$$\mathbf{y}_0^+ = \begin{bmatrix} \mathbf{y}_0 \\ \mathbf{y}_1 \\ \vdots \\ \mathbf{y}_N \end{bmatrix} \quad \text{and} \quad \mathbf{O}_N = \begin{bmatrix} \mathbf{C} \\ \mathbf{CA} \\ \vdots \\ \mathbf{CA}^{N-1} \end{bmatrix}$$

for some N. If $\mathbf{A}$ is singular, this problem is distinct from that of determining $\mathbf{x}_0$ from past data. (This is called a constructability problem. See Appendix at the end of the chapter.)

Since $\mathbf{x}_0$ is uniquely solved out by

$$\mathbf{x}_0 = (\mathbf{O}_N'\mathbf{O}_N)^{-1}\mathbf{O}_N'\mathbf{y}_0^+$$

when O_N has full rank, we call the system observable if

$$\operatorname{rank} \mathbf{O}_N = n,$$

where $\dim \mathbf{x}_0 = n$, or equivalently, the observability gramian is positive definite for $N = n$

$$\mathbf{O}_N'\mathbf{O}_N = \sum_{i-0}^{N-1} (\mathbf{A}')^i\mathbf{C}'\mathbf{C}\mathbf{A}^i > 0. \tag{13}$$

We also say that $(\mathbf{A}, \mathbf{C})$ is an observable pair if the system is observable. We follow Hautus (1969) in saying that the pair $(\mathbf{A}, \mathbf{C})$ is observable if the matrix $\left[\begin{smallmatrix}\lambda\mathbf{I}-\mathbf{A}\\ \mathbf{C}\end{smallmatrix}\right]$ has rank n for every eigenvalue of $\mathbf{A}$, i.e., we call an eigenvalue λ of matrix $\mathbf{A}$ $(\mathbf{A}, \mathbf{C})$-observable if no nonzero vector $\mathbf{z}$ exists such that $A\mathbf{z} = \lambda\mathbf{z}$ and $\mathbf{C}\mathbf{z} = 0$. If λ is not an eigenvalue of $\mathbf{A}$, then $|\lambda\mathbf{I} - \mathbf{A}| \neq 0$, i.e., rank $\left[\begin{smallmatrix}\lambda\mathbf{I}-\mathbf{A}\\ \mathbf{C}\end{smallmatrix}\right] = n$, where n is the dimension of state vector. If λ is an eigenvalue of $\mathbf{A}$, then $|\lambda\mathbf{I} - \mathbf{A}| = 0$. But rank n condition implies that there does not exist $\mathbf{q} \neq 0$ such that $\lambda\mathbf{q} = \mathbf{A}\mathbf{q}$ and $\mathbf{C}\mathbf{q} = 0$, i.e., $(\mathbf{A}, \mathbf{C})$ must be an observable pair.

To see this last point, note the following. If $\mathbf{q} \neq 0$ such that $\lambda\mathbf{q} = \mathbf{A}\mathbf{q}$ and $\mathbf{C}\mathbf{q} = 0$, then

$$\begin{bmatrix} \mathbf{C} \\ \mathbf{CA} \\ \vdots \\ \mathbf{CA}^{n-1} \end{bmatrix}\mathbf{q} = 0 \quad \text{i.e., rank} \begin{bmatrix} \mathbf{C} \\ \mathbf{CA} \\ \vdots \\ \mathbf{CA}^{n-1} \end{bmatrix} < n, \quad \text{i.e.,}$$

$(\mathbf{A}, \mathbf{C})$ is not an observable pair. Conversely, if $(\mathbf{A}, \mathbf{C})$ is not an observable pair, choose a basis in which $\mathbf{C} = (\mathbf{C}_1 \;\; \mathbf{0})$ and $\mathbf{A} = \left(\begin{smallmatrix}\mathbf{A}_1 & \mathbf{0}\\ \mathbf{A}_{21} & \mathbf{A}_2\end{smallmatrix}\right)$. Then $\mathbf{q} = \left(\begin{smallmatrix}\mathbf{0}\\ \mathbf{q}_2\end{smallmatrix}\right)$

where $\mathbf{A}_2\mathbf{q}_2 = \lambda\mathbf{q}_2$ is such that $\mathbf{Aq} = \binom{\mathbf{0}}{\mathbf{A}_2\mathbf{q}_2} = \lambda\binom{0}{\mathbf{q}_2}$ and $\mathbf{Cq} = 0$, i.e., there exists a nonzero vector such that $\mathbf{Aq} = \lambda\mathbf{q}$ and $\mathbf{Cq} = 0$. We call the pair $(\mathbf{A}, \mathbf{C})$ detectable if all unstable eigenvalues of $\mathbf{A}$ are observable of the pair $(\mathbf{A}, \mathbf{C})$. Then either $\mathbf{A}$ is asymptotically stable or the sum (13) becomes unbounded as N goes to infinity.

If the matrix $\mathbf{A}$ is asymptotically stable, then the infinite sum in (13) exists. The sum, denoted by $\mathbf{G}_0$, satisfies the equation

$$\mathbf{A}'\mathbf{G}_O\mathbf{A} - \mathbf{G}_O = -\mathbf{C}'\mathbf{C}. \tag{14}$$

It has the unique symmetric positive definite solution if $(\mathbf{A}, \mathbf{C})$ is an observable pair and $\mathbf{A}$ is asymptotically stable.

Equations (14) and (17) are called Lyapunov equations. See Aoki (1987c, p. 251) for further facts about the Lyapunov equation.

Reachability

The dynamic system

$$\mathbf{x}_{k+1} = \mathbf{A}\mathbf{x}_k + \mathbf{B}\mathbf{u}_k,$$

where $\mathbf{u}_k$ is an exogenous (input) vector, can be solved forward in time from the initial time 0 as

$$\mathbf{x}_N = \mathbf{A}^N\mathbf{x}_0 + \mathbf{\Omega}_N\mathbf{u}_{\bar{N}-1},$$

where

$$\mathbf{\Omega}_N = [\mathbf{B}, \mathbf{AB}, \ldots, \mathbf{A}^{N-1}\mathbf{B}]$$

and

$$\mathbf{u}_{\bar{N}-1} = \begin{bmatrix} \mathbf{u}_{N-1} \\ \mathbf{u}_{N-2} \\ \vdots \\ \mathbf{u}_0 \end{bmatrix}.$$

Then any state can be reached from the origin in N steps if

$$\mathbf{x}_N = \mathbf{\Omega}_N\mathbf{u}_{\bar{N}-1} \tag{15}$$

has a solution $\mathbf{u}_{\bar{N}-1}$. If the matrix $\mathbf{\Omega}_N$ has full row rank, the solution is given by

$$\mathbf{u}_{\bar{N}-1} = \mathbf{\Omega}'_N(\mathbf{\Omega}_N\mathbf{\Omega}'_N)^{-1}\mathbf{x}_N.$$

The system is called reachable when rank $\mathbf{\Omega}_N = \dim \mathbf{x}_t$ for sufficiently large N. When $\mathbf{A}$ is singular, this notion is distinct from that of controllability which requires that $\mathbf{x}_N = 0$ from arbitrary $\mathbf{x}_0$ for some N. (This means that the origin can be reached from an arbitrary initial condition. We do not use controllability in this book.)

The rank condition can be equivalently stated as the positive definiteness of the reachability gramian for some N,

$$\boldsymbol{\Omega}_N\boldsymbol{\Omega}_N' = \sum_{i=0}^{N-1} \mathbf{A}^i\mathbf{B}\mathbf{B}'(\mathbf{A}')^i > 0. \tag{16}$$

We also say that the pair $(\mathbf{A}, \mathbf{B})$ is reachable or that $(\mathbf{A}, \mathbf{B})$ is a reachable pair when the system is reachable. An alternative characterization of reachability is given by Hautus (1969). The pair $(\mathbf{A}, \mathbf{B})$, where $\mathbf{A}$ is $n \times n$ and is reachable if and only if the matrix $[\lambda\mathbf{I} - \mathbf{A}, \mathbf{B}]$, has rank n for every eigenvalue of $\mathbf{A}$. A corollary of this is that the pair $(\mathbf{A}, \mathbf{B})$ is reachable if and only if every vector $\mathbf{y}$, such that $\mathbf{y}'\mathbf{A} = \lambda\mathbf{y}'$ and $\mathbf{y}'\mathbf{B} = 0$, is zero for all eigenvalues of $\mathbf{A}$. We say an eigenvalue λ of matrix $\mathbf{A}$ is $(\mathbf{A}, \mathbf{B})$-reachable if no nonzero vector $\mathbf{y}$ exists such that $\mathbf{y}'\mathbf{A} = \lambda\mathbf{y}'$ and $\mathbf{y}'\mathbf{B} = 0$. In other words, the system is reachable if all eigenvalues of $\mathbf{A}$ are $(\mathbf{A}, \mathbf{B})$-reachable. An alternative statement is that if $(\mathbf{A}, \mathbf{B})$ is stabilizable—i.e., all unstable eigenvalues of $\mathbf{A}$ are reachable ones of the $(\mathbf{A}, \mathbf{B})$ pair—then either $\mathbf{A}$ is stable or the sum (16) is unbounded as N goes to infinity.

When A is asymptotically stable, the limits of the right-hand side of (16) exist as N goes to infinity. Denote it by $\mathbf{G}_\Omega$. It satisfies the equation

$$\mathbf{A}\mathbf{G}_\Omega\mathbf{A}' - \mathbf{G}_\Omega = -\mathbf{B}\mathbf{B}', \tag{17}$$

and it has the unique symmetric positive definite solution if $(\mathbf{A}, \mathbf{B})$ is a reachable pair and all eigenvalues of A have magnitude less than one.

Uniqueness is easy to see. Suppose there are two solutions: G_1 and G_2. Let $\boldsymbol{\Delta} = \mathbf{G}_1 - \mathbf{G}_2$. Then $\mathbf{A}\boldsymbol{\Delta}\mathbf{A}' = \boldsymbol{\Delta}$. Iterating this $\boldsymbol{\Delta} = \mathbf{A}^N\boldsymbol{\Delta}(\mathbf{A}')^N \to 0$ if $\mathbf{A}$ is asymptotically stable. Generate a sequence of $\mathbf{G}$s by $\mathbf{G}_{k+1} = \mathbf{A}\mathbf{G}_k\mathbf{A}' + \mathbf{B}\mathbf{B}'$, i.e., $\mathbf{G}_k = \mathbf{A}^k\mathbf{G}_0\mathbf{A}'^k + \Sigma_{j=0}^{k-1}\mathbf{A}^j\mathbf{B}\mathbf{B}'(\mathbf{A}')^j$. If A is asymptotically stable, we see that the first term vanishes and the sum converges because $\|\mathbf{A}^j\| \leqslant c\alpha^j$ for some c, and $\alpha < 1$. This sum is $\mathbf{G}_\Omega$.

Perfect Observability

In some cases, it is important that the initial condition $\mathbf{x}_0$ is uniquely determined from $\mathbf{y}^t$, irrespective of the disturbance or sequences of inputs. This requires a system property different from observability. When $\mathbf{u}$s are known, then $\mathbf{x}_0$ can be recovered uniquely from $\mathbf{y}^t$ if and only if the system is observable. From the relationship $\mathbf{y}^t = \mathbf{O}_t\mathbf{x}_0 + \mathbf{J}_t\mathbf{u}^t$, if $\mathbf{O}_t'\mathbf{O}_t$ is nonsingular, then $\mathbf{x}_0 = (\mathbf{O}'\mathbf{O})^{-1}\mathbf{O}'(\mathbf{y}^t - \mathbf{J}\mathbf{u}^t)$ if $\mathbf{u}^t$ is known and $\mathbf{O}'\mathbf{O}$ is nonsingular. When $\mathbf{u}^t$ is unknown, clearly $\mathbf{x}_0$ is still uniquely determined as before if $\mathbf{O}'\mathbf{J} = 0$; i.e., if $R(\mathbf{J}) \cap R(\mathbf{O}) = \{0\}$, the column vectors of $\mathbf{J}$ and $\mathbf{O}$ are linearly independent. This condition is also necessary; see Rappaport and Silverman (1971). This stronger notion of observability is called *perfect observability*. They also show that a perfectly observable system is invertible. An innovation model we discuss later is invertible. Given the original model

$$\begin{cases} \mathbf{x}_{t+1} = \mathbf{A}\mathbf{x}_t + \mathbf{B}\mathbf{e}_t \\ \mathbf{y}_t = \mathbf{C}\mathbf{x}_t + \mathbf{e}_t, \end{cases}$$

its inverse model is

$$\begin{cases} \mathbf{e}_t = \mathbf{y}_t - \mathbf{C}\mathbf{z}_t \\ \mathbf{z}_{t+1} = (\mathbf{A} - \mathbf{B}\mathbf{C})\mathbf{z}_t + \mathbf{B}\mathbf{y}_t. \end{cases}$$

Note that the roles of $\mathbf{e}_t$ and $\mathbf{y}_t$ are reversed in the original and inverse systems.

Observability and Reachability of Nonlinear Dynamic Systems

For a nonlinear dynamic system $\mathbf{x}_{t+1} = \mathbf{f}_t(\mathbf{x}_t, \mathbf{u}_t)$, the notion of local reachability (about a point or trajectory) can be defined in terms of a linearized dynamics (about a point or trajectory). Suppose $\{\mathbf{x}_t^0\}$ is a trajectory generated by $\{\mathbf{u}_t^0\}$. Then $\delta\mathbf{x}_{t+1} = \mathbf{A}_t(\mathbf{x}_t^0, \mathbf{u}_t^0)\delta\mathbf{x}_t + \mathbf{B}_t(\mathbf{x}_t^0, \mathbf{u}_t^0)\,\delta\mathbf{u}_t$ is a perturbed dynamics about the *reference path–reference control* pair. This time-varying linearized dynamics is reachable if the corresponding gramian is positive definite, for all t, $\mathbf{G}_t = \boldsymbol{\Omega}_t\boldsymbol{\Omega}_t' > 0$, where $\boldsymbol{\Omega}_t = [\mathbf{B}_t^0, \mathbf{A}_t^0\,\mathbf{B}_{t-1}^0, \ldots, \boldsymbol{\phi}_{t,1}\,\mathbf{B}_0^0]$ and $\boldsymbol{\phi}_{t,\tau} = \mathbf{A}_t^0 \ldots \mathbf{A}_{\tau+1}^0$ is the transition matrix. See Aoki (1976, pp. 85, 101) for example. A local observability notion is similarly defined.

Exercise 4

Let n be the dimension of a state vector, and suppose that rank of $\mathbf{O}_n$ is r which is less than n. Choose a coordinate system to render $\mathbf{A}$ lower triangular and $\mathbf{C} = [\mathbf{C}_1, \mathbf{0}]$ where $\mathbf{C}$ is $n \times r$ as in the preceding example. Similarly, when $\boldsymbol{\Omega}_n$ has rank less than n, the dynamic matrix $\mathbf{A}$ can be made upper triangular and $\mathbf{B}' = [\mathbf{B}_1', \mathbf{0}]$ where $\mathbf{B}_1$ is $r \times m$ where m is the dimension of the control vector.

Hint: Note that the observability matrix can be partitioned into $\mathbf{O}_n = (\bar{\mathbf{O}}_n, \mathbf{0})$ where $\bar{\mathbf{O}}_n$ has rank r. One way of choosing a coordinate is to use a singular value decomposition of $\mathbf{O}_n$ as $\mathbf{O}_n = \mathbf{U}\boldsymbol{\Sigma}\mathbf{V}'$ where $\mathbf{U}$ is $np \times np$, $\boldsymbol{\Sigma}$ is $np \times r$, and $\mathbf{V}$ is $r \times r$, where p is the dimension of the vector y.

1.5 Reference Path and Variational Dynamics

Most dynamic models, be they deterministic or stochastic, are usually too complex to be analyzed directly. In such cases, some types of variational or perturbation analysis often provide useful information about the system behavior near "benchmark" or "reference" time paths. There is no need to assign any normative sense or interpretation to the reference path even though that may certainly be possible in some cases. Reference time paths may merely represent the paths followed by the standard or somewhat simplified version of the models to facilitate analysis, or they may represent "ensemble" average in a sense of mean or statistical average of a (large) number of models or situations. Optimization problems with nonlinear dynamics and nonquadratic criterion functions may be approximated by a sequence of simpler problems with a linear dynamics and quadratic performance indices by treating the latter as (a sequence of) benchmark models. See Aoki (1964) for an illustration. For example, the assumption of a representative economic agent or a constant saving rate may be used to provide a "benchmark" model and a model with the actual agent characteristics can be treated as a deviation from the ensemble average embodied in the representative agent. Some examples of this type will be found in this section as well as in Aoki (1980b, 1981).

The next two examples illustrate the use of the reference path. In the second example, the reference path is chosen slightly differently from the ensemble average path to simplify analysis somewhat. (This choice is not crucial. We could have used the ensemble average path as the reference path just as easily.) More will be said on approximations later in this book.

Elementary Example (Solow's Growth Model). Consider

$$Y_t = Ak_t^{\alpha} N_{\tau}^{1-\alpha}, \qquad A > 0.$$

$$N_{t+1} = (1 + \lambda)N_t$$

$$K_{t+1} = sY_t + (1 - \delta)K_t, \qquad 0 < \delta, \alpha, s < 1.$$

The model in intensive or per capita form is

$$y_t = Ak_t^{\alpha} \quad \text{where} \quad y_t = Y_t/N_t\text{: output-labor ratio}$$

$$k_t = K_t/N_t\text{: capital–labor ratio.}$$

The dynamics become

$$\begin{aligned} k_{t+1} &= \frac{s}{1+\lambda} y_t + \frac{1-\delta}{1+\lambda} k_t \\ &= \frac{sA}{1+\lambda} k_t^\alpha + \frac{1-\delta}{1+\lambda} k_t \\ &\triangleq g(k_t). \end{aligned}$$

A stationary (steady) state of the model is defined to be $\bar{k}$ where $\bar{k} = g(\bar{k})$, which is $\bar{k} = \{sA/(\delta + \lambda)\}^{1/(1-\alpha)}$. Linearization about $\bar{k}$ produces a linear difference equation

$$\delta k_{t+1} = g'(\bar{k}) \delta k_t \tag{18}$$

where

$$\begin{aligned} \delta k_t &= k_t - \bar{k} \\ g'(\bar{k}) &= \{\alpha(\lambda + \delta) + 1 - \delta\}/(1 + \lambda). \end{aligned}$$

This is a particular application of the notion of a reference path when it is constant through time. The motion near the steady state is then stable if and only if $|g'(\bar{k})| < 1$, and if so, the convergence is monotone. The stability of the steady state can be stated then as $\alpha < 1$. Note that the asymptotic stability condition is independent of the saving rate s. The rate s affects only $\bar{k}$.

In this example, the reference path is really a steady state. In the next example, the reference path is not a steady state but is really a genuine time path.

Example 4: Dynastic Growth Model. This example is a simplified version of Aoki (1980a), which examines the implication of different saving rates in a context of deterministic growth models. The distribution of the equilibrium capital stock holdings among the different household types becomes proportional to the difference between the saving rate and the average rate. There are M types of households. Type i household holds capital K_t^i and works N_t units of labor, and its saving rate is s^i, assumed constant. Its income is given by

$$Y_t^i = r_t K_t^i + w_t N_t$$

where

$$K_{t+1}^i = (1 - \delta) K_t^i + s^i Y_t^i, \quad i = 1, \ldots, M.$$

Note that N_t is assumed to be the same for all types. The labor grows at a common rate for all types of households

$$N_{t+1} = \gamma N_t, \qquad \gamma > 1.$$

The average output and the average capital stock are defined by

$$Y_t = \Sigma_i Y_t^i/M, \qquad K_t = \Sigma_i K_t^i/M.$$

On the per capita basis, the capital held by type i households grows by

$$k_{t+1}^i = \frac{1-\delta}{\gamma} k_t^i + \frac{\mathrm{s}^i}{\gamma} y_t^i \tag{19}$$

where

$$y_t^i = r_t k_t^i + w_t$$

while the average per capita capital stock grows according to

$$\begin{aligned} k_{t+1} &= \frac{1-\delta}{\gamma} k_t + \frac{1}{\gamma} \Sigma \mathrm{s_i} y_t^i \\ &= \frac{1-\delta}{\gamma} k_t + \frac{\bar{\mathrm{s}}}{\gamma} y_t + \frac{1}{\gamma M} \Sigma(\mathrm{s_i} - \bar{\mathrm{s}})(y_t^i - y_t), \end{aligned} \tag{20}$$

where

$$\bar{\mathrm{s}} = \Sigma \mathrm{s_i}/M$$

and

$$y_t = r_t k_t + w_t.$$

The third term in (20) can be rewritten as

$$\begin{aligned} \xi_t &= \frac{1}{\gamma M} \Sigma(\mathrm{s_i} - \bar{\mathrm{s}})(y_t^i - y_t) \\ &= \frac{r_t}{\gamma M} \Sigma_i \, (\delta \mathrm{s_i})(k_t^i - k_t), \end{aligned}$$

where

$$\delta \mathrm{s_i} = \mathrm{s_i} - \bar{\mathrm{s}}.$$

Equation (20) describes the growth of "ensemble average" capital stock, which we use as the reference growth path. Since $k_t^i - k_t$ will be proportional to $\delta \mathrm{s_i}$, ξ_t is of second-order smallness. We choose a reference path defined by

$$k_{t+1}^0 = \frac{1-\delta}{\gamma} k_t^0 + \frac{\bar{\mathrm{s}}}{\gamma} f(k_t^0), \tag{21}$$

which differs from (20) by ξ_t.

This is the Solow one-sector growth model with a constant saving rate $\bar{\mathrm{s}}$. It has a well-defined steady state, k_∞^0.

Now subtract (21) and (19) to define the variational or deviational dynamics for $\delta k_t^i = k_t^i - k_t^0$,

$$\delta k_{t+1}^i = \frac{1-\delta}{\gamma}\delta k_t^i + \frac{\delta \mathrm{s_i}}{\gamma} y_t^i + \frac{\bar{\mathrm{s}}}{\gamma}\delta y_t^i, \tag{22}$$

where

$$\begin{aligned}\delta y_t^i &= y_t^i - y_t^0 = y_t^i - f(k_t^0) = f(k_t^i) - f(k_t^0)\\ &= f'(k_t^0)\delta k_t^i\\ &= r_t^0 \delta k_t^i,\end{aligned}$$

and

$$\delta \mathrm{s_i} = \mathrm{s_i} - \bar{\mathrm{s}}.$$

Since δy_t^i is of the order $\delta \mathrm{s_i}$, to the first order of smallness in $\delta \mathrm{s_i}$, Eq. (22) is approximated by

$$\delta k_{t+1}^i = \rho_t \delta k_t^i + \frac{\delta \mathrm{s^i}}{\gamma} y_t^0 \tag{23}$$

where

$$\rho_t = \frac{1-\delta}{\gamma} + \frac{\bar{\mathrm{s}} r_t^0}{\gamma}.$$

Here r_t^0 is the rental price along the reference path. Write (23) jointly as the vector $\boldsymbol{\delta}\mathbf{k}_t = (\delta k_t', \ldots, \delta k_t^M)'$

$$\boldsymbol{\delta}\mathbf{k}_{t+1} = \rho_t \boldsymbol{\delta}\mathbf{k}_t + \frac{\boldsymbol{\delta}\mathbf{s}}{\gamma} y_t^0$$

where

$$\boldsymbol{\delta}\mathbf{s} = (\delta s_1, \ldots, \delta s_M)'.$$

For simplicity assume that $f(k) = k^\alpha$. Then $r_t^0 = \alpha(k_t^0)^{\alpha-1} \to \alpha(\gamma + \delta - 1)/\bar{\mathrm{s}}$, as $t \to \infty$, and $(1 - \delta + \bar{\mathrm{s}} r_\infty^0)/\gamma = \alpha + (1-\delta)(1-\alpha)/\gamma$. Thus $1 - \delta < \gamma$ implies $\rho_t \to \rho_\infty < 1$, and hence $\boldsymbol{\delta}\mathbf{k}_t \to \boldsymbol{\delta}\mathbf{k}_\infty$ where

$$\boldsymbol{\delta}\mathbf{k}_\infty = \frac{\boldsymbol{\delta}\mathbf{s}}{\gamma(1-\rho_\infty)} y_\infty^0, \tag{24}$$

which shows that $\boldsymbol{\delta}\mathbf{k}_\infty$ is proportional to $\boldsymbol{\delta}\mathbf{s}$ as asserted.

From (21), noting that the difference between r_t and r_t^0 is at most of the order $\boldsymbol{\delta}\mathbf{s}$, we can approximate the limiting value of ξ_t by

$$\begin{aligned}\xi_t \to \xi_\infty &\simeq \frac{r_\infty^0}{\gamma M}\boldsymbol{\delta}\mathbf{s}'\boldsymbol{\delta}\mathbf{k}_\infty\\ &= \frac{r_\infty^0}{\gamma M}\frac{\boldsymbol{\delta}\mathbf{s}'\boldsymbol{\delta}\mathbf{s}}{\gamma(1-\rho_\infty)} y_\infty^0\end{aligned}$$

which is of the order $\Sigma_i \delta s_i^2$. This indicates that the effect of differing saving rates on the equilibrium capital stock is of the second order of smallness in δs.

Use of (20) as the reference path replaces (24) by $(I - \Phi_\infty)^{-1}\Delta \mathbf{s} y_\infty/\gamma$ where $\Phi_\infty = [(1-\delta)\mathbf{I} + r_\infty \mathbf{s}]/\gamma$ where $\mathbf{S} = \operatorname{diag}(s_1 \ldots s_M)$ and $\delta\mathbf{s}'\delta\mathbf{s}/\rho_\infty$ by $\delta\mathbf{s}'(I - \Phi_\infty)^{-1}\delta\mathbf{s}$ in ξ_∞. Nothing of significance changes.

1.6 Transfer Function Representation

Dynamic models make explicit time relations of relevant variables. Another way to state them is to specify lag structures among the variables. Lag structures are often stated by polynomials in lag operators: L in the econometric literature, z^{-1} or q^{-1} in the systems literature, $Lx_t = x_{t-1}$ or $q^{-1}x_t = x_{t-1}$, for example.

A typical dynamic model is represented by

$$A(q^{-1})y_k = B(q^{-1})x_k + C(q^{-1})e_k, \tag{25}$$

where y_k is the observed output or response, x_k is the input or control variable, and e_k is the noise disturbance, where $A(\sigma)$, $B(\sigma)$, and $C(\sigma)$ are polynomial matrices in σ. Multiplying by the inverse of $A(q^{-1})$ from the left, Eq. (12) can be put into

$$y_k = P(q^{-1})x_k + Q(q^{-1})e_k, \tag{26}$$

where

$$P(q^{-1}) = A^{-1}(q^{-1})B(q^{-1})$$

and

$$Q(q^{-1}) = A^{-1}(q^{-1})C(q^{-1}).$$

For example, $y_t + 5y_{t-1} + 2y_{t-2} = 3e_t + 7e_{t-1}$ may be rendered as $A(q^{-1})y_t = B(q^{-1})e_t$, where $A(q^{-1}) = 1 + 5q^{-1} + 2q^{-2}$, and $B(q^{-1}) = 3 + 7q^{-1}$. Since this alternative description of a dynamic relation is usually employed in connection with an infinite-horizon decision problem, the initial conditions are usually not mentioned explicitly. The same transfer function description is employed for stationary stochastic processes. Equation (25) expresses the fact that y_k consists of responses of a linear dynamic system to x_k and to a noise process, i.e.,

$$y_k = \bar{y}_k + \tilde{y}_k$$

where

$$A(q^{-1})\bar{y}_k = B(q^{-1})x_k$$

and

$$A(q^{-1})\tilde{y}_k = C(q^{-1})e_k.$$

Such a transfer function representation may arise from a time-domain description of dynamics of the form

$$(27)\qquad \begin{cases} \mathbf{z}_{k+1} = \mathbf{F}\mathbf{z}_k + \mathbf{G}\mathbf{x}_k + \mathbf{J}\mathbf{e}_k \\ \mathbf{y}_k = \mathbf{H}\mathbf{z}_k + \mathbf{D}\mathbf{e}_k. \end{cases}$$

In the transform notation, (27) can be put into a form of (26) (ignoring contributions due to initial conditions)

$$(28)\qquad \mathbf{y}_k = \mathbf{H}(q\mathbf{I} - \mathbf{F})^{-1}\mathbf{G}\mathbf{x}_k + [\mathbf{D} + \mathbf{H}(q\mathbf{I} - \mathbf{F})^{-1}\mathbf{J}]\mathbf{e}_k,$$

i.e.,

$$\mathbf{P}(q^{-1}) = \mathbf{H}(q\mathbf{I} - \mathbf{F})^{-1}\mathbf{G}$$

and

$$\mathbf{Q}(q^{-1}) = \mathbf{D} + \mathbf{H}(q\mathbf{I} - \mathbf{F})^{-1}\mathbf{J}.$$

These transfer functions can be put into matrix fraction representation

$$\mathbf{H}(q\mathbf{I} - \mathbf{F})^{-1}\mathbf{G} = \frac{1}{d(q)}\mathbf{T}_1(q)$$

and

$$\mathbf{D} + \mathbf{H}(q\mathbf{I} - \mathbf{F})^{-1}\mathbf{J} = \frac{1}{d(q)}\mathbf{T}_2(q),$$

where

$$d(q) = \det(q\mathbf{I} - \mathbf{F}).$$

This polynomial is the characteristic polynomial of the matrix $\mathbf{F}$. Zeros of the polynomial $d(z)$ are called poles of the system or the transfer function. They are the eigenvalues of the dynamic matrix $\mathbf{F}$. Note that the two transfer functions in (28) have the same denominator, but the numerator polynomials are different. The zeros of the determinants of the numerator polynomial, when they are square, are called transfer function zeros. They are calculated as the eigenvalues of matrix pencils

$$\begin{pmatrix} z\mathbf{I} - \mathbf{F} & \mathbf{G} \\ -\mathbf{H} & \mathbf{0} \end{pmatrix} = z\begin{bmatrix} \mathbf{I} & \mathbf{0} \\ \mathbf{0} & \mathbf{0} \end{bmatrix} - \begin{bmatrix} \mathbf{F} & -\mathbf{G} \\ \mathbf{H} & \mathbf{0} \end{bmatrix}$$

and

$$\begin{pmatrix} z\mathbf{I} - \mathbf{F} & \mathbf{J} \\ -\mathbf{H} & \mathbf{D} \end{pmatrix} = z\begin{bmatrix} \mathbf{I} & \mathbf{0} \\ \mathbf{0} & \mathbf{0} \end{bmatrix} - \begin{bmatrix} \mathbf{F} & -\mathbf{J} \\ \mathbf{H} & -\mathbf{D} \end{bmatrix}$$

respectively because of the matrix identity

$$\det\begin{pmatrix} \mathbf{A} & \mathbf{B} \\ \mathbf{C} & \mathbf{D} \end{pmatrix} = \det \mathbf{A} \det(\mathbf{D} - \mathbf{C}\mathbf{A}^{-1}\mathbf{B}),$$

for example

$$\begin{vmatrix} z\mathbf{I}-\mathbf{F} & \mathbf{J} \\ -\mathbf{H} & \mathbf{D} \end{vmatrix} = |z\mathbf{I}-\mathbf{F}||\mathbf{D}+\mathbf{H}(z\mathbf{I}-\mathbf{F})^{-1}\mathbf{J}|$$
$$= |\mathbf{T}_2(z)|.$$

Roots with unit magnitudes of the polynomial (matrices) $\mathbf{A}(\cdot)$ or $\mathbf{C}(\cdot)$, or equivalently in $d(\cdot)$ or $\mathbf{T}_2(\cdot)$, require special care in modeling and analysis since the presence of such roots signal either the presence of random walk components or noninvertability of the polynomial (matrices). We return to the unit root or the unit zero problems later in the chapter.

Decision rules may themselves be dynamic (i.e., be examples of dynamic models) since optimal decisions in some problems require knowledge of the values of some key variables not just now but at several points in the past.

1.7 Inverse of Dynamic Relations

Some notions of inverse of a given dynamic system are often found in various applications. In the field of time series, an AR (autoregressive) model is "inverted" to produce a MA (moving average) (Wold decomposition) representation, or a given Wold decomposition is transformed into an AR model. An "inverse" filter is used to whiten a stochastic sequence with nonconstant spectral densities and so on. In optimal controls, a time-optimal problem is to find a sequence of control to bring a dynamic system from a given state to another (the origin, say) in the smallest number of time periods. This problem also requires the notions of the inverse system.

This section discusses deterministic inverse systems that have the transfer functions related to the inverse of the original transfer functions. To put this problem into a proper framework, we need the notion of minimum-phase polynomials and their inverses.

Minimum-Phase Polynomial

We associate a finite sequence of (complex) members $a_0, a_1, \ldots a_M$ with a polynomial in q^{-1}

$$A(q^{-1}) = a_0 + a_1 q^{-1} + \cdots + a_M q^{-M}.$$

This is nothing but the z-transform of the finite sequence. This is the same way in which a transfer function is associated with the impulse response sequences, $\{h_i\}$, $i = 0, 1, \ldots,$ where $h_0 = 1$, and $h_i = \mathbf{CA}^{i-1}\mathbf{B}$, $i = 1, 2, \ldots,$ because then $H(q^{-1}) = 1 + h_1 q^{-1} + h_2 q^{-2} + \cdots = 1 + \mathbf{C}(z\mathbf{I} - \mathbf{A})^{-1}\mathbf{B}$. Another way to associate a polynomial with these coefficients would be to construct it after reversing the sequence, i.e.,

$$B(q^{-1}) = a_M + a_{M-1} q^{-1} + \cdots + a_0 q^{-M}. \tag{29}$$

Note that $B(q^{-1})$ is related to $A(q)$ by $B(q) = q^M A(q^{-1})$. The two other related polynomials are

$$\bar{A}(z) = A^*(1/z),$$

where * denotes complex conjugate, and

$$A^R(z) = z^{-M}\bar{A}(z) = z^{-M}A^*(z^{-1}).$$

By setting $z = e^{j\omega}$, we note that

$$A^*(e^{i\omega}) = \bar{A}(e^{-i\omega}),$$

i.e.,

$$|A(e^{i\omega})|^2 = A(e^{i\omega})\bar{A}(e^{-i\omega}).$$

From

$$A^R(e^{i\omega}) = e^{-jM\omega}A^*(e^{-j\omega})$$

and

$$[A^R(e^{i\omega})]^* = e^{jM\omega}A(e^{i\omega}),$$

note that

$$|A(e^{i\omega})|^2 = |A^R(e^{i\omega})|^2.$$

Thus, the sequence $(a_0, a_1, \ldots, a_M)$ and its reversed sequence $(a_M, a_{M-1}, \ldots a_0)$ have the same magnitude of the associated polynomals.

A sequence $\{a_k\}_0^M$ is said to be a minimum-phase sequence, and the associated polynomial $A(z^{-1})$ a minimum-phase polynomial, when the z-transform $A(z^{-1})$ has all its zeros inside the unit circle in the complex z-plane;

$$\begin{aligned} A(z^{-1}) &= a_0 + a_1 z^{-1} + \cdots + a_M z^{-M} \\ &= a_0(1 - z_1 z^{-1})(1 - z_2 z^{-1})\ldots(1 - z_M z^{-1}) \end{aligned}$$

with $|z_i| < 1$, $i = 1, \ldots, M$. An AR model

$$y_k + a_1 y_{k-1} + \cdots + a_M y_{k-M} = u_k$$

has a stable transfer function $G(z^{-1}) = (1 + a_1 z^{-1} + \cdots + a_M z^{-M})^{-1}$, i.e., a representation

$$Y(z^{-1}) = G(z^{-1}) U(z^{-1})$$

is stable if and only if $1 + a_1 z^{-1} + \cdots a_M z^{-M}$ has zeros inside the unit circle in the complex plane, i.e., $G(z^{-1})^{-1}$ is a minimum-phase polynomial.

A rational function with a minimum-phase numerator and a denominator polynomial is said to be a minimum-phase rational function. A transfer function $G(z^{-1})$ is said to be of minimum phase in the same way. Because the denominator has all zeros inside the unit circle, all the poles of $G(z^{-1})$ are inside the unit circle, i.e., the dynamic system with the transfer function $G(z)$ is stable. Because all the zeros if $G(z^{-1})$ are inside the unit circle, $1/G(z^{-1})$ is also a stable transfer function.

Given a minimum-phase polynomial, consider an associated polynomial, with zeros at z_i, $i = 1 \ldots M$ and with zeros at $1/z_i^*$, $i = 1 \ldots M$. These zeros are the mirror images of z_i relative to the unit circle because they lie on the same line segment from the origin, because $z_i = |z_i| e^{i\theta_i}$ implies that $1/z_i^* = (1/|z_i|) e^{i\theta_i}$, and because the product of the magnitudes is 1. Associate with $A(z^{-1}) = \Pi_{i=1}^M (1 - z_i/z)$, a polynomial $B(z^{-1}) = \Pi_{i=1}^M (-z_i^* + z^{-1})$. This polynomial has all its zeros outside the unit disk. Both are monic polynomials and are related by

$$B(z^{-1}) = \left(\prod_{i=1}^{M} \frac{-z_i^* + z^{-1}}{1 - z_i z^{-1}} \right) A(z^{-1}). \tag{30}$$

Note that each factor has a unit magnitude on the unit circle

$$\left| \frac{-z_i^* + z^{-1}}{1 - z_i z^{-1}} \right| = 1 \qquad \text{on } |z| = 1$$

because

$$\left| \frac{-z_i^* + e^{-i\omega}}{1 + z_i e^{-i\omega}} \right| = \left| \frac{e^{j\omega}(e^{-j\omega} - z_i^*)}{e^{i\omega} - z_i} \right| = 1,$$

and it is analytic inside the unit disk and has zeros also outside the unit disk.

The factor

$$\frac{1 - z_i z^{-1}}{z^{-1} - z_i^*} = \frac{|z_i^*|}{z_i} \frac{z - z_i}{1 - z_i^* z} \tag{31}$$

or its product is called a Blaschke product if $\Pi_{i=1}^{\infty}(1-|z_i|)<\infty$. In our case $|z_i|<1$, hence $A(z)$ and $B(z)$ differ by a Blaschke product. Note that $A(z)$ and $B(z)$ have the same magnitude $|A(z^{-1})| = |B(z^{-1})|$ on the unit circle.

The reason that $A(z^{-1})$ is called a minimum-phase polynomial can be indicated by considering two polynomials that differ by a single factor;

$$\left.\begin{aligned} A(z^{-1}) &= (1 - z_1 z^{-1})F(z) \\ &\text{and} \\ B(z^{-1}) &= (-z_1^* + z^{-1})F(z) \end{aligned}\right\} \tag{32}$$

where $F(z)$ is a polynomial with some zeros inside and possibly some outside the unit disk. The polynomials $A(z)$ and $B(z)$ have the same magnitude on $|z| = 1$ but have a different phase

$$\frac{A(e^{i\omega})}{B(e^{i\omega})} = \frac{1 - z_1 e^{-j\omega}}{-z_1^* + e^{-j\omega}} = \frac{e^{i\omega} - z_1}{e^{-j\omega} - z_1^*}\, e^{-j\omega}.$$

Let

$$\psi = \operatorname{Avg}(e^{i\omega} - z_1).$$

Then the angle is $2\psi - \omega = 2(\omega + \phi) - \omega = \omega + 2\phi \geqslant 0$ where ϕ is the angle between the radius connecting the origin to $e^{i\omega}$ and the line segment connecting z_1 to $e^{i\omega}$.

This shows that among all the polynomials with M zeros, the polynomial with all zeros inside the unit disk has the least phase lag. (This terminology is motivated by regarding the polynomial as the transfer function of a filter. Then the phase of the filter is the phase shift experienced by the signal in going through the filter.)

Another difference between $A(z)$ and $B(z)$ of (32) is that the partial sum of the squares of the coefficients is the largest for $A(z)$. Let

$$B(z^{-1}) = b_0 + b_1 z^{-1} + \cdots + b_M z^{-M}.$$

Then

$$a_k = b_k - z_1 b_{k-1}$$

and

$$b_k = -z_1^* b_k + b_{k-1}$$

for $0 < k < M$, from which

$$|a_k|^2 - |b_k|^2 = (1 - |z_1|^2)(|b_k|^2 - |f_{k-1}|^2)$$

and

$$\sum_{k=0}^{n} (|a_k|^2 - |b_k|^2) = (1 - |z_1|^2)|f_n|^2 \geqslant 0$$

for $0 \leqslant n \leqslant M - 1$.

The sum $\Sigma_{k=0}^{M} = 0$ because $f_M = 0$ since $F(z)$ is degree $M - 1$.

In physical signal processing, this implies that the filter with all zeros inside the unit disk has the most energy concentrated in the early part of the response, among all monic polynomials with m zeros, if one equates loosely the magnitude of the coefficient squared with the energy carried by the signal.

Inverse Systems

We consider only dynamic systems with scalar transfer functions here.

Associate with the system

$$\begin{aligned} \mathbf{x}_{k+1} &= \mathbf{A}\mathbf{x}_k + \mathbf{b}u_k \\ y_k &= \mathbf{c}\mathbf{x}_k + du_k \end{aligned} \tag{33}$$

an integer m, called the relative order of the system,

$$m = \min \{i;\ h_i \neq 0\}$$

where

$$h_0 = d$$

$$h_i = \mathbf{c}\mathbf{A}^{i-1}\mathbf{b},\ i = 1, 2, \ldots .$$

This integer is the delay of the output responses to an input sequence. It is also the difference in the degrees of the denominator polynomial and the numerator polynomial of the transfer function*

$$g(z) = d + \mathbf{c}(z\mathbf{I} - \mathbf{A})^{-1}\mathbf{b}. \tag{34}$$

Obviously $m = 0$ if $d \neq 0$. Note that y_k can be related to the current and past inputs by

$$y_k = \mathbf{c}\mathbf{A}^k\mathbf{x}_0 + \sum_{i=0}^{k} h_{k-i}u_i .$$

* If $d \neq 0$, then both the numerator and the denominator have degree n. If $d = 0$ and $h_1 = 0$ but $h_2 \neq 0$, then the numerator is an $(n - 2)$th-degree polynomial and so on.

In a dynamic system with the relative order m, $y_0, \dots y_{m-1}$ are not affected by the inputs, and

$$y_{k+m} = \mathbf{c}\mathbf{A}^m \mathbf{x}_k + h_m u_k, \qquad k \geqslant 0, \qquad h_m \neq 0.$$

Solve this for u_k,

$$u_k = (y_{k+m} - \mathbf{c}\mathbf{A}^m \mathbf{x}_k)/h_m. \tag{35}$$

Regarding u_k as the measurement $\hat{y}_k$ of the inverse system and y_{k+m} as the input to it, define

$$\hat{y}_k = u_k$$

$$\hat{u}_k = y_{m+k}.$$

Then (35) is the measurement equation for the inverse system

$$\hat{y}_k = \hat{\mathbf{c}}\hat{\mathbf{x}}_k + \hat{d}\hat{u}_k \tag{36}$$

where

$$\hat{\mathbf{c}} = -\mathbf{c}\mathbf{A}^m/h_m$$

and

$$\hat{d} = h_m^{-1}.$$

Substitute (35) in the dynamic equation (33)

$$\begin{aligned} \hat{\mathbf{x}}_{k+1} &= \mathbf{A}\hat{\mathbf{x}}_k + \mathbf{b}(y_{k+m} - \mathbf{c}\mathbf{A}^m \hat{\mathbf{x}}_k)/h_m \\ &= \hat{\mathbf{A}}_m \hat{\mathbf{x}}_k + \hat{\mathbf{b}} y_{k+m}, \end{aligned} \tag{37}$$

where

$$\hat{\mathbf{A}}_m = \mathbf{A} - \mathbf{b}\mathbf{c}\mathbf{A}^m/h_m$$

and

$$\hat{\mathbf{b}} = \mathbf{b}/h_m.$$

The system defined by (36) and (37) is the inverse system of (33) in the sense that

$$g(z)\hat{g}(z) = z^{-m}$$

where

(38) $$\hat{g}(z) = \hat{d} + \hat{\mathbf{c}}(z\mathbf{I} - \hat{\mathbf{A}}_m)\hat{\mathbf{b}}.$$

To verify (38), apply the matrix identity

$$(\mathbf{A} + \mathbf{bc})^{-1} = \mathbf{A}^{-1} - \frac{\mathbf{A}^{-1}\mathbf{bcA}^{-1}}{1 + \mathbf{cA}^{-1}\mathbf{b}}$$

when $\mathbf{A}^{-1}$ exists, use the identity relation

$$(z\mathbf{I} - \hat{\mathbf{A}}_m)^{-1} = (z\mathbf{I} - \mathbf{A})^{-1} + \frac{\mathbf{bcA}^m}{h_m}$$

$$= (z\mathbf{I} - \mathbf{A})^{-1} - \frac{(z\mathbf{I} - \mathbf{A})^{-1}\mathbf{bcA}^m(z\mathbf{I} - \mathbf{A})^{-1}}{h_m + \mathbf{cA}^m(z\mathbf{I} - \mathbf{A})^{-1}\mathbf{b}},$$

from which follow

$$\hat{\mathbf{c}}(z\mathbf{I} - \hat{\mathbf{A}}_m)^{-}\hat{\mathbf{b}} = \frac{-\mathbf{cA}^m(z\mathbf{I} - \mathbf{A})^{-1}\mathbf{b}/h_m}{h_m + \mathbf{cA}^m(z\mathbf{I} - \mathbf{A})^{-1}\mathbf{b}}$$

and

$$\hat{d} + \hat{\mathbf{c}}(z\mathbf{I} - \hat{\mathbf{A}}_m)^{-1}\hat{\mathbf{b}} = 1/[h_m + \mathbf{cA}^m(z\mathbf{I} - \mathbf{A})^{-1}\mathbf{b}] = z^{-m}/g(z).$$

Example 5. To illustrate the relations of the transfer functions of the original and its inverse, suppose that

$$g(q^{-1}) = 1 + a_1 q^{-1} + a_2 q^{-2} + \dots$$
$$= 1 + \mathbf{c}(q\mathbf{I} - \mathbf{A})^{-1}\mathbf{b},$$

i.e., $a_i = \mathbf{cA}^{i-1}\mathbf{b}$, $i = 1, 2, \dots$. Suppose that the inverse system has the transfer function

$$h(q^{-1}) = h_0 + h_1 q^{-1} + h_2 q^{-2} + \dots .$$

Then $g(q^{-1})h(q^{-1}) = 1$ by definition because $\hat{y} = g(q^{-1})\hat{u}$ and $\hat{u} = h(q^{-1})\hat{y}$. Collecting coefficients of equal powers in q^{-1}, the coefficients in $h(q^{-1})$ are recursively determined:

$$h_0 = 1$$
$$h_1 + h_0 a_1 = 0$$
$$h_2 + h_1 a_1 + h_0 a_2 = 0$$
$$\vdots$$

or

$$\begin{aligned} h_1 &= a_1 = -\mathbf{cb} \\ h_2 &= -h_1 a_1 - a_2 \\ &= a_1^2 - a_2 \\ &= (\mathbf{cb})^2 - \mathbf{cAb} \\ h_3 &= -h_2 a_1 - h_1 a_2 - a_3 \\ &= -[(\mathbf{cb})^2 - \mathbf{cAb}]\mathbf{cb} + \mathbf{cbcAb} - \mathbf{cA}^2\mathbf{b} \\ &= -(\mathbf{cb})^3 + 2\mathbf{cb}(\mathbf{cAb}) - \mathbf{cA}^2\mathbf{b} \text{ etc.} \end{aligned}$$

Alternatively, these coefficients can be generated as the transfer function of a dynamic system with

$$\begin{aligned} \hat{\mathbf{A}} &= \mathbf{A} - \mathbf{bc} \\ \hat{\mathbf{b}} &= \mathbf{b} \\ \hat{\mathbf{c}} &= -\mathbf{c} \end{aligned}$$

or

$$\hat{d} = 1,$$

because

$$1 + \hat{\mathbf{c}}(z\mathbf{I} - \hat{\mathbf{A}})^{-1}\hat{\mathbf{b}} = 1 - \mathbf{c}(z\mathbf{I} - \mathbf{A} + \mathbf{bc})^{-1}\mathbf{b}$$

$$= \frac{1}{1 + \mathbf{c}(z\mathbf{I} - \mathbf{A})^{-1}\mathbf{b}} = \frac{1}{g(z^{-1})}.$$

1.8 Infinite-Horizon Quadratic Cost-Minimization Problems

A basic or prototype control problem is to minimize a quadratic cost

$$\sum_0^{\infty} W_k$$

where

$$\begin{aligned} W_k &= \mathbf{x}_x'\mathbf{Q}\mathbf{x}_k + \mathbf{u}_k'\mathbf{R}u_k \\ \mathbf{Q}' &= \mathbf{Q} > 0 \quad \text{and} \quad \mathbf{R}' = \mathbf{R} > 0, \end{aligned}$$

subject to a linear dynamic equation

$$\mathbf{x}_{k+1} = \mathbf{A}\mathbf{x}_k + \mathbf{B}u_k.$$

Here, $\mathbf{A}, \mathbf{B}, \mathbf{Q}$, and $\mathbf{R}$ are all constant matrices. Dynamic programming formulation of this problem is to define a cost (or value) functional $J(\mathbf{x})$ as the minimum of the cost when the initial condition of the dynamic system is $\mathbf{x}$. Bellman's functional equation then becomes

$$J(\mathbf{x}) = \min_{\mathbf{u}} [\mathbf{x}'\mathbf{Q}\mathbf{x} + \mathbf{u}'\mathbf{R}\mathbf{u} + J(\mathbf{A}x + \mathbf{B}\mathbf{u})]. \tag{39}$$

The function $J(\mathbf{x})$ is quadratic in $\mathbf{x}$

$$J(\mathbf{x}) = \mathbf{x}'\mathbf{\Pi}\mathbf{x}.$$

Then (39) shows that the second-order condition is satisfied. Differentiating the expression in the square bracket of (39), the first-order condition becomes

$$\mathbf{R}\mathbf{u} + \mathbf{B}'\mathbf{\Pi}(\mathbf{A}\mathbf{x} + \mathbf{B}\mathbf{u}) = 0,$$

or we solve it for $\mathbf{u}$

$$\mathbf{u} = \mathbf{K}\mathbf{x},$$

where

$$\mathbf{K} = -[\mathbf{R} + \mathbf{B}'\mathbf{\Pi}\mathbf{B}]^{-1}\mathbf{B}'\mathbf{\Pi}\mathbf{A}, \tag{40}$$

which shows that the optimal control is proportional to $\mathbf{x}$ with a constant feedback matrix. When (40) is substituted back into (39), then $\mathbf{\Pi}$ is determined as the solution of the algebraic Riccati equation

$$\mathbf{\Pi} = \mathbf{F}'\mathbf{\Pi}\mathbf{F} + \mathbf{K}'\mathbf{R}\mathbf{K} + \mathbf{Q}, \tag{41}$$

where $\mathbf{F} = \mathbf{A} - \mathbf{B}\mathbf{K}$ describes the dynamics under the optimal control rule. When (40) is substituted in, (41) becomes

$$\mathbf{\Pi} = \mathbf{A}'\mathbf{\Pi}\mathbf{A} - \mathbf{A}'\mathbf{\Pi}\mathbf{B}[\mathbf{R} + \mathbf{B}'\mathbf{\Pi}\mathbf{B}]^{-1}\mathbf{B}'\mathbf{\Pi}\mathbf{A} + \mathbf{Q}. \tag{42}$$

With an infinite horizon and a constant transition law of dynamics, it is easy to see that the feedback control law must also be constant, since the same infinite-horizon control problem presents itself each period.

If $(\mathbf{A}, \mathbf{B})$ is reachable, then there is a sequence of controls to bring the initial state vector to zero in a finite step and keep the state vector there.* This means that the control cost J is bounded from above, and hence the optimal control cost for the infinite-horizon problem is finite even without any discount factor. This argument shows that $\mathbf{\Pi}$ is finite. Clearly $\mathbf{\Pi}$ is symmetric positive semidefinite. Partition $\mathbf{Q} = \mathbf{C}'\mathbf{C}$. We next show that if $(\mathbf{A}, \mathbf{C})$ is observable, then F is asymptotically stable. A simple way to establish this fact

---

* Strictly speaking, this requires the condition of perfect (output) controllability. See Aoki (1976, Section 3.2.3).

is to use $J(\mathbf{x})$ as a Lyapunov function in the stability theory. We capture the essence of the argument here. Note that the cost difference at two consecutive time points is

$$\mathrm{J}(\mathbf{x}_{k+1}) - \mathrm{J}(\mathbf{x}_k) = \mathbf{x}_k'(\mathbf{F}'\mathbf{\Pi}\mathbf{F} - \mathbf{\Pi})\mathbf{x}_k,$$

is nonpositive since

$$\mathbf{F}'\mathbf{\Pi}\mathbf{F} - \mathbf{\Pi} = -\mathbf{K}'\mathbf{R}\mathbf{K} - \mathbf{Q} \leqslant -\mathbf{C}'\mathbf{C}.$$

Let $\mathbf{x}_0$ be an eigenvector of $\mathbf{F}$, $\mathbf{F}\mathbf{x}_0 = \lambda\mathbf{x}_0$. Then $\mathbf{x}_k = \mathbf{F}^k\mathbf{x}_0 = \lambda^k\mathbf{x}_0$ and $\mathrm{J}(\mathbf{x}_{k+1}) - \mathrm{J}(\mathbf{x}_k) = (|\lambda|^{2k} - 1)\mathbf{x}_0'\mathbf{\Pi}\mathbf{x}_0 \leqslant -\mathbf{x}_0'\mathbf{C}'\mathbf{C}\mathbf{x}_0$. This expression is strictly negative if $\mathbf{C}\mathbf{x}_0 \neq 0$. $\mathbf{C}\mathbf{x}_0 = 0$ implies that λ is not an observable eigenvalue, a situation excluded if $(\mathbf{F}, \mathbf{C})$ is an observable pair. We see that

$$\begin{aligned}\mathrm{J}(\mathbf{x}_N) - \mathrm{J}(\mathbf{x}_0) &= \mathrm{J}(\mathbf{x}_N) - \mathrm{J}(\mathbf{x}_{N-1}) + \mathrm{J}(\mathbf{x}_{N-1}) - \ldots + \mathrm{J}(\mathbf{x}_1) - \mathrm{J}(\mathbf{x}_0)\\ &\leqslant -\mathbf{x}_0'\mathbf{C}'\mathbf{C}\mathbf{x}_0 - \mathbf{x}_1'\mathbf{C}'\mathbf{C}\mathbf{x}_1 \ldots - \mathbf{x}_{N-1}'\mathbf{C}'\mathbf{C}\mathbf{x}_N\\ &\leqslant -\mathbf{x}_0'(\mathbf{C}'\mathbf{C} + \mathbf{F}'\mathbf{C}'\mathbf{C}\mathbf{F} + \ldots + (\mathbf{F}')^{-N}\mathbf{C}'\mathbf{C}\mathbf{F}^N)\mathbf{x}_0\\ &< 0\end{aligned}$$

if $(\mathbf{F}, \mathbf{C})$ is an observable pair, i.e., $\mathrm{J}(\mathbf{x}_N) < \mathrm{J}(\mathbf{x}_0)$ as N goes to infinity. Since $\mathrm{J}(\mathbf{x}) = c_i$, $i = 1, 2$, $c_1 < c_2$ are two closed and bounded sets, which are strictly nested, in the n-dimensional Euclidean space, we see that $\|\mathbf{x}_N\| \to 0$ as N goes to infinity, i.e., $\mathbf{F}$ is asymptotically stable. Conversely, suppose $\mathbf{A}$ is not asymptotically stable and let $\mathbf{x}_0$ be an eigenvector of $\mathbf{A}$ such that

$$\mathbf{x}_0'\left[\sum_{j=0}^{k-1}(\mathbf{A}')^j\mathbf{C}'\mathbf{C}\mathbf{A}^i\right]\mathbf{x}_0 = \sum_{j=0}^{k-1}|\lambda|^{2j}\mathbf{x}_0'\mathbf{C}'\mathbf{C}\mathbf{x}_0.$$

Thus if $\mathbf{C}\mathbf{x}_0 \neq 0$, then the sum remains bounded as k goes to infinity if and only if $\mathbf{C}\mathbf{x}_0 = 0$. If the system is detectable, then $|\lambda| \geqslant 1$ is an observable eigenvalue of the pair. Thus $\mathbf{A}\mathbf{x}_0 = \lambda\mathbf{x}_0$, $\mathbf{C}\mathbf{x}_0 = 0$ implies that $\mathbf{x}_0$ is zero. This is a contradiction.

Once it is established that the optimal control is of the form $\mathbf{u} = \mathbf{K}\mathbf{x}$ for some constant $\mathbf{K}$, the optimal value of $\mathbf{K}$ is not hard to determine by a simple variational argument. In this infinite-horizon problem, if a constant feedback gain matrix $\mathbf{K}$ is employed, then the dynamics is

$$\mathbf{z}_{k+1} = \mathbf{F}\mathbf{z}_k$$

where

$$\mathbf{F} = \mathbf{A} + \mathbf{B}\mathbf{K},$$

and the cost expression J becomes

$$J = \mathbf{x}_0'\mathbf{\Pi}\mathbf{x}_0,$$

where

$$\mathbf{\Pi} = \sum_{k=0}^{\infty} \mathbf{F}'^k(\mathbf{Q} + \mathbf{K}'\mathbf{R}\mathbf{K})\mathbf{F}^k,$$

provided the infinite sum converges, i.e., if $\mathbf{F}$ is stable. Then the Lyapunov equation

$$-\mathbf{F}'\mathbf{\Pi}\mathbf{F} + \mathbf{\Pi} = \mathbf{Q} + \mathbf{K}'\mathbf{R}\mathbf{K}$$

can be solved with a unique symmetric positive definite matrix $\mathbf{\Pi}$ as its solution.

The optimal $\mathbf{K}$ from a class of constant feedback matrices with stable dynamics can then be obtained by variational arguments: Let $\mathbf{K}^*$ be the optimal and consider $\mathbf{K} = \mathbf{K}^* + \varepsilon\mathbf{K}_1$ where ε is a small positive number and $\mathbf{K}_1$ is arbitrary. For sufficiently small ε, $\mathbf{A} + \mathbf{B}\mathbf{K}$ is stable and the solution to the Lyapunov equation, $\mathbf{\Pi}(\varepsilon)$, exists. Then $d\mathbf{\Pi}/d\varepsilon$ is zero at $\varepsilon = 0$ by the fact that $\mathbf{K}^*$ is optimal. The first-order condition becomes

$$\mathbf{K}_1'(\mathbf{R}\mathbf{K}^* + \mathbf{B}'\mathbf{\Pi}\mathbf{F}^*) = 0$$

or

$$\mathbf{R}\mathbf{K}^* + \mathbf{\Pi}\mathbf{F}^* = 0$$

since $\mathbf{K}_1$ is arbitrary. Solving it, we obtain

$$\mathbf{K}^* = -(\mathbf{R} + \mathbf{B}'\mathbf{\Pi}\mathbf{B})^{-1}\mathbf{B}'\mathbf{\Pi}\mathbf{A}$$

as the optimal constant gain matrix.

Since we know that optimal gain is constant and unique for an infinite-horizon problem under some technical conditions that ensure existence and uniqueness, this $\mathbf{K}^*$ is indeed optimal. We later show that the same variational argument gives the optimal filter gain.

1.9 Hamiltonian Formulation

Differentiating the inside of the square bracket, the first condition of optimality of (39) is

$$\mathbf{R}\mathbf{u}_t + \mathbf{B}'\mathbf{\Pi}\mathbf{x}_{t+1} = 0$$

or

$$\mathbf{u}_t = -\mathbf{R}^{-1}\mathbf{B}'\mathbf{p}_{t+1}$$

where we define

$$\mathbf{p}_{t+1} = \mathbf{\Pi}\mathbf{x}_{t+1}.$$

From (42), this vector is governed by

$$\begin{aligned} \mathbf{p}_t &= \mathbf{\Pi x}_t \\ &= [\mathbf{A}'\mathbf{\Pi}(\mathbf{A} - \mathbf{BK}) + \mathbf{Q}]\mathbf{x}_t \\ &= \mathbf{Qx}_t + \mathbf{A}'\mathbf{\Pi x}_{t+1} \\ &= \mathbf{Qx}_t + \mathbf{A}'\mathbf{p}_{t+1}. \end{aligned} \tag{43}$$

The model dynamics become

$$\mathbf{x}_{t+1} = \mathbf{Ax}_t - \mathbf{BR}^{-1}\mathbf{B}'\mathbf{p}_{t+1}. \tag{44}$$

Note that time runs backward for $\mathbf{p}_t$,

$$\begin{bmatrix} \mathbf{x}_{t+1} \\ \mathbf{p}_t \end{bmatrix} = \begin{bmatrix} \mathbf{A} & -\mathbf{D} \\ \mathbf{Q} & \mathbf{A}' \end{bmatrix} \begin{bmatrix} \mathbf{x}_t \\ \mathbf{p}_{t+1} \end{bmatrix},$$

where $\mathbf{D} = \mathbf{BR}^{-1}\mathbf{B}'$. This form is known as the Hamiltonian formulation. It is related to the Pontryagin's maximum principle formulation which forms the Hamiltonian

$$H_t = \mathbf{x}_t'\mathbf{Qx}_t + \mathbf{u}_t'\mathbf{Ru}_t + \mathbf{p}_{t+1}'(\mathbf{Ax}_t + \mathbf{Bu}_t)$$

from which the dynamic equations are derived as

$$\mathbf{x}_{t+1} = \partial H_t / \partial \mathbf{p}_{t+1}$$

and

$$\begin{aligned} \mathbf{p}_t &= \partial H_t / \partial \mathbf{x}_t \\ &= \mathbf{Qx}_t + \mathbf{A}'\mathbf{p}_{t+1}. \end{aligned}$$

The optimal $\mathbf{u}_t$ minimizes H_t, i.e., $\mathbf{Ru}_t + \mathbf{B}'\mathbf{p}_t = 0$ or $\mathbf{u}_t = -\mathbf{R}^{-1}\mathbf{B}'\mathbf{p}_t$. With a finite horizon, this gives rise to a two-point boundary value problem. With the infinite horizon, a suitable transversality condition must be imposed. Although the Hamiltonian H_t has been introduced in the context of this quadratic cost-minimization problem subject to linear dynamics, this formulation applies to optimization problems with nonlinear dynamics and nonquadratic cost expressions as well. The next example illustrates this point.

Example 6: Optimal Investment Plan.* A firm faces given sequences of prices for the good it produces, $\mathbf{p}_t$, for investment good, $\mathbf{p}_t^I$, and for labor, w_t. It

* This example is a discrete-time version of the model described in Aoki and Leijonhufvud (1988).

wishes to maximize the present value of the earning stream

$$J = \sum_{t=0}^{\infty} \beta^t (p_t x_t - p_t^I I_t - w_t L_t)$$

subject to

$$K_{t+1} = (1 - \delta) K_t + I_t$$

and

$$x_t = \mathrm{f}(K_t, L_t).$$

Form the Hamiltonian,

$$H_t = \beta^t (p_t x_t - p_t^I I_t - w_t L_t) + \lambda_{t+1}[(1 - \delta) K_t + I_t].$$

The first-order condition for maximizing H_t with respect to w_t is

$$p_t \mathrm{f}_L(K_t, L_t) = w_t,$$

which determines the optimal L_t, given P_t, w_t, and K_t. Maximization of H_t with respect to I_t yields

$$\lambda_{t+1} = \beta^t p_t^I.$$

The dynamic equation for this adjoint (dual) variable is

$$\begin{aligned} \lambda_t &= \partial H_t / \partial K_t \\ &= \beta^t p_t F_k + (1 - \delta) \lambda_{t+1} \end{aligned}$$

or in terms of p_t^I,

$$p_t^I = V_t + \rho p_{t-1}^I,$$

where

$$V_t = p_t F_k(K_t, L_t)/(1 - \delta),$$

and

$$\rho = [\beta(1 - \delta)]^{-1} > 1.$$

Note that this equation is a perfect foresight version of a rational expectation model in which the future prices are all known. As such it must be solved forward in time.

Solve (43) for p_{t+1} and substitute it in (44) to write the dynamics as

$$\begin{bmatrix} \mathbf{x}_{t+1} \\ \mathbf{p}_{t+1} \end{bmatrix} = \mathbf{\Phi} \begin{bmatrix} \mathbf{x}_t \\ \mathbf{p}_t \end{bmatrix}$$

where

$$\mathbf{\Phi} = \begin{bmatrix} \mathbf{A} + \mathbf{D}\mathbf{A}^{-T}\mathbf{Q} & -\mathbf{D}\mathbf{A}^{-T} \\ -\mathbf{A}^{-T}\mathbf{Q} & \mathbf{A}^{-T} \end{bmatrix}$$

where

$$\mathbf{D} = \mathbf{B}\mathbf{R}^{-1}\mathbf{B}'.$$

This $2n \times 2n$ matrix is a symplectic matrix, i.e., eigenvalues are in pairs of λ and λ^{-1}. Any matrix that satisfies the definition

$$\mathbf{\Phi}'\mathbf{J} = \mathbf{J}\mathbf{\Phi}^{-1}$$

where

$$\mathbf{J} = \begin{bmatrix} \mathbf{0} & \mathbf{I} \\ -\mathbf{I} & \mathbf{0} \end{bmatrix}$$

possesses this characteristic. Here we have

$$\mathbf{\Phi}^{-1} = \begin{bmatrix} \mathbf{A}^{-1} & \mathbf{A}^{-1}\mathbf{D} \\ \mathbf{Q}\mathbf{A}^{-1} & \mathbf{A}' + \mathbf{Q}\mathbf{A}^{-1}\mathbf{D} \end{bmatrix}$$

Symplectic matrices have many interesting properties. These will enable us to show that the observability and reachability ensure the unique minimal symplectic positive definite solution. By putting $\mathbf{\Phi}$ into the real Schur decomposition, the solution of the Riccati equation Π is obtained. See Aoki (1987C, A.10) for further detail.

1.10 Finite-Horizon Quadratic Cost-Minimization Problem

Dynamic programming formulation becomes a little more cumbersome with finite-horizon problems because the optimal feedback control matrix is now changing with time. Also a final (valuation) cost term may be added to the cost expression. We give two slightly different versions of quadratic cost-minimization problems.

The corresponding closed-loop controls can be obtained by dynamic programming calculations. The calculations determine the last control, u_{N-1}, and proceed to determine $u_{N-2}, u_{N-3}, \ldots, u_0$ in that order. The calculations can be organized as follows. This example also serves to introduce the notations used in dynamic programming solutions for stochastic systems. Consider

(45) $$J = \mathbf{x}_N'\mathbf{\Gamma}\mathbf{x}_N + \sum_{k=0}^{N-1} W_k$$

where

(46) $$W_k = \mathbf{x}_k'\mathbf{Q}_k\mathbf{x}_k + \mathbf{u}_k'\mathbf{R}_k\mathbf{u}_k$$

subject to

(47) $$\mathbf{x}_{k+1} = \mathbf{A}_k\mathbf{x}_k + \mathbf{B}_k\mathbf{u}_k, \qquad k = 0, 1, \ldots, N$$

where*

(48) $$\mathbf{Q}_k' = \mathbf{Q}_k \geqslant 0, \qquad \mathbf{R}_k' = \mathbf{R}_k > 0, \qquad \mathbf{\Gamma}' = \mathbf{\Gamma} \geqslant 0.$$

The first term in (45) may be thought of as a proxy for the costs to be incurred from time N on, since the costs will be quadratic in $\mathbf{x}_N$. Aoki (1973) provides some numerical evaluations of approximating the "cost-to-go" from N onward by this type of terminal valuation term.

Define

$$\begin{aligned}\gamma_N &= \mathbf{x}_N'\mathbf{\Gamma}\mathbf{x}_N + \mathbf{u}_{N-1}'\mathbf{R}\mathbf{u}_{N-1}\\ &= (\mathbf{A}\mathbf{x}_{N-1} + \mathbf{B}\mathbf{u}_{N-1})'\mathbf{\Gamma}(\mathbf{A}\mathbf{x}_{N-1} + \mathbf{B}\mathbf{u}_{N-1}), \ + \mathbf{u}_{N-1}'\mathbf{R}\mathbf{u}_{N-1}.\end{aligned}$$

This is minimized by

$$\mathbf{u}_{N-1}^* = -\mathbf{K}_{N-1}\mathbf{x}_{N-1}$$

where

$$\mathbf{K}_{N-1} = (\mathbf{R} + \mathbf{B}'\mathbf{\Gamma}\mathbf{B})^{-1}\mathbf{B}'\mathbf{\Gamma}\mathbf{A}$$

and the minimum is

$$\begin{aligned}\gamma_N^* &= \min_{u_{N-1}} \gamma_N\\ &= \mathbf{x}_{N-1}'\mathbf{\Pi}_{N-1}\mathbf{x}_N\end{aligned}$$

where

$$\mathbf{\Pi}_{N-1} = \mathbf{A}'[\mathbf{\Gamma} - \mathbf{\Gamma}\mathbf{B}(\mathbf{R} + \mathbf{B}'\mathbf{Q}\mathbf{B})^{-1}\mathbf{B}'\mathbf{\Gamma}]\mathbf{A}.$$

The control $\mathbf{u}_{N-2}$ minimizes

$$\gamma_{N-1} = \gamma_N^* + \lambda_{N-1}$$

where

$$\lambda_{N-1} = \mathbf{x}_{N-1}'\mathbf{Q}\mathbf{x}_{N-1} + \mathbf{u}_{N-2}'\mathbf{R}\mathbf{u}_{N-2},$$

i.e.,

$$\mathbf{u}_{N-2} = -\mathbf{K}_{N-2}\mathbf{x}_{N-2}$$

---

* If a quadratic function is objectionable since it implies underachieving a desired level or path, and overachieving them is penalized equally, we may wish to use piecewise quadratic functions and assign different penalties to deviations on either side. (See B. Friedman 1975.) This is, however, a detail of technicalities. Basic analytical tools will be the same.

where

$$\mathbf{K}_{N-2} = [\mathbf{R} + \mathbf{B}'(\mathbf{Q} + \mathbf{\Pi}_{N-1})]^{-1}\mathbf{B}'(\mathbf{Q} + \mathbf{\Pi}_{N-1})\mathbf{A}$$

which results in

$$\begin{aligned}\gamma_{N-1}^* &= \min_{u_{N-2}} (\gamma_N^* + \lambda_{N-1}) \\ &= \mathbf{x}_{N-2}'\mathbf{\Pi}_{N-2}\mathbf{x}_{N-2}\end{aligned}$$

where

$$\begin{aligned}\mathbf{\Pi}_{N-2} = \mathbf{A}'\{&(\mathbf{Q} + \mathbf{\Pi}_{N-1}) \\ &- (\mathbf{Q} + \mathbf{\Pi}_{N-1})\mathbf{B}[\mathbf{R} + \mathbf{B}'(\mathbf{Q} + \mathbf{\Pi}_{N-1})\mathbf{B}]^{-1}\mathbf{B}'(\mathbf{Q} + \mathbf{\Pi}_{N-1})\}\mathbf{A}.\end{aligned} \tag{49}$$

In general

$$\begin{aligned}\mathbf{u}_k &= -\mathbf{K}_k\mathbf{x}_k \\ \mathbf{K}_k &= [\mathbf{R} + \mathbf{B}'(\mathbf{Q} + \mathbf{\Pi}_{k+1})\mathbf{B}]^{-1}\mathbf{B}'(\mathbf{Q} + \mathbf{\Pi}_{k+1})\mathbf{A}\end{aligned}$$

and

$$\begin{aligned}\gamma_k^* &= \min_{u_k} (\gamma_{k+1}^* + \lambda_k) \\ &= \mathbf{x}_k'\mathbf{\Pi}_k\mathbf{x}_k\end{aligned}$$

where

$$\begin{aligned}\mathbf{\Pi}_k = \mathbf{A}'\{&(\mathbf{Q} + \mathbf{\Pi}_{k+1}) \\ &- (\mathbf{Q} + \mathbf{\Pi}_{k+1})\mathbf{B}[\mathbf{R} + \mathbf{B}'(\mathbf{Q}' + \mathbf{\Pi}_{k+1})\mathbf{B}]^{-1}\mathbf{B}'(\mathbf{Q} + \mathbf{\Pi}_{k+1})\}\mathbf{A}.\end{aligned}$$

Example 7: Scalar Control Problem. Suppose we have a scalar deterministic control system described by the first-order scalar difference equation

$$x_{i+1} = ax_i + bu_i, \qquad u_i \in (-\infty, \infty), \qquad 0 \leqslant i \leqslant N-1 \tag{49}$$

where x, a, b, and u are all taken to be scalar quantities and a and b are known and x_i is exactly observed. The critierion function is taken to be

$$J = x_N^2.$$

This control problem has a very simple optimal control policy. Since

$$J = x_N^2 = (ax_{N-1} + bu_{N-1})^2,$$

clearly an optimal control variable at time $N-1$, denoted by u_{N-1}^*, is given by

$$u_{N-1}^* = -ax_{N-1}/b,$$

$u_0^*, u_1^*, \ldots, u_{N-2}^*$ are arbitrary and $\min J = 0$. Actually in this example we can choose any one or several of the N control variables $u_0, u_1, \ldots, u_{N-1}$ to appropriately minimize J. For the purpose of later comparisons we will choose $u_i^* = -ax_i/b$, $i = 0, 1, \ldots, N-1$.

We see that this control policy requires, among other things, that

(a) a and b of (49) be exactly known and
(b) x_{N-1} be exactly observable.

When both of these conditions are not satisfied, the optimal control problem of even such a simple problem is no longer trivial. Optimal control problems without assumptions (a) and/or (b) will be discussed later.

In some cases, we may want to include cross product terms in J, so that

$$W_k = (\mathbf{x}_k' \quad \mathbf{u}_k') \begin{pmatrix} \mathbf{Q}_k & \mathbf{S}_k \\ \mathbf{S}_k' & \mathbf{R}_k \end{pmatrix} \begin{pmatrix} \mathbf{x}_k \\ \mathbf{u}_k \end{pmatrix}, \tag{52a}$$

where

$$\begin{pmatrix} \mathbf{Q}_k & \mathbf{S}_k \\ \mathbf{S}_k' & \mathbf{R}_k \end{pmatrix} \geqslant 0, \qquad \mathbf{R}_k > 0.$$

In some cases, we may want to use **y**s instead of **x**s.

$$W_k = (\mathbf{y}_k' \quad \mathbf{u}_k') \begin{pmatrix} \mathbf{N}_k & \mathbf{L}_k \\ \mathbf{L}_k & \mathbf{M}_k \end{pmatrix} \begin{pmatrix} \mathbf{y}_k \\ \mathbf{u}_k \end{pmatrix}, \qquad \mathbf{N}_k \geqslant 0, \qquad \mathbf{M}_k > 0. \tag{52b}$$

All of these cases can be treated uniformly by introducing an artificial observation equation

$$\bar{\mathbf{y}}_k = \mathbf{C}_k \mathbf{x}_k + \mathbf{D}_k \mathbf{u}_k \tag{53}$$

and rewriting (45) as

$$J = \mathbf{x}_N' \mathbf{\Gamma} \mathbf{x}_N + \sum_{K=0}^{N-1} \bar{\mathbf{y}}_k' \bar{\mathbf{y}}_k. \tag{54}$$

In the case of (46), the matrices

$$\mathbf{C}_k = \begin{pmatrix} \mathbf{Q}_k^{1/2} \\ \mathbf{0} \end{pmatrix} \quad \text{and} \quad \mathbf{D}_k = \begin{pmatrix} \mathbf{0} \\ \mathbf{R}_k^{1/2} \end{pmatrix}$$

will put it in the form of (53) and (54). Note that $\mathbf{C}_k' \mathbf{D}_k = \mathbf{0}$ for (46).

When the cross-product terms are present in W_k as in (52a), the choice of matrices

$$\mathbf{C}_k = \begin{pmatrix} (\mathbf{Q}_k - \mathbf{S}_k\mathbf{R}_k^{-1}\mathbf{S}_k')^{1/2} \\ \mathbf{R}_k^{-1/2}\mathbf{S}_k' \end{pmatrix} \quad \text{and} \quad \mathbf{D}_k = \begin{pmatrix} \mathbf{0} \\ \mathbf{R}_k^{1/2} \end{pmatrix}$$

will accomplish the same purpose. Note that, by assumption,

$$\begin{pmatrix} \mathbf{Q}_k & \mathbf{S}_k \\ \mathbf{S}_k & \mathbf{R}_k \end{pmatrix} \geqslant 0,$$

hence the square root of $\mathbf{Q}_k - \mathbf{S}_k\mathbf{R}_k^{-1}\mathbf{S}_k'$ is well defined. Finally, if W_k is given by (52b) and if the target vector is related to $\mathbf{z}$ and $\mathbf{x}$ by

$$\mathbf{y}_k = \mathbf{H}_k\mathbf{x}_k + \mathbf{g}_k\mathbf{u}_k,$$

then

$$W_k = (\mathbf{x}_k' \quad \mathbf{u}_k')\begin{pmatrix} \mathbf{H}_k' \\ \mathbf{G}_k' \end{pmatrix}\begin{pmatrix} \mathbf{N}_k & \mathbf{L}_k \\ \mathbf{L}_k' & \mathbf{M}_k \end{pmatrix}(\mathbf{H}_k \quad \mathbf{G}_k)\begin{pmatrix} \mathbf{x}_k \\ \mathbf{u}_k \end{pmatrix}.$$

All we need to do to reduce it to the form in (54) is to define

$$\begin{pmatrix} \mathbf{Q}_k & \mathbf{S}_k \\ \mathbf{S}_k' & \mathbf{R}_k \end{pmatrix} = \begin{pmatrix} \mathbf{H}_k' \\ \mathbf{G}_k' \end{pmatrix}\begin{pmatrix} \mathbf{N}_k & \mathbf{L}_k \\ \mathbf{L}_k' & \mathbf{M}_k \end{pmatrix}(\mathbf{H}_k \quad \mathbf{G}_k)$$

and construct $\mathbf{C}_k$ and $\mathbf{D}_k$ accordingly.

Without loss of generality, then, we may say that the problem is to minimize (54) subject to (47) and (53) (we drop the overbar from the ys):

$$\text{minimize}\, \mathbf{x}_N'\mathbf{\Gamma}\mathbf{x}_N + \sum_{k=0}^{N-1} W_k$$

subject to

$$\mathbf{x}_{k+1} = \mathbf{A}_k\mathbf{x}_k + \mathbf{B}_k\mathbf{u}_k$$
$$\mathbf{y}_k = \mathbf{C}_k\mathbf{x}_k + \mathbf{D}_k\mathbf{u}_k$$

where

$$W_k = \mathbf{y}_k'\mathbf{y}_k.$$

Optimal Control Law

Denote by $\mathbf{J}_{k,N}(\mathbf{x}_k, \Gamma; \{\mathbf{u}_{k,N}\})$ the value of

$$\mathbf{x}_N'\mathbf{\Gamma}\mathbf{x}_N + \sum_{N=k}^{N-1} W_k$$

where $\{\mathbf{u}_{k,N}\}$ stands for $\mathbf{u}_k, \mathbf{u}_{k+1}, \ldots, \mathbf{u}_{N-1}$.

Quadratic Welfare Function

Let the optimal cost starting from the state x_k at time k be denoted by superscript *,

$$J^*_{k,N}(\mathbf{x}_k, \mathbf{\Gamma}) = \min_{\{\mathbf{u}_{k,N}\}} J_{k,N}(\mathbf{x}_k, \mathbf{\Gamma}; \{\mathbf{u}_{k,N}\}).$$

The subscripts k and N indicate that $[k, N]$ is the control horizon. Then by the principle of optimality of Bellman, $J^*_{k,N}$ satisfies the functional equation

$$J^*_{k,N}(\mathbf{x}_k, \mathbf{\Gamma}) = \min_k [\|\mathbf{y}_k\|^2 + J^*_{k+1,N}(\mathbf{x}_{k+1}, \mathbf{\Gamma})]. \tag{57}$$

A standard dynamic programming backward induction argument gives us the sequence of optimal instruments $\{\mathbf{u}^*_{k,N}\}$.

Consider time $N-1$ so that $\mathbf{u}_{N-1}$ is the only instrument yet to be determined. Since

$$\begin{aligned}\mathrm{J}_{N-1,N}(\mathbf{x}_{N-1}; \mathbf{\Gamma}; \mathbf{u}_{N-1}) &= \mathbf{x}'_k \mathbf{\Gamma} \mathbf{x}_N + \mathbf{y}'_{N-1}\mathbf{y}_{N-1} \\ &= \mathbf{x}'_N \mathbf{\Gamma} \mathbf{x}_N + \|\mathbf{C}_{N-1}\mathbf{x}_{N-1} + \mathbf{D}_{N-1}\mathbf{u}_{N-1}\|^2,\end{aligned}$$

we obtain

$$\mathbf{x}^*_{N-1} = -(\mathbf{B}'_{N-1}\mathbf{\Gamma}\mathbf{B}_{N-1} + \mathbf{D}'_{N-1}\mathbf{D}_{N-1})^+(\mathbf{B}'_{N-1}\mathbf{\Gamma}\mathbf{A}_{N-1} + \mathbf{D}'_{N-1}\mathbf{C}_{N-1})\mathbf{x}_{N-1}$$

and

$$J^*_{N-1,N}(\mathbf{x}_{N-1}, \mathbf{\Gamma}; \mathbf{u}^*_{N-1}) = \mathbf{x}'_{N-1}\mathbf{\Gamma}_{N-1,N}(\mathbf{\Gamma})\mathbf{x}_{N-1}$$

where

$$\begin{aligned}\mathbf{\Pi}_{N-1,N}(\mathbf{\Gamma}) = \mathbf{A}'_{N-1}\mathbf{\Gamma}\mathbf{A}_{N-1} &- (\mathbf{B}'_{N-1}\mathbf{\Gamma}\mathbf{A}_{N-1} + \mathbf{D}'_{N-1}\mathbf{C}_{N-1})'(\mathbf{B}'_{N-1}\mathbf{\Gamma}\mathbf{B}_{N-1} \\ &+ \mathbf{C}'_{N-1}\mathbf{C}_{N-1})^+(\mathbf{B}'_{N-1}\mathbf{\Gamma}\mathbf{A}_{N-1} + \mathbf{D}'_{N-1}\mathbf{C}_{N-1}).\end{aligned}$$

By mathematical induction, we can easily show that the solution to (7) is given by

$$\mathrm{J}^*_{k,N}(\mathbf{x}_k, \mathbf{\Gamma}; \{\mathbf{u}^*_{k,N}\}) = \mathbf{x}'_k \mathbf{\Pi}_{k,N}(\mathbf{\Gamma})\mathbf{x}_k, \tag{58}$$

where $\mathbf{\Pi}_{k,N}(\mathbf{\Gamma})$ satisfies what is called the algebraic matrix Riccati equation

$$\begin{aligned}\mathbf{\Pi}_{j,N}(\mathbf{\Gamma}) &= \mathbf{A}'_j\mathbf{\Pi}_{j+1,N}(\mathbf{\Gamma})\mathbf{A}_j + \mathbf{C}'_j\mathbf{C}_j \\ &- [\mathbf{B}'_j\mathbf{\Pi}_{j+1,N}(\mathbf{\Gamma})\mathbf{A}_j + \mathbf{D}'_j\mathbf{C}_j]'[\mathbf{B}'_j\mathbf{\Pi}_{j+1,N}(\mathbf{\Gamma})\mathbf{B}_j + \mathbf{D}'_j\mathbf{D}_j]^+ \\ &\cdot [\mathbf{B}'_j\mathbf{\Pi}_{j+1,N}(\mathbf{\Gamma})\mathbf{A}_j + \mathbf{D}'_j\mathbf{C}_j], \qquad \mathbf{\Pi}_{N,N}(\mathbf{\Gamma}) = \mathbf{\Gamma},\end{aligned} \tag{59}$$

and

$$\text{(60)}\quad \begin{aligned} \mathbf{u}_j^* &= -[\mathbf{B}_j'\mathbf{\Pi}_{j+1,N}(\mathbf{\Gamma})\mathbf{B}_j + \mathbf{D}_j'\mathbf{D}_j]^+ [\mathbf{B}_j'\mathbf{\Pi}_{t+1,N}(\mathbf{\Gamma})\mathbf{A}_j + \mathbf{D}_j'\mathbf{C}_j]\mathbf{x}_j, \\ & k \leqslant j \leqslant N-1. \end{aligned}$$

Note that the optimal instrument is linear in $\mathbf{x}_k$ with time-varying weights assigned to the components of it. The computation of these optimal combinations of the components of $\mathbf{x}_k$ requires that the values of the matrices $\mathbf{A}, \mathbf{B}, \mathbf{C}$, and $\mathbf{D}$ are known. This concludes the derivation of the sequence of optimal instruments for the discrete-time dynamic systems.

Denote $\mathrm{J}_{k,N}^*(\mathbf{x}, 0)$ by $\mathrm{V}(\mathbf{x}, k)$ for short. Along the optimal trajectory from $\mathbf{x}_k$ to $\mathbf{x}_N$,

$$\begin{aligned} \Delta\mathrm{V}(\mathbf{x}_k, k) &= \mathrm{V}(\mathbf{x}_{k+1}, k+1) - \mathrm{V}(\mathbf{x}_k, k) \\ &= \min_{u_{k+1,N}} \left(\sum_{k+1}^{N-1} \mathbf{y}_l'\mathbf{y}_l + \mathbf{x}_N'\mathbf{\Gamma}\mathbf{x}_N \right) \\ &\quad - \min_{(u_{k,N})} \left(\sum_{k}^{N-1} \mathbf{y}_l'\mathbf{y}_l + \mathbf{x}_N'\mathbf{\Gamma}\mathbf{x}_N \right) \\ &= -\|\mathbf{C}\mathbf{x}_k + \mathbf{D}\mathbf{u}_k\|^2 \leqslant 0. \end{aligned}$$

On every interval $[k, m]$, $k \leqslant m$, perfect observability implies that there is at least one l, $k \leqslant l \leqslant m$ such that $y_l \neq 0$, i.e.,

$$\Delta\mathrm{V}(\mathbf{x}_l, l) < 0$$

By the Lyapunov theorem, the system is therefore asymptotically stable.

In the usual control problem, the kth period has the cost

$$\begin{aligned} W_k &= \mathbf{x}_k'\mathbf{Q}\mathbf{x}_k + \mathbf{u}_k'\mathbf{R}\mathbf{u}_k, \\ \mathbf{Q}' &= \mathbf{Q} \geqslant 0 \quad \text{and} \quad \mathbf{R}' = \mathbf{R} > 0. \end{aligned}$$

Partition the matrices as

$$\mathbf{Q} = \mathbf{L}'\mathbf{L} \quad \text{and} \quad \mathbf{R} = \mathbf{H}'\mathbf{H}.$$

Then define

$$\mathbf{y}_k = \begin{bmatrix} \mathbf{L} \\ \mathbf{0} \end{bmatrix} \mathbf{x}_k + \begin{bmatrix} \mathbf{0} \\ \mathbf{H} \end{bmatrix} \mathbf{u}_k.$$

Then $\mathbf{S}_k = \mathbf{y}_k'\mathbf{y}_k$ and the matrices

$$\mathbf{C} = \begin{bmatrix} \mathbf{L} \\ \mathbf{0} \end{bmatrix} \quad \text{and} \quad \mathbf{D} = \begin{bmatrix} \mathbf{0} \\ \mathbf{H} \end{bmatrix}$$

are such that $\mathbf{C}'\mathbf{D} = 0$. Furthermore, note the positions of the zero submatrices in the observability matrix

$$\mathbf{O}_n = \begin{bmatrix} \mathbf{C} \\ \mathbf{CA} \\ \vdots \\ \mathbf{CA}^{n-1} \end{bmatrix} + \begin{bmatrix} \mathbf{L} \\ \mathbf{O} \\ \mathbf{LA} \\ 0 \\ \vdots \\ \mathbf{LA}^{n-1} \\ 0 \end{bmatrix}$$

and

$$\mathbf{J}_n = \begin{bmatrix} \mathbf{C} & \\ \mathbf{CA} & \vdots \\ \vdots & \\ \mathbf{CA}^{n-2}\mathbf{B} & \mathbf{D} \end{bmatrix} = \begin{bmatrix} \mathbf{0} & \mathbf{0} & \mathbf{0} \\ \mathbf{H} & \mathbf{0} & \mathbf{0} \\ \mathbf{LB} & \mathbf{0} & \vdots \\ \mathbf{0} & \mathbf{H} & \\ \vdots & & \mathbf{0} \\ \mathbf{LA}^{n-2}\mathbf{B} & \vdots & \mathbf{0} \\ \mathbf{0} & & \mathbf{H} \end{bmatrix}.$$

The column vectors of $\mathbf{O}_n$ and $\mathbf{J}_n$ are linearly independent, i.e., in this problem, the observability implies perfect observability.

1.11 Square Root Formulation

Let $\mathbf{\Gamma} = \mathbf{M}_N'\mathbf{M}_N$, where $\mathbf{M}_N$ is full rank. Stack the $\mathbf{y}$ vectors as

$$\mathbf{y}^t = \begin{pmatrix} \mathbf{y}_t \\ \vdots \\ \mathbf{y}_0 \end{pmatrix}.$$

Then the control cost functional has a convenient expression

$$J_{0,N} = \|\mathbf{y}^{N-2}\|^2 + \left\| \begin{pmatrix} \mathbf{y}_{N-1} \\ \mathbf{M}\mathbf{x}_N \end{pmatrix} \right\|^2, \tag{61}$$

which shows that the last-period control $\mathbf{x}_{N-1}$ affects only the last term in $J_{0,N}$.

Noting that

$$\begin{pmatrix} \mathbf{y}_{N-1} \\ \mathbf{M}_N \mathbf{x}_N \end{pmatrix} = \begin{bmatrix} \mathbf{C}_{N-1} & \mathbf{D}_{N-1} \\ \mathbf{M}_N \mathbf{A} & \mathbf{M}_N \mathbf{B} \end{bmatrix} \begin{bmatrix} \mathbf{x}_{N-1} \\ \mathbf{u}_{N-1} \end{bmatrix},$$

the optimal $\mathbf{x}_{N-1}$ will be linear in $\mathbf{u}_{N-1}$. To obtain this relation simply, we apply an orthogonal transformation to change the matrix into a lower block triangular form

$$\mathbf{S}_N \begin{pmatrix} \mathbf{y}_{N-1} \\ \mathbf{M}_N \mathbf{x}_{N-1} \end{pmatrix} = \begin{pmatrix} \mathbf{M}_{N-1} & \mathbf{0} \\ \mathbf{G}_{N-1} & \mathbf{F}_{N-1} \end{pmatrix} \begin{pmatrix} \mathbf{x}_{N-1} \\ \mathbf{u}_{N-1} \end{pmatrix},$$

where $\mathbf{F}_{N-1}$ has full rank and where $\mathbf{S}'_N \mathbf{S}_N = \mathbf{I}$. Applying the orthgonal transformation leves the norm of the vector the same, i.e., the magnitude of the last term in (61) remains the same, i.e., the contribution of the last term to $J_{0,N}$ is not altered by this transformation but now we can write the optimality condition for $\mathbf{x}_{N-1}$ by inspection as

$$\mathbf{G}_{N-1}\mathbf{x}_{N-1} + \mathbf{F}_{N-1}\mathbf{x}_{N-1} = 0.$$

Denoting the last term of (61) as γ_N, its minimal value, γ_N^*, is

$$\gamma_N^* = \left\| \begin{pmatrix} \mathbf{M}_{N-1}\mathbf{x}_{N-1} \\ 0 \end{pmatrix} \right\|^2.$$

Rewrite J_{0N} as

$$J_{0N} = \|\mathbf{y}^{N-3}\|^2 + \lambda_{N-2}$$

where

$$\lambda_{N-2} = \left\| \begin{pmatrix} \mathbf{y}_{N-2} \\ \mathbf{M}_{N-1}\mathbf{x}_{N-1} \end{pmatrix} \right\|^2.$$

Then u_{N-2} influences only λ_{N-2}.

In general

$$J_{0,N} = \|\mathbf{y}^{k-1}\|^2 + \lambda_k$$

where

$$\lambda_k = \left\| \begin{pmatrix} \mathbf{y}_k \\ \mathbf{M}_{k+1}\mathbf{x}_{k+1} \end{pmatrix} \right\|^2.$$

Find an orthogonal matrix $\mathbf{S}_{N+1}$ such that

$$\mathbf{S}_{k+1} \begin{bmatrix} \mathbf{y}_k \\ \mathbf{M}_{k+1}\mathbf{x}_{k+1} \end{bmatrix} = \begin{pmatrix} \mathbf{M}_k & \mathbf{0} \\ \mathbf{G}_k & \mathbf{F}_k \end{pmatrix} \begin{pmatrix} \mathbf{x}_k \\ \mathbf{u}_k \end{pmatrix}, \tag{62}$$

where $\mathbf{F}_k$ is full rank, from which λ_k is minimized by $\mathbf{u}_k$ satisfying

$$\mathbf{G}_k\mathbf{x}_k + \mathbf{F}_k\mathbf{u}_k = 0$$

and

$$\lambda_k^* = \|\mathbf{M}_k\mathbf{x}_k\|^2.$$

By deriving the recursion equation for $\mathbf{M}_k'\mathbf{M}_k$, we next show that it satisfies the same recursion equation as $\mathbf{\Pi}_{k,N}(\mathbf{\Gamma})$. Equation (62) shows that

$$\mathbf{S}_{k+1}\begin{bmatrix} \mathbf{C}_k & \mathbf{D}_k \\ \mathbf{M}_{k+1}\mathbf{A} & \mathbf{M}_{k+1}\mathbf{B} \end{bmatrix} = \begin{bmatrix} \mathbf{M}_k & \mathbf{0} \\ \mathbf{G}_k & \mathbf{F}_k \end{bmatrix}$$

from which

$$\begin{pmatrix} \mathbf{M}_k' & \mathbf{G}_k' \\ \mathbf{0} & \mathbf{F}_k' \end{pmatrix}\begin{pmatrix} \mathbf{M}_k & \mathbf{0} \\ \mathbf{G}_k & \mathbf{F}_k \end{pmatrix} = \begin{pmatrix} \mathbf{C}_k' & \mathbf{A}'\mathbf{M}_{k+1} \\ \mathbf{D}_k' & \mathbf{B}'\mathbf{M}_{k+1} \end{pmatrix}\begin{pmatrix} \mathbf{C}_k & \mathbf{D}_k \\ \mathbf{M}_{k+1}\mathbf{A} & \mathbf{M}_{k+1}\mathbf{B} \end{pmatrix}$$

or

$$\mathbf{\Pi}_k = \mathbf{A}'\mathbf{\Pi}_{k+1}\mathbf{A} + \mathbf{Q} - \mathbf{G}_k'\mathbf{G}_k$$

$$\mathbf{F}_k'\mathbf{G}_k = \mathbf{B}'\mathbf{\Pi}_{k+1}\mathbf{A}$$

because $\mathbf{C}_k'\mathbf{D}_k = 0$ and

$$\mathbf{F}_k'\mathbf{F}_k = \mathbf{R} + \mathbf{B}'\mathbf{\Pi}_{k+1}\mathbf{B}.$$

Identify $(\mathbf{R} + \mathbf{B}'\mathbf{\Pi}_{k+1}\mathbf{B})^{1/2}$ with $\mathbf{F}_k'$. Then

$$\mathbf{G}_k'\mathbf{G}_k = \mathbf{A}'\mathbf{\Pi}_{k+1}\mathbf{B}(\mathbf{R} + \mathbf{B}'\mathbf{\Pi}_{k+1}\mathbf{B})^{-1}\mathbf{B}'\mathbf{\Pi}_{k+1}\mathbf{A}.$$

Collecting terms,

$$\mathbf{\Pi}_k = \mathbf{A}'\mathbf{\Pi}_{k+1}\mathbf{A} - \mathbf{A}'\mathbf{\Pi}_{k+1}\mathbf{B}(\mathbf{R} + \mathbf{B}'\mathbf{\Pi}_{k+1}\mathbf{B})^{-1}\mathbf{B}'\mathbf{\Pi}_{k+1}\mathbf{A} + \mathbf{Q}.$$

When $\mathbf{C}_k'\mathbf{D}_k \neq 0$,

$$\mathbf{F}_k'\mathbf{G}_k = \mathbf{D}_k'\mathbf{C}_k + \mathbf{B}'\mathbf{\Pi}_{k+1}\mathbf{A}$$

or

$$\mathbf{G}_k = (\mathbf{R} + \mathbf{B}'\mathbf{\Pi}_{k+1}\mathbf{B})^{-1/2}(\mathbf{D}_k'\mathbf{C}_k + \mathbf{B}'\mathbf{\Pi}_{k+1}\mathbf{A})$$

and

$$\mathbf{G}_k'\mathbf{G}_k = (\mathbf{A}'\mathbf{\Pi}_{k+1}\mathbf{B} + \mathbf{C}_k'\mathbf{D}_k)(\mathbf{D}_k'\mathbf{D}_k + \mathbf{B}'\mathbf{\Pi}_{k+1}\mathbf{B})^{-1}$$
$$\times (\mathbf{D}_k'\mathbf{C}_k + \mathbf{B}'\mathbf{\Pi}_{k+1}\mathbf{A}).$$

or

$$\mathbf{\Pi}_k = \mathbf{A}'\mathbf{\Pi}_{k+1}\mathbf{A} - (\mathbf{A}'\mathbf{\Pi}_{k+1}\mathbf{B} + \mathbf{C}_k'\mathbf{D}_k)(\mathbf{D}_k'\mathbf{D}_k + \mathbf{B}'\mathbf{\Pi}_{k+1}\mathbf{B})^{-1} \times (\mathbf{D}_k'\mathbf{C}_k + \mathbf{B}'\mathbf{\Pi}_{k+1}\mathbf{A}).$$

Since $\mathbf{\Pi}_N = \mathbf{\Gamma}$, $\mathbf{M}_k'\mathbf{M}_k$ and $\mathbf{\Pi}_{k,N}(\Gamma)$ coincide if the Riccati equation has the unique symmetric positive infinite matrix solution.

Appendices

Intertemporal Maximization Problem

This appendix illustrates the types of conditions needed to ensure well-posed utility maximization problems.

Consider maximizing $\Sigma\beta^t u(c_t)$ where $u(c_t)$ is unbounded if c is taken to be in an unbounded region in some Euclidean space. The corresponding functional equation is

$$\gamma^*(k) = \sup_c [u(c) + \beta\gamma^*(k_+)],$$

where k_+ is the next state to which k is moved when a feasible c is chosen.

Write the right-hand side as $(T\gamma^*)(k)$, i.e., $\gamma^*(k) = (T\gamma^*)(k)$, and note that $\gamma^*(k) \geqslant u(c) + \beta\gamma^*(k)$, since $k_+ \neq k$ in general.

Assume that $\Sigma_0^\infty \beta^t u(c_t)$ is finite for all feasible consumption streams. Suppose that it is majorized by $V(k)$ for all feasible consumption streams and assume that $(TV)(k) \geqslant V(k)$. Clearly $\gamma^*(k) \leqslant V(k)$. Then $\gamma^*(k) \leqslant (TV)(k)$ and iterating we obtain $\gamma^*(k) \leqslant (T^n V)(k)$, for all $n \geqslant 0$.

Assume further that the limit exists and denote it by $V^*(k)$. Thus $\gamma^*(k) \leqslant V^*(k)$. Note that $V^*(k) = \lim (T^{n+1} V)(k) = TV^*(k)$. To establish $\gamma^* = V^*$, we proceed as follows. Iterating $\gamma^*(k) \geqslant u(c) + \beta\gamma^*(k)$, we obtain $\gamma^*(k) \geqslant \Sigma_1^n \beta^s u(c_s) + \beta^{n+1}\gamma^*(k)$ and taking the limit $\gamma^*(k) \geqslant V^*(k) + \overline{\lim}\, \beta^n \gamma^*(k)$. Thus if $\overline{\lim}\, T^n V^*(k) \geqslant 0$, then $\gamma^*(k) \geqslant V^*(k)$ follows, i.e., $\gamma^*(k) = V^*(k)$. Starting from a guess which majorizes the discounted sum of the utility stream, we can thus arrive at $\gamma^*(k)$ under the stated conditions above.

Linear Algebraic Equations

We often encounter linear algebraic equations as the first-order condition of some optimization problem, as a part of defining system properties, or as a set of an overdetermined system of equations for unknown systems parameters. Consider

(A1) $$\mathbf{Qx} = \mathbf{b}$$

where $\mathbf{Q}$ is $m \times n$, $m \geqslant n$, for example.

The solution to (A1) has the general form

$$\mathbf{x} = \mathbf{Q}^{+}\mathbf{b} + (\mathbf{I} - \mathbf{Q}^{+}\mathbf{Q})\mathbf{c}$$

where c is arbitrary and where $\mathbf{Q}^{+}$ is the Moore-Penrose pseudoinverse. This pseudoinverse is defined as

$$\mathbf{Q}^{+} = \mathbf{V}\boldsymbol{\Sigma}^{-1}\mathbf{U}'$$

where $\mathbf{Q} = \mathbf{U}\boldsymbol{\Sigma}\mathbf{V}'$ is its singular-value decomposition, where $\mathbf{V}$ is $m \times p$, $\boldsymbol{\Sigma}$ is $p \times p$, $\mathbf{U}$ is $n \times p$, and $p = \operatorname{rank}\mathbf{Q}$. Such a solution exists if and only if

$$\mathbf{Q}\mathbf{Q}^{+}\mathbf{b} = \mathbf{b}$$

i.e., if $\mathbf{b}$ is in the range space of $\mathbf{Q}$.

Matrix-Fraction Description

A transfer function $\mathrm{G}(z)$ put in the form

$$\mathbf{G}(\mathbf{z}) = \mathbf{D}^{-1}(z)\mathbf{N}(z)$$

where $\mathbf{D}(z)$ and $\mathbf{N}(z)$ are matrix polynomials is called (left) matrix-fraction description. For example,

$$\mathbf{G}(z) = \mathbf{C}(\lambda\mathbf{I} - \mathbf{A})^{-1}\mathbf{B} = \frac{1}{d(z)}\mathbf{C}\operatorname{adj}(\lambda\mathbf{I} - \mathbf{A})\mathbf{B}$$

where $d(z) = |\lambda\mathbf{I} - \mathbf{A}|$ is a matrix fraction description by setting $\mathbf{D}(z) = d(z)\mathbf{I}$.

There are many pairs of $\mathbf{D}(z)$ and $\mathbf{N}(z)$ for a given $\mathbf{G}(z)$. A matrix $\mathbf{W}(z)$ is said to be unimodular when $|\mathbf{W}(z)|$ is a nonzero constant independent of λ. $\mathbf{W}(z)$ is a common right divisor if $\mathbf{D}(z) = \bar{\mathbf{D}}(z)\mathbf{W}(z)$ and $\mathbf{N}(z) = \bar{\mathbf{N}}(z)\mathbf{W}(z)$. When $\mathbf{D}(z)$ and $\mathbf{N}(z)$ are right coprime (i.e., they only have unimodular common right divisors), $\mathbf{D}(z)$ and $\mathbf{N}(z)$ are said to be irreducible. See Kailath (1980) for extraction of right divisors.

The Smith-McMillan Forms

By elementary row operations, we mean operations involving only (1) multiplication of a row by a constant, (2) interchanges of two rows, and (3) addition of a multiple of one row to another. By elementary row and column operations, any $p \times q$ polynomial matrix $\mathbf{P}(z)$ can be put into

$$\mathbf{U}(z)\mathbf{P}(z)\mathbf{V}(z) = \boldsymbol{\Lambda}(z)$$

where $\mathbf{U}(z)$ and $\mathbf{V}(z)$ are unimodular matrices and $\mathbf{\Lambda}(z)$ is the $p \times q$ matrix

$$\mathbf{\Lambda}(z) = \begin{bmatrix} \lambda_1(z) & & & & 0 \\ & \ddots & & & \vdots \\ & & \lambda_r(z) & & 0 \\ & & & \ddots & \\ 0 & \cdots & 0 & \cdots & 0 \end{bmatrix}$$

where r = rank of $P(z)$ (a.e.) and λ's are monic polynomials arranged so that $\lambda_i(z)$ divides $\lambda_{i+1}(z)$, $i = 1 \ldots i - 1$. The matrix $\mathbf{\Lambda}(z)$ is said to be the Smith form. Starting with a rational matrix $\mathbf{G}(z)$

$$\mathbf{G}(z) = \mathbf{N}(z)/d(z),$$

where $d(z)$ is the monic least common multiple of the denominator of the elements of $\mathbf{G}(z)$, put $d(z)\mathbf{G}(z) = \mathbf{N}(z)$ into the Smith form $\mathbf{\Lambda}(z)$ so that we obtain

$$\mathbf{N}(z) = \mathbf{U}_1(z)\mathbf{\Lambda}(z)\mathbf{U}_2(z),$$

or

$$d(z)\mathbf{G}(z) = \mathbf{U}_1(z)\mathbf{\Lambda}(z)\mathbf{U}_2(z),$$

which can be put into

$$\mathbf{U}_1^{-1}(z)\mathbf{G}(z)\mathbf{U}_2^{-1}(z) = \frac{1}{d(z)}\mathbf{\Lambda}(z) = \operatorname{diag}\{\lambda_i(z)/d(z)\}$$

and cancel out all common factors from $\lambda_i(z)$ and $d(z)$, i.e.,

$$\lambda_i(z)/d(z) = \varepsilon_i(z)/\psi_i(z),$$

where $\varepsilon_i(z)$ and $\psi_i(z)$ are coprime, $i = 1 \ldots r$, and where $r = \operatorname{rank} \mathbf{G}(z)$. We have expressed $\mathbf{G}(z)$ as

$$\mathbf{G}(z) = \mathbf{U}_1(z) \operatorname{diag}[\varepsilon_i(z)/\psi_i(z)]\mathbf{U}_2(z).$$

This is the Smith-McMillan form. By construction $\psi_i(z)$ is divisible by $\psi_{i+1}(z)$ and $\psi_1(z) = d(z)$. The sum of the degrees of polynomials $\psi_i(z)$

$$n^* = \sum_i \deg \psi_i(z)$$

is called the McMillan degree.

When $d(z)$ has no repeated roots, the method of Gilbert (see Aoki 1987C) can express $\mathbf{G}(z)$ as

$$\mathbf{G}(z) = \sum_1^r \frac{\mathbf{R}_i}{\lambda - \lambda_i}, \qquad r = \operatorname{rank} \mathbf{G}(z)$$

and

$$n^* = \sum_i \text{rank } \mathbf{R}_i.$$

The McMillan degree can be shown to be a generalization of n^* obtained here to $d(z)$ with repeated roots, i.e., n^* is the dimension of the minimal realization state vector.

Example 8.

$$\mathbf{G}(z) = \frac{1}{d(z)} \begin{bmatrix} z & z(z-.5)^2 \\ -z(z-.5)^2 & -.25z(z-.5)^2 \end{bmatrix}$$

where

$$d(z) = (z-.5)^2(z-1)^2.$$

Then

$$\mathbf{G}(z) = \begin{bmatrix} 1 & 0 \\ -(z-1)^2 & 1 \end{bmatrix} \begin{bmatrix} z/(z-1)^2(z-.5)^2 & 0 \\ 0 & z^2/(z-1) \end{bmatrix} \begin{bmatrix} 1 & (z-.5)^2 \\ 0 & 1 \end{bmatrix}$$

is in the Smith-McMillan form. The McMillan degree is 5.

CHAPTER 2

Stochastic Models

2.1 Introduction

When transition equations or the observation equations for the state vector are not deterministic, we have stochastic models. In the time domain, a model in state space representation is specified by two equations:

$$\mathbf{x}_{t+1} = \mathbf{f}(\mathbf{x}_t, \mathbf{u}_t, \xi_t, t),$$

and

$$\mathbf{y}_t = \mathbf{h}(\mathbf{x}_t, \mathbf{u}_t, \boldsymbol{\eta}_t, t),$$

where $\mathbf{x}_t$ is the unobserved state vector, $\mathbf{u}_t$ is the decision vector, and $\boldsymbol{\xi}_t$s and $\boldsymbol{\eta}_t$s are exogenous disturbances. These are stochastic difference equations. The first is the dynamic transition equation which describes how $\mathbf{x}_t$ evolves with time, and the second observation equation relates $\mathbf{x}_t$ to the current observation $\mathbf{y}_t$.

When production functions are stochastic in the social planner's problem and rates of return are stochastic in the optimal saving problems in Chapter 1, these problems become stochastic control problems and their state equations are examples of stochastic models discussed in this chapter.

Linear Stochastic Models

A linear state space model consists of the dynamic or transition equation

$$\mathbf{x}_{t+1} = \mathbf{A}\mathbf{x}_t + \mathbf{B}\mathbf{u}_t + \mathbf{C}\xi_t$$

and the observation equation

$$\mathbf{y}_t = \mathbf{H}\mathbf{x}_t + \mathbf{G}\mathbf{u}_t + \mathbf{D}\eta_t.$$

Sometimes some of the matrices may also be time-varying.

Reachability

For stochastic dynamics, several notions of reachability can be introduced by requiring that the gramian is positive definite or, equivalently, that the reachability matrix is of full rank, in some probabilistic sense. For nonlinear stochastic dynamics, then, a system is locally stochastically controllable if the rank of the reachability matrix of the linearized dynamics is full for every state after some time for almost all realization of stochastic disturbances. A weaker notion is to require that this full rank condition holds for some but not all noise realization, etc.

Markovian Structure

When state vector is directly observed, stochastic models specify the conditional probability distribution for state vector $p_t(d\mathbf{x}_t|\mathbf{x}^{t-1}, \mathbf{u}^{t-1})$, $t = 1, 2, \ldots$ where $\mathbf{x}_t \in X$, X is a state space, and $\mathbf{u}_t \in U$, where U is the action or control space. Here $p_t(\mathrm{S}|\mathbf{x}^{t-1}, \mathbf{u}^{t-1})$ is measurable in $\mathbf{x}_0, \ldots, \mathbf{x}_{t-1}$, for all S in the σ-field of subsets of X and all $\mathbf{u}_0, \mathbf{u}_1 \ldots \mathbf{u}_{t-1}$ in U. (We assume that the spaces X and U are complete, separable metric spaces.) The model induces a family of probability distributions $\mu(\cdot\, |\mathbf{u}^{t-1})$, on $X^t = X \times X \times \ldots \times X$ parametrized by a sequence $\mathbf{u}_0, \ldots, \mathbf{u}_{t-1}$,

$$\mu(\mathrm{S}|\mathbf{u}^{t-1}) = \int \ldots \int p_0(d\mathbf{x}_0) p_1(d\mathbf{x}_1|\mathbf{x}_0, \mathbf{u}_0) \ldots p_t(d\mathbf{x}_t|\mathbf{x}^{t-1}, \mathbf{u}^{t-1})$$

for $\mathrm{S} = \{x^t | x_\tau \in \mathrm{S}_s, s = 0, \ldots, t\}$. In Markovian models, the conditional distribution simplifies to

$$P_t(\mathrm{S}|\mathbf{x}^{t-1}, \mathbf{u}^{t-1}) = P_t(\mathbf{x}_{t-1}, \mathrm{S}; \mathbf{u}_{t-1})$$

where the right side is the one-step transition probability from $\mathbf{x}_{t-1}$ to S.

When $\mathbf{x}_t$ is not observed but $\mathbf{y}_t$ is, let $\mathbf{s}_t = (\mathbf{x}_t', \mathbf{y}_t')'$. Stochastic models specify a conditional probability distribution

$$p_t(\mathrm{S} \times \mathrm{V}|\mathbf{y}^{t-1}, \mathbf{u}^{t-1}) = \int_{\mathrm{V}} p_t^{(1)}(\mathrm{S}|\mathbf{y}^{t-1}, \mathbf{y}_t; \mathbf{u}^{t-1}) p_t^{(2)}(d\mathbf{y}_t|\mathbf{y}^{t-1}; \mathbf{u}^{t-1}).$$

Admissible controls π are determined by the conditional probabilities $q_t(d\mathbf{u}_t|\mathbf{y}^{t-1}, \mathbf{u}^{t-1})$ and nonrandomized admissible controls as given by a sequence of functions $\mathbf{u}_t = \pi_t(\mathbf{y}^{t-1})$. More will be said later on Markov processes with incomplete observations.

2.2 Singular and Nonsingular Probability Distributions

In stochastic problems, the reachability condition is related to the existence of nonsingular probability distribution functions for the state vector. We illustrate this fact when the exogenous disturbances are normally distributed.

Suppose that ξ_t is a mean zero, serially independent random variable and has the covariance $\operatorname{cov} \xi_t = \sigma^2 \mathbf{I}_r$. Then the dynamic equation

$$\mathbf{x}_{t+1} = \mathbf{A}\mathbf{x}_t + \mathbf{B}\xi_t$$

implies that $\mathbf{x}_t = \mathbf{A}^t\mathbf{x}_0 + \mathbf{\Omega}_t \xi_{t-1}^-$ where

$$\mathbf{\Omega}_t = [\mathbf{B}\mathbf{A}\mathbf{B}\ldots\mathbf{A}^{t-1}\mathbf{B}],$$

and ξ_{t-1}^- stacks $\xi_{t-1}, \xi_{t-2}, \ldots, \xi_0$ in that order. Note that the covariance matrix of $\mathbf{\Omega}_t\xi_{t-1}^-$ is $\sigma^2\mathbf{\Omega}_t\mathbf{\Omega}_t'$. If $\mathbf{\Omega}_t\mathbf{\Omega}_t'$ is nonsingular, then the conditional distribution of $\mathbf{x}_t$ and $\mathbf{x}_0$ is a normal one with mean $\mathbf{A}^t\mathbf{x}_0$ and covariance $\sigma^2\mathbf{\Omega}_t\mathbf{\Omega}_t'$. Thus, if $(\mathbf{A}, \mathbf{B})$ is a reachable pair, then $\mathbf{x}_t$ has the nonsingular distribution, and if $(\mathbf{A}, \mathbf{B})$ is not a reachable pair, the probability mass for $\mathbf{x}_t$ is confined to some subspace in the n-dimensional Euclidean space, where $\dim \mathbf{x}_t = n$. See Appendix on singular distribution. As t goes to infinity, $\sigma^2\mathbf{\Omega}_t\mathbf{\Omega}_t'$ converges to $\sigma^2\mathbf{G}_\Omega$ where $\mathbf{G}_\Omega$ satisfies the Lyapunov equation, $\mathbf{G}_\Omega = \mathbf{A}\mathbf{G}_\Omega\mathbf{A}' + \mathbf{B}\mathbf{B}'$. If $\mathbf{A}$ is stable, then $\mathbf{A}^t\mathbf{x}_0 \to 0$, and the limit of $\mathbf{x}_t$ is normally distributed with mean zero and covariance $\sigma^2\mathbf{G}_\Omega$. When ξ_t is independently and identically distributed (henceforth i.i.d.) $N(0, \mathbf{I})$, and when $\mathbf{A}$ is asymptotically stable, $\mathbf{x}_t$ has a stationary normal distribution $N(0, \boldsymbol{\pi})$ where $\boldsymbol{\pi} = \mathbf{A}\boldsymbol{\pi}\mathbf{A}' + \mathbf{B}\mathbf{B}'$. The null space of $\boldsymbol{\pi}$ is zero if $(\mathbf{A}, \mathbf{B})$ is a reachable pair.

Later, we relax the condition on $\{\xi_t\}$ and assume it to be a martingale difference process to be explained shortly. We will show that the reachability condition will ensure the almost sure convergence of sample covariances. Since we encounter martingale and martingale difference processes when we discuss convergence properties of estimation and adaptive control schemes, this is as good a place as any to introduce them.

2.3 Martingales

The martingales are ubiquitous. They arise in analysis and estimation of dynamic systems and in macroeconomic and microeconomic intertemporal decision-making problems under uncertainty. This is partly because dual variables in stochastic optimization problems can be shown to be martingales under certain technical conditions. A brief account of how martingales arise as dual variables in a discrete time stochastic control problem, e.g., as shadow prices in optimal saving is given later. (See Foldes [1978].) One of the ways we discuss stability of stochastic dynamic systems is by introducing a stochastic Lyapunov function which is an expectation-decreasing martingale (also called supermartingale). A set of powerful theorems are available for establishing convergence for martingales. See Malliaris and Broh (1981) for a short survey of martingale applications in financial decision making.

This section is a brief introduction to this important class of stochastic processes. A basic setup, in addition to the probability triplet of the sample space, its σ-field $\mathscr{F}$, and a probability measure, is a notion of adapted stochastic processes. Given an increasing sequence of sub-σ-field of $\mathscr{F}$ indexed as $\mathscr{F}_0 \subset \mathscr{F}_1 \subset \ldots$, a sequence of random variables $\{x_n\}$ is an adapted process if for each $n \geqslant 0$, x_n is $\mathscr{F}_n$-measurable.* A stochastic process x is a predictable process if x_n is $\mathscr{F}_{n-1}$-measurable. In sequential decision problems, $\mathscr{F}_n$ is often the decision maker's information set available at time n. The choice variables are taken to be $\mathscr{F}_{n-1}$ measurable, i.e., predictable, i.e., known variables, at time n. The sequence $\{x_n\}$ is a martingale if $E|x_n| < \infty$ for all n and if

$$E(x_{n+1}|\mathscr{F}_n) = x_n, \qquad n \geqslant 0.$$

When the equality is replaced by $\geqslant$, the sequence is a submartingale, and with $\leqslant$, it is called a supermartingale.

At first sight, one may suspect that martingales are too special to be common or be of practical interest. This suspicion turns out to be unfounded and actually martingales are inseparable with optimization problems as dual variables. Loosely put, Lagrange multipliers are martingales, as shown in Rockafellar and Wets (1976) or Pliska (1982) under varying sets of assumptions.

Let $\{x_n\}$ be a sequence of mean zero random variables and finite variances. Let

$$y_n = x_1 + x_2 + \ldots + x_n.$$

* Let $\{x_n, \mathscr{F}_n, n \geqslant 1\}$ be a submartingale, and x_n is $\mathscr{G}_n$ measurable (Chow and Teicher, 1978, p. 228) and $\mathscr{G}_n \subset \mathscr{F}_n$, then $E(X_{n+1}|\mathscr{G}_n) = E(E(X_{n+1}|\mathscr{F}_n)|\mathscr{G}_n) \geqslant E(x_n|\mathscr{G}_n) = x_n$. Thus, $\{x_n, \mathscr{G}_n\}$ is also a submartingale.

If $\{y_n\}$ is a martingale, then $E(x_{n+1}|x_1 \dots x_n) = 0$, implying that $\{x_i\}$ is a sequence of orthogonal random variables. On the other hand, if $\{x_i\}$ is a sequence of independent random variables, then $\{y_n\}$ is a martingale. The martingale assumption is, thus, somewhere between that of independence and orthogonality. Laws of large numbers for martingales may then be expected to be of such a nature that they fall somewhere between the laws of large numbers for independent random variables and those for orthogonal random variables. This is indeed true as we will illustrate later.

Many examples of martingales are related to a fair gambling situation by a slight change of terminology. One example can be rephrased as a control system

$$x_{i+1} = x_i + f_i(\xi_i, u_i).$$

Assume that its control policy is given by $u_i = \Phi_i(x_{i-1})$. Suppose that ξ_is are independent random variables and that $E[f_i(\xi_i, \Phi_i(x_{i-1}))|x_i] = 0$ for all x_i. Then $E(x_{i+1}|x^i) = x_i$ and $\{x_i\}$ is a martingale. Cumulative gains or losses from a fair gamble is also a martingale. As another simple example, fix a random variable X with $E|X| < \infty$, and consider its conditional expectation, $X_n = E(X|y_1, \dots, y_n)$ where $\{y_i\}$ is another sequence of random variables. Then $\{X_n\}$ is a sequence of conditional expectation (orthogonal projection) of a fixed random variable relative to an increasing sequence of measurements, and it is thus a martingale. A special case of this is the conditional mean of the initial condition vector of some dynamic system.

Let

$$\mathbf{y}^n = (\mathbf{y}_0', \mathbf{y}_1', \dots, \mathbf{y}_n')'$$

and

$$\mathbf{u}_n = E(\mathbf{x}_0|\mathbf{y}^n).$$

Then

$$\begin{aligned} \mathbf{u}_{n+1} &= E(\mathbf{x}_0|\mathbf{y}^n, \mathbf{y}_{n+1}) \\ &= E(\mathbf{x}_0|\mathbf{y}^n, \mathbf{y}_{n+1} - \mathbf{y}_{n+1|n}) \\ &= E(\mathbf{x}_0|\mathbf{y}^n) + \mathbf{K}_n(\mathbf{y}_{n+1} - \mathbf{y}_{n+1|n}) \end{aligned}$$

for some constant $\mathbf{K}_n$.

Then

$$\begin{aligned} E(\mathbf{u}_{n+1}|\mathbf{y}^n) &= E(\mathbf{x}_0|\mathbf{y}^n) \\ &= \mathbf{u}_n, \end{aligned}$$

i.e., $\{\mathbf{u}_n\}$ is a martingale.

In prediction theory, a sequence of σ-algebra $\mathscr{F}_n$ is generated as successive refinements of our predictions based on a larger information set as n increases.

If X_n is measurable with respect to $\mathscr{F}_{n-1}$, then X_n is called predictable. A predictable martingale is shown to be a constant. Any sequence of random variables $\{X_n\}$ where X_n is measurable with respect to $\mathscr{F}_n$ can be split into a martingale and a predictable process as we show later. Here are some more examples of martingales.

Example 1. Let $x_1, x_2, \ldots$ be independent random variables with mean 0 and $S_n = x_1 + \ldots + x_n$. We already know that S_n is a martingale. For any $t > 0$, $Z_n = \exp(itS_n)/E(e^{itS_n})$ is also a martingale. Let $f_k(x_1, \ldots, x_k)$ be any bounded Borel measurable function. Define

$$\xi_1 = x_1$$

$$\xi_{t+1} = x_{k+1} f_k(x_1, \ldots, x_k)$$

and

$$y_n = \xi_1 + \ldots + \xi_n.$$

Then $\{y_n\}$ is a martingale. We verify the definitional relation directly as

$$\begin{aligned} E(y_{n+1} | y_1, \ldots, y_n) &= E(y_{n+1} | x_1, \ldots, x_n) \\ &= E(\xi_{n+1} | x_1, \ldots, x_n) + E(y_n | x_1, \ldots, x_n) = y_n, \end{aligned}$$

since y_n is a measurable function of $x_1, \ldots, x_n$. The first term is zero because $E(x_{n+1} f_n(x_1 \ldots x_n) | x_1, \ldots, x_n) = f_n(x_1 \ldots x_n) E(x_{n+1} | x_1 \ldots x_n) = 0$.

Example 2. Let

$$y_{n+1} = \prod_{k=1}^{n} (1 + x_k q)$$

where $x_1, x_2 \ldots$ are independent, mean zero, random variables and q is a constant.

Let

$$\begin{aligned} \xi_n &= y_n - y_{n-1} \\ &= x_{n-1} q \, y_{n-1}. \end{aligned}$$

Since $E(\xi_n | x_1 \ldots x_{n-1}) = 0$, $\{y_n\}$ is a martingale. Here ξ_n is called a martingale difference.

Example 3. Many optimization problems can be phrased as

$$\max E(X \cdot Z)_N$$

where

$$(X \cdot Z)_N = \sum_{0}^{N} X_n \Delta Z_n = X_0 Z_0 + X_1(Z_1 - Z_0) + \ldots + X_N(Z_N - Z_{N-1})$$

where

$$\Delta Z_n = Z_n - Z_{n-1}, n \geqslant 1,$$

and

$$\Delta Z_0 = Z_0,$$

where $\{X_n\}$ is a predictable process and Z_n is an adapted process. The maximization is over a class of predictable processes for which the expectation is finite. When $E(X \cdot Z)_N$ is finite, $(X \cdot Z)_N$ is a (super)martingale, if Z is a (super)martingale and X is nonnegative. This expression $(X \cdot Z)_N$ is called martingale transform (Kopp [1984, p. 44]). Martingale transforms provide less trivial examples of martingales. In mathematical programming, the basic duality relation between primal minimization problems and their dual maximization ones can be extended to stochastic control problems using martingales and martingale transforms. See this chapter's Appendix for more detail on the martingale transforms.

Example 4. Next we discuss the maximum likelihood estimation problem of an unknown parameter. For some optimal adaptive control problems, we know that the optimal control policy synthesis can be separated from the optimal estimation process of the unknown process parameters and/or noise statistics. Thus parameter estimations are important in themselves as well as being a part of control problems in many cases. Maximum likelihood or (extended) least squares estimates are often used when exact Bayesian conditional expectation estimates are not available or are too cumbersome to work with.

Suppose we have a system where its unknown parameter θ is assumed to be either θ_1 or θ_2. Consider a problem of constructing the maximum likelihood estimate of θ given a set of n observations at time n, y^n. Suppose that $p(y^n|\theta_i)$ is defined for all $n = 0, 1, \ldots$ and $i = 1, 2$.

Form the ratio

(1) $$z_n = \frac{p(y^n|\theta_2)}{p(y^n|\theta_1)}.$$

The probability density $p(y^n|\theta)$, when regarded as a function of θ for fixed y^n, is a likelihood function, and z_n is called the likelihood ratio. Since $\theta = \theta_1$ or θ_2 in the example, the maximum likelihood estimate of θ is θ_2 if $z_n > 1$, θ_1 if $z_n < 1$, and undecided for $z_n = 1$. Thus, the stochastic process $\{z_n\}$ of (1) describes the time history of the estimate of θ. To study the behavior of the sequential estimate of θ, one must study the behavior of $\{z_n\}$ as $n \to \infty$. Since p is a density function, the denominator is nonzero with a probability one. Let us assume that $p(y^n|\theta_2) = 0$ whenever $p(y^n|\theta_1) = 0$, since otherwise we can decide θ to be θ_2 immediately.

Suppose θ_1 is the true parameter value. Then

$$p(y_{n+1}|y^n) = \frac{p(y^{n+1}|\theta_1)}{p(y^n|\theta_1)}$$

and

$$\begin{aligned} E(z_{n+1}|y^n) &= \int z_{n+1} p(y_{n+1}|y^n) dy_{n+1} \\ &= \int \frac{p(y^{n+1}|\theta_2)}{p(y^n|\theta_1)} dy_{n+1} \\ &= \frac{p(y^n|\theta_2)}{p(y^n|\theta_1)} = z_n \quad \text{with probability 1.} \end{aligned}$$

Then, since z^n are random variables that are functions of y^n,

$$E(z_{n+1}|y^n) = E(z_{n+1}|y^n, z^n) = z_n \quad \text{with probability 1.} \tag{2}$$

Taking the conditional expectation of (2) with respect to z^n,

$$\begin{aligned} E(E(z_{n+1}|y^n)|z^n) &= E(E(z_{n+1}|y^n, z^n)|z^n) \\ &= E(z_{n+1}|z^n) \\ &= E(z_n|z^n) \\ &= z_n, \end{aligned}$$

i.e., the sequence of likelihood ratios, $\{z_n\}$, is a martingale.

For examples of more complicated martingales, see Rényi (1970).

2.4 Martingale Differences

In rational expectations models, how expectations are revised in response to new information is important. Expectation revisions are martingale differences, and estimation errors are martingale differences. Solutions of rational expectations models (which are described in Chapter 8) are intimately connected with the notion of martingale differences, i.e., of $\{\Delta X_n\}$ where $\Delta X_n = X_n - X_{n-1}$ for some martingale $\{X_n\}$. The martingale difference ΔX_n is adapted and $\Delta X_{n|n-1} = 0$, where the subscript $n|n-1$ means the expectation of ΔX_n is conditional on the information set at time $n-1$.

Let $e_t^0 = y_t - y_{t|t-1}$ where $y_{t|t-1} = E(y_t|\Omega_{t-1})$ with some information set Ω_{t-1}. Then $E(e_t^0|\Omega_{t-1}) = 0$. The estimate of y_{t+j} formed at time $t-1$ is revised at t, incorporating new information that becomes available from the formation of the previous estimate $y_{t+j|t-1}$. The expectation revision process is defined by

$$e_t^j = y_{t+j|t} - y_{t+j|t-1}, \qquad j = 1, 2, \ldots .$$

Because of the relationship

$$E_{t-1}(y_{t+j|t}) = E_{t-1}E_t(y_{t+j})$$
$$= E_{t-1}(y_{t+j}),$$

we note that

$$E(e_t^j \mid \Omega_{t-1}) = 0,$$

i.e., the expectation revision process e_t^j is also a martingale difference.

The next example shows that martingales arise in intertemporal optimization problems as adjoint variables.

Example 5: Optimal Saving Plan. The following is a simplified account of Foldes (1978), which shows that the shadow price of consumption associated with an optimal plan is a martingale. Let $w > 0$ be the initial endowment. A feasible plan is a nonnegative adapted process $\{k_t\}$, i.e., k_t is Ω_t-measurable where Ω_t is at the σ-algebra of information (observation) available at time t such that $\Omega_t \subset \Omega_{t+1}$ (no forgetting of past information).

Output produced by a random production function, $\mathrm{F}(k_t, t)$, is allocated between consumption and investment. The problem is to maximize $\Phi(k) = E\Sigma_0^\infty U(c_t, t)$ over K, a convex set of all capital plans, subject to

$$k_0 + c_0 = w,$$

and

$$k_t + c_t = \mathrm{F}(k_{t-1}, t), \qquad t = 1, 2, \ldots,$$

where k stands for the sequence $\{k_t\}$. The sequences of cs and ks are all random sequences. To avoid technical complications, assume that

$$\Phi^* = \sup[\Phi(k) \colon k \in K] < \infty.$$

Denote the marginal product of capital by $f_t > 0$ and the marginal utility by u_t.

Define the product

$$r_t = r_{t-1} \cdot f_t$$
$$= f_1 \ldots f_t > 0,$$

with

$$r_0 = 1,$$

and call it the interest process $\mathbf{r} = \{r_t\}$, and define

$$\tilde{k}_t = k_t / r_t,$$

and

$$\tilde{c} = c_t / r_t.$$

These measure k_t and c_t in reduced units, net of random compound interest. Define the shadow price by multiplying the martingale utility by r_t,

$$y_t = r_t u_t.$$

Note that y_t is Ω_t-measurable, i.e., $\{y_t\}$ is adapted.

Denote by * optimal plans or sequences such as $\mathbf{k}^* \in K$ and the corresponding consumption plan c^*. Denote the directional derivative of $\Phi(\mathbf{k})$ at $\mathbf{k}^*$ in the direction $\delta\mathbf{k}$ such that $\mathbf{k}^* + \delta\mathbf{k} \in K$ by

$$D\Phi(\mathbf{k}^*, \delta\mathbf{k}) = \lim_{\alpha\downarrow 0} \frac{1}{\alpha} [\Phi(\mathbf{k}^\alpha) - \Phi(\mathbf{k}^*)]$$

where

$$\mathbf{k}^\alpha = \mathbf{k}^* + \alpha\delta\mathbf{k}.$$

Denote by U_t^α the utility of the capital plan, $k_t^\alpha = k_t^* + \alpha\delta k_t$.

Noting that

$$\begin{aligned} U_t^\alpha - U_t^* &= U_t(c_t^* + \alpha\delta c_t) - U_t(c_t^*) \\ &= u_t^* \alpha\delta c_t > -\infty \end{aligned}$$

and that the constraint becomes

$$k_t^* + \alpha\delta k_t + c_t^* + \alpha\delta c_t = F(k_{t-1}^* + \alpha\delta k_{t-1}, t)$$

or

$$\delta c_t = f_t^* \delta k_{t-1} - \delta k_t,$$

we obtain

$$\begin{aligned} D\Phi &= \lim_{\alpha\downarrow 0} \frac{1}{\alpha} E\sum_0^\infty (U_t^\alpha - U_t^*) \\ &= E\sum_0^\infty (f_t^* \delta k_{t-1} - \delta k_t)u_t^* > -\infty. \end{aligned}$$

In reduced units (see Method of Partial Summation in Appendix),

$$\begin{aligned} D\Phi &= E\sum_0^\infty (\delta\tilde{k}_{t-1} - \delta\tilde{k}_t)y_t \\ &= \lim_{T\to\infty} \sum_0^T E(\delta\tilde{k}_{t-1} - \delta\tilde{k}_t)y_t \\ &= \lim_{T\to\infty} E\left[\sum_0^{T-1} \delta\tilde{k}_t(y_{t+1} - y_t) - \delta\tilde{k}_t y_t\right]. \end{aligned}$$

Taking the expectation as $E = EE_t$ where E_t is the expectation conditional on Ω_t, and noting that

$$E_t\delta\tilde{k}_t y_{t+1} = \delta\tilde{k}_t\, y_{t+1|t}$$

because $\delta\tilde{k}_t$ is Ω_t-measurable, the (directional) derivative is given by

$$D\Phi = \lim_{T\to\infty} E\left\{\sum_0^{T-1} \delta\tilde{k}_t(y_{t+1|t} - y_t) - \delta\tilde{k}_T y_T\right\}.$$

Now, $\mathbf{k}^*$ is optimal if and only if $\Phi(\mathbf{k}^*) > -\infty$ and $D\Phi(\mathbf{k}^*, \delta\mathbf{k}) \leqslant 0$ for all $\mathbf{k}^* + \delta\mathbf{k} \in K$. Denote this by $\mathbf{k}^1$. It is easy to see that if y_t is a martingale and the transversality condition is met, then $\mathbf{k}^*$ is optimal. Note first that if $\{y_t\}$ is a martingale, then

$$D\Phi = \lim_{t\to\infty} E(-\delta\tilde{k}_T y_T)$$
$$= \lim_{T\to\infty} E\{\tilde{k}_T - \tilde{k}_T^1) y_T.$$

If $\lim_{T\to\infty} \tilde{k}_T y_T = 0$, then $D\Phi \leqslant 0$ since $\tilde{k}_T^1 y_T \geqslant 0$ and $\mathbf{k}^*$ is optimal. Foldes then establishes that $\{y_t\}$ is a supermartingale, i.e., $y_{t+1|t} \leqslant y_t$. Then for any variation such that $\delta k_t < 0$, for all $t \geqslant 0$ (which is certainly feasible), each term $\delta\tilde{k}_t(y_{t+1|t} - y_t) \geqslant 0$ and $-\delta\tilde{k}_T y_T \geqslant 0$, implying that $D\Phi > 0$ and contradicting that $\mathbf{k}^*$ is optimal. Thus $\delta k_t(y_{t+1|t} - y_t) = 0$ a.s. or $y_{t+1|t} = y_t$ a.s., i.e., $\{y_t\}$ is a martingale. Variations of the functional equation of the dynamic program

$$V(k_{t-1}^*) = E_{t-1}[U(c_t^*, t) + V(k_t^*)]$$

where

$$V(k_t^*) = \max E_t \sum_t^\infty U(c_\tau, \tau)$$

and

$$k_t^* = F(k_{t-1}^*) - c_t^*$$

is

$$V'(k_{t-1}^*)\delta k_{t-1} = E_{t-1}[U'(c_t^* + \delta c_t) + \beta V'(k_t^*)\delta k_t]$$

where

$$\delta c_t = f_t^* \delta k_{t-1} - \delta k_t = r_t(\delta\tilde{k}_{t-1} - \delta\tilde{k}_t)$$

where $\delta k_{t-1} = r_{t-1}\delta\tilde{k}_{t-1}$ and $\delta k_t = r_t\delta\tilde{k}_t$. The preceding variational equation becomes

$$(V'(k_{t-1}^*)r_{t-1} - y_{t-1})\delta\tilde{k}_{t-1} = E_{t-1}\{[-y_t + V'(k_t^*)r_t]\delta\tilde{k}_t\}$$

where $y_t = u_t r_t$. Since $\delta\tilde{k}_t$ is arbitrary, $V'(k_t^*)r_t = y_t$ must hold, i.e., $V'(k_t^*)r_t$ is a martingale since y_t is.

Martingale Convergence Theorem. Convergence of martingales has been thoroughly investigated in the mathematical literature. We cite here one source that we will frequently refer to elsewhere in the book. Neveu (1975, p. 33) has shown* that when x_t and y_t are both finite positive random

---

* Neveu actually considers $E(x_{t+1}|\mathscr{F}_t) \leqslant (1 + r_t)x_t + y_t$, where r_t is positive and $\Sigma_t r_t < \infty$, a.s. By measuring x_t and y_t in "reduced units" by dividing them with $(1 + r_1)\cdots(1 + r_{t-1})$, the above inequality is restated as

$$E(x'_{t+1}|\mathscr{F}_t) \leqslant x'_t + y'_t$$

where

$$x'_t = x_t/(1 + r_1)\cdots(1 + r_{t-1})$$

and

$$y'_t = y_t/(1 + r_1)\cdots(1 + r_{t-1}).$$

variables that are measurable with respect to an increasing σ-field $\mathscr{F}_t$ and satisfy

$$E(x_{t+1}|\mathscr{F}_t) \leqslant x_t + y_t,$$

then x_t has a finite limit a.s., when

$$\Sigma_t y_t < \infty \ a.s.$$

Now consider x_t which satisfies the inequality

$$\begin{aligned} E(x_t|\mathscr{F}_{t-1}) &\leqslant x_{t-1} + \alpha_t - \beta_t \\ &\leqslant x_{t-1} + \alpha_t, \end{aligned}$$

where $\alpha_t > 0, \beta_t > 0$. We see that $x_t \to x$ finite, a.s., as t goes to infinity if $\Sigma \alpha_t < \infty$, a.s. Apply this to $T_t = x_t + b_t$ and $y_t = \alpha_t$ where $b_t = \Sigma_1^t \beta_s$. Then

$$E(x_t + b_t|\mathscr{F}_{t-1}) \leqslant x_{t-1} + b_{t-1} + \alpha_t,$$

and $x_t + b_t \to x' < \infty$. Since $x_t \to x < \infty$ a.s., we have $b_t \to b' < \infty$. Thus, Solo (1986) establishes a version of the martingale convergence theorem.*

Theorem (Solo (1986)). *Let T_{t-1}, α_t, and β_t be nonnegative random variables, measurable with respect to an increasing sequence of σ-algebra $\mathscr{F}_t$ such that*

$$E(T_t|\mathscr{F}_{t-1}) \leqslant T_{t-1} + \alpha_t - \beta_t. \tag{3}$$

If $\Sigma_1^\infty \alpha_t < \infty$ a.s., then $\Sigma_1^\infty \beta_t < \infty$ a.s., and $T_t \to T$ a.s.

---

* Let $Y = \Sigma_{t=1}^\infty y_t$, and suppose that $EY < \infty$. Define $Y_t = E(Y|\mathscr{F}_t)$. It is a martingale and $Y_t \to Y$ a.s., and $Y_t - \Sigma_1^{t-1} y_s \geqslant 0$, a.s.

Then $V_t = x_t + Y_t - \Sigma_1^{t-1} y_s$ is a positive supermartingale since

$$\begin{aligned} E(x_{t+1} + Y_{t+1} - \Sigma_1^t y_s|\mathscr{F}_t) &\leqslant x_t + y_t + Y_t - \Sigma_1^t y_s \\ &= x_t + Y_t - \Sigma_1^{t-1} y_s \\ &= V_t \end{aligned}$$

and $x_t + Y_t - \Sigma_1^{t-1} y_s \to L$ a.s. Since $Y_t - \Sigma_1^{t-1} y_s \to 0$, $x_t \to L$ a.s.

2.5 Bounds on $\{x_t\}$

Consider a dynamic model

$$\mathbf{x}_t = \mathbf{A}\mathbf{x}_{t-1} + \mathbf{e}_t \tag{4}$$

where e_t is an n-dimensional martingale difference sequence with respect to an increasing sequence of σ-fields $\{\mathscr{F}_t\}$, such that

$$E(\|\mathbf{e}_t\|^2 | \mathscr{F}_{t-1}) = \sigma^2$$

and

$$\sup_t E(\|\mathbf{e}_t\|^\alpha | \mathscr{F}_{t-1}) < \infty \qquad \text{for some } \alpha > 2. \tag{5}$$

Later in Chapter 6 we need to investigate the behavior of $\mathbf{V}_t = \Sigma_{\tau=1}^t \mathbf{x}_\tau \mathbf{x}_\tau'$ as t goes to infinity. Since

$$\lambda_{\max}(\mathbf{V}_t) \leqslant \operatorname{tr} \sum_1^t \mathbf{x}_\tau \mathbf{x}_\tau' = \sum_1^t \|\mathbf{x}_\tau\|^2 \leqslant n\lambda_{\max}(\mathbf{V}_t)$$

where $n = \dim \mathbf{x}_\tau$, one way to conduct this investigation is to examine the behavior of $\|\mathbf{x}_t\|$. Also, asymptotic behavior of the state vector governed by (4) is of independent interest. For simpler exposition, assume that the largest eigenvalue of $\mathbf{A}$, $\bar{\lambda}$, is of single multiplicity and that $\bar{\lambda} \leqslant 1$. Unlike deterministic systems $|\bar{\lambda}| < 1$ does not guarantee that $\|\mathbf{x}_t\|$ goes to zero exponentially fast. The next example due to Chow and Lai [1973] illustrates this.

Example 6: Fluctuation in $\{x_t\}$**.** This scalar example is an analogue of the law of the iterated logarithm. Consider

$$x_{t+1} = ax_t + bn_{t+1}$$

where $|a| < 1$. Its reduced form is

$$x_t = a^t x_o + \sum_{k=1}^t h_k n_{t-k} = a^t x_o + \sum_1^t h_{t-k} n_k$$

where

$$h_k = a^k b, \qquad k \geqslant 0$$

is the impulse response sequence. For simpler exposition, set x_o to zero without loss of generality. When $|a| < 1$, we have $\Sigma h_k^2 = b^2/(1 - a^2) < \infty$.

When ns are mean zero independent random variables, Lai (1986) has shown that

$$\overline{\lim}\,(2\ln t)^{-1/2}\,\Sigma\, h_{t-k}n_k = (\Sigma\, h_k^2)^{1/2} \quad \text{a.s.}$$

This example shows that the state variable of a stochastic dynamic system with $|a| < 1$ need not be bounded almost surely, while $x_t \to 0$ exponentially in asymptotically stable deterministic systems. It is easy to establish that the left side is bounded from above by the right side. Let $\sigma^2 = \Sigma\, h_k^2$. Since $\Sigma_1^t\, h_{t-k}n_k$ is $N(O, \sigma_t^2)$ where $\sigma_t^2 = \Sigma_1^t\, h_k^2$,

$$\mathrm{P}\left[\sum_{k=1}^{t} h_k n_{t-k} \geqslant b_t\right] \leqslant K b_t^{-1} \exp[-(2\sigma^2)^{-1} b_t^2]$$
$$= (K/\sigma)(\gamma \ln t)^{-1/2} t^{-\gamma/2}$$

where $b_t = \sigma(\gamma \ln t)^{1/2}$ for any $\gamma > 2$, and K is a positive constant, where we use the well-known inequality that

$$\int_b^\infty e^{-x^2/2\sigma^2}\,dx \leqslant \frac{K}{b} e^{-b^2/2\sigma^2}.$$

Therefore $\Sigma_{t=1}^{\infty} \mathrm{P}[\Sigma_{k=1}^{t} h_k n_{t-k} \geqslant b_t] < \infty$ and by the Borel-Cantelli lemma

$$\overline{\lim}\,(2\ln t)^{-1/2} \sum_{k=1}^{t} h_k n_{t-k} \leqslant \sigma, \quad \text{a.s.}$$

The reverse inequality can also be established by the Borel-Cantelli lemma, see Chow and Lai (1973, Th. 6).

Going back to an n-dimensional system

$$\mathbf{x}_t = \mathbf{A}^t \mathbf{x}_0 + \sum_1^t \mathbf{A}^{t-\tau} \mathbf{e}_\tau,$$

we bound the norm of $\mathbf{x}_t$ by

$$\|\mathbf{x}_t\| \leqslant \|\mathbf{A}^t\|\,\|\mathbf{x}_0\| + \sum_1^t \|\mathbf{A}^{t-\tau}\|\,\|\mathbf{e}_\tau\|$$

where

$$\|\mathbf{A}^t\| \sim c\bar{\lambda}^t$$

for some positive constant c as $t \to \infty$. When $\bar{\lambda}$ has multiplicity ρ this expression is multiplied by $t^{\rho-1}$. See Varga (1962, p. 65). Lai and Wei (1985) have established that

Theorem. *Let $\{\mathbf{e}_t\}$ be a martingale difference as just introduced.*

(a) *If $\bar{\lambda} < 1$, then $\|\mathbf{x}_t\| = o(t^{\beta})$ for every $\alpha\beta > 1$ and*
(b) *if $\bar{\lambda} = 1$, then $\|\mathbf{x}_t\| = 0(t^{1/2}\sqrt{\ln \ln t})$ a.s.*

To see (a), note from (5) that

$$\sum_{1}^{\infty} \mathrm{P}\{\|\mathbf{e}_t\|^{\alpha} > \varepsilon t^{\alpha\beta} | \mathscr{F}_{t-1}\} < \infty$$

for every $\varepsilon > 0$ and $\alpha\beta > 1$. By the conditional Borel-Cantelli lemma, Stout (1975, p. 55), the event $\{\|\mathbf{e}_t\|^{\alpha} > \varepsilon t^{\alpha\beta}\}$ occurs at most finitely often, i.e., $\|\mathbf{e}_t\| = o(t^{\beta})$. Combine this with the bound on $\|\mathbf{A}^t\|$ to see that $\|\mathbf{x}_t\| = o(t^{\beta})$, where $\beta > 1/\alpha$.

To derive (b), change the coordinate system to put $\mathbf{A}$ into the real Schur form as*

$$\mathbf{z}_t = \mathbf{D}\mathbf{z}_{t-1} + \mathbf{u}_t$$

where

$$d = \begin{bmatrix} 1 & \mathbf{0} \\ \mathbf{\Psi} & \mathbf{F} \end{bmatrix}, \qquad \mathbf{u}_t = \mathbf{G}\mathbf{e}_t,$$

and where the $\bar{\lambda}$ of $\mathbf{F}$ is less than one. Partition $\mathbf{z}_t$ conformably as

$$\mathbf{z}_t = \begin{bmatrix} z_t^1 \\ \mathbf{z}_t^2 \end{bmatrix}.$$

Here $z_t^1 = z_o^1 + \Sigma_{\tau=1}^{t} u_{\tau}$ is a martingale. The conclusion (b) follows from

(6) $$\sum_{1}^{t} \mathbf{u}_{\tau} = O[t^{1/2}(\ln \ln t)^{1/2}]$$

and

$$\|\mathbf{z}_t^1\| = O[t^{1/2}(\ln \ln t)^{1/2}] \quad \text{a.s.}$$

To see (6), use the truncation

$$\mathbf{u}_t' = \mathbf{u}_t I\left[|\mathbf{u}_t| \leqslant \varepsilon\left(\frac{t}{2 \ln \ln t}\right)^{1/2}\right].$$

By the conditional Borel-Cantelli lemma, $\Sigma_{\tau} \mathrm{P}\{\mathbf{u}_{\tau} \neq \mathbf{u}_{\tau}' | \mathscr{F}_{\tau-1}\} < \infty$ and $\mathrm{P}[\{\mathbf{u}_{\tau}' \neq \mathbf{u}_{\tau}$ infinitely often$\}] = 0$. Then use Stout (1974, Theorem 5.4.1)

* Alternatively, Jordan canonical forms may be used.

Note that

$$\mathbf{D}^t = \begin{bmatrix} 1 & 0 \\ \boldsymbol{\Psi}_t & \mathbf{F}^t \end{bmatrix}$$

where $\boldsymbol{\Psi}_{t+1} = \boldsymbol{\Psi}_t + \mathbf{F}^t\boldsymbol{\Psi}$ where $\|\boldsymbol{\Psi}_t\| \to \|(I-\mathbf{F})^{-1}\boldsymbol{\Psi}\| = 0(1)$

$$\|\mathbf{z}_t^2\| \leqslant \|\boldsymbol{\Psi}_t\|\,\|z_0^1\| + \|\mathbf{F}^t\|\,\|\mathbf{z}_0^2\| + \left|\sum_\tau \begin{bmatrix} \boldsymbol{u}_t^1 \\ \boldsymbol{\Psi}_{t-\tau}\boldsymbol{u}_\tau^1 + \mathbf{F}^{t-\tau}\boldsymbol{u}_\tau^2 \end{bmatrix}\right|.$$

Note that

$$\left\|\sum_t \boldsymbol{\Psi}_{t-\tau}\mathbf{u}_\tau^1 + \mathbf{F}^{t-\tau}\mathbf{u}_\tau^2\right\| \leqslant \left\|\sum_\tau \boldsymbol{\Psi}_{t-\tau}\mathbf{u}_\tau^1\right\| + \left\|\sum_\tau \mathbf{F}^{t-\tau}\mathbf{u}_\tau^2\right\|$$
$$= O(t^{1/2}(\ln\ln t)^{1/2}).$$

Bounds on $\Sigma_t x_t x_t'$

We next establish

Fact. Under the assumption as before,

(a) $\lambda_{\max}(V_t) = O(t^2 \ln\ln t)$ a.s. when $\bar{\lambda} = 1$, and
(b) $\lambda_{\max}(V_t) = O(t)$ a.s. when $\bar{\lambda} < 1$.

Fact (a) follows immediately from theorem (b) since $\lambda_{\max}(V_t) \leqslant \hbar V_t = \Sigma_{\tau=1}^t \|\mathbf{x}_\tau\|^2 \leqslant n\lambda_{\max}(V_t)$.

Fact (b) is established as follows. Let $\varepsilon_\tau = \|\mathbf{e}_t\|^2 - E(\|\mathbf{e}_t\|^2|\mathscr{F}_{t-1})$. This is a martingale difference and $E(|\varepsilon_t|^{\alpha/2}|\mathscr{F}_{t-1}) < \infty$ for some $\alpha > 2$, from which $\Sigma_1^t \varepsilon_\tau = o(t)$ follows (Chow [1965]), and

$$\sum_1^t \|\mathbf{e}_t\|^2 = \sum_1^t E(\|\mathbf{e}_t\|^2|\mathscr{F}_{\tau-1}) + o(t) = O(t) \tag{7}$$

where (5) is used. In

$$\sum_{t=1}^s \|\mathbf{x}_t\|^2 = \sum_{t=1}^s \left(\|\mathbf{A}^t\|\,\|\mathbf{x}_0\| + \sum_{\tau=1}^t \|\mathbf{A}^{t-\tau}\|\,\|\mathbf{e}_\tau\|\right)^2,$$

the first term is negligible compared with the second. By the Schwarz inequality

$$\sum_{t=1}^t \left(\sum_{\tau=1}^t \|\mathbf{A}^{t-\tau}\|\,\|\mathbf{e}_\tau\|\right)^2 \leqslant \left(\sum_1^\infty \|\mathbf{A}^{t-\tau}\|\right)^2 \sum_{\tau=1}^t \|\mathbf{e}_\tau\|^2 = O(t) \quad \text{a.s.}$$

where (7) is used, and we conclude that

$$\sum_1^t \|\mathbf{x}_\tau\|^2 = O(t).$$

From (4),

$$\begin{aligned}\sum_1^t \mathbf{x}_\tau \mathbf{x}_\tau' &= \mathbf{A}\left(\sum_1^t \mathbf{x}_{\tau-1}\mathbf{x}_{\tau-1}'\right)\mathbf{A}' + \sum_1^t \mathbf{e}_\tau \mathbf{e}_\tau' \\ &\quad + \sum_1^t (\mathbf{A}\mathbf{x}_{\tau-1}\mathbf{e}_\tau' + \mathbf{e}_\tau \mathbf{x}_{\tau-1}'\mathbf{A}') \\ &= \mathbf{A}\left(\sum_1^t \mathbf{x}_\tau \mathbf{x}_\tau' - \mathbf{x}_t\mathbf{x}_t' + \mathbf{x}_0\mathbf{x}_0'\right)\mathbf{A}' + \sum_1^t \mathbf{e}_\tau \mathbf{e}_\tau' \\ &\quad + \sum_1^t (\mathbf{A}\mathbf{x}_{\tau-1}\mathbf{e}_\tau' + \mathbf{e}_\tau \mathbf{x}_{\tau-1}'\mathbf{A}').\end{aligned}$$

Since $\|\mathbf{x}_t\| = o(t^\beta)$,

$$\|\mathbf{A}(\mathbf{x}_t\mathbf{x}_t' - \mathbf{x}_0\mathbf{x}_0')\mathbf{A}'\| = o(t) \quad \text{a.s.},$$

and

$$\left\|\sum_t^t (\mathbf{A}\mathbf{x}_{\tau-1}\mathbf{e}_\tau' + \mathbf{e}_\tau \mathbf{x}_{\tau-1}'\mathbf{A}')\right\| = o(t) \quad \text{a.s.}$$

Let

$$\mathbf{\Pi}_t = t^{-1}\sum_1^t \mathbf{x}_\tau\mathbf{x}_\tau', \qquad \mathbf{\Delta}_t = t^{-1}\sum_1^t \mathbf{e}_\tau\mathbf{e}_\tau'.$$

Then

$$\mathbf{\Pi}_t - \mathbf{A}\mathbf{\Pi}_t\mathbf{A}' = \mathbf{\Delta}_t + o(t) \quad \text{a.s.}$$

From Fact (b) and (7), $\mathbf{\Pi}_t$ and $\mathbf{\Delta}_t$ are both relatively compact matrix sequences. The limiting equation is a Lyapunov equation $\mathbf{\Pi} = \mathbf{A}\,\mathbf{\Pi}\,\mathbf{A}' + \mathbf{\Delta}$ which has a unique solution $\mathbf{\Pi} = \Sigma\, \mathbf{A}^i\mathbf{\Delta}(\mathbf{A}')^i$. One can show analogously to (7) that

$$\sum_1^t \mathbf{e}_\tau\mathbf{e}_\tau' = \sum_1^t E(\mathbf{e}_\tau\mathbf{e}_\tau' | \mathscr{F}_{\tau-1}) + o(t) \quad \text{a.s.}$$

Thus, with probability 1, $\mathbf{\Delta}$ is a limit point of $\{t^{-1}\Sigma_1^t\, \mathbf{e}_\tau\mathbf{e}_\tau', t \geqslant 1\}$ if and only if $\mathbf{\Delta}$ is a limit point of $\{t^{-1}\Sigma_1^t\, E(\mathbf{e}_\tau\mathbf{e}_\tau'|\mathscr{F}_{\tau-1}), \tau \geqslant 1\}$. Thus, when $\bar{\lambda} < 1$,

$$\overline{\lim}\, t^{-1}\lambda_{\min}(V_t) > 0 \quad \text{a.s.}$$

if and only if

$$\underline{\lim}\, \lambda_{\min}\left[\sum_{\tau=0}^{\infty} \mathbf{A}^{\tau} \boldsymbol{\Delta}_t (\mathbf{A}')^{\tau}\right] > 0 \quad \text{a.s.}$$

Suppose that

$$\underline{\lim}\, \lambda_{\min}\left[\sum_{\tau=0}^{\infty} \mathbf{A}^{\tau} \boldsymbol{\Delta}_t(\omega) (\mathbf{A}')^{\tau}\right] > 0.$$

Then there exists a $k = k(\omega) \geqslant n$ such that

$$\lambda_{\min}\left[\sum_{\tau=0}^{k} \mathbf{A}^{\tau} \boldsymbol{\Delta}(\omega) (\mathbf{A}')^{\tau}\right] > 0.$$

This is equivalent to

$$\lambda_{\min}\left[\sum_{\tau=0}^{n-1} \mathbf{A}^{\tau} \boldsymbol{\Delta}(\omega) (\mathbf{A}')^{\tau}\right] > 0$$

and to the fact that $[\mathbf{A}, \boldsymbol{\Delta}^{1/2}(\omega)]$ is a reachable pair. Since $\{\boldsymbol{\Delta}_t(\omega)\}$ is relatively compact for almost all ω, if this statement is true for every limit point $\Delta(\omega)$ of $\{\boldsymbol{\Delta}_t(\omega)\}$, then

$$\underline{\lim}\, \lambda_{\min}\left[\sum_{\tau=0}^{n-1} \mathbf{A}^{\tau} \boldsymbol{\Delta}_t(\omega) (\mathbf{A}')^{\tau}\right] > 0.$$

Lai and Wei (1985) established that $\underline{\lim}\, t^{-1} \lambda_{\min}(V_t) > 0$ without assuming $\bar{\lambda} < 1$, if $(\mathbf{A}, \boldsymbol{\Delta}^{1/2})$ is a reachable pair.

Example 7: (Lai and Wei [1982]). Specialize ε_t to be i.i.d. random variables with mean zero and variance $\sigma^2 > 0$. In the least squares estimate problem for an AR(2) process

$$y_t = \beta_1 y_{t-1} + \beta_2 y_{t-2} + \varepsilon_t, \qquad y_0 = y_{-1} = 0,$$

define its state space model by

$$\mathbf{x}_t = (y_t, y_{t-1})', \qquad \mathbf{e}_t = (\varepsilon_t, 0)', \quad \text{and}$$

$$\mathbf{A} = \begin{bmatrix} \beta_1 & \beta_2 \\ 1 & 0 \end{bmatrix}.$$

Consider

$$V_t = \begin{bmatrix} \sum_{2}^{t-1} y_\tau^2 & \sum_{2}^{t-1} y_\tau y_{\tau-1} \\ \sum_{2}^{t-1} y_\tau y_{\tau-1} & \sum_{1}^{t-1} y_\tau^2 \end{bmatrix}.$$

Assume that $\beta_1 = 1$ and $\beta_2 = 0$. Then $y_t = S_t$ and

$$\lambda_{\max}(V_t) + \lambda_{\min}(V_t) = \sum_2^{t-1} S_\tau^2 + \sum_1^{t-2} S_\tau^2$$
$$= 2\sum_1^{t-1} S_{\tau-1}^2 + \sum_1^{t-1} \varepsilon_\tau^2 + 2\sum_{t=1}^{t-1} S_{\tau-1}\varepsilon_\tau$$

and

$$\lambda_{\max}(V_t)\cdot\lambda_{\min}(V_t) = \left(\sum_2^{t-1} S_\tau^2\right)\left(\sum_1^{t-1} S_{\tau-1}^2\right) - \left(\sum_2^{t-1} S_\tau S_{\tau-1}\right)$$
$$= \left(\sum_2^{t-1} \varepsilon_\tau^2\right)\left(\sum_1^{t-1} S_{\tau-1}^2\right) - \left(\sum_2^{t-1} \varepsilon_\tau S_{\tau-1}\right)^2.$$

In the preceding, Strassen's law of the iterated logarithm yields

$$\overline{\lim} \sum_1^t S_\tau^2/(t^2 \ln\ln t) = \text{constant},$$

and Lai and Wei (1982, lemma 2, cor 2) established that

$$\left(\sum_o^{t-1} \varepsilon_\tau S_{\tau-1}\right)^2 = o\left[\left(\sum_1^t S_{\tau-1}^2\right)\left(\ln \sum_1^t S_{\tau-1}^2\right)^{1+\delta}\right]$$
$$= o\left[\left(\sum_1^t S_{\tau-1}^2\right)(\ln t)^{1+\delta}\right] \quad \text{a.s.}$$

for every $\delta > 0$. Since $\Sigma_2^{t-1}\varepsilon_\tau^2 \sim t\sigma^2$ a.s., we have $\lambda_{\max}(V_t) \sim 2\Sigma_1^t S_\tau^2 = O(t^2 \ln\ln t)$ and $\lambda_{\min}(V_t) \sim \frac{1}{2}t\sigma^2$. From (b) of Theorem, $\|\mathbf{x}_t\|^2 = O(t \ln\ln t)$. Thus $\lambda_{\max}(V_t) = \mathrm{O}(t^2 \ln\ln t)$.

Example 8: (Lai and Wei [1985]). An $AR(p)$ model

$$y_t = \beta_1 y_{t-1} + \ldots + \beta_p y_{t-p} + \varepsilon_t$$

has a state space form

$$\mathbf{x}_t = \mathbf{A}\mathbf{x}_{t-1} + \mathbf{e}_t$$

where

$$\mathbf{x}_t = (y_t y_{t=1} \ldots y_{t-p})', \qquad \mathbf{e}_t = (1 \quad 0 \ldots 0)'\varepsilon_t$$

and

$$\mathbf{A} = \begin{bmatrix} \beta_1 \ldots \beta_{p-1} & \beta_p \\ \mathbf{I}_{p-1} & \mathbf{0} \end{bmatrix}.$$

Condition (5) is satisfied if

$$\sup_t E(|\varepsilon_t|^\alpha | \mathscr{F}_{t-1}) < \infty \quad \text{a.s.}$$

for some $\alpha > 0$. The pair $(\mathbf{A}, \mathbf{\Delta}^{1/2})$ is reachable, i.e., $\underline{\lim}\, \lambda_{\min}[\Sigma_{i=0}^{n-1} \mathbf{A}^i \mathbf{\Delta}_t (\mathbf{A}')^i] > 0$ if $\underline{\lim}\, t^{-1} \Sigma_1^t E(\varepsilon_\tau^2 | \mathscr{F}_{\tau-1}) > 0$ a.s.

2.6 Stability of Stochastic Systems

It is well known that the nature of stability of deterministic dynamic systems can be answered if we can construct Lyapunov functions with certain specified properties. See, for example LaSalle and Lefschetz (1961) or Hahn (1963). As an immediate application of the martingale convergence theorem in the preceding section, we examine stability property of stochastic systems via Lyapunov functions.

Generally speeaking, given a dynamic system with state vector $\mathbf{x}$, the stability of its equilibrium point (which is taken to be the origin without loss of generality) can be shown by constructing a positive definite continuous function of $\mathbf{x}$, $V(\mathbf{x})$, called a Lyapunov function, such that its time derivative or its first difference along a system trajectory is nonpositive definite in continuous-time dynamic models and in discrete-time dynamic models, respectively. A monotonically decreasing behavior of $V(\mathbf{x})$ along the trajectory implies a similar behavior for the norm of the state vector $\mathbf{x}$, $\|\mathbf{x}\|$, i.e., $\|\mathbf{x}\| \to 0$ as $t \to \infty$, which is in correspondence with our intuitive notion that the origin is asymptotically stable. To be a Lyapunov function, $V(x)$ is chosen to be continuous in an n-dimensional vector $\mathbf{x}$, such that $V(0) = 0$, $V(\mathbf{x}) > 0$, $\mathbf{x} \neq 0$, is finite for any finite $\|\mathbf{x}\|$, and $V(\mathbf{x}) \to \infty$ as $\|\mathbf{x}\| \to \infty$, and the sets $\{\mathbf{x}; V(\mathbf{x}) = c_i\}$, $c_1 \leqslant c_2 \leqslant \ldots$, should be nested.

Stability of the origin is implied by the behavior of $V(\mathbf{x})$ such that, for any set of i discrete sampling points in time, $0 \leqslant n_1 < \ldots < n_i$,

(8) $$V(\mathbf{x}_{n_1}) \geqslant V(\mathbf{x}_{n_2}) \geqslant \ldots \geqslant V(\mathbf{x}_{n_i}).$$

This behavior of $V(\mathbf{x})$ may be intuitively understood by interpreting $V(\mathbf{x})$ as a "generalized" distance of the system that must not increase with time for stable dynamic systems.

Now consider a discrete-time stochastic dynamic system described by

(9) $$\mathbf{x}_{k+1} = \mathbf{D}_k(\mathbf{x}_k, \xi_k), \qquad k = 0, 1, \ldots$$

where $\mathbf{x}_k$ is the n-dimensional state vector and ξ_k is a random variable (generally a vector). A control system described by

(10) $$\mathbf{x}_{k+1} = F_k(\mathbf{x}_k, \mathbf{u}_k, \xi_k), \qquad k = 0, 1, \ldots$$

can be regarded as a dynamic system described by (9) for a given control policy, say $\mathbf{u}_k = \boldsymbol{\phi}_k(\mathbf{x}_k)$, where $\boldsymbol{\phi}_k$ is a known function of $\mathbf{x}_k$:

$$\begin{aligned}\mathbf{x}_{k+1} &= \mathbf{F}_k(\mathbf{x}_k, \phi_k(\mathbf{x}_k), \xi_k) \\ &\triangleq \mathbf{D}_k(\mathbf{x}_k, \xi_k).\end{aligned}$$

Only one trajectory results from a given initial state $\mathbf{x}_0$, for a deterministic system. A collection of trajectories is possible from a given $\mathbf{x}_0$ for a stochastic system, depending on realizations of the stochastic process $\{\xi_k\}$.

Using the same intuitive arguments as before, consider a positive definite continuous function of $\mathbf{x}$, $V(\mathbf{x})$, which may be regarded again as representing the system's generalized distance. Because $\{\mathbf{x}_n\}$ is now a stochastic process, one must now consider the behavior of $V(\mathbf{x})$ for the class of all possible realizations of trajectories. One may, for example, replace condition (8) for the behavior of Lyapunov functions of deterministic systems by

$$EV(\mathbf{x}_{n_1}) \geqslant EV(\mathbf{x}_{n_2}) \geqslant \dots \geqslant EV(\mathbf{x}_{n_i}) \tag{11}$$

for any given set of i time instants, $0 \leqslant n_1 < n_2 < \dots < n_i$. Namely, (11) requires that the average behavior of $V(\mathbf{x})$ over possible realizations of trajectories behaves like the Lyapunov function of a deterministic system.

Intuitively speaking, stable stochastic systems should be such that $E(V(\mathbf{x}_n))$ remains bounded. In terms of $\{\mathbf{x}_n\}$, stable stochastic systems should be such that, for the majority of sample sequences $\{\mathbf{x}_n\}$, i.e., for the set of sample sequences with probability arbitrarily close to 1, $\|\mathbf{x}_n\|$ remains bounded.

For an asymptotically stable stochastic system, not only should $E(V(\mathbf{x}_n))$ remain bounded for all n, but actually $E(V(\mathbf{x}_n))$ should decrease monotonically to zero with probability 1.

Equation (11) is implied by the relation that, given any particular realization of $\mathbf{x}_0, \mathbf{x}_2, \dots, \mathbf{x}_n$, $V(\mathbf{x})$ satisfies the inequality

$$E(V(\mathbf{x}_n) | \mathbf{x}_0, \mathbf{x}_1, \dots, \mathbf{x}_{n-1}) \leqslant V(\mathbf{x}_{n-1}) \tag{12}$$

since $E(V(\mathbf{x}_n)) = E(E(V(\mathbf{x}_n) | \mathbf{x}_0, \mathbf{x}_1, \dots, \mathbf{x}_{n-1}))$, where the outer E refers to the expectation operation over possible $\mathbf{x}_0 \mathbf{x}_1, \dots, \mathbf{x}_{n-1}$. Inequality (12) can be interpreted to mean that the behavior of $\{\mathbf{x}_n\}$ is such that, given past behavior of $V(\mathbf{x})$ for a realized (sample) trajectory $\mathbf{x}_0, \mathbf{x}_1, \dots, \mathbf{x}_{n-1}$, the expected value of $V(\mathbf{x})$ at the next time instant, $E(V(\mathbf{x}_n) | \mathbf{x}_0, \dots, \mathbf{x}_{n-1})$, is not greater than the previous value of $V(\mathbf{x})$, $V(\mathbf{x}_{n-1})$.

Then the system can be regarded as stable in some stochastic sense, since the conditional expected generalized distance is not increasing with time.

Inequality (12) is precisely the definition that $V(\mathbf{x}_n), n = 0, 1, \ldots,$ is an expectation-decreasing martingale or a supermartingale.

The idea of discussing the stability of stochastic systems by suitably extending the deterministic Lyapunov theory seems to have appeared first in the papers by Bertram and Sarachik (1959) and Kats and Krazovskii (1960). The realization that such stochastic Lyapunov functions are supermartingales seems to be due to Bucy (1965) and Kushner (1965).

In the following paragraphs, we shall increase the precision of the statements in this section and discuss the problems of stability of stochastic systems. The exposition is based in part on Bucy and Kushner.

We consider a stochastic system described by (9) such that

$$\mathbf{D}_k(0, \xi_k) = 0 \quad \text{for all } k, \tag{13}$$

i.e., the origin is taken to be the equilibrium point of the system. The solution of (9) is denoted by $\mathbf{x}_k, k = 0, 1, \ldots,$ or by

$$\mathbf{x}(k, \mathbf{x}_0), \qquad k = 0, 1, \ldots$$

when it is desired to indicate the dependence of the solution on the initial state vector $\mathbf{x}_0$. Note that $\mathbf{x}_0$ is generally a random variable.

We now give definitions of stability and asymptotic stability in a way that parallels the definitions for deterministic systems.

Definition of stability. The origin is stable with probability 1 if and only if, for any $\delta > 0$ and $\varepsilon > 0$, there is a $\rho(\delta, \varepsilon) > 0$ such that, if $\|\mathbf{x}_0\| \leqslant \rho(\delta, \varepsilon)$, then

$$P[\sup_n \|\mathbf{x}_n\| \geqslant \varepsilon] \leqslant \delta.$$

Definition of asymptotic stability. The origin is asymptotically stable with probability 1 if and only if it is stable with probability 1 and

$$\|\mathbf{x}(n, \mathbf{x}_0)\| \to 0 \quad \text{with probability 1 as } n \to \infty \tag{14}$$

for $\mathbf{x}_0$ in some neighborhood of the origin.

If (14) is true for all $\mathbf{x}_0$ in the state space, then we say that the origin is asymptotically stable in the large. Let us now indicate how we prove the stability of a stochastic system if a positive definite continuous scalar function $V(\cdot)$ is defined on the space of state vectors such that $\{V(\mathbf{x}(n, \mathbf{x}_0))\}$ is an expectation-decreasing martingale in some region about the origin.

Such a proof can be made to depend essentially on the semimartingale inequality Doob (1953): for any $\lambda > 0$,

$$\text{(15)} \qquad P[\sup_n V(\mathbf{x}_n) \geqslant \lambda] \leqslant EV(\mathbf{x}_0)/\lambda.$$

To simplify the arguments, we show the proof for positive definite scalar function $V(\mathbf{x}_n)$ which forms positive supermartingale in the whole state vector space. It is a minor technical complication to extend the arguments to include $\{V(\mathbf{x}_n)\}$, which is a positive supermartingale only in some region about the origin. We assume that

$$EV(\mathbf{x}_0) < \infty.$$

Consider a positive definite continuous function of n-dimensional vector $\mathbf{x}$, $V(\mathbf{x})$, such that $V(0) = 0$, $V(\mathbf{x})$ finite for any finite $\|\mathbf{x}\|$, and $V(\mathbf{x}) \to \infty$ as $\|\mathbf{x}\| \to \infty$, and suppose that the sequence $\{V(\mathbf{x}(n, x_0))\}$ is an expectation-decreasing martingale for all $\mathbf{x}_0$. Then we will show that the origin is stable. From (15) for any $\varepsilon > 0$,

$$\text{(16)} \qquad P[\sup_n V(\mathbf{x}_n) \geqslant \varepsilon] \leqslant EV(\mathbf{x}_0)/\varepsilon.$$

Given such a scalar-valued function $V(\mathbf{x})$, it is always possible to find continuous positive nondecreasing functions α and β of real variables such that

$$\alpha(0) = 0, \qquad \beta(0) = 0$$

$$\alpha(\|\mathbf{x}\|) \to \infty, \qquad \beta(\|\mathbf{x}\|) \to \infty \quad \text{as} \quad \|\mathbf{x}\| \to \infty$$

and

$$\alpha(\|\mathbf{x}\|) \leqslant V(\mathbf{x}) \leqslant \beta(\|\mathbf{x}\|).$$

For example, such α and β can be constructed by

$$\beta(\|\mathbf{x}\|) \triangleq \max_{\|\mathbf{y}\| \leqslant \|\mathbf{x}\|} V(\mathbf{y})$$

and

$$\alpha(\|\mathbf{x}\|) \triangleq \min_{\mathbf{y}} V(\mathbf{y})$$

where

$$\|\mathbf{x}\| \leqslant \|\mathbf{y}\| \leqslant b(c(\|\mathbf{x}\|))$$

where

$$c(\|\mathbf{x}\|) \triangleq \min_{\|\mathbf{y}\| = \|\mathbf{x}\|} V(\mathbf{y})$$

and where the function b is chosen so that, for any $a > 0$,

$$\mathrm{V}(\mathbf{x}) > \alpha \quad \text{for} \quad \|\mathbf{x}\| > \mathrm{b}(a).$$

Since $\mathrm{V}(\mathbf{x}) \to \infty$ as $\|\mathbf{x}\| \to \infty$, it is always possible to choose such $b(a)$. Choose $\rho(\delta, \varepsilon)$ such that if

$$\|\mathbf{x}_0\| \leqslant \rho(\delta, \varepsilon)$$

then

$$EV(\mathbf{x}_0) \leqslant E\beta(\|\mathbf{x}_0\|) \leqslant \varepsilon\delta. \tag{17}$$

From (16) and (17),

$$\mathrm{P}[\sup_n \mathrm{V}(\mathbf{x}_n) \geqslant \varepsilon] \leqslant \delta. \tag{18}$$

Since

$$\mathrm{P}[\sup_n \alpha(\|\mathbf{x}\|) \geqslant \varepsilon] \leqslant \mathrm{P}[\sup_n \mathrm{V}(\mathbf{x}_n) \geqslant \varepsilon]$$

and since $\alpha(\cdot)$ has a unique inverse in the neighborhood of the origin, the origin is stable with probability 1.

Condition (17) can be satisfied in other ways. For example, if $\mathbf{x}_0$ is the random variable such that

$$\mathrm{P}[\|\mathbf{x}_0\| \leqslant M] = 1$$

for some finite $M > 0$, then

$$E\mathrm{V}(\|\mathbf{x}_0\|) \leqslant \beta(a)[1 - \mathrm{P}(\|\mathbf{x}_0\| \geqslant a)] + \beta(M)\mathrm{P}(\|\mathbf{x}_0\| \geqslant a)$$

or

$$\mathrm{P}[\|\mathbf{x}_0\| \leqslant a] \leqslant [E\mathrm{V}(\|\mathbf{x}_0\|) - \beta(a)]/\beta(M)$$

Choose $\rho(\delta, \varepsilon) > 0$ sufficiently small so that $\beta(\rho) + \rho\beta(M) \leqslant \varepsilon\delta$. Then for $\mathbf{x}_0$ satisfying $\mathrm{P}(\|\mathbf{x}_0\| \geqslant \rho) \leqslant \rho$ we have

$$E\mathrm{V}(\|\mathbf{x}_0\|) \leqslant \varepsilon\delta.$$

From this inequality and (16), (18) follows.

Thus we have proved, with a slightly different definition of stability, that the origin is stable if and only if, for any $\delta > 0$ and $\varepsilon > 0$, there exists a $\rho(\delta, \varepsilon) > 0$ such that for every x_0 satisfying $P(\|\mathbf{x}_0\| \geqslant \rho) \leqslant \rho$, and $P(\|\mathbf{x}_0\| \leqslant M) = 1$, (18) holds.

The criterion for asymptotic stability is given by the following.

Suppose that there exists a continuous nonnegative function $\gamma(\cdot)$ of real numbers such that it vanishes only at zero and

$$(19) \quad E[V(\mathbf{x}(n, \mathbf{x}_0))|\mathbf{x}_0, \ldots, \mathbf{x}_{n-1}] - V(\mathbf{x}(n-1, \mathbf{x}_0)) \leqslant -\gamma(\|\mathbf{x}(n-1, \mathbf{x}_0)\|) < 0$$

for all x_0. Then the origin is asymptotically stable with probability 1. As mentioned earlier in connection with the definition of the asymptotic stability, in order to show the asymptotic stability, it is necessary to show $EV(\mathbf{x}_n) \to 0$ as $n \to \infty$. Letting

$$V_n \triangleq V(\mathbf{x}_n)$$

and

$$\gamma_n \triangleq \gamma(\|\mathbf{x}_n\|)$$

(19) is written as

$$E(V_n|\mathbf{x}_0, \mathbf{x}_1, \ldots, \mathbf{x}_{n-1}) \leqslant V_{n-1} - \gamma_{n-1}.$$

Now apply Solo's theorem. We can also proceed directly.

Taking the expectation of this with respect to the random variables $x_0, \ldots, x_{n-1}$,

$$E(V_n) \leqslant EV_{n-1} - E\gamma_{n-1}$$

or

$$EV_{n+1} - EV_0 \leqslant -\sum_{i=0}^{n} E\gamma_i < 0$$

or

$$(20) \qquad 0 \leqslant \sum_{i=0}^{n} E\gamma_i \leqslant EV_0 \leqslant E\beta_0 < \infty$$

for every n. Equation (20) implies that

$$E\gamma_n \to 0 \quad \text{as} \quad n \to \infty.$$

Thus, $\gamma_n \to 0$ in probability.

Since it is possible to pick a subsequence of $\{\gamma_n\}$ such that the subsequence converges almost surely, let $\{\gamma_{n_i}\}$ be such a subsequence. Then, since γ is continuous and vanishes only at zero, we have $\|x(n_i, x_0)\| \to 0$ with probability 1.

Since $0 \leqslant EV_n \leqslant EV_0 < \infty$ by the semimartingale convergence theorem (see Doob [1953, p. 324]),

$$\lim_{n \to \infty} V_n \triangleq V_\infty$$

exists with probability 1. But

$$0 \leqslant V_n \leqslant \beta_n. \tag{21}$$

Then taking the limit of (21) on the subsequence $\{\beta_{n_i}\}$,

$$0 \leqslant V_\infty \leqslant \beta(0) = 0.$$

Therefore,

$$V_\infty = 0 \quad \text{with probability 1}$$

or

$$\overline{\lim}\, \alpha_n = \alpha(\overline{\lim}\, \|x(n, x_0)\|) = 0$$

hence,

$$\|x(n, x_0)\| \to 0 \quad \text{with probability 1.}$$

Examples. The following examples are taken from Kushner (1965). Consider a scalar system

$$x_{k+1} = (a + \xi_k)x_k \tag{22}$$

where the ξ_n are independent and identically distributed with

$$E\xi_n = 0, \qquad E\xi_n^2 = \sigma^2. \tag{23}$$

Choose

$$\mathrm{V}(x) = x^2. \tag{24}$$

Then

$$E(V_n | x_0, \ldots, x_{n-1}) = (a^2 + \sigma^2)x_{n-1}^2.$$

Therefore, if $(a^2 + \sigma^2) < 1$, then

(25) $$E(V_n|x_0, \ldots, x_{n-1}) - V_{n-1} = x_{n-1}^2(a^2 + \sigma^2 - 1) < 0.$$

hence the origin is asymptotically stable. This example can be extended to systems described by vector difference equations immediately.

From the basic semimartingale inequality (15),

$$c\mathrm{P}[\sup_n V_n \geqslant c] \leqslant \int V_0 = Ex_0^2$$

(26) or

$$\mathrm{P}[\sup_n |x_n| < c^{1/2}] > 1 - Ex_0^2/c$$

gives a useful probability expression on the magnitude of x_n. Now for the same system, if $\mathrm{V}(x)$ is chosen to be

(27) $$\mathrm{V}(x) = x^{2r}$$

for some positive integer $r > 1$, then

$$E(V_n|x_0, \ldots, x_{n-1}) = x_{n-1}^{2r}[E(a + \xi_0)^{2r} - 1].$$

Thus, for those positive integers r such that

(28) $$E(a + \xi_0)^{2r} < 1,$$

$\{x_n^{2r}\}$ is still an expectation-decreasing martingale and is still asymptotically stable. Now, instead of Eq. (26), one has

$$\mathrm{P}[\sup_n x_n^{2r} \geqslant c] \leqslant Ex_0^{2r}/c$$

(29) or

$$\mathrm{P}[\sup_n |x_n| \geqslant c^{1/2r}] \geqslant Ex_0^{2r}/c.$$

Another inequality that is sometimes useful is

$$\mathrm{P}\left[\max_{i \leqslant j \leqslant n} \left|\sum_{i=1}^{j} z_i\right| \geqslant \lambda\right] \leqslant \frac{1}{\lambda^2} \sum_{i=1}^{n} \operatorname{var} z_i,$$

where zs are independent random variables with mean zero and finite variances.

2.7 Transfer Functions and Spectral Densities

Besides difference equations, transfer functions are used to represent dynamic models. The latter is particularly convenient in dealing with stationary stochastic processes and stationary decision rules. In transfer function descriptions of dynamics, lag operators are used to relate variables at different time instants. In the systems literature, often z^{-1} or q^{-1} are used instead of the lag operator L; e.g., $y_t + 5y_{t-1} + 3y_{t-2} = e_t$ may be rendered as $(1 + 5q^{-1} + 3q^{-2})y_t = e_t$. Since transfer function description is usually employed in problems of infinite horizon, initial conditions are seldom mentioned explicitly.

Since the one-sided z-transform of a sequence $\{a_k\}_0^\infty$ is defined to be*

$$A(z) = \sum_{k=0}^{\infty} a_k z^{-k}, \tag{30}$$

its kth member of the sequence can be recovered by the inversion formula

$$a_k = \frac{1}{2\pi j}\oint A(z)z^{k-1}\,\mathrm{d}z = \frac{1}{2\pi}\int_{-\pi}^{\pi} A(e^{i\omega})e^{ik\omega}\,\mathrm{d}\omega. \tag{31}$$

The definitional relation of the z-transform (30) and its inversion formula (31) take on the form of a Fourier coefficient calculation when $z = e^{i\omega}$ is substituted in.

In the transfer function representation, a model takes the form

$$\mathbf{A}(q^{-1})\mathbf{y}_t = \mathbf{B}(q^{-1})\mathbf{u}_t + \mathbf{C}(q^{-1})\mathbf{e}_t, \tag{32}$$

where $\mathbf{u}_t$ is a control or instrument vector and where $\mathbf{e}_t$ is some exogenous noise disturbance. For example, in the absence of exogenous disturbances, an ARMA model relates a sequence $\{\mathbf{y}_k\}$ to $\{\mathbf{u}_k\}$ by

$$\mathbf{A}_0\mathbf{y}_k + \mathbf{A}_1\mathbf{y}_{k-1} + \ldots + \mathbf{A}_n\mathbf{y}_{k-n} = \mathbf{B}_0\mathbf{u}_k + \ldots + \mathbf{B}_n\mathbf{u}_{k-n},$$

where $\mathbf{A}_0$ is usually assumed to be nonsingular. This relationship is expressed in shorthand notation

$$\mathbf{A}(z^{-1})\mathbf{y}_k = \mathbf{B}(z^{-1})\mathbf{u}_k,$$

which is a special case of (32), where

$$\mathbf{A}(z^{-1}) = \mathbf{A}_0 + \mathbf{A}_1 z^{-1} + \ldots + \mathbf{A}_n z^{-n}$$

and

$$\mathbf{B}(z^{-1}) = \mathbf{B}_0 + \mathbf{B}_1 z^{-1} + \ldots + \mathbf{B}_n z^{-n}.$$

---

* In the systems literature this is usually written as $A(z^{-1})$.

Similar expressions are obtained when **u** is absent from (32). In (32) there are two transfer functions: one from **u** to **y**, and the other from **e** to **y**. The transfer function relation

$$\mathbf{y}_k = \mathbf{T}(z^{-1})\mathbf{u}_k$$

is obtained by

$$\mathbf{A}(z^{-1})\mathbf{T}(z^{-1}) = \mathbf{B}(z^{-1}),$$

i.e., $\mathbf{T}(z^{-1}) = \mathbf{A}^{-1}(z^{-1})\mathbf{B}(z^{-1})$ when the indicated inverse exists except at a finite number of z values. The coefficient matrices on $\mathbf{T}$ are determined by

$$\mathbf{T}(z^{-1}) = \mathbf{T}_0 + \mathbf{T}_1 z^{-1} + \dots$$

where

$$(33)\qquad \begin{aligned} &\sum_{i=0}^{j} \mathbf{A}_i \mathbf{T}_{j-i} = \mathbf{B}_j, \qquad && j = 0, 1, \dots, n \\ &\sum_{j=0}^{n} \mathbf{A}_i \mathbf{T}_{j-i} = 0, \qquad && j > n. \end{aligned}$$

The matrix $\mathbf{T}_0$ is called the impact multiplier and $\mathbf{T}_i, i = 1, 2, \dots$ are called dynamic multipliers. Their sum is the total multiplier, i.e., it expresses the long-run effects of once-and-for-all shifts in the exogenous variables. These multipliers can be more easily calculated by putting models in state space form first, as we shall show shortly. Equation (33) makes clear that these coefficient matrices are not all independently determined. For example, with $n = 2$, the column vectors of 6 by 2 matrices which stack $\mathbf{T}_j, \mathbf{T}_{j+1}$, and $\mathbf{T}_{j+2}, j = 1, 2, \dots$ must all lie in the same null space of the matrix $[\mathbf{A}_2 \mathbf{A}_1 \mathbf{A}_0]$. More generally, arrange the coefficient matrices of the transfer matrix $\mathbf{T}(z^{-1})$ into a Hankel matrix, i.e., a matrix with its (i, j)th submatrix being $\mathbf{T}_{i+j-1}$. The rank of this Hankel matrix is finite (when the spectral density of the processes in the right-hand of (32) or its special cases are rational) by the Kronecker's theorem, i.e., the column vectors of the Hankel matrix are not all independent. See Aoki (1987c, Chapter 5) for further detail.

The process $\{\mathbf{y}_k\}$ has zero mean and is wide-sense stationary when $\{\mathbf{u}_k\}$ is. Its covariance matrices are

$$\mathbf{R}_k = E(\mathbf{y}_{l+k}\mathbf{y}_l') = \begin{cases} \displaystyle\sum_{j=0}^{\infty} \mathbf{T}_{j+k} \boldsymbol{\Sigma} \mathbf{T}_j', & k \geqslant 0 \\ \displaystyle\sum_{j=0}^{\infty} \mathbf{T}_j \boldsymbol{\Sigma} \mathbf{T}_{j-k}', & k < 0 \end{cases}$$

where $\boldsymbol{\Sigma} = \operatorname{cov} \mathbf{u}$. These relations will be discussed further later.

A stochastic process $\{\mathbf{y}_k\}$ can often be thought of as passing another stochastic process $\{\mathbf{e}_k\}$ through a linear dynamic system with transfer matrix $\mathbf{F}(z)$

$$\mathbf{y}_k = \mathbf{F}(z)\mathbf{e}_k \tag{34}$$

where

$$\mathbf{F}(z) = \frac{1}{d(z^{-1})}(\mathbf{F}_0 + \mathbf{F}_1 z^{-1} + \ldots + \mathbf{F}_n z^{-n})$$

and where $d(\sigma)$ is a polynomial in σ. When $d(\sigma) = 1$, (34) merely means that

$$\mathbf{y}_k = \sum_{i=0}^{n} \mathbf{T}_i \mathbf{e}_{k-i}.$$

When $d(\sigma) \neq 1$ then $\mathbf{F}(z)$ in effect is an infinite matrix sequence in z^{-1} obtained by formal expansion. The sequences of exogenous disturbances may not be uncorrelated.

ARMA models for variables in levels may be transformed into models for the differences, and models for differenced variables can be integrated to give models for levels. Such changes of model representations are useful in models for variables that are cointegrated or variables with unit roots. See papers in the special issue of *Journal of Economic Dynamics and Control* (1988) for examples of such variables and models. To illustrate, consider the model $\mathbf{A}(L)\mathbf{y}_t = \mathbf{C}(L)\mathbf{e}_t$ to be the same as $\mathbf{F}(L)\Delta\mathbf{y}_t = \mathbf{C}(L)\mathbf{e}_t - \mathbf{F}_k\mathbf{y}_{t-k}$, where

$$\mathbf{A}(L) = \mathbf{F}(L)(1 - L) + \mathbf{F}_k L^k$$

where

$$\mathbf{A}(L) = \mathbf{I} - \mathbf{A}_1 L - \mathbf{A}_2 L^2 - \ldots - \mathbf{A}_k L^k$$

and where

$$\mathbf{F}(L) = \mathbf{I} - \mathbf{F}_1 L - \mathbf{F}_2 L^2 - \ldots - \mathbf{F}_{k-1} L^{k-1}$$

and

$$\mathbf{F}_i = -\mathbf{I} + \mathbf{A}_1 + \ldots + \mathbf{A}_i, \qquad i = 1, \ldots, k.$$

Conversely, given an ARMA model for $\Delta\mathbf{y}_t$, $\theta(L)\Delta\mathbf{y}_t = \phi(L)\mathbf{e}_t$, its integrated expression is

$$\theta(L)y_t = \phi(1)\mathbf{w}_t + \phi^*(L)\mathbf{e}_t$$

where

$$\mathbf{w}_t = \sum_{s=1}^{t} \mathbf{e}_s, \quad \text{and} \quad \phi^*(L) = \sum \phi_j^* L^j$$

where

$$\phi_j^* = -\sum_{i=j+1}^{\infty} \phi_j$$

if the sum converges.

The spectral density matrix of a mean-zero weakly stationary (vector-valued) process $\{\mathbf{y}_t\}$ is defined to be the two-sided z-transform of (auto-) covariance matrices

$$\mathbf{S}_{\mathbf{yy}}(z) = \sum_{-\infty}^{\infty} \mathbf{R}_{\mathbf{yy}}(k) z^{-k} \tag{35}$$

where

$$\mathbf{R}_{\mathbf{yy}}(k) = E(\mathbf{y}_{l+k}\mathbf{y}_l'), \qquad k \geqslant 0,$$
$$\mathbf{R}_{\mathbf{yy}}(k) = \mathbf{R}_{\mathbf{yy}}'(-k), \qquad k < 0,$$

and

$$z = e^{i\omega}.$$

This is the same except for trivial notational differences, such as the (auto-)covariance generating function which is defined by

$$\mathbf{G}_{\mathbf{yy}}(z) = \sum_{-\infty}^{\infty} \mathbf{R}_{\mathbf{yy}}(k) z^{k}$$

in the statistical literature with $z = e^{-j\omega}$ to obtain its frequency domain expression. In both notations, the spectral density at zero frequency is equal to

$$\mathbf{S}_{\mathbf{yy}}(1) = \mathbf{G}_{\mathbf{yy}}(1) = \sum_{-\infty}^{\infty} \mathbf{R}_{\mathbf{yy}}(k)$$
$$= E\mathbf{y}_0\mathbf{y}_0' + \sum_{k=1}^{\infty} E(\mathbf{y}_0\mathbf{y}_k' + \mathbf{y}_k\mathbf{y}_0')$$

if the series converges.*

When $\{\mathbf{u}_t\}$ and $\{\mathbf{y}_t\}$ are jointly weakly stationary and $\mathbf{y}_t$ is related to $\mathbf{u}_t$ by

$$\mathbf{y}_t = \sum_{-\infty}^{t} \mathbf{H}_i \mathbf{u}_{t-i},$$

* When $\{\mathbf{y}_t\}$ has a stable state space representation, the series converges because $E|\mathbf{y}_0\mathbf{y}_k'| < \lambda^k$ for some $|\lambda| < 1$. Sufficient conditions in a more general context are discussed in Ibragimov and Linnik (1971, Section 18.5).

its z-transform is

(36) $$\mathbf{Y}(z) = \mathbf{H}(z)\mathbf{U}(z)$$

in both engineering and statistics notation, even though

$$\mathbf{H}(z) = \Sigma\, \mathbf{H}_i z^{-i}$$

in the engineering notation and

$$\mathbf{H}(z) = \Sigma\, \mathbf{H}_i z^{i}$$

in the generating function notation. $\mathbf{H}(z)$ is stable if its poles all lie inside the unit circle in the engineering notation and all its poles lie outside in the statistics notation.

In statistical terminology, a generating function $\theta(z) = \Sigma\, \theta_j z^j$ is realizable if $\theta_j = 0$ for $j < 0$. It is said to be stable if realizable and $\Sigma_0^\infty \theta_j z^j$ converges in $|z| \leqslant 1$, i.e., analytic inside the unit disk in the complex plane. It is invertible if $\theta(z)^{-1}$ is stable. A generating function has a canonical factorization

$$g(z) = \theta_1(z)\Delta\theta_2'(z^{-1})$$

if θ_1 and θ_2 are canonical, i.e., stable and invertible, and Δ is independent of z.

In the engineering notation, the spectral density matrix $\mathbf{S}(z)$ has a canonical factorization

$$\mathbf{S}(z) = \mathbf{H}(z)\Delta\mathbf{H}'(1/z)$$

if $\mathbf{H}(z)$ corresponds to a transfer function of a realizable system, i.e., causal and stable dynamic system, i.e.,

$$\mathbf{H}(z) = \sum_0^\infty \mathbf{H}_i z^{-i}$$

with poles inside the unit disk, i.e., analytic in $|z| > 1$. It is invertible if $\mathbf{H}(z)^{-1}$ is stable. To lessen the possibility of confusion, we sometimes use q^{-1}, as in

$$\mathbf{H}(q^{-1}) = \Sigma\, \mathbf{H}_i q^{-i},$$

or explicitly indicate the associated polynomials.

Given $\mathbf{S}_{yy}$, one can always obtain the factorization $\mathbf{H}(e^{i\omega})\Delta\mathbf{H}(e^{-j\omega})$ in which both poles and zeros of $\mathbf{H}(z)$ lie inside the unit disk since all poles and zeros have mirror images about the unit circle (except when they happen to lie on the unit circle). Since the inverse of $\mathbf{H}(z)$ is also stable because the zeros of the numerator all lie within the unit disk, the dynamics with transfer function $\mathbf{H}^{-1}(z)$ driven by the $\mathbf{y}$-sequence is stable and will produce a white noise sequence. This system is known as the whitening filter.

The one-sided z-transform of the input and output sequences of a dynamic system are related by (36) where

$$\mathbf{Y}(z) = \sum_{0}^{\infty} \mathbf{y}_k z^{-k}$$

and

$$\mathbf{U}(z) = \sum_{0}^{\infty} \mathbf{u}_k z^{-k},$$

and $\mathbf{H}(z)$ is the transfer function of the dynamics

$$\mathbf{H}(z) = \sum_{0}^{\infty} \mathbf{H}_k z^{-k}$$

where $\{\mathbf{H}_k\}$ is the impulse response sequence. Because the dynamics are assumed to be causal, $\mathbf{H}_{-k} = 0$ for $k > 0$.

Corresponding to the cross-covariance matrix

$$\mathbf{R}_{\mathbf{yu}}(k) = E(\mathbf{y}_{l+k}\mathbf{u}_l'),$$

the cross-spectral density matrix, $\mathbf{S}_{\mathbf{yu}}(z)$, is defined as its z-transform in the same way

(37)
$$\mathbf{S}_{\mathbf{yu}}(z) = \sum_{-\infty}^{\infty} \mathbf{R}_{\mathbf{yu}}(k) z^{-k}.$$

In terms of $\mathbf{Y}(z)$ and $\mathbf{U}(z)$, we note that (37) is given by

$$\begin{aligned}\mathbf{S}_{\mathbf{yu}}(z) &= E[\mathbf{Y}(z)\mathbf{U}'(z^{-1})]\\ &= \mathbf{H}(z)\mathbf{S}_{\mathbf{uu}}(z)\end{aligned}$$

and

$$\begin{aligned}\mathbf{S}_{\mathbf{uy}}(z) &= E[\mathbf{U}(z)\mathbf{Y}'(z^{-1})]\\ &= \mathbf{S}_{\mathbf{uu}}(z)\mathbf{H}'(z^{-1}).\end{aligned}$$

In deriving the preceding, we use the relation

$$\begin{aligned}\mathbf{S}_{\mathbf{yu}}(-l) &= E[\mathbf{y}(k)\mathbf{u}'(k+l)] = E[\mathbf{u}(k+l)\mathbf{y}'(k)]'\\ &= \mathbf{S}_{\mathbf{uy}}'(l).\end{aligned}$$

From the definitional equation

$$\mathbf{S}_{\mathbf{yu}}(z) = \sum_{-\infty}^{\infty} \mathbf{R}_{\mathbf{yu}}(k) z^{-k},$$

we obtain

$$\begin{aligned}\mathbf{S}_{\mathbf{yu}}(1/z) &= \sum_{-\infty}^{\infty} \mathbf{R}_{\mathbf{yu}}(\mathrm{k})z^{k} \\ &= \sum_{-\infty}^{\infty} \mathbf{R}_{\mathbf{yu}}(-l)z^{-l} \\ &= \sum_{-\infty}^{\infty} \mathbf{R}'_{\mathbf{uy}}(l)z^{-l} \\ &= \mathbf{S}'_{\mathbf{uy}}(z).\end{aligned}$$

Thus

(38)
$$\begin{aligned}\mathbf{S}_{\mathbf{yy}}(z) &= \mathbf{S}_{\mathbf{yu}}(z)\mathbf{H}'(z^{-1}) \\ &= \mathbf{H}(z)\mathbf{S}_{\mathbf{uu}}(z)\mathbf{H}'(z^{-1}).\end{aligned}$$

A special relation of much interest is

(39)
$$\begin{aligned}E(\mathbf{y}_k\mathbf{y}'_k) = \mathbf{R}_{\mathbf{yy}}(0) &= \frac{1}{2\pi}\int_{-\pi}^{\pi} \mathbf{S}_{\mathbf{yy}}(z)\frac{\mathrm{d}z}{z} \\ &= \frac{1}{2\pi}\int_{-\pi}^{\pi} \mathbf{S}_{\mathbf{yy}}(e^{i\omega})\mathrm{d}\omega\end{aligned}$$

which, by (38), becomes

$$\mathbf{R}_{\mathbf{yy}}(0) = \frac{1}{2\pi}\int_{-\pi}^{\pi} \mathbf{H}(e^{i\omega})\mathbf{S}_{\mathbf{uu}}(e^{i\omega})\mathbf{H}(e^{-j\omega})\,\mathrm{d}\omega.$$

In particular, if $\{\mathbf{u}_k\}$ is a white noise sequence, then $\mathbf{R}_{\mathbf{uu}}(u) = \boldsymbol{\Delta}$ is the only nonzero covariance matrix and hence $\mathbf{S}_{\mathbf{uu}}(e^{j\omega}) = \boldsymbol{\Delta}$. Eq. (39) reduces to

(40)
$$\begin{aligned}E\mathbf{y}_k\mathbf{y}'_k &= \frac{1}{2\pi}\int_{-\pi}^{\pi} \mathbf{H}(e^{i\omega})\boldsymbol{\Delta}\mathbf{H}(e^{-j\omega})\mathrm{d}\omega \\ &= \frac{1}{2\pi}\sum_{k=0}^{\infty} \mathbf{H}_k\boldsymbol{\Delta}\mathbf{H}'_k.\end{aligned}$$

This is a particular case of the Parseval identity.

Parseval Relation

We have a discrete version of the Parseval identity for those sequences with $\Sigma(a_k)^2 < \infty$. Slightly generalizing, we note that

$$\begin{aligned}\sum_{k=0}^{\infty} \mathbf{a}'_k\mathbf{Q}\mathbf{a}_k &= \frac{1}{2\pi j}\oint \mathbf{A}'(1/z)\mathbf{Q}\mathbf{A}(z)\mathrm{d}z/z \\ &= \frac{1}{2\pi}\int_{-\pi}^{\pi} \mathbf{A}'(e^{-j\omega})\mathbf{Q}\mathbf{A}(e^{i\omega})\mathrm{d}\omega\end{aligned}$$

which is valid for all (vector-valued) sequences with finite **Q**-metric if $\mathbf{Q}' = \mathbf{Q} \geqslant 0$, i.e., if $\Sigma\, \mathbf{a}_k' \mathbf{Q} \mathbf{a}_k < \infty$.

With y and u scalar, (39) reduces to

$$\sigma_y^2 = \frac{\sigma_u^2}{2\pi} \int_{-\pi}^{\pi} \mathrm{H}(e^{i\omega})\mathrm{H}'(e^{-j\omega})d\omega$$

$$= \frac{\sigma_u^2}{2\pi} \sum_{k=0}^{\infty} \mathrm{h}_k^2.$$

If a filter input is a white noise sequence $\{u_k\}$ with transfer function $\mathrm{F}(z)$ and the output sequence $\{s_k\}$, the cross-spectral density is equal to

$$\mathbf{S}_{su}(z) = \mathrm{F}(z)\mathbf{S}_{uu}(z),$$
$$= \sigma_u^2 \mathrm{F}(z)$$

i.e.,

$$\mathbf{R}_{su}(k) = \sigma_u^2 \mathrm{f}_k, \qquad k \geqslant 0$$

i.e.,

$$\mathrm{f}_k = \mathbf{R}_{su}(k)/\sigma_u^2, \qquad k \geqslant 0$$

or

$$\mathrm{F}(z) = \frac{1}{\sigma_u^2}[\mathbf{S}_{su}(z)]_+$$

where $[\,]_+$ selects only the part of $[\,\cdot\,]$ with negative powers of z. A practical way of carrying this operation is via partial fraction (Laurent series) expansion, as in Hansen and Sargent (1980, p. 35).

2.8 State Space Models

A mean zero and covariance stationary sequence $\{\mathbf{y}_k\}$ can be represented as a state model

(41)
$$\begin{cases} \mathbf{s}_{k+1} = \mathbf{A}\mathbf{s}_k + \xi_k \\ \mathbf{y}_k = \mathbf{C}\mathbf{s}_k \end{cases}$$

where the dynamic matrix **A** is asymptotically stable, and where

$$E\xi_k = 0,$$
$$E\xi_k \xi_l' = \mathbf{Q}\delta_{kl}.$$

The covariance matrices of the **y**-sequence are

$$\boldsymbol{\Lambda}_l = E\mathbf{y}_{k+l}\mathbf{y}'_k = \begin{cases} \mathbf{C}\mathbf{A}^l\boldsymbol{\Pi}\mathbf{C}', & l \geqslant 0 \\ \mathbf{C}\boldsymbol{\Pi}\mathbf{A}'^{-l}\mathbf{C}', & l < 0 \end{cases}$$

where $\boldsymbol{\Pi} = \operatorname{cov}(\mathbf{s}_k)$ satisfies

$$\boldsymbol{\Pi} = \mathbf{A}\boldsymbol{\Pi}\mathbf{A}' + \mathbf{Q}. \tag{42}$$

The spectral density matrix for the y-sequence is put into the factored form

$$\begin{aligned} \mathbf{S}_y &= \sum_{-\infty}^{\infty} \boldsymbol{\Lambda}_l z^{-1} \\ &= \mathbf{C}\left(\boldsymbol{\Pi} + \sum_{1}^{\infty} z^{-l}\mathbf{A}^l\boldsymbol{\Pi} + \sum_{-\infty}^{-1} \boldsymbol{\Pi}\mathbf{A}'^{-l}z^{-l}\right)\mathbf{C}' \\ &= \mathbf{C}[\boldsymbol{\Pi} + \mathbf{A}(z\mathbf{I} - \mathbf{A})^{-1}\boldsymbol{\Pi} + \boldsymbol{\Pi}(z^{-1} - \mathbf{A}')^{-1}\mathbf{A}']\mathbf{C}' \\ &= \mathbf{C}(z\mathbf{I} - \mathbf{A})^{-1}\mathbf{Q}(z^{-1}\mathbf{I} - \mathbf{A})^{-1}\mathbf{C}'. \end{aligned} \tag{43}$$

Note from the dynamics that the z-transform of the y-sequence is given by

$$\mathbf{Y}(z) = \mathbf{C}(z\mathbf{I} - \mathbf{A})^{-1}\boldsymbol{\Xi}(z) = \mathbf{H}(z)\boldsymbol{\Xi}(z),$$

where $\boldsymbol{\Xi}(z)$ is the z-transform of the ξ-sequence and $\mathbf{H}(z)$ is the transfer function of (41) $\mathbf{H}(z) = \mathbf{C}(z\mathbf{I} - \mathbf{A})^{-1}$. Since ξs are uncorrelated, their spectral density matrix is a constant matrix

$$\mathbf{S}_\xi(z) = \mathbf{Q},$$

and hence we have factored it as

$$\mathbf{S}_y(z) = \mathbf{H}(z)\mathbf{S}_\xi(z)\mathbf{H}'(z^{-1}). \tag{44}$$

Now modify (41) by including a measurement noise process

$$\mathbf{y}_k = \mathbf{C}\mathbf{s}_k + \boldsymbol{\eta}_k \tag{45}$$

where

$$\operatorname{cov}\begin{pmatrix} \boldsymbol{\xi} \\ \boldsymbol{\eta} \end{pmatrix}_k = \begin{pmatrix} \mathbf{Q} & \mathbf{S}' \\ \mathbf{S} & \mathbf{R} \end{pmatrix} \geqslant 0.$$

This matrix can be factored as

$$\begin{pmatrix} \mathbf{Q} & \mathbf{S} \\ \mathbf{S}' & \mathbf{R} \end{pmatrix} = \begin{pmatrix} \mathbf{B} \\ \mathbf{I} \end{pmatrix} \boldsymbol{\Delta}(\mathbf{B}'\mathbf{I}), \qquad \boldsymbol{\Delta} \geqslant 0, \tag{46}$$

for appropriate $\mathbf{B}$ and $\boldsymbol{\Delta}$ so that the model is rewritten as

$$\begin{aligned} \mathbf{z}_{k+1} &= \mathbf{A}\mathbf{z}_k + \mathbf{B}\mathbf{e}_k \\ \mathbf{y}_k &= \mathbf{C}\mathbf{z}_k + \mathbf{e}_k \end{aligned} \tag{47}$$

where

$$\operatorname{cov} \mathbf{e}_k = \boldsymbol{\Delta}$$

This system has the transfer matrix

$$\mathbf{H}(z) = \mathbf{I} + \mathbf{C}(z\mathbf{I} - \mathbf{A})^{-1}\mathbf{B}.$$

The covariance matrix $\boldsymbol{\Pi}$ now satisfies

$$\boldsymbol{\Pi} = \mathbf{A}\boldsymbol{\Pi}\mathbf{A}' + \mathbf{B}\boldsymbol{\Delta}\mathbf{B}'$$

instead of (42). The matrices $\mathbf{B}$ and $\boldsymbol{\Delta}$ are obtained by solving a Riccati equation. See Aoki (1987c, Chapter 7) for further computational details.

For $l \geqslant 0$, the model (47) shows that

$$\mathbf{y}_{k+l} = \mathbf{C}\mathbf{A}^l\mathbf{z}_k + \sum_{i=1}^{l} \mathbf{H}_i\mathbf{e}_{k+l-i}.$$

Therefore, noting $\mathbf{H}_k = \mathbf{C}\mathbf{A}^{k-1}\mathbf{B}$,

$$\boldsymbol{\Lambda}_l = E\mathbf{y}_{l+k}\mathbf{y}_k'$$

and

$$= \mathbf{C}\mathbf{A}^{l-1}\mathbf{M}, \qquad l \geqslant 1$$

where $\mathbf{M} = \mathbf{A}\boldsymbol{\Pi}\mathbf{C}' + \mathbf{B}\boldsymbol{\Delta}$,

$$\boldsymbol{\Lambda}_{-l} = \boldsymbol{\Lambda}_l' \quad \text{for} \quad l < 0$$

and

$$\boldsymbol{\Lambda}_0 = \mathbf{C}\boldsymbol{\Pi}\mathbf{C}' + \boldsymbol{\Delta}.$$

Then the spectral density matrix has the decomposition

$$\mathbf{S}_y(z) = \mathbf{H}(z)\boldsymbol{\Delta}\mathbf{H}'(z^{-1})$$

where

$$\mathbf{H}(z) = \mathbf{I} + \mathbf{C}(z\mathbf{I} - \mathbf{A})^{-1}\mathbf{B}.$$

This last equality is verified by direct calculation after noting that $\mathbf{BS} = \mathbf{B}\boldsymbol{\Delta}\mathbf{B}'$ from (46).

Exercise 1

Given a state space model $\mathbf{x}_{t+1} = \mathbf{A}\mathbf{x}_t + \mathbf{u}_t$ and $\mathbf{y}_t = \mathbf{C}\mathbf{x}_t$, choose a new coordinate so that the model is given by $\boldsymbol{\xi}_{t+1} = \mathbf{A}\boldsymbol{\xi}_t + \boldsymbol{v}_t$, $\mathbf{y}_t = \mathbf{C}\boldsymbol{\xi}_t + \mathbf{e}_t$.

Exercise 2

Express $\text{vec}\,\mathbf{\Pi}$ of the covariance equation $\mathbf{\Pi} = \mathbf{A\Pi A}' + \mathbf{B\Delta B}'$ in terms of $\text{vec}\,\mathbf{\Delta}$ by solving the Lyaparnov equation. What are the conditions on $\mathbf{A}$ for this procedure to be valid?

Exercise 3

Suppose you solve the Riccati equation $\mathbf{\Pi} = \mathbf{A\Pi A}' + \mathbf{P}(\mathbf{\Pi})$ iteratively, where $\mathbf{P}(\mathbf{\Pi}) = (\mathbf{M} - \mathbf{A\Pi C}')(\mathbf{\Lambda}_0 - \mathbf{C\Pi C}')^{-1}(\mathbf{M} - \mathbf{A\Pi C}')'$ by $\mathbf{\Pi}_{n+1} = \mathbf{A\Pi}_n\mathbf{A}' + \mathbf{P}(\mathbf{\Pi}_n)$. Show that $\mathbf{Z}_n = \mathbf{\Pi}_n - \mathbf{\Pi}$ is governed by

$$\text{vec}\,\mathbf{Z}_{n+1} = (\mathbf{F} \otimes \mathbf{F})\,\text{vec}\,\mathbf{Z}_n$$

where

$$\mathbf{F} = \mathbf{A} - \mathbf{BC}.$$

2.9 Predictor Model Representation

Suppose that a process $\{\mathbf{y}_k\}$ is generated by

$$\mathbf{y}_k + \mathbf{A}_1\mathbf{y}_{k-1} + \ldots + \mathbf{A}_n\mathbf{y}_{k-n} = \mathbf{B}_0\mathbf{x}_k + \mathbf{B}_1\mathbf{x}_{k-1} + \ldots + \mathbf{B}_n\mathbf{x}_{k-n} + \mathbf{e}_k + \mathbf{C}_1\mathbf{e}_{k-1} + \ldots + \mathbf{C}_n\mathbf{e}_{k-n}$$

where $\mathbf{e}$s are zero-mean and independent and $\{\mathbf{x}_k\}$ is a deterministic sequence. This is called an ARMAX model, where X refers to the presence of exogenous $\mathbf{x}$s in the model. The values of $\mathbf{y}_{-1}, \mathbf{y}_{-2}, \ldots, \mathbf{y}_{-n}, \mathbf{x}_{-1}, \mathbf{x}_{-2}, \ldots, \mathbf{x}_{-n}, \mathbf{e}_{-1}, \ldots, \mathbf{e}_{-n}$ are given as initial data. The conditional expectation, given $\mathbf{y}_0, \mathbf{y}_1, \ldots, \mathbf{y}_{k-1}$, written as $\mathbf{y}_{k|k-1}$, is

$$\begin{aligned}\mathbf{y}_{k|k-1} &= -\mathbf{A}_1\mathbf{y}_{k-1} - \ldots - \mathbf{A}_n\mathbf{y}_{k-n} \\ &= \mathbf{B}_0\mathbf{x}_k + \ldots + \mathbf{B}_n\mathbf{x}_{k-n} + \mathbf{C}_1\mathbf{e}_{k-1} + \ldots + \mathbf{C}_n\mathbf{e}_{k-n}.\end{aligned}$$

Here the conditioning variables are $\mathbf{e}_s, s \leqslant k-1$, which are assumed available.

When $\{\mathbf{y}_k\}$, $\{\mathbf{x}_k\}$, and $\{\mathbf{e}_k\}$ are related by

$$\mathbf{y}_k = \mathbf{f}_k(\mathbf{y}^{k-1}, \mathbf{x}^{k-1}) + \mathbf{e}_k, \qquad k = 0, 1, \ldots,$$

where $\mathbf{y}^{k-1}$ and $\mathbf{x}^{k-1}$ denote the set of $\mathbf{y}_{k-1}, \mathbf{y}_{k-2}, \ldots, \mathbf{y}_0$ and $\mathbf{x}_{k-1}, \mathbf{x}_{k-2}, \ldots, \mathbf{x}_0$, the term $\mathbf{f}(\mathbf{y}^{k-1}, \mathbf{x}^{k-1})$ is the expected value of $\mathbf{y}_k$ given $\mathbf{y}^{k-1}$ and $\mathbf{x}^{k-1}$, which is the next one-step–ahead predictor in the mean square sense, if $\mathbf{e}_k$ is a sequence of mean zero white noises.

For example, the representation

$$\mathbf{y}_k = \mathbf{P}(\mathbf{q}^{-1})\mathbf{x}_{k-1} + \mathbf{Q}(\mathbf{q}^{-1})\mathbf{e}_k, \qquad k \geqslant 0$$

with initial conditions

$$\mathbf{x}_k = 0, \qquad \mathbf{y}_k = 0, \qquad \mathbf{e}_k = 0, \qquad k < 0,$$

can be put into the predictor model representation

$$\begin{aligned} \mathbf{y}_k &= [\mathbf{I} - \mathbf{Q}^{-1}(q^{-1})]\mathbf{y}_k + \mathbf{Q}^{-1}(q^{-1})\mathbf{P}(q^{-1})\mathbf{x}_{k-1} + \mathbf{e}_k \\ \mathbf{x}_k &= 0, \qquad \mathbf{y}_k = 0, \qquad k \geqslant 0, \end{aligned}$$

where

$$\mathbf{f}_k(\mathbf{y}^{k-1}, \mathbf{x}^{k-1}) = [\mathbf{I} - \mathbf{Q}^{-1}(q^{-1})]\mathbf{y}_k + \mathbf{Q}^{-1}(q^{-1})\mathbf{P}(q^{-1})\mathbf{x}_{k-1}.$$

Example 9: Covariance Matrices in State Space Models. Given a state space model

$$\begin{cases} \mathbf{z}_{k+1} = \mathbf{A}\mathbf{z}_t + \mathbf{G}\mathbf{e}_t \\ \mathbf{y}_t = \mathbf{C}\mathbf{z}_t + \mathbf{e}_t, \end{cases}$$

where $\mathbf{e}_t$ is serially uncorrelated and uncorrelated with $\mathbf{z}_t$, the state vectors have covariance matrices

$$\mathbf{\Pi}_k = \operatorname{cov} \mathbf{z}_k,$$

which satisfies the recursion

$$\mathbf{\Pi}_{k+1} = \mathbf{A}\,\mathbf{\Pi}_k\,\mathbf{A}' + \mathbf{G}\mathbf{\Delta}\mathbf{G}'$$

and

$$\operatorname{cov}(\mathbf{y}_k, \mathbf{y}_{k-j}) = \begin{cases} \mathbf{H}\,\mathbf{\Pi}_k\,\mathbf{H}' + \mathbf{G}\mathbf{\Delta}\mathbf{G}', & j = 0 \\ \mathbf{H}\mathbf{A}^j\mathbf{\Pi}_{k-j}\mathbf{H}' + \mathbf{H}\mathbf{A}^{j-1}\mathbf{C}\mathbf{\Delta}\mathbf{G}', & j = 1, \ldots, k > 0, \end{cases}$$

where $\mathbf{\Delta} = \operatorname{cov} \mathbf{e}_k$. If $\mathbf{A}$ is stable, then $\mathbf{\Pi}_k$ has the limit $\mathbf{\Pi}$ as $k > \infty$, where $\mathbf{\Pi}$ is the unique solution of

$$\mathbf{\Pi} = \mathbf{A}\,\mathbf{\Pi}\,\mathbf{A}' + \mathbf{G}\mathbf{\Delta}\mathbf{G}'.$$

Let $\mathbf{\Lambda}_l = E(\mathbf{y}_{t+l}\mathbf{y}_t')$ and $l \geqslant 0$. Then $\mathbf{\Lambda}_0 = \mathbf{C}\,\mathbf{\Pi}\,\mathbf{C}' + \mathbf{\Delta}$.

Example 10: Variance Calculation. Given

$$\hat{\mathbf{y}} = \mathbf{H}(q)\hat{\mathbf{e}},$$

where

$$\mathbf{H}(q) = \mathbf{I} + \mathbf{C}(q\mathbf{I} - \mathbf{A})^{-1}\mathbf{B}$$

and $\mathbf{e}$ is a mean zero weakly stationary process with covariance matrix $\mathbf{\Delta}$, the covariance of $\mathbf{y}$ is calculated as

$$\frac{1}{2\pi j}\oint \mathbf{H}(z)\mathbf{\Delta}\mathbf{H}'(1/z)dz/z.$$

The two cross-product terms in the integrand contribute nothing to the integral. One way to see that

$$\frac{1}{2\pi j}\oint \mathbf{C}(z\mathbf{I}-\mathbf{A})^{-1}\mathbf{B}\mathbf{\Delta}\mathbf{B}'(z^{-1}\mathbf{I}-\mathbf{A}')^{-1}\mathbf{C}'dz/z = \mathbf{CZC}'$$

where $\mathbf{Z}$ is

$$\mathbf{Z}-\mathbf{AZA}'=\mathbf{D} \tag{48}$$

where $\mathbf{D}=\mathbf{B\Delta B}'$ is to expand $(z\mathbf{I}-\mathbf{A})^{-1}$ as $z^{-1}(\mathbf{I}+z^{-1}\mathbf{A}'+z^{-2}\mathbf{A}'^2+\dots)$ and collect terms independent of z as $\mathbf{C}[\mathbf{B\Delta B}'+\mathbf{AB\Delta B}'\mathbf{A}'+\dots]\mathbf{C}'$. Define the expression inside the parentheses as $\mathbf{Z}$. When the matrix $\mathbf{A}$ is asymptotically stable, the infinite sum is well defined and satisfies the Lyapunov equation shown previously. This may be solved as $\operatorname{vec}\mathbf{Z}=(\mathbf{I}-\mathbf{A}\otimes\mathbf{A})^{-1}$ $(\mathbf{B}\otimes\mathbf{B})\operatorname{vec}\Delta$ and $\operatorname{vec}(\mathbf{CZC}')$ evaluated as

$$(\mathbf{C}\otimes\mathbf{C})\operatorname{vec}\mathbf{Z}=(\mathbf{C}\otimes\mathbf{C})(\mathbf{I}-\mathbf{A}\otimes\mathbf{A})^{-1}(\mathbf{B}\otimes\mathbf{B})\operatorname{vec}\mathbf{\Delta}.$$

To see this, directly expand the transfer function as

$$\mathbf{y}_t=\mathbf{e}_t+\mathbf{CBe}_{t-1}+\mathbf{CABe}_{t-2}+\dots$$

and

$$\begin{aligned}\operatorname{cov}\mathbf{y}_t &= \mathbf{\Delta}+\mathbf{CB\Delta B}'\mathbf{C}'+\mathbf{CAB\Delta B}'\mathbf{A}'\mathbf{C}'+\dots\\ &= \mathbf{\Delta}+\mathbf{CZC}'\end{aligned}$$

where

$$\mathbf{Z}=\mathbf{B\Delta B}'+\mathbf{AB\Delta B}'\mathbf{A}'+\dots .$$

To illustrate the residue calculation to evaluate the integral, suppose that $\mathbf{A}$ is 2×2. The Lyapunov equation (48) determines $\mathbf{Z}$ via

$$\mathbf{v}_i'\mathbf{Zu}_j=\xi_{ij}=\mathbf{v}_i'\mathbf{Du}_j/(1-\lambda_i\lambda_j),\qquad i\to j=1,2$$

where

$$\mathbf{A}=[\mathbf{u}_1\mathbf{u}_2]\mathbf{\Lambda}\begin{bmatrix}\mathbf{v}_1'\\ \mathbf{v}_2'\end{bmatrix}$$

where

$$\mathbf{\Lambda}=\operatorname{diag}(\lambda_1,\lambda_2)$$

$$\begin{pmatrix}\mathbf{v}_1'\\ \mathbf{v}_2'\end{pmatrix}=(\mathbf{u}_1\mathbf{u}_2)^{-1}.$$

The integral

$$\mathbf{Z} = \frac{1}{2\pi j}\oint(z\mathbf{I} - \mathbf{A})^{-1}\mathbf{D}(z^{-1}\mathbf{I} - \mathbf{A}')^{-1}dz/z$$

is calculated by evaluating the residues at the poles which are the eigenvalues of $\mathbf{A}$. The residue calculus yields

$$\mathbf{Z} = (\mathbf{u}_1\mathbf{u}_2)(\xi_{ij})\begin{pmatrix}\mathbf{u}_1' \\ \mathbf{u}_2'\end{pmatrix},$$

noting that

$$(z\mathbf{I} - \mathbf{A})^{-1} = (z - \lambda_1)^{-1}\mathbf{u}_1\mathbf{v}_1' + (z - \lambda_2)^{-1}\mathbf{u}_2\mathbf{v}_2',$$

the residue at $z = \lambda_1$ is

$$\mathbf{u}_1\mathbf{v}_1'\mathbf{D}\left(\frac{1}{\lambda_1^{-1} - \lambda_1}\mathbf{v}_1\mathbf{u}_1' + \frac{1}{\lambda_1^{-1} - \lambda_2}\mathbf{v}_2\mathbf{u}_2'\right)\frac{1}{\lambda_1}$$
$$= \zeta_{11}\mathbf{u}_1\mathbf{u}_1' + \zeta_{12}\mathbf{u}_1\mathbf{u}_2'.$$

The residue at the pole $z = \lambda_2$ is equal to $\zeta_{12}\mathbf{u}_2\mathbf{u}_1' + \zeta_{22}\mathbf{u}_2\mathbf{u}_2'$. The solution of the Lyapunov equation can be restated as

$$\begin{pmatrix}\mathbf{v}_1' \\ \mathbf{v}_2'\end{pmatrix}\mathbf{Z}(\mathbf{v}_1\mathbf{v}_2) = (\zeta_{ij})$$

or

$$\mathbf{Z} = \begin{pmatrix}\mathbf{v}_1' \\ \mathbf{v}_2'\end{pmatrix}^{-1}(\zeta_{ij})(\mathbf{v}_1\mathbf{v}_2)^{-1}$$
$$= (\mathbf{u}_1\mathbf{u}_2)(\zeta_{ij})\begin{pmatrix}\mathbf{u}_1' \\ \mathbf{u}_2'\end{pmatrix},$$

which is the same as the result of the residue calculation. Thus these two approaches give the same result, as they should.

2.10 Unit Zeros

Zeros on the unit circle in the complex plane either in the denominator or the numerator of a transfer function present special problems. The former (i.e., poles of the transfer function on the unit circle) are called unit roots, and the latter (i.e., zeros of the transfer functions) are called unit zeros.

When a transfer function is a matrix, it is written in one representation called left matrix fraction description as

$$\mathbf{T}(z) = \mathbf{D}(z)^{-1}\mathbf{N}(z)$$

where

$$\mathbf{D}(z) = d(z)\mathbf{I}.$$

The zeros of the transfer function are defined as the zeros of $\det \mathbf{N}(z)$.

A state space model

$$\mathbf{x}_{t+1} = \mathbf{A}\mathbf{x}_t + \mathbf{B}\mathbf{u}_t$$
$$\mathbf{y}_t = \mathbf{C}\mathbf{x}_t + \mathbf{D}\mathbf{u}_t$$

has the transfer function

$$\mathbf{T}(z) = \mathbf{D} + \mathbf{C}(z\mathbf{I} - \mathbf{A})^{-1}\mathbf{B} = \frac{1}{d(z)}\mathbf{N}(z)$$

where $d(z) = |z\mathbf{I} - \mathbf{A}|$.

Its zeros are calculated as the eigenvalues of a matrix pencil

$$\lambda\begin{bmatrix} \mathbf{I} & 0 \\ 0 & 0 \end{bmatrix} - \begin{bmatrix} \mathbf{A} & -\mathbf{B} \\ \mathbf{C} & -\mathbf{D} \end{bmatrix} = \begin{bmatrix} \lambda\mathbf{I} - \mathbf{A} & \mathbf{B} \\ -\mathbf{C} & \mathbf{D} \end{bmatrix}$$

since

$$\det\begin{bmatrix} \lambda\mathbf{I} - \mathbf{A} & \mathbf{B} \\ -\mathbf{C} & \mathbf{D} \end{bmatrix} = |\lambda\mathbf{I} - \mathbf{A}|\,|\mathbf{D} + \mathbf{C}(\lambda\mathbf{I} - \mathbf{A})^{-1}\mathbf{B}|$$
$$= \det \mathbf{N}(z).$$

A transfer function with a zero on the unit circle does not have a stable inverse system. Some unusual features of such a system are not commonly appreciated. Exceptions are Whittle (1963, p. 43) and Hannan (1970, p. 134) where a series

(49) $$y_t = e_t - e_{t-1}$$

is used to highlight problems of constructing constant coefficient one-step predictors, where e_t are mean zero serially uncorrelated with variance σ^2. We follow Hannan here. Another view of the same problem is given as an application of the Kalman filter in Chapter 6.

One way to see that a unit zero presents a problem in (one-step) forecasting has been pointed out by Hannan. Suppose one builds a regression model for (49). The regression coefficients in

$$\hat{y}_t = a_1 y_{t-1} + a_2 y_{t-2} + \ldots + a_N y_{t-N}$$

satisfy

$$\begin{bmatrix} 2 & -1 & 0 \ldots & & 0 \\ -1 & 2 & -1 \ldots & & \\ & & & & \\ 0 \ldots & & & -1 & 2 \end{bmatrix}\begin{bmatrix} a_1 \\ \vdots \\ \\ a_N \end{bmatrix} = \begin{bmatrix} -1 \\ 0 \\ \vdots \\ 0 \end{bmatrix}$$

or

$$a_j = -(1 - j/N), \qquad j = 1 \ldots N.$$

In other words, the one-step prediction is not a linear combination of past values $\Sigma_{j=1}^{\infty} a_j y_{t-j}$ because $\Sigma\, a_j^2$ diverges.

When the process has a unit zero, the correlation between the future and past canonical variables has unit magnitude. Hannan and Poskitt (1986) have shown that the number of canonical correlations between the future and past canonical variables of unit magnitude is equal to the number of zeros of the determinant of the canonical factor in the rational spectral density matrix on the unit circle.

Let $y_{t+1}^{+} = y_{t+1}$ and $\mathbf{y}_t^{-} = (y_t, y_{t-1}, \ldots, y_{t-N})'$. Then

$$\begin{aligned}\mathbf{R} &= E(\mathbf{y}_t^{-}\,\mathbf{y}_t^{-\prime}) \\ &= \mathbf{L}\mathbf{L}'\end{aligned}$$

where $\mathbf{L}$ is the Cholesky factor.

Define

$$d_t^{+} = \mathbf{L}^{-1} y_{t+1}$$

and

$$\mathbf{d}_t^{-} = \mathbf{L}^{-1}\mathbf{y}_t^{-}.$$

Denote the singular value decomposition of $\mathbf{L}^{-1}\mathbf{H}\mathbf{L}^{-T}$ where $-T$ stands for the inverse of the transposed matrix, and $\mathbf{H} = E(y_{t+1}\mathbf{y}_t^{-\prime})$ by $\mathbf{P}\mathbf{\Gamma}\mathbf{Z}'$. The future and the past canonical variables are defined by

$$\mathbf{u}_t^{+} = \mathbf{P}'\mathbf{d}_t^{+}$$

and

$$\mathbf{u}_t^{-} = \mathbf{Z}'\mathbf{d}_t^{-}.$$

Then the canonical correlation matrix is

$$E(\mathbf{u}_t^{+}\,\mathbf{u}_t^{-\prime}) = \mathbf{\Gamma}.$$

From (49), $H = -\sigma^2(1\,0 \ldots 0)$, i.e., $\Gamma = 1$. This implies that a linear function of the future is perfectly predictable from the past. It is straightforward to calculate that

$$\operatorname{var}(y_t - \hat{y}_t) = \sigma^2(N+1)/N \to \sigma^2 \quad \text{as} \quad N \to \infty.$$

Indeed construct an approximation to e_t by

$$\hat{e}_t^N = y_t + b_1 y_{t-1} + \ldots + b_{N-1} y_{t-N+1}$$

where

$$b_j = 1 - j/N.$$

Then

$$e_t - \hat{e}_t^N = (1 - a_1)e_{t-1} + (a_1 - a_2)e_{t-2} + \dots$$
$$+ (a_{N-2} - a_{N-1})e_{t-N+1} + a_{N-1}e_{t-N}$$

and hence $\hat{e}_t^N$ converges in mean square to e_t.

Reverse the time in (49) and consider $y_t = \eta_t - \eta_{t+1}$ where η_t is mean zero and serially uncorrelated with $E(\eta_t^2) = \sigma^2$.

Then

$$\hat{\eta}_{t+1}^N = \sum_0^{N-1} b_j y_{t+j+1}$$
$$b_0 = 1$$

converges in mean square to η_{t+1}. Since $E(\hat{e}_t^N \hat{\eta}_{t+1}^N) = -1, E(e_t \eta_{t+1}) = -1$, i.e., $\eta_{t+1} = -e_t$ a.s. Yet another way to see this is to iterate y_t as

$$\begin{aligned} y_t &= e_t - e_{t-1} \\ &= e_t - y_{t-1} - e_{t-2} \\ &= e_t - y_{t-1} - y_{t-2} - \dots - y_{t-N} - e_{t-N}. \end{aligned}$$

If the process starts at $t = 0$, then

$$y_N = -(y_{N-1} + \dots + y_0) + e_N - e_0.$$

In this sense the effects of y_0 persist.

2.11 Decomposition of Dynamic Representation

Some economic variables have dynamic modes with eigenvalues equal to one or very close to one so that $|\lambda| < 1$ and $|\lambda| = 1$ are statistically hard to distinguish. Sometimes they are modeled by specifying how the first difference Δy_t evolves with time, i.e., one of the eigenvalues is constrained to be one. This is how the model

$$\Delta y_t = A(q^{-1})e_t + \mu, \tag{50}$$

where μ is a constant, is often posited in the literature where e_t is a mean zero i.i.d. white noise sequence, where

$$A(q^{-1}) = 1 + a_1 q^{-1} + a_2 q^{-1} + \dots,$$

and the coefficients are constrained by $\Sigma_{i=0}^{\infty} a_i^2 < \infty$ so that $A(q^{-1})e_t$ has a well-defined variance,

$$\operatorname{var} \Delta y_t = \left(\sum a_i^2 \right) \sigma_e^2.$$

The coefficients in the Wold decomposition representation (50) are the impulse (dynamic multiplier) responses. For example, a_3 in $\Delta y_t = A(q^{-1})e_t$ tells us how much the shock three periods earlier, e_{t-3}, still affects Δy_t. This class of models has been proposed by Beveridge and Nelson (1981), for example.

By rewriting (50) as

$$\Delta y_t = \mu + A(1)e_t + [A(q^{-1}) - A(1)]e_t,$$

where

$$A(1) = \sum_0^\infty a_i < \infty$$

is assumed, one can integrate this equation. First define

$$\Delta y_{1t} = \mu + A(1)e_t \tag{51}$$

and

$$\Delta y_{2t} = [A(q^{-1}) - A(1)]e_t. \tag{52}$$

The dynamics are decomposed into a part generating y_{1t} and the other y_{2t}. Equation (51) immediately yields y_{1t}. If the difference operator is defined by

$$\Delta y_{1t} = y_{1t+1} - y_{1t} = (1 - q^{-1})y_{1t+1},$$

then

$$\begin{aligned} y_{1t+1} &= \frac{\mu}{1-q^{-1}} + \frac{A(1)}{1-q^{-1}}e_t \\ &= \mu + A(1)(e_t + e_{t-1} + e_{t-2} + \dots) \end{aligned}$$

or

$$y_{1t} = \mu + A(1)(e_{t-1} + e_{t-2} + \dots) \tag{53}$$

is the integral of past disturbances.

If $\Delta y_{1t} = y_{1t} - y_{1t-1} = (1 - q^{-1})y_{1t}$, then instead of (53), the integral is given by

$$y_{1t} = A(1)(e_t + e_{t-1} + \dots). \tag{54}$$

In either form, y_{1t} is the part of y_t that is called a random walk with drift. For a simpler explanation, set μ to zero. In (53), $y_{1t} = y_{1t|t-1}$ is a predictable process. In (54), $y_{1t} = y_{1t|t-1} + A(1)e_t$.

We use the definition $\Delta y_t = y_t - y_{t-1}$. Note that the variance of y_{1t} tends to become unbounded as t goes to infinity. The second term defined by (52) has a well-defined variance for all t. This can be seen as follows. From $\Delta y_{2t} = (1 - q^{-1})y_{2t}$,

$$y_{2t} = \frac{A(q^{-1}) - A(1)}{1 - q^{-1}} e_t. \tag{55}$$

When the spectral density function of Δy_t is rational, it can be factored in terms of a rational transfer function, and Δy_t can be regarded as an output of a finite-dimensional state space model driven by a white noise sequence. The impulse responses are also characterized by finite parameter combinations. When $a_i = \mathbf{h}'\mathbf{F}^{i-1}\mathbf{g}, i \geqslant 1, a_0 = 1$,

$$A(q^{-1}) = 1 + \mathbf{h}'(\mathbf{I} - \mathbf{F}q^{-1})^{-1}\mathbf{g}\mathbf{q}^{-1}.$$

Then

$$\frac{A(q^{-1}) - A(1)}{1 - q^{-1}} = -H(q^{-1}) \tag{56}$$

where $H(z) = \mathbf{h}'(\mathbf{I} - \mathbf{F}q^{-1})^{-1}(\mathbf{I} - \mathbf{F})^{-1}\mathbf{g}$. To see this directly, let $A(q^{-1}) = \Sigma_0^\infty a_i q^i$ and express

$$\begin{aligned} A(q^{-1}) - A(1) &= a_1(q^{-1} - 1) + a_2(q^{-2} - 1) + a_3(q^{-3} - 1) + \dots \\ &= (q^{-1} - 1)[a_1 + a_2(q^{-1} + 1) + a_3(q^{-2} + q^{-1} + 1) + \dots] \end{aligned}$$

since $\{a_i\}$ is absolutely convergent, and note that

$$\begin{aligned} a_1 + a_2 + a_3 + \dots &= \mathbf{h}'(\mathbf{I} + \mathbf{F} + \mathbf{F}^2 + \dots)\mathbf{g} \\ &= \mathbf{h}'(\mathbf{I} - \mathbf{F})^{-1}\mathbf{g}. \end{aligned}$$

All the terms are similar expressed in closed forms as

$$\begin{aligned} a_2 + a_3 + a_4 + \dots &= \mathbf{h}'\mathbf{F}(\mathbf{I} - \mathbf{F})^{-1}\mathbf{g}, \\ a_3 + a_4 + \dots &= \mathbf{h}'\mathbf{F}^2(\mathbf{I} - \mathbf{F})^{-1}\mathbf{g} \text{ etc.} \end{aligned}$$

Thus

$$\begin{aligned} A(q^{-1}) - A(1) &= (q^{-1} - 1)\mathbf{h}'(\mathbf{I} + \mathbf{F} + \mathbf{F}^2 + \dots)(\mathbf{I} - \mathbf{E}q^{-1})^{-1}\mathbf{g} \\ &= (q^{-1} - 1)\mathbf{h}'(\mathbf{I} - \mathbf{F})^{-1}(\mathbf{I} - \mathbf{F}z^{-1})^{-1}\mathbf{g} \\ &= (q^{-1} - 1)\mathbf{h}'(\mathbf{I} - \mathbf{E}q^{-1})^{-1}\mathbf{\Psi} \end{aligned}$$

where

$$\mathbf{\Psi} = (\mathbf{I} - \mathbf{F})^{-1}\mathbf{g}.$$

From (55) and (56)

$$y_{2t} = -H(q^{-1})e_t = -(a_1 e_t + a_2 e_{t-1} + \dots).$$

We note that $\operatorname{cov} y_{2t} = \Sigma_1^\infty a_i^2 < \infty$, and y_{2t} is weakly stationary.

There is an alternative way to decompose y_t into two components. In the state representation, the eigenvalue $\rho = 1$ is not imposed in the model. When (random) trends have one-dimensional dynamics, the model takes the form

$$\begin{pmatrix} \tau \\ \mathbf{x} \end{pmatrix}_{t+1} = \begin{pmatrix} \rho & b\mathbf{h}' \\ \mathbf{0} & \mathbf{F} \end{pmatrix} \begin{pmatrix} \tau \\ \mathbf{x} \end{pmatrix}_t + \begin{pmatrix} b \\ \mathbf{G} \end{pmatrix} e_t, \tag{57}$$

$$y_t = c\tau_t + \mathbf{h}'\mathbf{x}_t + e_t.$$

Such a (block) triangular representation results from putting the dynamic matrix in Schur decomposition form in which eigenvalues of magnitude larger than some critical values are separated from the rest. Equation (57) shows that $y_{t|t-1}$, the predictable component of y_t at time $t-1$, consists of trend part

$$y_{1t|t-1} = c\tau_t$$

and cyclical component part

$$y_{2t|t-1} = \mathbf{h}'\mathbf{x}_t.$$

By defining $y_{1t} = y_{1t|t-1}$ and $y_{2t} = y_{2t|t-1} + e_t$, we decompose y_t as the sum of y_{1t} and y_{2t} where the former is a predictable process and the latter is a martingale. That this decomposition is possible for any adapted process is guaranteed by the Doob decomposition theorem. (See Kopp [1984, p. 66], for example.) The lag transform is given by

$$\begin{aligned}\hat{\mathbf{y}} &= \left\{(c, \mathbf{h}')\begin{bmatrix} 1-\rho q^{-1} & -\mathbf{b}\mathbf{h}'q^{-1} \\ 0 & \mathbf{I}-\mathbf{F}q^{-1}\end{bmatrix}^{-1}\begin{bmatrix}\mathbf{b} \\ \mathbf{G}\end{bmatrix} q^{-1} + \mathbf{I}\right\}\hat{\mathbf{e}} \\ &= [1 + (1-\rho q^{-1})^{-1} b q^{-1}][1 + \mathbf{h}'(\mathbf{I} - \mathbf{F}q^{-1})^{-1}\mathbf{G}q^{-1}]\hat{e}.\end{aligned}$$

Now suppose that $\rho = 1$. This decomposition of the transfer function matrix into low-frequency component (first factor) and high-frequency component (second factor) suggests the decomposition

$$\hat{y}_1 = c(1-\rho q^{-1})^{-1} b q^{-1}[1 + \mathbf{h}'(\mathbf{I} - \mathbf{F}q^{-1})^{-1}\mathbf{G}q^{-1}]\hat{e}$$

and

$$\hat{y}_2 = [1 + \boldsymbol{\omega}'(\mathbf{I} - \mathbf{F}q^{-1})^{-1}\mathbf{G}q^{-1}]\hat{e}$$

where

$$\omega' = \mathbf{h}' - b\mathbf{h}'(\mathbf{I} - \mathbf{F})^{-1}.$$

When $\rho = 1$,

$$\hat{y}_1 = \frac{cbq^{-1}}{1 - q^{-1}}[1 + \mathbf{h}'(\mathbf{I} - \mathbf{F}q^{-1})^{-1}\mathbf{G}q^{-1}]\hat{e}$$

or

$$y_{1t} - y_{1t-1} = bc[e_{t-1} + \mathbf{h}'\mathbf{G}e_{t-2} + \mathbf{h}'\mathbf{F}\mathbf{G}e_{t-3} + \ldots]$$

and

$$y_{2t} = e_t + \mathbf{h}'\mathbf{G}e_{t-1} + \mathbf{h}'\mathbf{F}\mathbf{G}e_{t-2} + \ldots .$$

Note the difference in the timing if e. In Beveridge and Nelson $\Delta y_{1t} = y_{1t} - y_{1t-1}$ contains $e_t, e_{t-1}, \ldots$ (i.e., $\Delta y_{1t} = e_t + \lambda_1 e_{t-1} \ldots$), while in the state space model, Δy_{1t} contains only $e_{t-1}, e_{t-2}, \ldots$.

In other words, the trend component of the state space model y_{1t} lies in the subspace spanned by $e_{t-1}, \ldots$, i.e., is a predictable process in the sense of Doob decomposition of an arbitrary stochastic process (Kopp [1984, p. 66]), and y_{2t} is a martingale process. Beveridge-Nelson decomposition does not have this "canonical" property. For further analysis and details on this and related topics, see Aoki (1988a–d).

Appendix

Evaluating Integrals

Given a spectral density matrix $\mathbf{S}(z)$ of a time series $\{y_t\}$, the variance is calculated as

$$Ey_t^2 = \frac{1}{2\pi j}\oint \mathbf{S}(z)\frac{dz}{z}.$$

Since

$$S(z) = G(z)G'(1/z)$$

where

$$G(z) = B(z)/A(z),$$

the evaluation of the integral amounts to calculation of residues of the polynomial $A(z)$,

$$A(z) = a_0 z^n + \ldots + a_n, \qquad a_0 > 0.$$

By construction of factors by the spectral density factorization, both polynomials $A(z)$ and $B(z)$ can be chosen to be of minimal phase polynomials, i.e., have zeros inside the unit disk. Alternatively, let $\{y_t\}$ be an output sequence of a dynamic system driven by white noises which is stable and of minimum phase. The transfer function of the system is of the form given by $G(z)$. We suppose, then, that $A(z)$ and $B(z)$ are minimum phase polynomials from now on. Designing Wiener filters with continuous time data requires evaluation of variances by integrating spectral density matrices in Laning and Battin (1056) or in Newton, Gould, and Kaiser (1957) and in Åstrom (1970) for discrete-time systems. The same type of integrals need to be evaluated in calculating the cost associated with the so-called minimum variance control problems in which the variance of state vector is being controlled.

Evaluation of the integral is carried out sequentially by calculating the residues and sequentially reducing the degrees of the denominator polynomials. In this process, it is convenient to use the associated reverse polynomial of $A(z)$ defined thus:

$$A^R(z) = z^n A(1/z)$$
$$= a_0 + a_1 z + \ldots + a_n z^n.$$

Consider a sequence of polynomials defined by

(A1) $$A_{k-1}(z) = z^{-1}[A_k(z) - \alpha_k A_k^R(z)],$$

where

$$A_k(z) = a_{k0} z^k + a_{k1} z^{h-1} + \ldots + a_{kk},$$

where

$$B_{k-1}(z) = z^{-1}[B_k(z) - \beta_k A_k^R(z)], \qquad k = n, n-1, \ldots, 1$$

and similarly for

$$B_k(z) = b_{k0} z^k + b_{k1} z^{k-1} + \ldots + b_{kk},$$

(A2)
$$\alpha_k = a_{kk}/a_{k0},$$
$$\beta_k = b_{kk}/a_{k0},$$
$$A_n(z) = A(z),$$

and

$$B_n(z) = B(z).$$

The polynomials $A_0(z)$ and $B_0(z)$ are constants, a_{00} and b_{00}. The coefficients of these polynomials are related by (comparing the like powers of z)

$$\text{(A3)} \quad \begin{aligned} a_{k-1,i} &= a_{ki} - \alpha_k a_{k,k-i}, \\ b_{k-1,i} &= b_{ki} - \beta_k a_{kk-i}, \qquad i = 0, 1, \ldots, k-1. \end{aligned}$$

By assumption, $A(z)$ is stable and a_0 can be chosen to be positive. By mathematical induction, we show that if $A_k(z)$ is stable, then $A_{k-1}(z)$ is stable and $a_{k-1,0} > 0$. This shows that all $A_k(z)$ are stable and a_{k0} are positive for $k = 1, 2, \ldots, n$. The polynomials $A(z)$ and $A^R(z)$ are related by essentially Blaschke factors, because

$$A(z) = a_0 \prod_{i=1}^{n} (z - \gamma_i), \qquad |\gamma_i| < 1,$$

and

$$A^R(z) = a_0 \prod_{i=1}^{n} (1 - \gamma_i z)$$

imply that

$$\begin{aligned} A(z) &= A^R(z) \prod_{i=1}^{n} \frac{z - \gamma_i}{1 - \alpha_i z} \\ &= A^R(z) \prod_{i=1}^{n} \frac{z - \gamma_i}{1 - \bar{\alpha}_i z}. \end{aligned}$$

In the preceding, the polynomials have real coefficients, and α_i can be replaced by $\bar{\alpha}_i$ in the denominator.

Since the Blaschke factors are less than one in magnitude for $|z| < 1$,

$$a_{kk} = |A_k(0)| < |A_k^R(0)| = a_{k0}.$$

Hence, $|\alpha_k| < 1$, and from (A2) and (A3)

$$\text{(A4)} \quad \begin{aligned} a_{k-1,0} &= a_{k0} - \alpha_k a_{kk} \\ &= (a_{k0}^2 - a_{kk}^2)/a_{k0} > 0. \end{aligned}$$

Also,

$$|A_k(z)| \geqslant |A_k^R(z)|$$

for $|z| \geqslant 1$, and because $|\alpha_k|$ is less than one,

$$|A_k(z)| > |\alpha_k| |A_k^R(z)|$$

for $|z| \geqslant 1$. From the polynomial recursion (A1),

$$|z||A_{k-1}(z)| = |A_k(z) - \alpha_k A^*(z)|$$
$$\geqslant |A_k(z)| - |\alpha_k||A^*(z)| > 0 \quad \text{for} \quad |z| \geqslant 1,$$

showing that $A_{k-1}(z)$ has no root outside the unit disk, i.e., $A_{k-1}(z)$ is stable.

We can argue the converse result that $A_{k-1}(z)$ is stable and $a_0^{k-1} > 0$ implies that $A_k(z)$ is stable and a_0^k is stable in a similar way. The inequality (A4) shows that $a_{k-1,0}$ implies that $|\alpha_k| < 1$. The recursion (A1) shows that

$$A_k(z^{-1}) - \alpha_k A_k^R(z^{-1}) = z^{-1} A_{k-1}(z^{-1}),$$

and multiplying the preceding by z^k creates

(A5) $$A_k^R(z) - \alpha_k A_k(z) = A_{k-1}^R(z).$$

Substitute $A_k^R(z)$ by (A1) to see that

(A6) $$A_k(z) = \frac{z}{1-\alpha_k^2} A_{k-1}(z) + \frac{\alpha_k}{1-\alpha_k^2} A_{k-1}^R.$$

Outside the unit disk (A6) shows that

$$|A_k(z)| \geqslant \left|\frac{z}{1-\alpha_k^2}\right| |A_{k-1}(z)| - \frac{|\alpha_k|}{1-\alpha_k^2} |A_{k-1}^R(z)| > 0$$

because

$$|\alpha_k| < 1$$

and

$$|A_{k-1}(z)| \geqslant |A_{k-1}^R(z)|.$$

This establishes that $A_k(z)$ has no pole outside the unit disk, i.e., $A_k(z)$ is stable.

The recursion (A5) and (A6) can be used in reducing the order of the integrands sequentially, a topic to which we turn next.

Now consider the integral

$$I_{k-1} = \frac{1}{2\pi j} \oint \frac{B_{k-1}(z) B_{k-1}(z^{-1})}{A_{k-1}(z) A_{k-1}(1/z)} \frac{dz}{z}.$$

The poles are $z = 0$ and the zeros of the polynomial $A_k(z), z_i, i = 1 \ldots k-1$. At zero z_i of $A_{k-1}(z_i)$, the term $A_{k-1}(z_i^{-1})$ contributes to the residue. From (A6),

$$A_k(z_i^{-1}) = \frac{z_i^{-1}}{1-\alpha_k^2} A_{k-1}(z_i^{-1})$$

or because

$$A_{k-1}^R(z_i^{-1}) = z_i^{-k+1} A_{k-1}(z_i) = 0,$$

we obtain

$$\frac{1}{A_{k-1}(z_i^{-1})} = \frac{1}{(1-\alpha_k^2) z_k A_k(z_i^{-1})}.$$

Noting from (A4) that $A_k^R(0)(1-\alpha_k^2) = A_{k-1}^R(0)$, we see that the residues at $z = 0$ of I_{k-1}, and that of

$$\frac{B_{k-1}(z) B_{k-1}^R(z^{-1})}{(1-\alpha_k^2) A_{k-1}(z) A_k^R(z)}$$

are the same. Thus

$$\begin{aligned} I_{k-1} &= \frac{1}{(1-\alpha_k^2)} \frac{1}{2\pi j} \oint \frac{B_{k-1}(z) B_{k-1}(z^{-1})}{A_{k-1}(z) A_k(z^{-1})} \frac{\mathrm{d}z}{z^2} \\ &= \frac{1}{2\pi j (1-\alpha_k^2)} \oint \frac{B_{k-1}(z) B_{k-1}(z^{-1})}{A_k(z) A_{k-1}(z^{-1})} \mathrm{d}z \end{aligned} \tag{A7}$$

where the change of variables from z to z^{-1} is made. Now the residues are contributed by the zeros of $A_k(z)$ and $z = 0$. Let z_j be a zero of $A_k(z)$. Then from (A1)

$$A_{k-1}(z_j^{-1}) = z_j A_k(z_j^{-1}),$$

and hence from (A7)

$$(1-\alpha_k^2) I_{k-1} = \frac{-1}{2\pi j} \oint \frac{B_{k-1}(z) B_{k-1}(z^{-1})}{A_k(z) A_{k-1}(z^{-1})} = \frac{-1}{2\pi j} \oint \frac{B_{k-1}(z) B_{k-1}(z^{-1})}{A_k(z) A_k(z^{-1})} \frac{\mathrm{d}z}{z}.$$

Also from (A1), the numerator of the integrand becomes

$$\begin{aligned} B_{k-1}(z) B_{k-1}(z^{-1}) &= [B_k(z) - \beta_k A_k^R(z)][B_k(z^{-1}) - \beta_k A_k^R(z^{-1})] \\ &= B_k(z) B_k(z^{-1}) - \beta_k [A_k^R(z) B_k(z^{-1}) \\ &\quad + B_k(z) A_k^R(z^{-1})] + \beta_k^2 A_k^R(z) A_k^R(z^{-1}). \end{aligned}$$

Note that

$$\begin{aligned} \frac{1}{2\pi j} \oint \frac{B_k(z) A_k^R(z^{-1})}{A_k(z) A_k(z^{-1})} \frac{\mathrm{d}z}{z} &= \frac{1}{2\pi j} \oint \frac{B_k(z) A_k(z)}{A_k(z) A_k^R(z)} \frac{\mathrm{d}z}{z} \\ &= \frac{1}{2\pi j} \oint \frac{B_k(z)}{A_k^R(z)} \frac{\mathrm{d}z}{z} \\ &= \frac{B_k(0)}{A^R(0)} = \frac{b_{kk}}{a_{k0}} = \beta_k. \end{aligned}$$

The integral becomes

$$\frac{1}{2\pi j}\oint \frac{A_k^R(z)B_k(z^{-1})}{A_k(z)A_k(z^{-1})}\frac{dz}{z} = \frac{1}{2\pi j}\oint \frac{B_k^R(z)}{A_k(z)}\frac{\mathrm{d}z}{z}$$

$$= \frac{1}{2\pi j}\oint \frac{B_k^R(z^{-1})}{A_k(z^{-1})}\frac{\mathrm{d}z}{z}$$

$$= \frac{1}{2\pi j}\oint \frac{B_k(z)}{A_k^R(z)}\frac{\mathrm{d}z}{z}$$

$$= \frac{B_k(0)}{A_k^R(0)} = \beta_k.$$

Finally,

$$\frac{1}{2\pi j}\oint \frac{A_k^R(z)A_k^R(z^{-1})}{A_k(z)A_k(z^{-1})}\frac{\mathrm{d}z}{z} = \frac{1}{2\pi j}\oint \frac{A_k^R(z)A_k(z)}{A_k(z)A_k^R(z)}\frac{\mathrm{d}z}{z} = 1.$$

Collecting terms together, the recursion for the integrals is obtained as

$$I_k = (1 - \alpha_k^2)I_{k-1} + \beta_k^2, \qquad k = 1, \ldots, n$$

where

$$I_0 = \beta_0^2.$$

Method of Partial Summation

Abel's method of partial summation (Lukacs [1975, p. 95]), Chow and Teicher [1978, p. 111]) is quite useful in obtaining order of magnitudes of weighted sums of random variables.

Lemma. *Let* $\{u_n\}$ *and* $\{v_n\}$, $n = 0, 1, \ldots$, *be two sequences of real numbers. Let* $U_n = \Sigma_{i=0}^n u_k$. *Then the "summation by part" formula is*

$$\sum_{i=1}^{n} u_i v_i = U_n v_n - U_0 v_1 - \sum_{i=1}^{n-1} U_i(v_{i+1} - v_i).$$

To see this rewrite the sum as

$$\sum_{i=1}^{n} u_i v_i = \sum_{i=1}^{n} (U_i - U_{i-1})v_i.$$

After multiplying v_i out, rewrite the second term as $-\Sigma_{i=1}^n U_{i-1}(v_i - v_{i-1}) - \Sigma_{i=1}^n U_{i-1}v_{i-1}$. Collect these terms together to derive the expression.

The next example (Lai and Robbins, 1979, lemma 2) shows how to use the method. Let $z_1, z_2, \ldots$ be i.i.d. random variables with finite mean and variance $< \infty$. Let $S_n = \Sigma_1^n z_i$, $S_0 = 0$. Then the sum $\Sigma_1^n z_i/i$ can be calculated by partial summation

$$\sum_1^n i^{-1} z_i = \sum_1^n i^{-1} \Delta S_i = n^{-1} S_n - \sum_1^{n-1} S_i[(i+1)^{-1} - i^{-1}].$$

For large $i, (i+1)^{-1} - i^{-1} \sim -i^{-2}$. By the strong law of large numbers $i^{-1} S_i \to Ez_1$, and hence

$$\lim_n \sum_1^n i^{-1} z_i / \ln n = Ez_1 \quad \text{a.s.}$$

Let ε_t be mean zero i.i.d. random variables with $E\varepsilon_t^2 = \sigma^2 < \infty$, and set $\varepsilon_t^2 = z_t$. Then $Ez_1 = \sigma^2$ and we obtain a useful

Fact 1. $\dfrac{\Sigma_1^n i^{-1} \varepsilon_i^2}{\ln n} \to \sigma^2$ a.s.

Another useful application of the method is

Fact 2. (Lai-Robbins [1979, cor. 1]).

$$\lim_n \sum_1^n \bar{\varepsilon}_i^2 / \ln n = \sigma^2 \quad \text{a.s.}$$

where

$$\bar{\varepsilon}_i = \sum_{j=1}^i j^{-1} \varepsilon_j.$$

To see this, let $S_i = \Sigma_1^i \varepsilon_j$, $S_0 = 0$. Then let $u_i = -i^{-1}$. Then $\Delta u_i \approx i^{-2}$ for large i. Apply the lemma to obtain

$$\sum_1^n i^{-2} S_i^2 = n^{-1} S_n + \sum_1^n (S_i^2 - S_{i-1}^2) i^{-1}$$

where

$$S_i^2 - S_{i-1}^2 = \varepsilon_i^2 + 2S_{i-1}\varepsilon_i.$$

The term $\Sigma\, i^{-1} \varepsilon_i^2 \sim \sigma^2 \ln n$ a.s. by Fact 1. The second term $\Sigma\, i^{-1} S_i \varepsilon_{i+1}$ is a martingale transform. Use a standard truncation argument to ensure its finite expectation. Then it is a martingale. By the strong law for martingales

(Neveu [1965, p. 150]) on the set where $[d_n = \infty], d_n = \sigma^2 \Sigma_1^n S_i^2/i^2$, we have

$$\Sigma^n \frac{S_i \varepsilon_{i+1}}{i} \sim o(\Sigma\, d_n^{1/2}(\ln d_n)^{1/2+\delta}), \delta > 0,$$

and hence the ratio $\Sigma^n i^{-1} S_i \varepsilon_{i+1}/\ln n$ converges to 0 a.s.

Fact 3. Let ε_t be i.i.d. mean zero and finite variance. Then

$$\sum_1^t \varepsilon_\tau/\tau < \infty, \quad \text{a.s.}$$

Then $\Sigma_\tau (E\phi(\varepsilon_\tau)/\phi(\tau)) < \infty \Rightarrow \Sigma\, \varepsilon_\tau/\tau$ converges a.s. where $\phi(x) = x^2$ by a theorem of Loève (Chow and Teicher [1978, p. 114]). Since $E\varepsilon_\tau^2 = \sigma^2$ and $\Sigma\, 1/\tau^2 < \infty$, Fact 3 is established.

Martingale Transform

Given $(\Omega, P, \mathscr{F})$, fix a finite sequence of σ-fields $\{\mathscr{F}_n\}_0^N$, with $\mathscr{F}_0$ generated by P-null sets and $\mathscr{F}_N = \mathscr{F}$. The optional σ-field Σ_0 is that generated on $T \times \Omega$ by all adapted processes, where $T = \{0, 1, \ldots, N\}$. A bounded measure on Σ_0 is defined by $m(A) = E[\Sigma_0^N I_A(n, \omega)]$. Let $L^p(S, \Sigma_0, m)$ denote the Lebesque space, $1 \leqslant p \leqslant \infty$. The space L^p consists of all adapted stochastic processes $\{X_n\}_0^N$ such that $|X_n(\omega)|^p$ is m-integrable over S. Any bounded linear functional f on L^p is in the dual space of L^p and of the form $\mathrm{f}(X) = E(\Sigma_0^N X_n Y_n)$ for some $Y \in L^q$, $1/p + 1/q = 1$. In particular for $p = 1$, the space of bounded adapted processes L_∞ is its dual space. Let $\mathscr{D}_p$ denote the set of predictable processes in L^p, i.e., $\mathscr{D}_p = \{X \in L^p : X_n \in \mathscr{F}_{n-1}, n = 1, \ldots N\}$. Let $\mathscr{D}_p^\perp = \{y \in L^q : E(\Sigma_0^N X_n Y_n) = 0 \text{ for all } X \in \mathscr{D}_p\}$. Define ΔX by $\Delta X_0 = 0, \Delta X_n = X_n - X_{n-1}, n > 1$.

We have

Proposition (Pliska)

A stochastic process $Y \in \mathscr{D}_p^\perp$ if and only if there exists a martingale process $M \in \mathscr{D}_q$ such that $Y = \Delta M$.

Since $\mathrm{f}(X) = E(X_0 Y_0 + \Sigma_1^N X_n \Delta Y_n)$, $\mathscr{D}_p^\perp$ consists of all martingales in L^p that are null at zero. To see this, for any $M \in L^q$, let $\mathrm{Y} = \Delta M$. Then $\mathrm{f}(X) = E(\Sigma_1^N X_n \Delta M_n)$ for any $X \in \mathscr{D}_p$ is a martingale transform in L^1 and hence is a martingale. Since $(X \cdot M)_0 = 0, \mathrm{f}(X) = E[(X \cdot M)_N] = E[(X \cdot M)_0] = 0$, i.e., $Y = \Delta M \in \mathscr{D}_p^\perp$, where $(X \cdot Y)_n = X_0 Y_0 + X_1(Y_1 - Y_0) + \ldots + X_n(Y_n - Y_{n-1})$. Conversely, $Y \in \mathscr{D}_p^\perp$ implies that Y_0 is 0, for $Y_0 \in \mathscr{F}_0$ implies that Y_0 is a constant and any nonzero constant Y_0 would lead to $E(\Sigma X_n Y_n) \neq 0$ for some $X \in \mathscr{D}_p$. Define M by $M_0 = 0, M_n = Y_n + M_{n-1}$. The process M is adapted. To show that it is a martingale, set $X_n = I_B, B \in \mathscr{F}_{n-1}, X_m = 0, m \neq n$. Then $X_n \in \mathscr{D}_p$ and $0 = E(\Sigma X_k Y_k) = E(I_B Y_n) = \int_B Y_n dP$. Since B is arbitrary, we have shown that $E(\Delta M_n | \mathscr{F}_{n-1}) = E(Y_n | \mathscr{F}_{n-1}) = 0$.

Basic Duality Relations

(P) Principal problem: Minimize $\mathrm{f}(X)$
subject to $X \in C \cap \mathscr{D}_p$
where C is a convex subset of L^p and $\mathscr{D}_p$ is a space of all predictable processes in L^p.

(D) Dual problem: Maximize $-\mathrm{f}^*(Y)$
subject to $Y \in C^* \cap \mathscr{D}_p^{\perp}$
where

$$\mathrm{f}^*(Y) = \sup_{X \in C} \{E(X \cdot Y) - \mathrm{f}(X)\}$$

and

$$C^* = \{Y \in L^q : \sup \mathrm{f}^q(Y) < \infty\}$$

and

$$\mathscr{D}_p^{\perp} = \{Y \in L^q : E(X \cdot Y) = 0 \quad \text{for all } X \in \mathscr{D}_p\}.$$

Note from the definition that $-\mathrm{f}^*(Y) \leqslant -E(X \cdot Y) + \mathrm{f}(X) \leqslant \mathrm{f}(X)$, i.e.,

$$-\mathrm{f}^*(Y) \leqslant \mathrm{f}(X)$$

for all $Y \in C^{\infty} \cap \mathscr{D}_p^{\perp}$ and for all $X \in C \cap \mathscr{D}_p$. The equality is obtained by

Theorem (Fenchel Duality). *Suppose that $C \cap \mathscr{D}_p$ contains points in the relative interior of C and $\mathscr{D}_p$, that the epigraph of* f *over C has a nonempty interior, and that the infirmum of the primal problem is finite. Then*

$$\inf(P) = \sup(D).$$

Multidimensional Normal Distributions

In this section certain useful facts on multidimensional normal distributions are listed for easy reference.

Partition of Random Vectors. Let $\mathbf{X}$ be an n-dimensional random vector with $N(\mathbf{m}, \mathbf{\Lambda})$. Assume that $\mathbf{\Lambda}$ is nonsingular. Partition $\mathbf{X}$ into two vectors $\mathbf{X}_1$ and $\mathbf{X}_2$ of k and $(n-k)$ dimensions each. Define

$$\begin{aligned} \mathbf{\Lambda}_{11} &= E(\mathbf{X}_1 - \mathbf{m}_1)(\mathbf{X}_1 - \mathbf{m}_1)', \qquad & \mathbf{m}_1 = E(\mathbf{X}_1) \\ \mathbf{\Lambda}_{22} &= E(\mathbf{X}_2 - \mathbf{m}_2)(\mathbf{X}_2 - \mathbf{m}_2)', \qquad & \mathbf{m}_2 = E(\mathbf{X}_2) \\ \mathbf{\Lambda}_{12} &= E(\mathbf{X}_1 - \mathbf{m}_1)(\mathbf{X}_2 - \mathbf{m}_2)'. \end{aligned}$$

If $\mathbf{\Lambda}_{12} = 0$, then

$$|\mathbf{\Lambda}| = |\mathbf{\Lambda}_{11}||\mathbf{\Lambda}_{22}|$$

$$\mathbf{\Lambda}^{-1} = \begin{pmatrix} \mathbf{\Lambda}_{11}^{-1} & 0 \\ 0 & \mathbf{\Lambda}_{22}^{-1} \end{pmatrix},$$

and the density function of x becomes

$$\begin{aligned} f(\mathbf{X}_1, \mathbf{X}_2) = &\frac{1}{(2\pi)^{k/2}|\mathbf{\Lambda}_{11}|^{1/2}} \exp(-\{\tfrac{1}{2}(\mathbf{X}_1 - \mathbf{m}_1)'\mathbf{\Lambda}_{11}^{-1}(\mathbf{X}_1 - \mathbf{m}_1)\}) \\ &\times \frac{1}{(2\pi)^{(n-k)/2}|\mathbf{\Lambda}_{22}|^{1/2}} \exp(-\{\tfrac{1}{2}(\mathbf{X}_2 - \mathbf{m}_2)'\mathbf{\Lambda}_{22}^{-1}(\mathbf{X}_2 - \mathbf{m}_2)\}). \end{aligned}$$

Therefore when $\mathbf{\Lambda}_{12} = 0, \mathbf{X}_1$ and $\mathbf{X}_2$ are independent and are distributed according to $N(\mathbf{m}_1, \mathbf{\Lambda}_{11})$ and $N(\mathbf{m}_2, \mathbf{\Lambda}_{22})$, respectively. Thus, we have

Lemma 1. *Two uncorrelated normally distributed random vectors are independent.*

Conditional Distributions. Generally,

$$\mathbf{\Lambda}_{12} \neq 0.$$

In this case, introduce random vectors $\mathbf{Y}_1$ and $\mathbf{Y}_2$ of k and $(n-k)$ dimensions each by

$$\begin{aligned} \mathbf{Y}_1 &= \mathbf{X}_1 - \mathbf{D}\mathbf{X}_2 \\ \mathbf{Y}_2 &= \mathbf{X}_2 \end{aligned}$$

where $\mathbf{D}$ is a $(k \times (n-k))$ matrix to be specified in a moment. Then

$$E[(\mathbf{Y}_1 - E\mathbf{Y}_1)(\mathbf{Y}_2 - E\mathbf{Y}_2)'] = \mathbf{\Lambda}_{12} - \mathbf{D}\mathbf{\Lambda}_{22}.$$

If $\mathbf{D}$ is chosen to be

$$D = \mathbf{\Lambda}_{12}\mathbf{\Lambda}_{22}^{-1},$$

then $\mathbf{Y}_1$ and $\mathbf{Y}_2$ are uncorrelated normally distributed random vectors, hence independent from Lemma 1.

Since $\mathbf{Y}_1$ and $\mathbf{Y}_2$ are normally distributed, their distributions are specified by computing their means $\boldsymbol{\mu}_1$ and $\boldsymbol{\mu}_2$ and covariance matrices $\boldsymbol{\Sigma}_1$ and $\boldsymbol{\Sigma}_2$ where

$$\begin{aligned}
\boldsymbol{\mu}_1 &= E\mathbf{Y}_1 = \mathbf{m}_1 - \boldsymbol{\Lambda}_{12}\boldsymbol{\Lambda}_{22}^{-1}\mathbf{m}_2 \\
\boldsymbol{\mu}_2 &= E\mathbf{Y}_2 = \mathbf{m}_2 \\
\boldsymbol{\Sigma}_1 &= E[(\mathbf{Y}_1 - E\mathbf{Y}_1)(\mathbf{Y}_1 - E\mathbf{Y}_1)'] \\
&= \boldsymbol{\Lambda}_{11} - \boldsymbol{\Lambda}_{12}\boldsymbol{\Lambda}_{22}^{-1}\boldsymbol{\Lambda}_{21}
\end{aligned}$$

and

$$\begin{aligned}
\boldsymbol{\Sigma}_2 &= E[(\mathbf{Y}_2 - E\mathbf{Y}_2)(\mathbf{Y}_2 - E\mathbf{Y}_2)'] \\
&= \boldsymbol{\Lambda}_{22}.
\end{aligned}$$

Then the joint density function of $(\mathbf{X}_1, \mathbf{X}_2)$ when $\boldsymbol{\Lambda}_{12} \neq 0$ is given by

$$f(\mathbf{X}_1, \mathbf{X}_2) = g(\mathbf{Y}_1) \cdot g(\mathbf{Y}_2)|J|$$

where

$$J = \left|\frac{\partial y_i}{\partial x_j}\right|, \qquad 1 \leqslant i, j \leqslant n.$$

Then the conditional probability density function of $\mathbf{X}_1$ on $\mathbf{X}$ is obtained from

$$f(\mathbf{X}_1|\mathbf{X}_2) = \frac{f(\mathbf{X}_1, \mathbf{X}_2)}{f(\mathbf{X}_2)}.$$

This has the normal distribution law

$$N(\mathbf{m}_1 + \boldsymbol{\Lambda}_{12}\boldsymbol{\Lambda}_{22}^{-1}(\mathbf{X}_2 - \mathbf{m}_2), \boldsymbol{\Sigma}_1).$$

Thus the conditional mean of a normally distributed random vector is linear in the conditioning vector $\mathbf{X}_2$:

$$E(\mathbf{X}_1|\mathbf{X}_2) = E(\mathbf{X}_1) + \boldsymbol{\Lambda}_{12}\boldsymbol{\Lambda}_{22}^{-1}(\mathbf{X}_2 - E(\mathbf{X}_2)).$$

Singular Distributions. When a covariance matrix Λ is positive semidefinite, then Λ^{-1} does not exist and the density function cannot be obtained from the inversion formula as has been done in previous sections.

The function $\phi(t) = E(e^{it\mathbf{X}})$ where $\mathbf{X}$ is a random vector, however, is still a characteristic function. Therefore, there exists a corresponding distribution function even when Λ^{-1} does not exist. (For necessary and sufficient condition for $\phi(t)$ to be a characteristic function see, for example, Cramèr (1946)).

This d.f. can be obtained as a limit of a d.f. with nonsingular $\Lambda_k \to \Lambda$. For example, let

$$\Lambda_k = \Lambda + \varepsilon_k t't, \qquad \varepsilon_k > 0.$$

Λ_k^{-1} now exists and the corresponding d.f. F_k can be found. As $\varepsilon_k \to 0$, a ch.f. with Λ_k converges at every t.

Then it can be shown that there exists a d.f. F with $\phi(t)$ as its ch.f. to which F_k converges at every continuity point of F. This limit d.f. is called a singular normal distribution.

Let

$$\operatorname{rank} \mathbf{\Lambda} = r < n.$$

Consider a linear transformation

$$\mathbf{Y} = \mathbf{C}(\mathbf{X} - \mathbf{m}). \tag{A8}$$

Then the covariance matrix $\mathbf{M}$ of y is given by

$$\mathbf{M} = E(\mathbf{YY}') = \mathbf{C\Lambda C}'.$$

Choose C as an orthogonal matrix such that $\mathbf{\Lambda}$ is diagonalized. Since

$$\operatorname{rank} \mathbf{M} = \operatorname{rank} \mathbf{\Lambda} = r,$$

only r diagonal elements of M are positive; the rest are all zero. Therefore,

$$E(y_i^2) > 0, \qquad 1 \leqslant i \leqslant r$$
$$E(y_j^2) = 0, \qquad r + 1 \leqslant j \leqslant n$$

by rearranging components of y, if necessary.

This implies

$$y_j = 0 \qquad \text{with probability 1}, \qquad r + 1 \leqslant j \leqslant n.$$

Then, from Eq. (A8),

$$\mathbf{X} = \mathbf{m} + \mathbf{C}'\mathbf{Y}.$$

It is seen, therefore, that random variables $x_1, \ldots, x_n$, with probability 1, can be expressed as linear combinations of r uncorrelated random variables $y_1, \ldots, y_r$. Since each y_i, $1 \leqslant i \leqslant r$, is a linear combination of $x_1, \ldots, x_n$, each y_i, $1 \leqslant i \leqslant r$, is normally distributed and is independent.

Lemma 2. *If n random variables are distributed normally with the covariance matrix of rank r, then they can be expressed as linear combinations of r independent and normally distributed random variables with probability* 1.

CHAPTER 3

Stochastic Control Problems

When we go beyond control problems of known deterministic systems in deterministic environments, many complications arise of a computational and conceptual nature. Even in possibly the simplest problem of controlling known deterministic systems in a noisy environment, we must now distinguish open-loop and closed-loop control laws.* When we consider controlling systems with randomly varying parameters, another kind of complication arises. Even the familiar quadratic cost minimization problems subject to linear dynamic constraint may become unbounded when parameter variability exceeds critical thresholds. Systems with unknown or imprecisely known parameters introduce further complications. This chapter deals with various complications of the first two types. All random variables involved are assumed to have known probability densities, and no unknown parameters are present in the system dynamics or in the system observation equations. The third type is extensively discussed in Chapter 7. In this chapter, we develop a systematic procedure for obtaining optimal control policies for discrete-time stochastic control systems, where the random variables involved

* Closed-loop feedback decision rules, and open-loop ones we later discuss, are contingency rules in that the actual values of decisions to be implemented are contingent on the additional information which will be gathered, and not known until the time the decisions are to be made.

are such that they all have known probability distribution functions, or at least have known first, second, and possibly higher moments.

We describe a method for deriving optimal control policies that is later extended to treat a much larger class of optimal control problems than those just mentioned, such as for systems with unknown parameters and dependent random disturbances. This method can also be extended to tackle problems with unknown parameters or random variables with only partially known statistical properties. Thus, we will be able to discuss optimal controls of parameter adaptive systems without too much extra effort.

The method presented in this chapter improves on the one proposed by Fel'dbaum. It uses the notion of information state advanced by Bellman and that of conditional expected cost-to-go (conditionally expected future costs to be incurred), rather than the expected costs used by Fel'dbaum.* This method is more concise and less cumbersome than Fel'dbaum's when applied to control problems. For example, the concept of sufficient statistics is incorporated in the method, and some assumptions on the systems that lead to simplified formulations are explicitly pointed out. Evaluations of various expectation operations necessary in deriving optimal control policies are all based on recursive calculations of certain conditional probabilities or probability densities. As a result, the expositions are simpler and most formulas are stated recursively, which makes them easier to implement by means of digital computers.

3.1 Markovian Structure

In problems of this chapter, the Markovian structure of the problem is embodies in recursions that relate $p_{t+1}(\mathbf{x}_{t+1} \mid \mathbf{y}^t, \mathbf{u}^t)$ to $p_t(\mathbf{x}_t \mid \mathbf{y}^{t-1}, \mathbf{u}^{t-1})$, and $p(\mathbf{y}_{t+1} \mid \mathbf{y}^t, \mathbf{u}^t)$ to $p(\mathbf{y}_t \mid \mathbf{y}^{t-1}, \mathbf{u}^{t-1})$. To derive these updating relations, we assume that the dynamic (transition) equation is now represented by

$$P_{t+1}(\mathbf{x}_{t+1} \in A \mid \mathbf{x}^t, \mathbf{u}^t) = P_{t+1}(\mathbf{x}_t, A; \mathbf{u}_t)$$

where the right side is the one-step transition probability from a state $\mathbf{x}_t$ to a set of states at the next time instant, when control $\mathbf{u}_t$ is used. This is a generalization of the deterministic state transition equation in the previous

* There are slightly different ways for defining expected conditional costs, whether inf is used or P-essential inf with respect to the probability measure induced by decision rules. In this book, classes of decision rules are sufficiently rich and regular so that we do not worry about measure-theoretic subtleties. See Striebel (1970, Chapter 4), Hinderer (1970), or Bertsekas and Shreve (1978) about universally measurable policies, or ε-optimal rules.

chapter. Any dynamic model satisfying this property is called Markovian. Markovian processes are discussed further in Chapter 7. The dynamics of the object to be controlled are stated by

$$\text{(1)} \qquad \mathbf{x}_{t+1} = \mathrm{f}(\mathbf{x}_t, \mathbf{u}_t, t, \xi_t),$$

where $\mathbf{x}_t$ is the state vector, $\mathbf{u}_t$ the decision vector, and ξ_t is an exogenous random vector that causes the transition from $\mathbf{x}_t$ to $\mathbf{x}_{t+1}$ to be random even if $\mathbf{u}_t$ is kept fixed. The state vector for the decision maker need not be the same as $\mathbf{x}_t$ when it is not directly observed. When it is necessary to distinguish, $\mathbf{x}_t$ is called process or model state vector. The random variable ξ_t may represent disturbances acting on the controlled object or may represent random changes in the dynamic characteristics of the object to be controlled. The random vector ξ_t is independent of $\mathbf{u}^t$ or $\mathbf{x}^t$.* When ξ_t is independent of $\mathbf{x}^{t-1}$ and $\mathbf{u}^{t-1}$, the transition probability density function is simply given by $p(\mathbf{x}_{t+1}|\mathbf{x}_t, \mathbf{u}_t)$. Knowing $P(\xi_t)$, one can calculate $P(\mathbf{x}_{t+1}|\mathbf{x}^t, \mathbf{u}^t) = P(\mathbf{x}_{t+1}|\mathbf{x}_t, \mathbf{u}_t)$ when the noises are independent. In this sense, control variables are "parametrizing" variables, i.e., the distribution of $\mathbf{x}$ conditional on $\mathbf{y}$ is parametrized by $\mathbf{u}$ rather than $\mathbf{u}$ being joiontly specified with $\mathbf{x}$, i.e., $P(\mathbf{x}\,|\,\mathbf{y};\mathbf{u})$, not $P(\mathbf{x},\mathbf{u}\,|\,\mathbf{y})$.

For example, when noises enter additively into the state transition equation

$$\mathbf{x}_{t+1} = \mathbf{f}(\mathbf{x}_t, \mathbf{u}_t) + \xi_t$$

where

$$\xi_t \sim N(0, \sigma^2),$$

then

$$P(\mathbf{x}_{t+1}|\mathbf{x}_t, \mathbf{u}_t) \sim N(\mathbf{f}(\mathbf{x}_t, \mathbf{u}_t), \sigma^2),$$

where the conditional mean is parametrized by $\mathbf{u}_t$.

Stochastic problems in which $\mathbf{x}_t$ is known, i.e., $\mathbf{y}_t = \mathbf{x}_t$, are almost the same as deterministic problems as we shall soon see. When ξ_t is absent from Eq. (1), then (1) represents a deterministic system. Control problems become stochastic, then, through noisy measurements of the state vector $\mathbf{x}_t$, i.e., $\mathbf{x}_t$ is

---

* When not independent, ξ is not part of underlying elementary or basic stochastic processes that induce probability distributions for state vectors and observations. Dependence can be removed from basic stochastic processes by including in the dynamics of the model those for the noise processes as well. For example, if $\xi_{t+1} = \rho\xi_t + \varepsilon_t$ and if ε is independent, then the latter is treated as basic processes, not the former.

only incompletely known from current and past $\mathbf{y}$'s, i.e., $\mathbf{y}^t$ when the relation between $\mathbf{x}_t$ and $\mathbf{y}_t$ is not deterministic but is given by

$$\mathbf{y}_t = \mathbf{h}(\mathbf{x}_t, \mathbf{u}_t, t, \boldsymbol{\eta}_t) \tag{2}$$

where $\boldsymbol{\eta}_t$ is an exogenous random vector.

Finite-dimensional sufficient statistics are assumed to be available in this book. For example, when the transition equation $P(\mathbf{x}_{t+1}|\mathbf{x}_t, \mathbf{u}_t, t)$ is Gaussian, it is completely specified by the conditional mean of $\mathbf{x}_{t+1}$, $E(\mathbf{x}_{t+1}|\mathbf{x}_t, \mathbf{u}_t, t)$ and the conditional covariance $\operatorname{cov}(\mathbf{x}_{t+1}|\mathbf{x}_t, \mathbf{u}_t, t)$, which is independent of conditioning variables.

Denote by Ω_t the information set available to a decision maker at time t. When $\Omega_t = \{\mathbf{x}^t, \mathbf{u}^{t-1}, t\}$, then $\mathbf{x}_t$ is observed exactly. When $\Omega_t = \{\mathbf{y}^t, \mathbf{u}^{t-1}, t\}$, $\mathbf{x}_t$ is not observed exactly where the current data y_t is related to $\mathbf{x}_t$ by Eq. (2). When $\mathbf{x}_t$ is not observed exactly and its value has to be inferred from $\mathbf{y}^t$, decision problems become more difficult even if the transition equations are known exactly.

One of our primary concerns is the derivation or construction of optimal control policies, in other words, obtaining methods for controlling dynamic systems in such a way that some conditionally expected cost indices are minimized. Loosely speaking, a control policy is a sequence of functions (mappings) that generates a sequence of control actions $\mathbf{u}_0, \mathbf{u}_1, \ldots$ according to some rule. The class of control policies to be considered throughout this book is that of closed-loop control policies,* i.e., control policies such that the control $\mathbf{u}_t$ at time t is to depend only on the past and current observations $\mathbf{y}^t$ and on the past control sequences $\mathbf{u}^{t-1}$, which are assumed to be also observed. A nonrandomized closed-loop control policy for an N-stage control process is a sequence of N control actions u_t, such that each $\mathbf{u}_t$ takes a value in the set of admissible control U_t, $\mathbf{u}_t \in U_t$, $0 \leqslant t \leqslant N-1$ depending on the past and current observations on the system $\mathbf{y}_0, \mathbf{y}_1, \ldots, \mathbf{y}_{t-1}, \mathbf{y}_t$ and on the past control vectors $\mathbf{u}_0, \ldots, \mathbf{u}_{t-1}$. Since past controls $\mathbf{u}_0, \ldots, \mathbf{u}_{t-1}$ really depend on $\mathbf{y}_0, \ldots, \mathbf{y}_{t-1}$, $\mathbf{u}_t$ depends on $\mathbf{y}_0, \ldots, \mathbf{y}_{t-1}, \mathbf{y}_t$. Thus a control policy $\pi(u)$ is a sequence of functions (mappings) $\boldsymbol{\pi}_0, \boldsymbol{\pi}_1, \ldots, \boldsymbol{\pi}_{N-1}$ such that the domain of π_t is defined to be the collection of all points

$$\mathbf{y}^t = (\mathbf{y}_0, \ldots, \mathbf{y}_t), \quad \text{with} \quad \mathbf{y}_s \in Y_s, \qquad 0 \leqslant s \leqslant t$$

where Y_s is the set in which the sth observation takes its value, and such that the range of $\boldsymbol{\pi}_t$ is U_t. Namely, $\mathbf{u}_t = \mathbf{u}_t(\mathbf{y}^t, \mathbf{u}^{t-1}) = \boldsymbol{\pi}_t(\mathbf{y}^t) \in U_t$. We take U_t not

* The so-called open-loop feedback decision rules are later considered as approximations.

to depend on $\mathbf{y}^t$ or $\mathbf{u}^{t-1}$. When the value of $\mathbf{u}_t$ is determined uniquely from $\mathbf{y}^t$ and $\mathbf{u}^{t-1}$, that is when the function $\boldsymbol{\pi}_t$ is deterministic, we say a control policy is nonrandomized. When $\boldsymbol{\pi}_t$ is a random transformation from $\mathbf{y}^t$ and $\mathbf{u}^{t-1}$ to a point in U_t, such that $\boldsymbol{\pi}_t$ is a probability distribution on U_t, a control policy is called randomized.

A nonrandomized optimal control policy, therefore, is a sequence of mappings from the space of observable quantities of the space of control vectors, given all the past and current observations, in such a way that the sequence minimizes the expected value of cost functions. A class of nonrandomized control policies is included in the class of randomized control policies since a nonrandomized control policy may be regarded as a sequence of probability distributions, each of which assigns probability mass 1 to a point in U_t, $0 \leqslant t \leqslant N-1$. It is the functional form of π_t that is chosen as optimal. To make this explicit, sometimes a subscript π_t is used to indicate the dependence of the argument of the form of the past and current control, e.g.,

$$p_{\pi_t}(\mathbf{x}_{t+1} | \mathbf{x}_t, \mathbf{y}^t) = p[\mathbf{x}_{t+1} | \mathbf{x}_t, \mathbf{u}_t = \pi_t(\mathbf{y}^t, \mathbf{u}^{t-1})].$$

The question of whether one can really find optimal control policies in the class of nonrandomized control policies is dicussed, for example, in Sworder and Aoki (1965). We later give heuristic reasonings for using nonrandomized controls. For more mathematical arguments the reader is referred to Gihman and Skorohod (1979, Chapter 1). For Markovian stochastic models with complete and separable metric spaces as state spaces and compact control spaces, sufficient conditions for the existence of nonrandomized controls are basically that the transition probability distributions preserve boundedness and continuity subject to certain measurability conditions and that the control costs are lower semicontinuous.

For randomized control policies,

$$p(\mathbf{y}_{t+1} | \mathbf{y}^t) = \int p(\mathbf{y}_{t+1} | \mathbf{u}_t, \mathbf{y}^t) p(\mathbf{u}_t | \mathbf{y}^t) \, d\mathbf{u}_t,$$

hence $p(\mathbf{y}_{t+1} | \mathbf{y}^t)$ is a functional depending on the form of the density function of $\mathbf{u}_t$, $p(\mathbf{u}_t | \mathbf{y}^t)$. When $\mathbf{u}_t$ is nonrandomized, $p(\mathbf{y}_{t+1} | \mathbf{y}^t)$ is a functional depending on the value of $\mathbf{u}_t$ and we write

$$p_{\pi_t}(\mathbf{y}_{t+1} | \mathbf{y}^t) = [p(\mathbf{y}_{t+1} | \mathbf{y}^t)]_{\mathbf{u}_t = \pi_t(\mathbf{y}^t)} = p[\mathbf{y}_{t+1} | \mathbf{y}^t, \mathbf{u}_t = \boldsymbol{\pi}_t(\mathbf{y}^t)]$$

or simply $p(\mathbf{y}_{t+1} | \mathbf{y}^t, \mathbf{u}_t)$.

The variables $\mathbf{u}_t$ or $\mathbf{u}^t$ are sometimes dropped from expressions such as $p(\cdot \,|\, \mathbf{y}^t, \mathbf{u}^t)$ or $p(\cdot \,|\, \mathbf{y}^t, \mathbf{u}^t)$ where no confusion is likely to occur.

Let

$$p_{\pi^{t+1}}(\mathbf{x}_t, \mathbf{y}^{t-1}) d(\mathbf{x}^t, \mathbf{y}^{t-1})$$

be the joint conditional probability that the sequence of the state vectors and observed vectors will lie in the elementary volume $d\mathbf{x}_0 \ldots d\mathbf{x}_t\, d\mathbf{y}_0 \ldots d\mathbf{y}_{t-1}$ around $\mathbf{x}^t$ and $\mathbf{y}^{t-1}$, given a sequence of control specified by π^{t-1}, where the notation

$$d(\mathbf{x}^t, \mathbf{y}^{t-1}) = d(\mathbf{x}_0, \ldots, \mathbf{x}_t, \mathbf{y}_0, \ldots, \mathbf{y}_{t-1})$$
$$\triangleq d\mathbf{x}_t \ldots d\mathbf{x}_t\, d\mathbf{y}_0 \ldots d\mathbf{y}_{t-1}$$

is used to indicate the variables with respect to the integrations carried out.

Let

$$p(\mathbf{y}_t \mid \mathbf{x}_t)\, d\mathbf{y}_t$$

be the conditional probability that the observation at time t lies in the elementary volume $d\mathbf{y}_t$ about $\mathbf{y}_t$, given $\mathbf{x}_t$. Finally, let

$$p_0(\mathbf{x}_0)\, d\mathbf{x}_0$$

be the probability that the initial condition is in the elementary volume about $\mathbf{x}_0$. Various probability densities are assumed to exist.

Separation of Estimation and Control

When state vector $\mathbf{x}_t$ is not available, i.e., when states are unobservable, it is not generally optimal to construct a control as a function of its estimate $\mathbf{x}_{t|t-1}$ or $\mathbf{x}_{t|t}$ as the case may be. A notable exception is found for a class of linear dynamic models with quadratic criteria. In this class of problems, optimal decisions can be shown to be linear functions of the most up-to-date estimate of the state unobserved state vector. Controls are functions of sufficient statistics that are $\mathbf{x}_{t|t}$ or $\mathbf{x}_{t|t-1}$ and they can be updated by Kalman filters for this class of problems. Instead of $\mathbf{u}_t = \mathbf{K}\mathbf{x}_t$, optimal $\mathbf{u}_t$ is $\mathbf{K}\mathbf{x}_{t|t-1}$ if $\mathbf{y}^{t-1}, \mathbf{u}^{t-1}$ is the information set, or $\mathbf{K}\mathbf{x}_{t|t}$ when y_t is additionally available for some constraint matrix $\mathbf{K}$. This fact is sometimes referred to as "separation principle," i.e., the problems of obtaining the best estimate and of synthesizing best control can be attacked separately rather than as a whole.

3.2 Infinite Horizon Problems*

In this section, all random variables involved are assumed to have *known* probability densities, and no unknown parameters are present in the system dynamics or in the system observation mechanisms.

* This section follows Gihman and Skorohod (1979, Section 7).

Consider a control system described by

(3) $$\mathbf{x}_{t-1} = \mathbf{f}_t(\mathbf{x}_t, \mathbf{u}_t, \xi_t), \qquad \mathbf{u}_t \in U_t, \qquad t = 0, 1, \ldots, N-1$$

where $p_0(\mathbf{x}_0)$ is given and observed by

(4) $$\mathbf{y}_t = \mathbf{h}_t(\mathbf{x}_t, \boldsymbol{\eta}_t), \qquad t = 0, 1, \ldots, N$$

where $\mathbf{x}_t$ is an n-dimensional state vector at tth time instant, $\mathbf{u}_t$ is a r-dimensional control vector at the tth time instant, U_t is the set in the r-dimensional Euclidean vector space and is called the admissible set of controls, ξ_t is a q-dimensional random vector at the tth time instant, and $\mathbf{y}_t$ is an n-dimensional random vector at the t the time instant.

The functional forms of $\mathbf{f}_t$ and $\mathbf{h}_t$ are assumed known for all t.

The vectors ξ_t and $\boldsymbol{\eta}_t$ are the random noises in the system dynamics and in the observation device, or they may be random parameters in the system. In this chapter, they are assumed to be mutually and serially independent, unless stated otherwise.* From now on, Eq. (3) is referred to as the dynamic equation and Eq. (4) is referred to as the vector observation equation or simply as the observation equation.

A Markov process is time-homogeneous if the transition probability is independent of time, $\mathrm{P}_t(\mathbf{x}, A; \mathbf{u}) = \mathrm{P}(\mathbf{x}, A; \mathbf{u})$. The first of two types of infinite horizon problems we treat here are discounted cost and average cost problems.

Discounted Cost Problem

Let $\underline{\mathbf{x}}$ and $\underline{\mathbf{u}}$ denote points $(\mathbf{x}_0, \mathbf{x}_1, \ldots)$ and $(\mathbf{u}_0, \mathbf{u}_1 \ldots)$.

Consider minimizing

$$c(\underline{\mathbf{x}}, \underline{\mathbf{u}}) = \sum_0^\infty \beta^t \mathrm{f}(\mathbf{x}_t, \mathbf{x}_{t+1}, \mathbf{u}_t)$$

---

* When the random variables in Eqs. (3) and (4) are independent, they are the basic or elementary underlying stochastic processes that induce probability measures for $\{\mathbf{x}_t\}$ and $\{\mathbf{y}_t\}$ processes, given a decision rule or control law. When ξ is not independent but is generated by $\xi_{t+1} = \rho\xi_t + \mathbf{e}_t$ where $\mathbf{e}_t$ is independent, for example, then e_t and $\mathbf{y}_t$ are the underlying basic stochastic processes. The state vector $\mathbf{x}_t$ need be augmented by ξ_t as

$$\mathbf{z}_{t+1} = \begin{bmatrix} \mathbf{x}_{t+1} \\ \xi_{t+1} \end{bmatrix} = \begin{bmatrix} \mathbf{f}(\mathbf{x}_t, \mathbf{u}_t, t, \xi_t) \\ \rho\xi_t + \mathbf{e}_t \end{bmatrix} = \mathrm{F}(\mathbf{z}_t, \mathbf{u}_t, t, \mathbf{e}_t)$$

which is the state transition equation for the augmented vector $\mathbf{z}_t$. Equation (4) needs be modified as $\mathbf{y}_t = \mathrm{H}(\mathbf{z}_t, \mathbf{u}_t, t, \boldsymbol{\eta}_t)$ in which ξ_t does not appear. This procedure of augmenting state vectors so that not all components are observed is used in Chapter 7.

where $0 < \beta < 1$ and $\{\mathbf{x}_t\}$ is a time-homogeneous Markov process. The function $\mathrm{f}(\mathbf{x}, \mathbf{y}, \mathbf{u})$ is assumed to be bounded and lower semicontinuous.

Noting that cost-to-go from time t onward is

$$C_t(\underline{\mathbf{x}}, \underline{\mathbf{u}}) = \sum_{s=t}^{\infty} \beta^{s-t} \mathrm{f}(\mathbf{x}_s, \mathbf{x}_{s+1}, \mathbf{u}_s),$$

and that because of time-homogeneity the origin of time is immaterial, the Bellman functional equation is

$$\rho(\mathbf{x}) = \inf_{\mathbf{u}} \int [\mathrm{f}(\mathbf{x}, \mathbf{y}, \mathbf{u}) + \beta\rho(\mathbf{y})] P(\mathbf{x}, d\mathbf{y}; \mathbf{u}) \tag{5}$$

where

$$\rho(\mathbf{x}) = \inf_{\pi} E_\pi [\mathrm{C}(\underline{\mathbf{x}}, \underline{\mathbf{u}}) \,|\, \mathbf{x}_0 = \mathbf{x}].$$

Equation (5) has a unique bounded solution, since the contrary assumption that $\bar{\rho}(\mathbf{x})$ is also a solution implies that

$$|\rho(\mathbf{x}) - \bar{\rho}(\mathbf{x})| \leqslant \sup_{u} \int \beta |\rho(\mathbf{y}) - \bar{\rho}(\mathbf{y})| P(\mathbf{x}, d\mathbf{y}; \mathbf{u}),$$

where we use the inequality

$$|\inf_{\mathbf{u}} g_1(\mathbf{u}) - \inf_{\mathbf{u}} g_2(\mathbf{u})| \leqslant \sup |g_1(\mathbf{u}) - g_2(\mathbf{u})|, \tag{6}$$

and

$$\sup_{\mathbf{x}} |\rho(\mathbf{x}) - \bar{\rho}(\mathbf{x})| \leqslant \beta \sup_{\mathbf{y}} |\rho(\mathbf{y}) - \bar{\rho}(\mathbf{y})|,$$

i.e., $\sup_{\mathbf{x}} |\rho(\mathbf{x}) - \bar{\rho}(\mathbf{x})| = 0$. The optimal control is stationary, $\mathbf{u} = \varphi(\mathbf{x})$, and satisfies

$$\rho(\mathbf{x}) = \int \{\mathrm{f}[\mathbf{x}, \mathbf{y}, \varphi(\mathbf{x})] + \beta\rho(\mathbf{y})\} \rho[\mathbf{x}, d\mathbf{y}; \varphi(\mathbf{x})], \tag{7}$$

and $\rho(\mathbf{x})$ is unique.

Value Iteration

To solve Eq. (5), an iterative scheme is used to generate $\{v_n(\mathbf{x})\}$ where $v_0(\mathbf{x})$ is an arbitrary bounded Borel function

$$v_{n+1}(\mathbf{x}) = \min_{\mathbf{u}} [\mathrm{F}(\mathbf{x}, \mathbf{u}) + \beta \int v_n(\mathbf{y}) P(\mathbf{x}, d\mathbf{y}; \mathbf{u})] \tag{8}$$

where

$$\mathrm{F}(\mathbf{x}, \mathbf{u}) = \int \mathrm{f}(\mathbf{x}, \mathbf{y}, \mathbf{u}) P(\mathbf{x}, d\mathbf{y}; \mathbf{u}).$$

From (7) and (8), noting (6),

$$|v_{n+1}(\mathbf{x}) - \rho(\mathbf{x})| \leqslant \beta \int |v_n(\mathbf{y}) - \rho(\mathbf{y})| P(\mathbf{x}, d\mathbf{y}; \mathbf{u}). \tag{9}$$

Set $\delta_n = \max_x |v_n(\mathbf{x}) - p(\mathbf{x})|$. Equation (6) implies that $\delta_{n+1} \leqslant \beta\delta_n$, i.e., $\delta_n \to 0$ as $n \to \infty$.

Average Cost Problem

This section uses the average cost

$$1/N \sum_{k=0}^{N-1} \mathrm{f}(\mathbf{x}_k, \mathbf{x}_{k+1}, \mathbf{u}_k)$$

instead of the sum of discounted cost.

Let

$$r_N(\mathbf{x}) = \inf_\pi E_\pi\left[1/N \sum_{k=0}^{N-1} \mathrm{f}(\mathbf{x}_k, \mathbf{x}_{k+1}; \mathbf{u}_k) | \mathbf{x}_0 = \mathbf{x}\right].$$

Then it satisfies the Bellman functional equation

$$\begin{aligned} r_{N+1}(\mathbf{x}) &= \inf_\pi E_\pi\left[\frac{1}{N+1}\mathrm{f}(\mathbf{x}_0, \mathbf{x}_1; \mathbf{u}_0)\right. \\ &\qquad \left. + \frac{N}{N+1}\frac{1}{N}\sum_{k=1}^{N} \mathrm{f}(\mathbf{x}_k, \mathbf{x}_{k+1}; \mathbf{u}_k) | \mathbf{x}_0 = \mathbf{x}\right] \\ &= \inf_\pi E_\pi\left\{\frac{1}{N}\mathrm{f}(\mathbf{x}_0, \mathbf{x}_1; \mathbf{u}_0)\right. \\ &\qquad \left. + \frac{N}{N+1}E_\pi\left[\frac{1}{N}\sum_{k=1}^{N} \mathrm{f}(\mathbf{x}_k, \mathbf{x}_{k+1}; \mathbf{u}_k) | \mathbf{x}_0, \mathbf{x}_1, \mathbf{u}_0\right] \Bigg| \mathbf{x}_0 = \mathbf{x}\right\} \\ &= \inf_u \int\left[\frac{1}{N+1}\mathrm{f}(\mathbf{x}, \mathbf{y}, \mathbf{u}) + \frac{N}{N+1} r_N(\mathbf{y})\right] P(\mathbf{x}, d\mathbf{y}; \mathbf{u}). \end{aligned}$$

Assume that $r_N(\mathbf{x}) \to a$ constant s, as $N \to \infty$ and that $[r_N(\mathbf{x}) - s]N \to c(x)$ for some $c(\cdot)$. Then, in the equation,

$$s + c(\mathbf{x}) = \inf_{\mathbf{u}} \int [\mathrm{f}(\mathbf{x}, \mathbf{y}, \mathbf{u}) + c(\mathbf{y})] P(\mathbf{x}, d\mathbf{y}; \mathbf{u}).$$

Note that $c(\mathbf{x})$ is determined up to an additive constant, i.e., $c(\mathbf{x}) + \alpha$ for any constant α also satisfies the preceding. See Gihman and Skorohod (1979, Section 7) for the existence of ϵ-optimal stationary policies.

Alternatively under certain technical conditions, the limit of the average cost exists as the ergodic average when the dynamic systems admit invariant measure. Doob (1953) has shown that if π is an invariant probability for the Markov process $\{\mathbf{x}_t\}$, then for any $\Phi \in L^1(X^{Z_t}, B(X^{Z_t}), P_\pi)$,

$$\lim 1/T \Sigma_1^T \Phi(\mathbf{x}_t, \mathbf{x}_{t+1}, \ldots) = E_\pi[\Phi | \Sigma_I](\mathbf{x}_0, \mathbf{x}_1, \ldots),$$

where Σ_I is the σ-field of invariant events, almost surely when $\mathbf{x}_0$ has initial distribution which is absolutely continuous with respect to π. When the set Σ_I is trivial, the right side becomes $E_\pi(\Phi)$. The measure π is then called ergodic. This theorem, however, says nothing about the asymptotic behavior of $\{\mathbf{x}_t\}$ when the initial condition lies outside the set of full π-measure. A Markov process is said to be positive Harris recurrent if

$$P_{\mathbf{x}}\{\{\mathbf{x}_t\} \in A \text{ i.o.}\} = 1$$

where $[\{x_t\} \in A \text{ i.o.}] = \bigcap_{\tau>0} \bigcup_{t=\tau}^{\infty} \{\mathbf{x}_t \in A\}$. See Nummelin (1984, p. 42). A Markov process with invariant measure is positive Harris recurrent if and only if

$$\lim 1/T \sum_{t=1}^{T} \mathrm{f}(\mathbf{x}_t) = \int f d\pi \quad \text{a.s. } P_{\mu_0}$$

for every initial distribution μ_0 in the set of probabilities on the Borel field on the state space X. A Harris recurrent Markov process has a unique ergodic invariant probability. When control u and observation y are Borel functions of a Markov process $\{\mathbf{x}_t\}$, then the ergodic average of the form

$$\lim 1/T \sum_{t=1}^{T} (y_t^2 + \rho u_t^2)$$

exists for every initial distribution. Meyn and Caines (1988) showed that even when a Markov process is not positive Harris recurrent, subsets of the state space on which the process is Harris recurrent satisfy a certain controllability condition to ensure that the probability density for $\mathbf{x}_t$ is strictly positive for some t.

Example 1: Optimal Savings under Uncertainty. Slightly different versions of optimal savings under uncertainty over an infinite horizon are reported in the literature. Levhari and Srinivasan (1969) treated k_t as wealth to be invested so that $k_{t+1} = \xi_t(k_t - c_t)$ is the next period wealth available for investment after c_t is consumed, and a random interest rate r_t is the source of uncertainty in maximizing the expected discounted stream of utilities $E\Sigma_t \beta^t u(c_t)$.

Investment is explicitly modeled in Brock and Mirman (1972). Let k_t be the capital stock at time t; then the same criterion function is to be maximized subject to

$$c_t + x_t = \mathrm{f}(k_{t-1}, \xi_t),$$

and

$$k_t = (1 - \delta)k_{t-1} + x_t,$$

where δ in the depreciation rate of capital stock per period and ξ_t is productivity shock. This basic social-planning optimization problem has been generalized in the literature in several ways. Long and Plosser (1983) specialize $u(\cdot)$ to a log–linear form but allow for leisure and different output goods. Hansen (1985) allows for straight-time and overtime in his production function.

Levhari and Srinivansan (1969) obtained optimal saving plans for the utility function

$$u(c) = \frac{1}{1-\alpha} c^{1-\alpha}, \qquad \alpha > 0$$

on the assumption that the stochastic disturbances, ξ_ts, are independent. We sketch how the solution procedure goes through without an independent assumption. The Bellman equation is

$$V(k_t) = \max_{c_{t+1}} [u(c_{t+1}) + \beta E_t(V(k_t^+))]$$

where

$$k_t^+ = \xi_{t+1}(k_t - c_{t+1}).$$

Differentiating the inside of the square bracket with respect to c_{t+1}, the first-order condition is

$$u'(c_{t+1}) = \beta E_t[\xi_{t+1} V'(k_t^+)]. \tag{10}$$

Solving it for c_{t+1}, postulate $c_{t+1} = \phi(k_t)$ as an optimal rule and differentiate the functional equation with respect to k_t to obtain

$$\begin{aligned} V'(k_t) &= u(\phi(k_t))\phi'(k_t) + \beta E_t\{V'(k_t^+)[1 - \phi'(k_t)]\xi_{t+1}\} \\ &= \beta E_t[\xi_{t+1} V'(k_t^+)] \end{aligned} \tag{11}$$

where (10) is used. From (10) and (11) we see that $u'(\phi(k_t)) = V'(k_t)$ and hence

$$u'(\phi(k_t)) = \beta E_t[\xi_{t+1} u'(\phi(k_t^+))] \tag{12}$$

where

$$k_t^+ = \xi_{t+1}(k_t - \phi(k_t)).$$

Try $\phi(k) = \lambda k$. Then

$$\phi(k_t^+) = \lambda\xi_{t+1}(1-\lambda)k_t$$

and hence (12) becomes

$$\lambda^{-\alpha}k^{-\alpha} = \beta\lambda^{-\alpha}(1-\lambda)^{-\alpha}k^{-\alpha}E_t(\xi_{t+1}^{1-\alpha})$$

or

$$(1-\lambda)^{\alpha} = \beta E_t(\xi_{t+1}^{1-\alpha}), \qquad t \geqslant 0.$$

We assume that $\beta E_t(\xi_{t+1}^{1-\alpha}) < 1$.

With the logarithmic utility function $U(c) = \ln c$, the coefficient of the optimal consumption rule $c_t = \lambda x_{t-1}$ is equal to $1 - \beta$ and is independent of the distribution of the stochastic noise. This is a special case. In the case of $U(c) = (1-\alpha)^{-1}c^{1-\alpha}$, $\alpha \geqslant 0$, $\lambda = 1 - \beta^{1/\alpha}[E_t(\xi_{t+1}^{1-\alpha})]^{1/\alpha}$, so the distribution of ξ matters.

Levhari and Srinivasan (1969) examine a case where ξ is independently distributed as a log-normal random variable with mean μ and variance σ^2. Then

$$(1-\lambda)^{\alpha} = \beta^{-(1-\alpha)}e^{-\alpha(1-\alpha)\sigma^2/2}.$$

Assuming that the right-hand side is less than one to ensure the existence of a feasible consumption rule with $0 < \lambda < 1$, the effect of increasing σ^2 while keeping $\bar{\xi}$ fixed is to increase λ if $\alpha < 1$ and to decrease λ if $\alpha > 1$.

With this policy k_t evolves according to the dynamic law

$$\begin{aligned} k_{t+1} &= \xi_{t+1}(k_t - \lambda k_t) \\ &= (1-\lambda)\xi_{t+1}k_t \\ &= (1-\lambda)^{t+1}k_0\prod_{s=1}^{t+1}\xi_s \\ &= (1-\lambda)^{t+1}k_0 w_{t+1} \quad \text{where} \quad w_{t+1} = \prod_{s=1}^{t+1}\xi_s. \end{aligned}$$

By evaluating the expectation successively, we obtain

$$V(k_0) = E\left[\sum_0^{\infty}\beta^t u(\phi(k_t))\right] = \frac{\lambda^{1-\alpha}}{1-\alpha}k_0^{1-\alpha}\sum_0^{\infty}(1-\lambda)^t = \frac{\lambda^{-\alpha}}{1-\alpha}k_0^{1-\alpha}.$$

The Levhari-Srinivasan model can be generalized to include serially correlated disturbances. Consider $x_{t+1} = \xi_{t+1}(x_t - c_t)$ where

$$\xi_{t+1} = \rho\xi_t + \varepsilon_t,$$

where ε_t is a mean zero i.i.d. random variable with finite variance. The optimal consumption rule is now time-varying

$$c_t = \lambda_t x_t$$

where λ_t must satisfy

$$(1 - \lambda_t)^\alpha = \beta E_t[\xi_{t+1}^{1-\alpha}(\lambda_t/\lambda_{t+1})^\alpha].$$

According to Foldes (1978), $\beta^t w_t c_t^{-\alpha} = \beta^t w_t k^{-\alpha} k_t^{-\alpha}$ is a martingale, where $w_{t+1} = \xi_{t+1} w_t$ and $w_0 = 1$. This can be directly verified.

The transversality condition (Foldes [1978, Th 2]) is satisfied if $0 < \lambda < 1$.

3.3 Sufficient Statistics

We have seen in previous sections that $\mathbf{u}_t$ is generally a function of $\mathbf{y}^t$ and $\mathbf{u}^{t-1}$ and not just $\mathbf{y}_t$. This dependence of $\mathbf{u}_t$ on $\mathbf{y}^t$ occurs through $p(\mathbf{x}_t \mid \mathbf{y}^t)$ and $p(\mathbf{y}_{t+1} \mid \mathbf{y}^t, \mathbf{u}^t)$ in computing γs. Intuitively speaking, if a vector $\mathbf{s}_t$, a function of $\mathbf{y}^t$ and $\mathbf{u}^{t-1}$, exists such that $p(\mathbf{x}_t \mid \mathbf{y}^t, \mathbf{u}^{t+1}) = p(\mathbf{x}_t \mid \mathbf{s}_t)$, then the dependence of $\mathbf{u}_t$ on past observation is summarized by $\mathbf{s}_t$ and optimal $\mathbf{u}_t$ will be determined given $\mathbf{s}_t$ and perhaps $\mathbf{y}_t$ without the need of additional knowledge of $\mathbf{y}^{t-1}$ or $\mathbf{u}^{t-1}$. Such a function of observations is called a sufficient statistic.

Means and variances of Gaussian random vectors are sufficient statistics.* Updating formulas for Gaussian probability densities can be obtained by noting recursions or updating relations for these sufficient statistics. For this purpose we use the orthogonal projection formulas we describe next.

Orthogonal Projection and Conditional Expectations

For Gaussian random vectors, conditional expectations are the same as orthogonal projections on the subspace spanned by the conditioning variables. The orthogonal projection, $\hat{E}(\mathbf{x} \mid \mathbf{y})$, is proportional to $\mathbf{y}$ and is such that $\mathbf{x} - \hat{E}(\mathbf{x} \mid \mathbf{y})$ is orthogonal to $\mathbf{y}$, i.e.,

$$\mathbf{x} - \hat{E}(\mathbf{x} \mid \mathbf{y}) \perp \mathbf{y}$$

---

* This notion is a slight modification of the notion of sufficient statistics because $\mathbf{s}_t$ depends on $\mathbf{y}^t$ and on $\mathbf{u}^{t-1}$. For example, in linear dynamics where ξ_t is normally distributed, $E(\mathbf{x}_{t+1} \mid \mathbf{y}^t, \mathbf{u}^t) = \mathbf{A}\mathbf{x}_{t|t} + \mathbf{B}\mathbf{u}_t$ where $\mathbf{x}_{t|t} = E(\mathbf{x}_t \mid \mathbf{y}^t, \mathbf{u}^{t-1})$ and $\mathbf{A} \operatorname{cov}(\mathbf{x}_t \mid \mathbf{y}^t, \mathbf{u}^{t-1})\mathbf{A}' + \operatorname{cov} \xi_t$ are sufficient. Note that the conditional mean depends on $\mathbf{u}_t(\mathbf{y}^t)$ for a given control law. Particular cost structure may admit different statistics, sufficient for the purpose of synthesizing control values. An example illustrating this point is found at the end of this section.

or

$$E\mathbf{x}\mathbf{y}' = E(\hat{E}(\mathbf{x}\,|\,\mathbf{y})\mathbf{y}') = \mathbf{K}\overline{\mathbf{y}\mathbf{y}'},$$

i.e.,

$$\mathbf{K} = E(\mathbf{x}\mathbf{y}')E(\mathbf{y}\mathbf{y}')^{-1}$$

if $\mathbf{x}$ and $\mathbf{y}$ are both mean zero.

When $\mathbf{x}$ and $\mathbf{y}$ are jointly normally distributed, then

$$\hat{E}(\mathbf{x}\,|\,\mathbf{y}) = \mathbf{m}_x + \mathbf{K}(\mathbf{y} - \mathbf{m}_y) \tag{13}$$

where $\mathbf{m}_x$ and $\mathbf{m}_y$ are means of $\mathbf{x}$ and $\mathbf{y}$, and

$$\mathbf{K} = E(\mathbf{x} - \mathbf{m}_x)(\mathbf{y} - \mathbf{m}_y)'[E(\mathbf{y} - \mathbf{m}_y)(\mathbf{y} - \mathbf{m}_y)']^{-1}.$$

The matrix $\mathbf{K}$ is the matrix of regression coefficients of $\mathbf{x}$ on $\mathbf{y}$. Equation (13) is the regression equation for $\mathbf{x}$.

Note that $\mathbf{x} - \hat{E}(\mathbf{x}\,|\,\mathbf{y})$ is uncorrelated with $\mathbf{y}$. When $\mathbf{x}, \mathbf{y}$, and $\mathbf{z}$ are mean zero and jointly normally distributed and when $\mathbf{y}$ and $\mathbf{z}$ are uncorrelated, we can think of $E(\mathbf{x}\,|\,\mathbf{y}, \mathbf{z})$ as $E[\mathbf{x}\,|\,\binom{\mathbf{y}}{\mathbf{z}}]$. Then the matrix $\operatorname{cov}\binom{\mathbf{y}}{\mathbf{z}}$ is block-diagonal and

$$\hat{E}(\mathbf{x}\,|\,\mathbf{y}, \mathbf{z}) = \hat{E}(\mathbf{x}\,|\,\mathbf{y}) + \hat{E}(\mathbf{x}\,|\,\mathbf{z}).$$

Since $\hat{E}(\mathbf{x}\,|\,\mathbf{y}, \mathbf{z}) = \hat{E}(\mathbf{x}\,|\,\mathbf{y}, \mathbf{z} - \hat{E}(\mathbf{z}\,|\,\mathbf{y}))$, because $\mathbf{y}$ and $\mathbf{z}$ and $\mathbf{y}$ and $\mathbf{z} - \hat{E}(\mathbf{z}\,|\,\mathbf{y})$ span the same subspace, we immediately derive a useful formula:

$$\hat{E}(\mathbf{x}\,|\,\mathbf{y}, \mathbf{z}) = \hat{E}(\mathbf{x}\,|\,\mathbf{y}) + \hat{E}(\mathbf{x}\,|\,\mathbf{x} - \hat{E}(\mathbf{z}\,|\,\mathbf{y})).$$

The next two examples apply the orthogonal projection formula to update conditional probability densities rather than by algebraically more tedious direct integration of the relation given in the Appendix.

Example 2. Consider a scalar control system with a dynamic equation

$$x_{t+1} = ax_t + bu_t, \qquad b \neq 0, \qquad u_t \in (-\infty, \infty) \tag{14}$$

$$y_t = x_t + \eta_t, \qquad 0 \leqslant t \leqslant N-1 \tag{15}$$

where a and b are known constants, where ηs are independent random variables with $E(\eta_t) = 0$, $\operatorname{var}(\eta_t) = r_t^2$, $0 \leqslant t \leqslant N-1$, and where x_0 is a random variable independent of everything else.

Let $E(x_t | y^t, u^{t-1}) = \mu_t$ and $\text{var}(x_t | y^t, u^{t-1}) = \sigma_t^2$, $0 \leqslant i \leqslant N$. Since x_{t+1}, y_{t+1} and y^t are jointly Gaussian

$$E(x_{t+1} | y^t, y_{t+1}, u^t) = E(x_{t+1} | y^t, u^t) + E(x_{t+1} | y_{t+1} - y_{t+1|t}, u^t)$$

where the notation

$$y_{t+1|t} = E(y_{t+1} | y^t, u^t)$$

is used.

From the dynamics,

$$E(x_{t+1} | y^t, u^t) = aE(x_t | y^t, u^t) + bu_t = a\mu_t + bu_t.$$

Since η_{t+1} is independent of y^t and u^t,

$$y_{t+1|t} = a\mu_t + bu_t,$$

i.e.,

$$\mu_{t+1} = a\mu_t + bu_t + K_{t+1}(y_{t+1} - y_{t+1|t})$$

because $E(x_{t+1} | y_{t+1} - y_{t+1|t})$ is proportional to $y_{t+1} - y_{t+1|t}$ with the constant of proportionality denoted by K_{t+1}. A quick way to determine K_{t+1} is to use the fact that K_{t+1} minimizes the conditional covariance $\text{cov}(x_{t+1} | y^{t+1}, u^t)$. From the dynamic equation and the expression for the conditional mean we obtain

$$x_{t+1} - \mu_{t+1} = a(1 - K_{t+1})(x_t - \mu_t) - K_{t+1}\eta_{t+1}$$

and

$$\sigma_{t+1}^2 = a^2(1 - K_{t+1})^2\sigma_t^2 + K_{t+1}^2 r_{t+1}^2.$$

The best K_{t+1} is then equal to

$$K_{t+1} = a^2\sigma_t^2/(a^2\sigma_t^2 + r_{t+1}^2),$$

and substitution of this expression in the preceding relation yields the recursion for the conditional variances

$$\sigma_{t+1}^2 = a^2\sigma_t^2 r_{t+1}^2/(a^2\sigma_t^2 + r_{t+1}^2)$$

or

$$1/\sigma_{t+1}^2 = 1/r_{t+1}^2 + 1/a^2\sigma_t^2, \qquad t \geqslant 1.$$

From $y_0 = x_0 + \eta_0$,

$$\mu_0 = E(x_0 \mid y_0) = (\alpha/\sigma^2 + y_0/r_0^2)/\sigma_0^2$$

where

$$1/\sigma_0^2 = 1/\sigma^2 + 1/r_0^2.$$

When J is given, λ_t is computable sincce $p(x_t \mid y^t)$ is known as a Gaussian probability density function with the conditional mean μ_t and the conditional variance σ_t^2. The conditional probability density $p(y_{t+1} \mid y^t, u^t)$ needed to compute $E(\gamma_{t+2}^* \mid y^t)$ is obtained as follows:

From Eq. (15),

$$x_t = y_t - \eta_t,$$

hence

$$\mu_t = E(x_t \mid y^t) = y_t - E(\eta_t \mid y^t)$$

or

$$E(\eta_t \mid y^t) = y_t - \mu_t.$$

Similarly,

$$\operatorname{var}(\eta_t \mid y^t) = \operatorname{var}(x_t \mid y^t) = \sigma_t^2.$$

From Eqs. (14) and (15),

$$y_{t+1} = ay_t + bu_t + \eta_{t+1} - a\eta_t$$

where y_{t+1} is a Gaussian random variable since it is a linear combination of Gaussian random variables. Now

$$\begin{aligned} E(y_{t+1} \mid y^t, u^t) &= ay_t + bu_t + E(\eta_{t+1} \mid y^t) - aE(\eta_t \mid y^t) \\ &= ay_t + bu_t - a(y_t - \mu_t) \qquad (16) \\ &= a\mu_t + bu_t \end{aligned}$$

because

$$E(\eta_{t+1} \mid y^t) = E(\eta_{t+1}) = 0.$$

Also,

$$[y_{t+1} - (a\mu_t - bu_t)]^2 = [-a(\eta_t - y_t + \mu_t) + \eta_{t+1}]^2,$$

hence

$$\operatorname{var}(y_{t+1} \mid y^t, u^t) = a^2\sigma_t^2 + r_{t+1}^2. \qquad (17)$$

Equations (16) and (17) determine $p(y_{t+1} \mid y^t, u^t)$ completely.

Example 3. The model is given by

$$\mathbf{x}_{t+1} = \mathbf{A}\mathbf{x}_t + \boldsymbol{\xi}_t \tag{18}$$

$$\mathbf{y}_t = \mathbf{C}\mathbf{x}_t + \boldsymbol{\eta}_t \tag{19}$$

where $\mathbf{x_0}$, $\boldsymbol{\xi}$s and $\boldsymbol{\eta}$s are all independent,

$$\mathbf{x_0} \sim N(\mathbf{0}, \boldsymbol{\Sigma}_0)$$

$$\boldsymbol{\xi}_t \sim N(\mathbf{0}, \mathbf{Q}),$$

and

$$\mathbf{y}_t \sim N(\mathbf{0}, \mathbf{R}).$$

Taking the conditional expectation of (18) with respect to $\mathbf{y}^t$ yields

$$\mathbf{x}_{t+1|t} = \mathbf{A}\mathbf{x}_{t|t}, \tag{20}$$

since $\boldsymbol{\xi}_t$ is independent of $\mathbf{y}^t$. From (18) and (20),

$$\boldsymbol{\Sigma}_{t+1|t} = \mathbf{A}\boldsymbol{\Sigma}_{t|t}\mathbf{A}' + \mathbf{Q}.$$

The conditional expectation of (19) with respect to $\mathbf{y}^{t-1}$ is

$$\mathbf{y}_{t|t-1} = \mathbf{C}\mathbf{x}_{t|t-1}, \tag{21}$$

because $\mathbf{y}_t$ is independent of $\mathbf{y}^{t-1}$. Now note that

$$\begin{aligned} \mathbf{x}_{t+1|t+1} &= \hat{E}(\mathbf{x}_{t+1} | \mathbf{y}^t, \mathbf{y}_{t+1}) \\ &= \hat{E}(\mathbf{x}_{t+1} | \mathbf{y}^t, \mathbf{y}_{t+1} - \mathbf{y}_{t+1|t}) \\ &= \mathbf{x}_{t+1|t} + \hat{E}(\mathbf{x}_{t+1} | \mathbf{y}_{t+1} - \mathbf{y}_{t+1|t}), \end{aligned} \tag{22}$$

where the second term is equal to

$$E[\mathbf{x}_{t+1}(\mathbf{y}_{t+1} - \mathbf{y}_{t+1|t})'] \operatorname{cov}(\mathbf{y}_{t+1|t} - \mathbf{y}_{t+1|t})^{-1}(\mathbf{y}_{t+1} - \mathbf{y}_{t+1|t}). \tag{23}$$

From (19) and (21),

$$\mathbf{y}_t - \mathbf{y}_{t|t-1} = \mathbf{C}(\mathbf{x}_t - \mathbf{x}_{t|t-1}) + \boldsymbol{\eta}_t$$

and hence

$$\operatorname{cov}(\mathbf{y}_t - \mathbf{y}_{t|t-1}) = \mathbf{C}\boldsymbol{\Sigma}_{t|t-1}\mathbf{C}' + \mathbf{R}.$$

Denote this by $\boldsymbol{\Delta}_t$. The first factor is

$$\begin{aligned} E[\mathbf{x}_t(\mathbf{y}_t - \mathbf{y}_{t|t-1})'] &= E[\mathbf{x}_t(\mathbf{x}_t - \mathbf{x}_{t|t-1})']\mathbf{C}' \\ &= \boldsymbol{\Sigma}_{t|t-1}\mathbf{C}' \end{aligned}$$

since $\boldsymbol{\eta}_t$ is independent of $\mathbf{x}_t$ and $\mathbf{x}_{t|t-1}$. Thus (23) is equal to $\mathbf{K}_{t+1}(\mathbf{y}_{t+1} - \mathbf{y}_{t+1|t})$ where

$$\mathbf{K}_{t+1} = \boldsymbol{\Sigma}_{t+1|t}\mathbf{C}'\boldsymbol{\Delta}_{t+1}^{-1}.$$

Combining (23) with (22) we obtain the recursion

$$\mathbf{x}_{t+1|t+1} = \mathbf{x}_{t+1|t} + \mathbf{K}_{t+1}(\mathbf{y}_{t+1} - \mathbf{y}_{t+1|t}),$$

and subtracting the preceding from $\mathbf{x}_{t+1}$, we calculate

$$\boldsymbol{\Sigma}_{t+1|t+1} = \boldsymbol{\Sigma}_{t+1|t} - \boldsymbol{\Sigma}_{t+1|t}\mathbf{C}'\boldsymbol{\Delta}^{-1}\mathbf{C}\boldsymbol{\Sigma}_{t+1|t}.$$

Summarizing, we obtain by assumption of normality the density functions

$$p(\mathbf{x}_t|\mathbf{y}^{t-1}) \sim N(\mathbf{x}_{t|t-1}, \boldsymbol{\Sigma}_t),$$
$$p(\mathbf{x}_{t+1}|\mathbf{x}_t) \sim N(\mathbf{A}\mathbf{x}_t, \mathbf{Q}),$$
$$p(\mathbf{y}_t|\mathbf{x}_t) \sim N(\mathbf{C}\mathbf{x}_t, \mathbf{R}),$$

and

$$p(\mathbf{x}_{t+1}|\mathbf{y}^t) \sim N(\mathbf{x}_{t+1|t}, \boldsymbol{\Sigma}_{t+1})$$

where

$$\mathbf{x}_{t+1|t} = \mathbf{A}\mathbf{x}_{t|t-1} + \mathbf{K}_t(\mathbf{y}_t - \mathbf{C}\mathbf{x}_{t|t-1}).$$

Example 4. The next example illustrates the notion of statistics for a given cost structure. The problem is to minimize

$$J = \mathbf{x}_T'\mathbf{Q}\mathbf{x}_T + \sum_0^{T-1} \mathbf{u}_t'\mathbf{R}\mathbf{u}_t \tag{24}$$

where $\mathbf{Q}' = \mathbf{Q} \geqslant 0$ and $\mathbf{R}' = \mathbf{R} > 0$. Here $\mathbf{x}_T$ or $\mathbf{H}\mathbf{x}_T$ where $\mathbf{H}'\mathbf{H} = \mathbf{Q}$ may model "miss" in achieving a set of target values at the end of a T-period plan for a social planner. (The model may be set up in deviational terms so that $\mathbf{H}\mathbf{x}_T = 0$ represents achieving target levels.) The dynamics and the observation scheme are

$$\mathbf{x}_{t+1} = \mathbf{A}\mathbf{x}_t + \mathbf{B}\mathbf{u}_t + \boldsymbol{\xi}_t$$

and

$$\mathbf{y}_t = \mathbf{C}\mathbf{x}_t + \boldsymbol{\eta}_t$$

where $\boldsymbol{\xi}_t$ and $\boldsymbol{\eta}_t$ are mean-zero mutually and serially independent Gaussian random variables. With the cost structure (24), the last control minimizes

$$\gamma_T = E[(\mathbf{x}_T'\mathbf{Q}\mathbf{x}_T + \mathbf{u}_{T-1}'\mathbf{R}\mathbf{u}_{T-1})|\mathbf{y}^{T-1}, \mathbf{u}^{T-1}].$$

Since $\mathbf{x}_T$ is Gaussian with conditional mean

$$\mathbf{x}_{T|T-1} = \mathbf{A}\mathbf{x}_{T-1|T-1} + \mathbf{B}\mathbf{u}_{T-1}$$

and covariance $\boldsymbol{\Sigma}_{T|T-1} = \mathbf{A}\boldsymbol{\Sigma}_{T-1|T-1}\mathbf{A}' + \operatorname{cov}\xi_{T-1}$, γ_T is minimized by minimizing $\mathbf{x}'_{T|T-1}\mathbf{A}\mathbf{x}_{T|T-1} + \mathbf{u}'_{T-1}\mathbf{R}\mathbf{u}_{T-1}$, i.e., by

$$\mathbf{u}^*_{T-1} = -(\mathbf{R} + \mathbf{B}'\mathbf{Q}\mathbf{B})^{-1}\mathbf{B}'\mathbf{Q}\mathbf{x}_{T|T-1}$$

and

$$\gamma^*_T = \min \gamma_T = \mathbf{x}'_{T|T-1}\mathbf{Q}_{T-1}\mathbf{x}_{T|T-1} + c_T$$

where c_T is a constant, independent of $\mathbf{u}$s and

$$\mathbf{Q}_{T-1} = \mathbf{A}'[\mathbf{Q} - \mathbf{Q}\mathbf{B}(\mathbf{R} + \mathbf{B}'\mathbf{Q}\mathbf{B})^{-1}\mathbf{B}'\mathbf{Q}]\mathbf{A} = \mathbf{A}'(\mathbf{Q}^{-1} + \mathbf{B}\mathbf{R}^{-1}\mathbf{B}')^{-1}\mathbf{A}.$$

In evaluating

$$E(\gamma^*_T|\mathbf{y}^{T-2}, \mathbf{u}^{T-2}) = \int \gamma^*_T p(\mathbf{y}_{T-1}|\mathbf{y}^{T-2}, \mathbf{u}^{T-2})\,d\mathbf{y}_{T-1},$$

the expression $\mathbf{x}_{T|T-1}$ in γ^*_T is replaced by

$$E(\mathbf{x}_{T|T-1}|\mathbf{y}^{T-2}, \mathbf{u}^{T-2}) = \mathbf{A}^2\mathbf{x}_{T-2|T-2} + \mathbf{A}\mathbf{B}\mathbf{u}_{T-2} = \mathbf{A}\mathbf{x}_{T-1|T-2}$$

and the first-order condition for the minimum of $\gamma_{T-1} = \lambda_{T-1} + E(\gamma^*_T|\mathbf{y}^{T-2}, \mathbf{u}^{T-2})$ is

$$\mathbf{R}\mathbf{u}_{T-2} + \mathbf{B}'\mathbf{A}'\mathbf{Q}_{T-1}\mathbf{A}(\mathbf{A}\mathbf{x}_{T-2|T-2} + \mathbf{B}\mathbf{u}_{T-2}) = 0$$

or

$$\mathbf{u}_{T-2} = -(\mathbf{R} + \mathbf{B}'\mathbf{Q}_{T-1}\mathbf{B})^{-1}\mathbf{B}'\mathbf{Q}_{T-1}\mathbf{A}\mathbf{x}_{T-2|T-2}$$

and

$$\gamma^*_{T-1} = \min_{u_{T-2}} \gamma_{T-1} = \mathbf{x}'_{T-2|T-2}\mathbf{Q}_{T-2}\mathbf{x}_{T-2|T-2} + c_{T-1}$$

where c_{T-1} is a constant

$$\begin{aligned}\mathbf{Q}_{T-2} &= \mathbf{Q}'[\mathbf{A}_{T-1} - \mathbf{Q}_{T-1}\mathbf{B}(\mathbf{R} + \mathbf{B}')^{-1}\mathbf{B}'\mathbf{Q}_{T-1}]\mathbf{A} \\ &= \mathbf{A}'(\mathbf{Q}^{-1}_{T-1} + \mathbf{B}\mathbf{R}^{-1}\mathbf{B}')^{-1}\mathbf{A},\end{aligned}$$

where $\mathbf{Q}^{-1}_{T-1}$ is assumed to exist. In general, we have

$$\gamma_t = \lambda_t + E(\gamma^*_{t+1}|\mathbf{y}^{t-1}, \mathbf{u}^{t-1})$$

where

$$\gamma_{t+1}^{*} = \mathbf{x}_{t|t}' \mathbf{Q}_t \mathbf{x}_{k|t} + c_{t+1}$$

and

$$\lambda_t = \mathbf{u}_{t-1}' \mathbf{R} \mathbf{u}_{t-1}.$$

Since $E(\mathbf{x}_{t|t} | \mathbf{y}^{t-1}, \mathbf{u}^{t-1}) = \mathbf{A}\mathbf{x}_{t-1|t-1} + \mathbf{B}\mathbf{u}_{t-1}$,

$$\gamma_t = \mathbf{u}_{t-1}' \mathbf{R} \mathbf{u}_{t-1} + E(\gamma_{t+1}^{*} | \mathbf{y}^{t-1}, \mathbf{u}^{t-1})$$

where

$$E(\gamma_{t+1}^{*} | \mathbf{y}^{t-1}, \mathbf{u}^{k-1}) = (\mathbf{A}\mathbf{x}_{t-1|t-1} + \mathbf{B}\mathbf{u}_{t-1})' \mathbf{Q}_t (\mathbf{A}\mathbf{x}_{t-1|t-1} + \mathbf{B}\mathbf{u}_{t-1}) + c_t.$$

The optimal decision is

$$\mathbf{u}_{t-1}^{*} = -(\mathbf{R} + \mathbf{B}'\mathbf{Q}_t\mathbf{B})^{-1} \mathbf{B}'\mathbf{Q}_t \mathbf{A} \mathbf{x}_{t-1|t-1}$$

and

$$\gamma_t^{*} = \mathbf{x}_{t-1|t-1}' \mathbf{Q}_{t-1} \mathbf{x}_{t-1|t-1} + c_t$$

where

$$\mathbf{Q}_{t-1} = \mathbf{A}'[\mathbf{Q}_t - \mathbf{Q}_t\mathbf{B}(\mathbf{R} + \mathbf{B}'\mathbf{Q}_t\mathbf{T})^{-1}\mathbf{B}'\mathbf{Q}_t]\mathbf{A} = \mathbf{A}'[\mathbf{Q}_t^{-1} + \mathbf{B}\mathbf{R}^{-1}\mathbf{B}']^{-1}\mathbf{A}.$$

We note that $\mathbf{u}_t^{*}$ is proportional to $\hat{\mathbf{x}}_{T|t}$ where

$$\hat{\mathbf{x}}_{T|t} = \mathbf{A}^{T-k} \mathbf{x}_{t|t}.$$

This is a sufficient statistic for the cost structure of this example. This $\hat{\mathbf{x}}_{T|t}$ is conditional expectation of $\mathbf{x}_T$, given $\mathbf{y}^t, \mathbf{u}^t$ on the assumption that $\mathbf{u}_{t+1} = \ldots = \mathbf{u}_{T-1} = 0$.

Let $\mathbf{s}_t = \mathbf{A}^{T-k} \mathbf{x}_{t|t}$. It is computed by

$$\begin{aligned} \mathbf{s}_{t+1} &= \mathbf{A}^{T-k-1} \mathbf{x}_{t+1|t+1} \\ &= \mathbf{s}_t + \mathbf{A}^{T-k-1}[\mathbf{B}\mathbf{u}_t + \mathbf{K}_{t+1}(\mathbf{y}_{t+1} - \mathbf{C}\mathbf{A}\mathbf{x}_{t|t} - \mathbf{C}\mathbf{B}\mathbf{u}_t)], \end{aligned}$$

it has a normal distribution with mean $\hat{\mathbf{x}}_{T|t} \mathbf{A}^{T-t-1} \mathbf{B}\mathbf{u}_t$ and the covariance matrix

$$\operatorname{cov} \hat{\mathbf{x}}_{T|t+1} = \mathbf{A}^{T-t-1} \mathbf{K}_{t+1} \operatorname{cov}(\mathbf{y}_{t+1} | \mathbf{y}^t, \mathbf{u}^t) \mathbf{K}_{t+1}' (\mathbf{A})^{T-t-1}.$$

Note that $\mathbf{x}_{t|t}$ and $\mathbf{s}_t$ jointly satisfy the Markovian properties, i.e., their updated equations involve only $\mathbf{y}_{t-1}$ and $\mathbf{u}_t$ not $\mathbf{y}^{t+1}$ and $\mathbf{u}^t$.

3.4 Finite Horizon Problems

The cost index is taken to be

$$J = \sum_{k=1}^{N} \mathrm{W}_k(\mathbf{x}_k, \mathbf{u}_{k-1}), \qquad W_k \geqslant 0. \tag{25}$$

This form of cost index is fairly general.

Although in a more general formulation the set of admissible controls at time k, U_k, will depend on $\mathbf{x}^k$ and $\mathbf{u}^{k-1}$. U_k is assumed to be independent of $\mathbf{x}^k, \mathbf{u}^{k-1}$ in this book.

From (25), using $E(\cdot)$ to denote the expectation operation, the expected value of J is evaluated as

$$EJ = E\left(\sum_{k=1}^{N} W_k\right) = \sum_{1}^{N} R_k$$

where

$$R_k \triangleq E(W_k).$$

We evaluate R_k not directly, as Feld'baum did, but as a sequence of its conditional means

$$E(W_k \mid \mathbf{y}^{k-1}, \mathbf{u}^{k-2}) = \int \mathrm{W}_k(\mathbf{x}_k, \mathbf{u}_{k-1}) p(\mathbf{x}_k, \mathbf{u}_{k-1} \mid \mathbf{y}^{k-1}, \mathbf{u}^{k-2})\, d\mathbf{x}_k\, d\mathbf{u}_{k-1},$$

and generate

$$p(\mathbf{x}_k \mid \mathbf{y}^k, \mathbf{u}^{k-1}) \quad \text{and} \quad p(\mathbf{y}_{k+1} \mid \mathbf{y}^k, \mathbf{u}_k), \qquad 0 \leqslant k \leqslant N-1$$

recursively.

Derivation of Optimal Control Policies

We will now derive a general formula to obtain optimal control policies. At this point, we must look for optimal control policies from the class of closed-loop randomized control policies and then show that nonrandomized policies can achieve the same minimum.

Last Stage

Consider the last stage of control, assuming $\mathbf{y}^{N-1}$ has been observed and $\mathbf{u}^{N-2}$ has been determined somehow and that only the last control variable, $\mathbf{u}_{N-1}$, remains to be specified. Since $\mathbf{u}_{N-1}$ appears only in W_N, EJ is minimized with respect to $\mathbf{u}_{N-1}$ by minimizing EW_N with respect to $\mathbf{u}_{N-1}$. Since

$$R_N = E(W_N) = E[E(W_N \mid \mathbf{y}^{N-1}, \mathbf{u}^{N-2})]$$

where the outer expectation is with respect to $\mathbf{y}^{N-1}$ and $\mathbf{u}^{N-2}$, R_N is minimized if $E(W_N|\mathbf{y}^{N-1}, \mathbf{u}^{N-2})$ is minimized for every $\mathbf{y}^{N-1}$ and $\mathbf{u}^{N-2}$.

One can write

$$E(W_N|\mathbf{y}^{N-1}, \mathbf{u}^{N-2}) = \int W_N(\mathbf{x}_N, \mathbf{u}_{N-1}) p(\mathbf{x}_N, \mathbf{u}_{N-1}|\mathbf{y}^{N-1}, \mathbf{u}^{N-2}) d(\mathbf{x}_N, \mathbf{u}_{N-1}). \tag{26}$$

By the chain rule, the probability density in (26) can be written as

$$\begin{aligned} p(\mathbf{x}_N, \mathbf{u}_{N-1}|\mathbf{y}^{N-1}, \mathbf{u}^{N-2}) &= p(\mathbf{u}_{N-1}|\mathbf{y}^{N-1}, \mathbf{u}^{N-2}) \\ &\quad \times p(\mathbf{x}_N|\mathbf{u}^{N-1}, \mathbf{y}^{N-1}) \end{aligned} \tag{27}$$

where

$$\begin{aligned} p(\mathbf{x}_N|\mathbf{u}^{N-1}, \mathbf{y}^{N-1}) &= \int p(\mathbf{x}_N|\mathbf{x}_{N-1}, \mathbf{u}^{N-1}, \mathbf{y}^{N-1}) \\ &\quad \times p(\mathbf{x}_{N-1}|\mathbf{u}^{N-1}, \mathbf{y}^{N-1})\, d\mathbf{x}_{N-1}. \end{aligned} \tag{28}$$

If the ξs and $\boldsymbol{\eta}$s are mutually independent and for each t, i.e., if $\mathbf{x}_0$, $\xi_0, \xi_1, \ldots, \xi_{N-1}, \boldsymbol{\eta}_0, \ldots, \boldsymbol{\eta}_{N-1}$ are all independent, then

$$\begin{aligned} p(\mathbf{x}_{t+1}|\mathbf{x}_t, \mathbf{y}^t, \mathbf{u}^t) &= p(\mathbf{x}_{t+1}|\mathbf{x}_t, \mathbf{u}_t) \\ p(\mathbf{y}_t|\mathbf{x}^t, \mathbf{y}^{t-1}, \mathbf{u}^{t-1}) &= p(\mathbf{y}_t|\mathbf{x}_t) \qquad 0 \leqslant t \leqslant k-1. \end{aligned}$$

We will use Eq. (28) throughout this section. Developments are quite similar when this Markov property does not hold if suitably augmented. See Chapter 7 for more general discussions of the Markov property.

In particular, in (27),

$$p(\mathbf{x}_N|\mathbf{x}_{N-1}, \mathbf{u}^{N-1}, \mathbf{y}^{N-1}) = p(\mathbf{x}_N|\mathbf{x}_{N-1}, \mathbf{u}_{N-1}) \tag{29}$$

and

$$p(\mathbf{x}_{N-1}|\mathbf{u}^{N-1}, \mathbf{y}^{N-1}) = p(\mathbf{x}_{N-1}|\mathbf{y}^{N-1}, \mathbf{u}^{N-2}) \tag{30}$$

since $\mathbf{u}_{N-1}$ affects $\mathbf{x}_N$ but not $\mathbf{x}_{N-1}$. Define

$$p(\mathbf{u}_{N-1}|\mathbf{y}^{N-1}, \mathbf{u}^{N-2}) \triangleq \rho_{N-1}(\mathbf{u}_{N-1}).$$

Therefore, if one asumes that (30) is available, then (26) can be written as

$$E(W_N|\mathbf{y}^{N-1}, \mathbf{u}^{N-2}) = \int \lambda_N \rho_{N-1}\, d\mathbf{u}_{N-1} \tag{31}$$

where

$$\begin{aligned} \lambda_N \triangleq \int &\mathrm{W}_N(\mathbf{x}_N, \mathbf{u}_{N-1}) p(\mathbf{x}_N|\mathbf{x}_{N-1}, \mathbf{u}_{N-1}) \\ &\times p(\mathbf{x}_{N-1}|\mathbf{y}^{N-1}, \mathbf{u}^{N-2}) d(\mathbf{x}_N, \mathbf{x}_{N-1}). \end{aligned} \tag{32}$$

In (32), the probability density $p(\mathbf{x}_N | \mathbf{x}_{N-1}, \mathbf{u}_{N-1})$ is obtainable from the known probability density function for ξ_{N-1} and the model dynamics.

The second probability density in Eq. (32), $p(\mathbf{x}_{N-1} | \mathbf{y}^{N-1}, \mathbf{u}^{N-2})$, is not generally directly available. It will be shown in the next section how it can be generated. For the moment, assume that it is available.

Thus λ_N is, in principle, computable as a function of $\mathbf{y}^{N-1}$ and $\mathbf{u}^{N-1}$. Therefore its minimum with respect to u_{N-1}, assumed to exist, can in principle be found. Denote this minimizing $\mathbf{u}_{N-1}$ by $\mathbf{u}^*_{N-1}$. Define

$$\min_{\rho_{N-1}} E(W_N | \mathbf{y}^{N-1}, \mathbf{u}^{N-2}) = \gamma_N{}^*. \tag{33}$$

Thus, the minimization of EW_N with respect to ρ_{N-1} is accomplished by that of $E(W_N | \mathbf{y}^{N-1}, \mathbf{u}^{N-2})$, which is achieved by taking $\rho_{N-1} = \delta(\mathbf{u}_{N-1} - \mathbf{u}^*_{N-1})$. Since λ_N is a function of $\mathbf{y}^{N-1}$ and $\mathbf{u}^{N-1}$, $\mathbf{u}^*_{N-1}$ is obtained as a function of $\mathbf{y}^{N-1}$ and $\mathbf{u}^{N-2}$ as desired.

In Eq. (31) the expression $\rho_{N-1}(\mathbf{u}_{N-1})$ represents a probability density function of $\mathbf{u}_{N-1} \in U_{N-1}$, where the functional form of the density function depends on the history of observation or on $\mathbf{y}^{N-1}$. The functional form of ρ_{N-1} specifies the probability $\rho_{N-1}(\mathbf{u}_{N-1})\, d\mathbf{u}_{N-1}$ with which a control in the neighborhood of a point u_{N-1} is used in the last control stage.

However, we have seen that this generality is not necessary, at least for the last control $\mathbf{u}_{N-1}$, and we can actually confine our search for the optimal $\mathbf{u}_{N-1}$ to the class of nonrandomized control policies; i.e., the value of the optimal control vector $\mathbf{u}_{N-1}$ will actually be determined, given $\mathbf{y}^{N-1}$, and it is not that merely the form of the probability density will be determined. We can see by similar arguments that the u_i are all nonrandomized, $0 \leqslant i \leqslant N-1$. Thus, we can remove $\mathbf{u}_{N-1}$ from the probability density function in Eq. (26), and we can deal with $p(\mathbf{x}_N | \mathbf{y}^{N-1})$ with the understanding that $\mathbf{u}_{N-1}$ is uniquely determined by $\mathbf{y}^{N-1}$.

When ξ_{N-1} is Gaussian, the dynamics are linear, and W_N is quadratic in $\mathbf{x}_N$ and $\mathbf{u}_{N-1}$, Eq. (32) is easily evaluated because the probability density of $\mathbf{x}_N$ conditional on $\mathbf{x}_{N-1}$ and $\mathbf{u}_{N-1}$ is normal with the conditional mean $\mathbf{x}_{N|N-1}$ and conditional variance $\mathbf{\Pi}_N$ which is equal to $\mathbf{A\Pi}_{N-1}\mathbf{A}' + \operatorname{cov} \xi_{N-1}$. Eq. (32) becomes

$$\lambda_N = \mathbf{x}'_{N|N-1}\mathbf{Q}\mathbf{x}_{N|N-1} + \mathbf{u}'_{N-1}\mathbf{R}\mathbf{u}_{N-1} + \operatorname{tr}[\mathbf{Q} \operatorname{cov}(\mathbf{x}_N | \mathbf{y}^{N-1}, \mathbf{u}^{N-1})].$$

---

* If $\mathbf{u}^*_{N-1}$ is not unique, then the following arguments must be modified slightly. By choosing any one control which minimizes λ_N and concentrating the probability mass one there, a nonrandomized control still results.

The Last Two Stages

Putting aside, for the moment, the question of how to evaluate $p(\mathbf{x}_{N-1}|\mathbf{y}^{N-1})$, let us proceed next to the consideration of optimal control policies for the last two stages of the process. Assume that $\mathbf{y}^{N-2}$ and $\mathbf{u}^{N-3}$ are given. The control variable $\mathbf{u}_{N-2}$ appears in W_{N-1} and W_N. Since

$$E[W_{N-1}(\mathbf{x}_{N-1}, \mathbf{u}_{N-2}) + W_N(\mathbf{x}_N, \mathbf{u}_{N-1})] = E[E(W_{N+1} + W_N|\mathbf{y}^{N-2}, \mathbf{u}^{N-3})]$$

where the outer expectation is with respect to $\mathbf{y}^{N-2}$, and since a choice of $\mathbf{u}_{N-2}$ transforms the problem into the last-stage situation just considered, EJ is minimized by choosing $\mathbf{u}_{N-2}$ such that it minimizes $E(W_{N-1} + W_N|\mathbf{y}^{N-2}, \mathbf{u}^{N-3})$ for every $\mathbf{y}^{N-2}$ and by following this $\mathbf{u}_{N-2}$ by $\mathbf{u}_{N-1}^*$. Analogous to (31) we have

$$E(W_{N-1}|\mathbf{y}^{N-2}, \mathbf{u}^{N-3}) = \int \lambda_{N-1}\rho_{N-2}\, d\mathbf{u}_{N-2} \tag{34}$$

where

$$\rho_{N-2}(\mathbf{u}_{N-2}) = p(\mathbf{u}_{N-2}|\mathbf{y}^{N-2}, \mathbf{u}^{N-3})$$

and where

$$\begin{aligned} \lambda_{N-1} \triangleq \int &\mathrm{W}_{N-1}(\mathbf{x}_{N-1}, \mathbf{u}_{N-2})p(\mathbf{x}_{N-1}|\mathbf{x}_{N-2}, \mathbf{u}_{N-2}) \\ &\times p(\mathbf{x}_{N-2}|\mathbf{y}^{N-2}, \mathbf{u}^{N-3})\, d(\mathbf{x}_{N-1}, \mathbf{x}_{N-2}). \end{aligned} \tag{35}$$

Also

$$E(W_N|\mathbf{y}^{N-2}, \mathbf{u}^{N-3}) = E[E(W_N|\mathbf{y}^{N-1}, \mathbf{u}^{N-2})|\mathbf{y}^{N-2}, \mathbf{u}^{N-3}]$$

since the information set at time $(N-2)$ is included in that at time $(N-1)$. This is seen also from

$$\begin{aligned} p(\cdot\,|\mathbf{y}^{N-2}, \mathbf{u}^{N-3}) = \int &p(\cdot\,|\mathbf{y}^{N-1}, \mathbf{u}^{N-2})p(\mathbf{y}_{N-1}|\mathbf{y}^{N-2}, \mathbf{u}^{N-2}) \\ &\times \rho(\mathbf{u}_{N-2})\, d(\mathbf{y}_{N-1}, \mathbf{u}_{N-2}). \end{aligned}$$

The optimal ρ_{N-2} is such that it minimizes $E(W_{N-1} + W_N^*)$ where the asterisk on W_N indicates that $\mathbf{u}_{N-1}^*$ is used for the last control. Now,

$$\begin{aligned} \min_{\rho_{N-2}} E(W_{N-1} + W_N^*|\mathbf{y}^{N-2}, \mathbf{u}^{N-3}) &= \min_{\rho_{N-2}} [E(W_{N-1}|\mathbf{y}^{N-2}, \mathbf{u}^{N-3}) \\ &\qquad + E(W_N^*|\mathbf{y}^{N-2}, \mathbf{u}^{N-3})] \\ &= \min_{\rho_{N-2}} E[W_{N-1} \\ &\qquad + E(W_N^*|\mathbf{y}^{N-1}, \mathbf{u}^{N-2})|\mathbf{y}^{N-2}, \mathbf{u}^{N-3}] \end{aligned}$$

$$
\begin{aligned}
(36) \qquad &= \min_{\rho_{N-2}} E[W_{N-1} \\
&\quad + \gamma_N^* | \mathbf{y}^{N-2}, \mathbf{u}^{N-3}] \\
&= \min_{\rho_{N-2}} \int \Bigg[\lambda_{N-1} \\
&\quad + \int \gamma_N^* p(\mathbf{y}_{N-1} | \mathbf{u}^{N-2}, \mathbf{y}^{N-2})\, d\mathbf{y}_{N-1} \Bigg] \\
&\quad \times \rho_{N-2}\, d\mathbf{u}_{N-2}
\end{aligned}
$$

where it is assumed that $p(\mathbf{y}_{N-1} | \mathbf{u}^{N-2}, \mathbf{y}^{N-2})$ is available. Defining γ_{N-1} by

$$\gamma_{N-1} = \lambda_{N+1} + \int \gamma_N^* p(\mathbf{y}_{N-1} | \mathbf{y}^{N-2}, \mathbf{u}^{N-2})\, d\mathbf{y}_{N-1},$$

Equation (36) is written as

$$\min_{\rho_{N-2}} E(W_{N-1} + W_N^* | \mathbf{y}^{N-2}, \mathbf{u}^{N-3}) = \min_{\rho_{N-2}} \int \gamma_{N-1} \rho_{N-2}\, d\mathbf{u}_{N-2}.$$

Comparing this with Eq. (31), it is seen that the optimal control is such that $\rho_{N-2}^* = \delta(\mathbf{u}_{N-2} - \mathbf{u}_{N-2}^*)$, where $\mathbf{u}_{N-2}^*$ is $\mathbf{u}_{N-2}$ which minimizes γ_{N-1}, and the control at the $(N-2)$th stage is also nonrandomized.

General Case

Generally, $E(\Sigma_{k+1}^N W_i)$ is minimized by minimizing $E(\Sigma_{k+1}^N W_i | \mathbf{y}^k, \mathbf{u}^{k-1})$ with respect to ρ_k for each $\mathbf{y}^k, \mathbf{u}^{k-1}$ and following it with $\rho_{k+1}^*, \ldots, \rho_{N-1}^*$. It should now be clear that arguments quite similar to those employed when deriving ρ_{N-1}^* and ρ_{N-2}^* can be used to determine ρ_k^*. Define

$$
\begin{aligned}
(37) \qquad \gamma_k &= \lambda_k + \int \gamma_{k+1}^* p(\mathbf{y}_k | \mathbf{y}^{k-1}, \mathbf{u}^{k-1})\, d\mathbf{y}_k, \qquad 1 \leqslant k \leqslant N \\
\gamma_{N+1}^* &\equiv 0
\end{aligned}
$$

where $p(\mathbf{y}_k | \mathbf{y}^{k-1}, \mathbf{u}^{k-1})$ is assumed available and where λ_k is given, assuming $p(\mathbf{x}_{k-1} | \mathbf{y}^{k-1}, \mathbf{u}^{k-2})$ is available, by

$$
\begin{aligned}
(38) \qquad \lambda_k = \int & \mathrm{W}_k(\mathbf{x}_k, \mathbf{u}_{k-1}) p(\mathbf{x}_k | \mathbf{x}_{k-1}, \mathbf{u}_{k-1}) \\
& \times p(\mathbf{x}_{k-1} | \mathbf{y}^{k-1}, \mathbf{u}^{k-2})\, d(\mathbf{x}_k, \mathbf{x}_{k-1}), \qquad 1 \leqslant k \leqslant N.
\end{aligned}
$$

Then the optimal control at time $k-1$, $\mathbf{u}_{k-1}^*$, is $\mathbf{u}_{k-1}$, which minimizes γ_k:

$$(39) \qquad \min_{u_{k-1}} \gamma_k = \gamma_k^*, \qquad 1 \leqslant k \leqslant N$$

By computing γ_k recursively, optimal control variables are derived in the order of $\mathbf{u}_{N-1}^*, \mathbf{u}_{N-2}^*, \ldots, \mathbf{u}_0^*$. Once the optimal control policy is derived, these optimal control variables are used, of course, in the order of time $\mathbf{u}_0^*$, $\mathbf{u}_1^*, \ldots, \mathbf{u}_{N-1}^*$. The conditional probability densities assumed available in connection with (37) and (38) are derived in the next section.

At each time k, $\mathbf{u}_0^*, \ldots, \mathbf{u}_{k-1}^*$ and $\mathbf{y}_0, \ldots, \mathbf{y}_k$ are *known*. Therefore, $\mathbf{u}_k^*$ is determined definitely since

$$\mathbf{u}_k^* = \pi_k(\mathbf{u}_0^*, \ldots, \mathbf{u}_{k-1}^*, \mathbf{y}_0, \ldots, \mathbf{y}_k)$$

and π is given as a deterministic function.

From (38), $\lambda_k = 0$ if $W_k = 0$. Therefore, if we have a final value problem, then $\lambda_k = 0$, $k = 1, 2, \ldots, N-1$, and, from (37), γ_ks are simply obtained by repeated operations of minimization with respect to u and integrations with respect to y.

From (37) and (39) we have

$$\begin{aligned} \gamma_k^* &= \min \gamma_k \\ &= \min_{u_{k-1}} \left[\lambda_k + \int \gamma_{k+1}^* p(\mathbf{y}_k \mid \mathbf{y}^{k-1}, \mathbf{u}^{k-1})\, d\mathbf{y}_k \right]. \end{aligned}$$

This is precisely the statement of the Bellman's principle of optimality applied to this problem where

$$\gamma_k^* = \min_{u_{k-1}, \ldots, u_{N-1}} E[W_k + W_{k+1} + \ldots + W_N \mid \mathbf{y}^{k-1}].$$

To see this simply, let us assume that the state vectors are perfectly observable, i.e.,

$$\mathbf{y}_i = \mathbf{x}_i, \qquad 0 \leqslant i \leqslant N-1.$$

Then, the key relation (37) reads

$$\gamma_k^*(\mathbf{x}^{k-1}, \mathbf{u}^{k-2}) = \min_{u_{k-1}} [\lambda_k + E(\gamma_{k+1}^* \mid \mathbf{x}^{k-1}, \mathbf{u}^{k-1})],$$

which is the result of applying the principle of optimality to

$$\gamma_k^* = \min_{u_{k-1}, \ldots, u_{N-1}} E[W_k + \ldots + W_N \mid \mathbf{x}^{k-1}].$$

We have the usual functional equation of the dynamic programming if the $\{\mathbf{x}_k\}$-process is a first-order Markov sequence, for example, if ξ_ks are all independent. Then

$$\gamma_k^*(\mathbf{x}_{k-1}) = \min_{u_{k-1}} [\lambda_k(\mathbf{x}_{k-1}, \mathbf{u}_{k-1}) + E(\gamma_{k+1}^*(\mathbf{x}_k) \mid \mathbf{x}_{k-1}, \mathbf{u}_{k-1})].$$

When the observations are not perfect, then the arguments of γ_k^* are generally $\mathbf{y}^{k-1}$ and $\mathbf{u}^{k-2}$. Thus the number of the arguments changes with k. γ_N^* is computed as a function of $\mathbf{y}^{N-1}$ and $\mathbf{u}^{N-2}$ and, at step k, $\mathbf{y}_k$ in γ_{k+1}^* is integrated out and the presence of $\mathbf{u}_{k-1}$ is erased by the minimization operation on $\mathbf{u}_{k-1}$ to obtain γ_k^* as a function of $\mathbf{y}^{k-1}$ and $\mathbf{u}^{k-2}$. When the information in $(\mathbf{y}^k, \mathbf{u}^{k-1})$ is summarized by a statistic, $\mathbf{s}_k$, and when it satisfies a certain condition, then the recursion relation for the general noisy observation case also reduces to the usual functional equation of dynamic programming

$$\gamma_k^*(\mathbf{s}_{k-1}) = \min_{u_{k-1}} [\lambda_k(\mathbf{s}_{k-1}, \mathbf{u}_{k-1}) + E(\gamma_k^*(\mathbf{s}_k) | \mathbf{s}_{k-1}, \mathbf{u}_{k-1})]$$

where $\mathbf{s}_k$ satisfies the relation

$$\mathbf{s}_k = \psi(\mathbf{s}_{k-1}, \mathbf{u}_{k-1})$$

for some function ψ. (This idea is taken up later. Similar observations are valid for recurrence equations in later chapters.)

Updating Conditional Probability Densities

Equations (37)–(39) constitute a recursive solution of optimal control policies. One must evaluate γs recursively, and this requires that the conditional densities $p(\mathbf{x}_t | \mathbf{y}^t, \mathbf{u}^{t-1})$ and $p(\mathbf{y}_{t+1} | \mathbf{y}^t, \mathbf{u}^t)$ are available.* They can be calculated recursively easily in some cases.

To indicate this let us derive these densities under the assumption that noise random vectors $\boldsymbol{\xi}$s and $\boldsymbol{\eta}$s are mutually and serially independent.

Consider a conditional density $p(\mathbf{x}_{t+1}, \mathbf{y}_{t+1} | \mathbf{y}^t, \mathbf{u}^t)$.

By the chain rule, remember that we are interested in control policies in the form of $\mathbf{u}_t = \boldsymbol{\phi}_t(\mathbf{y}^t, \mathbf{u}^{t-1})$, $0 \leqslant t \leqslant N-1$,

$$p(\mathbf{x}_{t+1}, \mathbf{y}_{t+1} | \mathbf{y}^t, \mathbf{u}^t) = p(\mathbf{y}_{t+1} | \mathbf{y}^t, \mathbf{u}^t) p(\mathbf{x}_{t+1} | \mathbf{y}^{t+1}, \mathbf{u}^t).$$

We can write, using (28),

$$\begin{aligned} p(\mathbf{x}_t, \mathbf{x}_{t+1}, \mathbf{y}_{t+1} | \mathbf{y}^t, \mathbf{u}^t) &= p(\mathbf{x}_t | \mathbf{y}^t, \mathbf{u}^{t-1}) p(\mathbf{x}_{t+1} | \mathbf{x}_t, \mathbf{y}^t, \mathbf{u}^t) p(\mathbf{y}_{t+1} | \mathbf{x}_t, \mathbf{x}_{t+1}, \mathbf{y}^t, \mathbf{u}^t) \\ &= p(\mathbf{x}_t | \mathbf{y}^t, \mathbf{u}^{t-1}) p(\mathbf{x}_{t+1} | \mathbf{x}_t, \mathbf{u}_t) p(\mathbf{y}_{t+1} | \mathbf{x}_{t+1}). \end{aligned} \tag{40}$$

---

* Alternatively, one can just as easily generate $p(\mathbf{x}_{t+1} | \mathbf{y}^t, \mathbf{u}^t)$ and $p(\mathbf{y}_{t+1} | \mathbf{y}^t, \mathbf{u}^t)$ recursively. They are related by

$$p(\mathbf{x}_{t+1} | \mathbf{y}^t, \mathbf{u}^t) = \int p(\mathbf{x}_{t+1} | \mathbf{x}_t, \mathbf{u}_t) p(\mathbf{x}_t | \mathbf{y}^t, \mathbf{u}^{t-1}) \, d\mathbf{x}_t.$$

Thus, from (40),

$$
\begin{aligned}
p(\mathbf{x}_{t+1}, \mathbf{y}_{t+1} | \mathbf{y}^t, \mathbf{u}^t) &= \int p(\mathbf{x}_t, \mathbf{x}_{t+1}, \mathbf{y}_{t+1} | \mathbf{y}^t, \mathbf{u}^t)\, d\mathbf{x}_t \\
&= \int p(\mathbf{x}_t | \mathbf{y}^t, \mathbf{u}^{t-1}) p(\mathbf{x}_{t+1} | \mathbf{x}_t, \mathbf{u}_t) \\
&\quad \times p(\mathbf{y}_{t+1} | \mathbf{x}_{t-1})\, d\mathbf{x}_t .
\end{aligned} \tag{41}
$$

Hence

$$
\begin{aligned}
p(\mathbf{x}_{t+1} | \mathbf{y}^{t+1}, \mathbf{u}^t) &= \frac{p(\mathbf{x}_{t+1}, \mathbf{y}_{t+1} | \mathbf{y}^t, \mathbf{u}^t)}{p(\mathbf{y}_{t+1} | \mathbf{y}^t, \mathbf{u}^t)} \\
&= \frac{p(\mathbf{y}_{t+1} | \mathbf{x}_{t+1}) \int p(\mathbf{x}_t | \mathbf{y}^t, \mathbf{u}^{t-1}) p(\mathbf{x}_{t+1} | \mathbf{x}_t, \mathbf{u}_t)\, d\mathbf{x}_t}{\int (\text{numerator})\, d\mathbf{x}_{t+1}}
\end{aligned} \tag{42}
$$

where the denominator of (42) gives $p(\mathbf{y}_{t+1} | \mathbf{y}^t, \mathbf{u}^t)$ and where $p(\mathbf{x}_{t+1} | \mathbf{x}_t, \mathbf{u}_t)$ and $p(\mathbf{y}_t | \mathbf{x}_t)$ are obtainable from the plant and observation equations and the density functions for $\boldsymbol{\xi}_t$ and $\boldsymbol{\eta}_t$.

The recursion formula is started from $p(\mathbf{x}_0 | \mathbf{y}_0)$, which may be computed by the Bayes formula

$$
p(\mathbf{x}_0 | \mathbf{y}_0) = \frac{p_0(\mathbf{x}_0) p(\mathbf{y}_0 | \mathbf{x}_0)}{\int p_0(\mathbf{x}_0) p(\mathbf{y}_0 | \mathbf{x}_0)\, d\mathbf{x}_0}
$$

where $p_0(\mathbf{x}_0)$ is assumed available as a part of the a priori information on the system.

Equation (42) is typical in that the recursion formulas for $p(\mathbf{x}_t | \mathbf{y}^t, \mathbf{u}^{t-1})$ and $p(\mathbf{y}_{t+1} | \mathbf{y}^t, \mathbf{u}^t)$ generally have this structure for general stochastic and adaptive control problems discussed in later chapters.

In the numerator of Eq. (42), $p(\mathbf{x}_{t+1} | \mathbf{x}_t, \mathbf{u}_t)$ is computed from the state equation and the known density function for $\boldsymbol{\xi}_t$, and $p(\mathbf{y}_{t+1} | \mathbf{x}_{t+1})$ is computed from the observation equation and the known density function for $\boldsymbol{\eta}_t$. The first factor $p(\mathbf{x}_t | \mathbf{y}^t, \mathbf{u}^{t-1})$ is available from the previous stage of the recursion formula. With suitable conditions

$$
p(\mathbf{x}_{t+1} | \mathbf{x}_t, \mathbf{u}_t) = p(\boldsymbol{\xi}_t) | J_\xi |
$$

and

$$
p(\mathbf{y}_t | \mathbf{x}_t) = p(\boldsymbol{\eta}_t) | J_\eta | \tag{43}
$$

where J_ξ and J_η are appropriate Jacobians and where the plant and the observation equations are solved for $\boldsymbol{\xi}_t$ and $\boldsymbol{\eta}_t$, respectively, and substituted in the right-hand sides.

When ξs and ηs enter into dynamic and observation equations linearly, then the probability densities in Eq. (42) can be obtained particularly simply from the probability densities for ξs and ηs. For example, suppose that

$$x_{t+1} = \mathrm{f}_t(x_t, u_t) + \xi_t$$

$$y_t = \mathrm{h}_t(x_t) + \eta_t$$

then

$$|J_\xi = |J_\eta| = 1,$$

$$\xi_t = x_{t+1} - \mathrm{f}_t(x_t, u_t),$$

and

$$\eta_t = y_t - \mathrm{h}_t(x_t)$$

are substituted in the right-hand sides of Eq. (43). Thus, if

$$p(\xi_t) = \frac{1}{(2\pi)^{1/2}\sigma_1} \exp\left(-\frac{\xi_t^2}{2\sigma_1^2}\right) \quad \text{and} \quad p(\eta_t) = \frac{1}{(2\pi)^{1/2}\sigma_2} \exp\left(-\frac{\eta_t^2}{2\sigma_2^2}\right),$$

then

$$p(x_{t+1}|x_t, u_t) = \frac{1}{(2\pi)^{1/2}\sigma_1} \exp\left[-\frac{1}{2\sigma_1^2}[x_{t+1} - f_t(x_t, u_t)]^2\right]$$

and

$$p(y_t|x_t) = \frac{1}{(2\pi)^{1/2}\sigma_2} \exp\left[-\frac{1}{2\sigma_2^2}(y_t - h_t(x_t))^2\right].$$

Equation (42) clearly indicates the kind of difficulties we will encounter time and again in optimal control problems.

Equation (42) can be evaluated explicitly by analytical methods only in a special class of problems. Although this special class contains useful problems of linear control systems with Gaussian random noises as will be discussed in later sections of this chapter, in a majority of cases, Eq. (42) cannot be integrated analytically. We must resort either to numerical evaluation, to some approximate analytical evaluations of Eq. (42), or to both. Numerical integrations of Eq. (42) are nontrivial by any means since the probability density function $p(x_t | y^t, u^{t-1})$ will not be any well-known probability density in general and cannot be represented conveniently analytically; hence it must be stored numerically.

In order to synthesize u_t^*, it is necessary to compute $p(x_t | y^t, u^{t-1})$ by (42) and then to compute λ_{t+1}, to generate $p(y_{t+1} | y^t, u^t)$ to evaluate $E(\gamma_{t+2}^* | y^t, u^t)$, to obtain γ_{t+1}, and finally to minimize γ_{t+1} with respect to u_t.

Note that the controller must generally remember y^t and u^{t-1} at time t in order to generate u_t^*.

Although some of the information necessary to compute u_t can be precomputed (i.e., generated off-line), all these operations must generally be done on the real-time basis if the control problem is the real-time optimization problem.

If k sampling times are needed to perform these operations, one must then either find the optimal control policy from the class of control policies such that

$$u_t = \pi_t(y^{t-k}, u^{t-1}), \qquad t = k, k+1, \ldots, N-1$$

where u_0^* through u_{k-1}^* must be chosen based on the a priori information only, or one must use approximations so that all necessary computations can be performed within one sampling time. In practice we may have to consider control policies with the constraints on the size of the memory in the controller and/or we may be forced to use control policies as functions of several statistical moments (such as mean or variance), instead of the probability density functions, and generate these statistics recursively. For example, u_t^* may have to be approximated from the last few observations and controls, say y_{t-1}, y_t, u_{t-2}, and u_{t-1}.

The effects of any suboptimal control policies on the system performance need be evaluated carefully either analytically or computationally, for example, by means of Monte Carlo simulations of system behaviors.

We will return to these points many times in the course of this book, in particular in Chapter 9, where some approximation techniques are discussed.

Example 5: System with Unknown Random Time Constant. Consider a one-dimensional control system

$$x_{t+1} = a_t x_t + u_t, \qquad 0 \leqslant t \leqslant N-1,\ u_t \in (-\infty, \infty)$$

where as are independently and identically distributed random variables with

$$\mathscr{L}(a_t) = N(\theta, \sigma^2)$$

for each θ where θ is the unknown parameter of the distribution function. It is assumed to have an a priori distribution function

$$\mathscr{L}(\theta) = N(\theta_0, \sigma_0^2).$$

The system is assumed to be perfectly observed:

$$y_t = x_t, \qquad 0 \leqslant t \leqslant N-1.$$

When the common mean of the random time constants is known, the problem reduces to that of a purely stochastic system. See Section 3.5.

By letting $\sigma_0 \to 0$, the solution of this problem reduces to that for the purely stochastic system as we will see shortly.

The criterion function is taken to be

$$J = x_N^2.$$

Now,

$$\begin{aligned}\gamma_N &= E(x_N^2 \mid x^{N-1})\\ &= \int x_N^2 p(x_N \mid x_{N-1}, u_{N-1}, a_{N-1}) p(a_{N-1} \mid x^{N-1}) d(x_N, a_{N-1})\\ &= \int (a_{N-1} x_{N-1} + u_{N-1})^2 p(a_{N-1} \mid x^{N-1})\, da_{N-1}.\end{aligned}$$

Because of the assumption of perfect observation, the knowledge of x^{N-1} is equivalent to that of a^{N-2}, since $a_t = (x_{t+1} - u_t)/x_t$ if $x_t \neq 0$. If $x_t = 0$ for some t, it is easy to see that $u_s = 0$, $s = t, t+1, \ldots, N-1$, is optimal from that point on.

Define

$$\begin{aligned}\hat{a}_t &= E(a_t \mid a^{t-1}), \qquad 1 \leqslant t \leqslant N-1\\ \hat{\sigma}_t^2 &= \operatorname{var}(a_t \mid a^{t-1}), \qquad 1 \leqslant t \leqslant N-1.\end{aligned}$$

We easily calculate the expected cost to be

$$\lambda_N = (\hat{a}_{N-1} x_{N-1} + u_{N-1})^2 + \hat{\sigma}_{N-1}^2 x_{N-1}^2,$$

from which the optimal control variable at time $N-1$ is given by

$$u_{N-1}^* = -\hat{a}_{N-1} x_{N-1},$$

since $\hat{a}_{N-1}$ and $\hat{\sigma}_{N-1}$ can be seen to be independent of u_{N-1}. The minimal value of γ_N is given as

$$\gamma_N^* = \hat{\sigma}_{N-1}^2 x_{N-1}^2.$$

Now $\lambda_t = 0$, $t < N$, and we have

$$\gamma_{N-1} = \int \gamma_N^* p(x_{N-1} \mid x^{N-2})\, dx_{N-1}.$$

As we will see later, $\hat{\sigma}_{N-1}$ is independent of u_{N-2}.

Therefore, the problem of finding an optimal control policy for the remaining $(N-1)$ control stages essentially remains unchanged.

Hence

$$u_t^* = -\hat{a}_t x_t, \qquad 0 \leqslant t \leqslant N-1$$

is the optimal control policy and

$$\min EJ = \gamma_1^* = \left(\prod_{i=0}^{N-1} \hat{\sigma}_t^2\right) x_0^2.$$

It now remains to compute the conditional mean and variance of a_t. Now, from the conditional independence assumption on θ,

$$p(a_t \mid a^{t-1}) = \int p(a_t \mid \theta) p(\theta \mid a^{t-1})\, d\theta$$

where, by assumption, the a priori probability density function is

$$p(a_t \mid \theta) = \text{const} \exp\left(-\frac{(a_t - \theta)^2}{2\sigma^2}\right).$$

To compute $p(\theta \mid a^{t-1})$, we use the recursion formula

$$p(\theta \mid a^s) = \frac{p(\theta \mid a^{s-1}) p(a_s \mid \theta, a^{s-1})}{\int p(\theta \mid a^{s-1}) p(a_s \mid \theta, a^{s-1})\, d\theta}, \qquad 1 \leqslant s \leqslant N-1$$

where its initial conditional probability density function is given by

$$\begin{aligned} p(\theta \mid a_0) &= \frac{p_0(\theta) p(a_0 \mid \theta)}{\int p_0(\theta) p(a_0 \mid \theta)\, d\theta} \\ &= \text{const} \exp\left(-\frac{(\theta - \mu_0)^2}{2\Sigma_0^2}\right), \end{aligned}$$

where

$$\begin{aligned} \mu_0 &= \frac{a_0/\sigma^2 + \theta_0/\sigma_0^2}{1/\sigma^2 + 1/\sigma_0^2}, \\ 1/\Sigma_0^2 &= 1/\sigma^2 + 1/\sigma_0^2. \end{aligned}$$

As in previous examples, θ has the sufficient statistics and we can write

$$p(\theta \mid a^s) = \text{const} \exp\left(-\frac{(\theta - \mu_s)^2}{2\Sigma_s^2}\right)$$

where the sufficient statistics (μ_t, Σ_t) are given by

$$\begin{aligned} \mu_s &= \frac{\mu_{s-1}/\Sigma_{s-1}^2 + a_s/\sigma^2}{1/\Sigma_{s-1}^2 + 1/\sigma^2}, \qquad 0 \leqslant s \leqslant N-1 \\ 1/\Sigma_s^2 &= 1/\Sigma_{s-1}^2 + 1/\sigma^2. \end{aligned}$$

Collecting these terms and carrying out the integral, we see that

$$p(a_t|a^{t-1}) = \text{const}\exp\left(-\frac{(a_t-\mu_{t-1})^2}{2(\sigma^2+\Sigma_{t-1}^2)}\right)$$

where

$$\hat{a}_t = \mu_{t-1}, \qquad 1 \leqslant i \leqslant N-1$$
$$\hat{a}_0 = \theta_0$$

and

$$\hat{\sigma}_t^2 = \sigma^2 + \Sigma_{t+1}^2.$$

Exercise 1

Consider a time-varying linear dynamics

$$\mathbf{x}_{t+1} = \mathbf{A}_t\mathbf{x}_t + \mathbf{B}_t\mathbf{u}_t + \boldsymbol{\xi}_t$$
$$\mathbf{y}_t = \mathbf{C}_t\mathbf{x}_t + \boldsymbol{\eta}_t$$

where $\boldsymbol{\xi}_t$ and $\boldsymbol{\eta}_t$ are mean zero, mutually, and serially uncorrelated Gaussian noises. For a quadratic cost function show that the optimal control is still of the form

$$\mathbf{u}_t^* = -\mathbf{\Lambda}_t\boldsymbol{\mu}_t$$

where $\boldsymbol{\mu}_t$ is $E(\mathbf{x}_t|\mathbf{y}^t, \mathbf{u}^{t-1})$ or

$$\boldsymbol{\mu}_{t+1} = \mathbf{D}_{t+1}\mathbf{u}_t + \mathbf{K}_{t+1}\mathbf{y}_{t+1}$$

where

$$\mathbf{D}_j = (\mathbf{I} - \mathbf{K}_j\mathbf{C}_j)(\mathbf{A}_{j-1} - \mathbf{B}_{j-1}\mathbf{\Lambda}_{j-1}).$$

This can be written as

$$\boldsymbol{\mu}_{t+1} = \mathbf{K}_{t+1}\mathbf{y}_{t+1} + \sum_{s=0}^{t}\mathbf{D}_{t+1}\ldots\mathbf{D}_{s+1}\mathbf{K}_s\mathbf{y}_s.$$

Therefore the control is a linear combination of current and past data

$$\mathbf{u}_t^* = -\mathbf{\Lambda}_t\mathbf{K}_t\mathbf{y}_t - \mathbf{\Lambda}_t\left(\sum_{s=0}^{t-1}\mathbf{D}_{t+1}\ldots\mathbf{D}_{s+1}\mathbf{K}_s\mathbf{y}_s\right), \qquad 0 \leqslant t \leqslant N-1,$$

$$\mathbf{u}_0^* = -\mathbf{\Lambda}_0\mathbf{K}_0\mathbf{y}_0,$$

which can be interpreted to mean that the optimal control is of the proportional plus integral type (see Rosenbrock [1963]), where the first term gives the control proportional to the measurement of the current state vector and the second term expresses the control due to the integral on the past state vector measurements. Thus, if $\|(\Pi_{k=j+1}^{i-1}\mathbf{D}_k)\mathbf{K}_j\| \ll \|\mathbf{K}_i\|$ for $i \gg j$, then the remote past measurements have negligible effects on the current control variables to be chosen. In the extreme case $\mathbf{D}_k = \mathbf{0}$, $\mathbf{u}_i^*$ depends only on $\mathbf{y}_i$, and past observations $\{\mathbf{y}^{i-1}\}$ have no effect on $\mathbf{u}_i^*$.

Intuitively speaking, if the measurements of the state vectors are exact and there are no unknown parameters in the problem statements as we are assuming now, then the control at time $\mathbf{t}$ will be a function of $\mathbf{x}_t$ alone, indicating $\mathbf{K}_j\mathbf{C}_j$ will be equal to $\mathbf{I}$ under the perfect measurements. For systems with poor measurements of the state vectors, it is intuitively reasonable that the controller makes use not only of the current measurements but also of past measurements in synthesizing optimal controls. Model accurate measurements by $\varepsilon\mathbf{R}_j$ where ε is a small positive scalar quantity to show that

$$\begin{aligned}\mathbf{\Gamma}_0 &= (\mathbf{\Sigma}_0^{-1} + 1/\mathbf{H}_0'\mathbf{R}_0^{-1}\mathbf{H}_i)^{-1}\\ &= \varepsilon(\mathbf{H}_0'\mathbf{R}_0^{-1}\mathbf{H}_0 + \varepsilon\mathbf{\Sigma}_0^{-1})^{-1}\\ &\approx \varepsilon(\mathbf{H}_0'\mathbf{R}_0^{-1}\mathbf{H}_0)^{-1}.\end{aligned}$$

In general,

$$\mathbf{\Gamma}_i = \varepsilon\mathbf{H}_i'\mathbf{R}_i^{-1}\mathbf{H}_i \approx \mathbf{I}.$$

Therefore

$$\mathbf{K}_i\mathbf{C}_i = \mathbf{\Gamma}_i\mathbf{H}_i' \approx \mathbf{I}.$$

Thus

$$-\mathbf{u}_i^* = \mathbf{\Lambda}_i\boldsymbol{\mu}_i \approx \varepsilon\mathbf{\Lambda}_i(\mathbf{H}_i'\mathbf{R}_i^{-1}\mathbf{H}_i)^{-1}\mathbf{H}_i'\mathbf{R}_i^{-1}\mathbf{y}_i.$$

When the accuracy of measurement is poor, use $\varepsilon\mathbf{R}_j^{-1}$ instead of $\mathbf{R}_j^{-1}$, where ε is a small positive scalar quantity to show that

$$\begin{aligned}\mathbf{\Gamma}_0 &= (\mathbf{\Sigma}_0^{-1} + \varepsilon\mathbf{H}_0'\mathbf{R}_0^{-1}\mathbf{H}_0)^{-1}\\ &\approx \mathbf{\Sigma}_0\\ \boldsymbol{\mu}_0 &\approx \varepsilon_0\mathbf{H}_0'\mathbf{R}_0^{-1}\mathbf{y}_0.\end{aligned}$$

In general,

$$\mathbf{\Gamma}_{i+1} = (\varepsilon\mathbf{H}_{i+1}'\mathbf{R}_{i+1}^{-1}\mathbf{H}_{i+1} + \mathbf{Q}_i + \mathbf{A}_i\mathbf{\Gamma}_i\mathbf{A}_i')^{-1}$$

or

$$\mathbf{\Gamma}_{i+1} = \mathbf{L}_i$$

where

$$\begin{aligned}\mathbf{L}_j &\triangleq \mathbf{Q}_j\mathbf{A}_j\mathbf{L}_{j+1}\mathbf{A}_j'\\ \mathbf{L}_{-1} &\triangleq \mathbf{\Sigma}_0.\end{aligned}$$

Thus

$$\begin{aligned}\mathbf{K}_i &= \varepsilon\mathbf{L}_{i-1}\mathbf{C}_i'\mathbf{R}_i^{-1},\\ \mathbf{I} - \mathbf{K}_i\mathbf{C}_i &\approx \mathbf{I},\end{aligned}$$

and consequently,

$$\boldsymbol{\mu}_i = \varepsilon \mathbf{L}_{i-1} \mathbf{C}'_{i-1} \mathbf{R}_{i-1}^{-1} \mathbf{y}_i + \varepsilon \sum_{k=1}^{i-2} (\mathbf{A}_k - \mathbf{B}_k \boldsymbol{\Lambda}_k) \boldsymbol{\Sigma}_0 \mathbf{H}'_0 \mathbf{R}_0^{-1} \mathbf{y}_0 .$$

Show that the integral part is of the same order as the proportional part unless $\|\Pi_{k=1}^{i-2} (\mathbf{A}_k - \mathbf{B}_k \boldsymbol{\Lambda}_k)\|$ is of the order ε or less, for example by satisfying the inequality $\|\mathbf{A}_k - \mathbf{B}_k \boldsymbol{\Lambda}_k\| \ll 1$ for $0 \leqslant k \leqslant i-2$.

Exercise 2

Consider a scalar control system

$$x_{t+1} = a_t x_t + b_t u_t + \xi_t, \qquad 0 \leqslant t \leqslant N-1, \qquad u_t \in (-\infty, \infty)$$

and the observation equation

$$y_t = h_t x_t + \eta_t, \qquad h_t \neq 0, \qquad 0 \leqslant i \leqslant N-1.$$

Take as the performance index a quadratic form in x and u,

$$J = \sum_{1}^{N} (v_t x_t^2 + \lambda_{t-1} u_{t-1}^2), \qquad v_t > 0, \qquad t_{t-1} > 0$$

where a_t and b_t are known deterministic model parameters and where ξs and ηs are assumed to be independent Gaussian random variables with

$$\begin{aligned} E(\xi_t) = E(\eta_t) &= 0, & 0 &\leqslant t \leqslant N-1 \\ E(\xi_t^2) &= q_t^2 > 0, & 0 &\leqslant t \leqslant N-1 \\ E(\eta_y^2) &= r_t^2 > 0, & 0 &\leqslant t \leqslant N-1 \\ E(\xi_t \eta_s) &= 0, & &\text{all } t \text{ and } s. \end{aligned}$$

Assume also that x_0 is Gaussian, independent of ξs and ηs with mean α and variance σ^2. Show that

$$p(x_t \mid y^t, u^{t-1}) = \text{const} \exp\left(-\frac{(x_t - \mu_t)^2}{2\sigma_t^2} \right)$$

and

$$p(x_{t+1} \mid y^{t+1}) = \text{const} \exp\left(\frac{(x_{t+1} - \mu_{t+1})^2}{2\sigma_{t+1}^2} \right)$$

where

$$\mu_{t+1} = a_t \mu_t + b_t u_t + K_{t+1} [y_{t+1} - h_{t+1}(a_t \mu_t + b_t u_t)]$$

where

$$K_{t+1} = \frac{h_{t+1}(q_t^2 + a_t^2 \sigma_t^2)}{h_{t+1}^2 (q_t^2 + a_t^2 \sigma_t^2) + r_{t+1}^2}$$

and where

$$1/\sigma_{t+1}^2 = h_{t+1}^2/r_{i+1}^2 + 1/(q_t^2 + a_t^2\sigma_t^2)$$
$$= \frac{h_{t+1}^2(q_t^2 + a_i^2\sigma_t^2) + r_{t+1}^2}{r_{t+1}^2(q_t^2 + a_t^2\sigma_t^2)}.$$

Next, derive that

$$\lambda_N = E(W_N \mid y^{N-1})$$
$$= D_N + \lambda_{N-1}u_{n-1}^2 + v_N(a_{N-1}\mu_{N-1} + b_{N-1}u_{N-1})^2$$

where

$$D_N = v_N(q_{N-1}^2 + a_{N-1}^2\sigma_{N-1}^2)$$

and hence the optimal control is

$$u_{N-1}^* = -\Lambda_{N-1}\mu_{N-1}$$

where

$$\Lambda_{N-1} = \frac{v_N b_{N-1} a_{N-1}}{t_{N-1} + v_N b_{N-1}^2}$$
$$\gamma_N^* = c_N + T_N \mu_{N-1}^2$$

where

$$T_N = \frac{v_N \lambda_{N-1} a_{N-1}^2}{(t_{N-1} + v_N b_{N-1}^2)}$$
$$C_N = D_N.$$

At the next-to-last stage, show that λ_{N-1} is given by

$$\lambda_{N-1} = D_{N-1} + \lambda_{N-2}u_{N-2}^2 + v_{N-1}(a_{N-2}\mu_{N-2} + b_{N-2}u_{N-2})^2.$$

The probability density $p(y_{N-1} \mid y^{N-2})$, necessary to evaluate $E(\gamma_N^* \mid y^{N-2})$, is the Gaussian probability density with mean $h_{N-1}(a_{N-2}\mu_{N-2} + b_{N-2} + b_{N-2}u_{N-2})$ and variance

$$h_{N-1}^2(a_{N-2}^2\sigma_{N-2}^2 + q_{N-2}^2) + r_{N-1}^2.$$

Next, show that

$$E(\mu_{N-1} \mid y^{N-2}) = a_{N-2}\mu_{N-2} + b_{N-2}u_{N-2}$$

and

$$\operatorname{var}(\mu_{N-1} \mid y^{N-2}) = K_{N-1}^2[h_{N-1}^2(a_{N-2}^2\sigma_{N-2}^2 + q_{N-2}^2) + r_{N-1}^2]$$

and hence

$$E(\gamma_N^* \mid y^{N-2}) = C_N + T_N\{(a_{N-2}\mu_{N-2} + b_{N-2}u_{N-2})^2$$
$$+ K_{N-1}^2[h_{N-1}^2(a_{N-2}^2\sigma_{N-2}^2 + q_{N-2}^2) + r_{N-1}^2]\}.$$

Derive

$$u_{N-2}^* = -\Lambda_{N-2}\mu_{N-2}$$

where

$$\Lambda_{N-2} = \frac{(v_{N-1} + T_N)b_{N-2}a_{N-2}}{t_{N-2} + (v_{N-1} + T_N)b_{N-2}^2}$$

and

$$\gamma_{N-1}^* = C_{N-1} + T_{N-1}\mu_{N-1}^2$$

where

$$T_{N-1} = \frac{(v_{N-1} + T_N)\lambda_{N-2}a_{N-2}^2}{\lambda_{N-2} + (v_{N-1} + T_N)b_{N-2}^2},$$

and in general

$$u_t^* = -\Lambda_t u_t$$

where

$$\Lambda_t = \frac{(v_t + T_{t+2})b_t a_t}{t_t + (v_{t+1} + T_{t+2})b_t^2}, \qquad 0 \leqslant i \leqslant N-1, \qquad T_{N+1} = 0$$

$$\gamma_t^* = C_t + T_t\mu_t^2,$$

where

$$C_t = D_t + C_{t+1} + T_{i+1}K_t^2[h_t^2(a_{t-1}^2\sigma_{t-1}^2 + q_{t-1}^2) + r_i^2]$$

$$C_{N+1} = 0$$

$$D_t = v_t(q_t^2 + a_{i-1}^2\sigma_{t-1}^2), \qquad 1 \leqslant i, \leqslant N,$$

and

$$T_t = \frac{(v_t + T_{t+1})\lambda_{t-1}a_{t-1}^2}{\lambda_{t-1} + (v_t + T_{t+1})b_{t-1}^2}, \qquad T_{N+1} = 0, \qquad 1 \leqslant i, \leqslant N.$$

When $\lambda_t = 0$, $1 \leqslant t \leqslant N$, the above reduces to

$$\mu_t = K_t y_t, \qquad 1 \leqslant i, \leqslant N$$

and

$$u_t^* = -a_t K_t y_t / b_t, \qquad 0 \leqslant i \leqslant N-1.$$

Exercise 3

Consider a simple scalar final-value control problem where the state vector is perfectly observed,

$$y_{t+1} = ay_t + u_t + r_t$$

where

$$u_t \in (-\infty, \infty)$$

and where ys are used instead of xs because of the assumption of perfect observation. The criterion function is taken to be $J = W_N(y_N)$. The noise r_ts are assumed to be independently distributed binomial random variables

$$r_t = \begin{cases} +c & \text{with probability } \theta \\ -c & \text{with probability } 1-\theta \end{cases}$$

for each θ, where θ is the unknown parameter with its a priori distribution given. Assume, for example, that*

$$r_t = \begin{cases} \theta_1 & \text{with probability } z_0 \\ \theta_2 & \text{with probability } 1-\theta, \qquad (0 \leqslant z_0 \leqslant 1). \end{cases}$$

From the model,

$$r_t = y_{t+1} - ay_t - u_t.$$

Therefore, at time t, the past realization of the noise sequence $(r_0, r_1, \ldots, r_{t-1})$ is available from the knowledge of y^t. Therefore, the joint probability** of $r_0, \ldots, r_{t-1}$ is given by

$$\mathrm{P}[r_0, r_1, \ldots, r_{t-1} | \theta] = \theta^k (1-\theta)^{t-k}$$

where k is the number of times $+c$ is observed and is known from y^t. The pair (t, k) is said to be sufficient for θ, or the number of times rs are $+c$ is the sufficient statistic for θ. Denote this number i by s_t.

To obtain an optimal control policy for the system one computes, as usual,

$$\begin{aligned} \gamma_N(y^{N-1}) &= E(W_N | y^{N-1}) \\ &= \int W_N(y_N) p_N(y^{N-1})\, dy_N \\ &= \int W_N^*(y_N) p(y_N | y_{N-1}, r_{N-1}, u_{N-1}) P(r_{N-1} | y^{N-1})\, d(y_N, r_{N-1}) \\ &= \int W_N(ay_{N-1} + r_{N-1} + u_{N-1}) P(r_{N-1} | r^{N-2})\, dr_{N-1}. \end{aligned}$$

* The form of the unknown parameter θ is not essential for the development. If θ is a continuous variable, $0 \leqslant \theta \leqslant 1$, then an a priori probability density may be taken, for example, to be of Beta type, i.e.,

$$p_0(\theta) = \frac{\Gamma(c+d+2)}{\Gamma(c+1)\Gamma(d+1)} \theta^c (1-\theta)^d, \qquad c, d > 0.$$

** In this example, the random variables are discrete and it is convenient to deal with probabilities rather than probability densities. It is, therefore, necessary to modify prior developments in an obvious way. Such modifications will be made without further comments.

The conditional probability one needs in evaluating γ_N, therefore, is $P(r_t | y^t)$ or $P(r_t | r^{t-1})$, $0 \leqslant t \leqslant N-1$. Note that

$$P[r_t = c | y^t] = P[r_t = c | r^{t-1}]$$
$$= P[r_t = c | \theta_1] z_t + P[r_t = c | \theta_2](1 - z_t)$$

where

$$z_t \triangleq P[\theta_1 | r^{t-1}] = P[\theta_1 | s_t], \qquad 0 \leqslant t \leqslant N-1$$

and where $s_t = (t+1)/c \, \Sigma_{j=0}^{t-1} r_j$ is the number of times $+c$ is observed. Show that

$$E(W_N | y^{N-1}) = E(W_N | s_{N-1}, y^{N-1}) = \gamma_N(y_{N-1}, z_{N-1})$$
$$= W_N(a y_{N-1} + c + u_{N-1}) \hat{\theta}_{N-1}$$
$$= W_N(a y_{N-1} + c + u_{N-1})(1 - \hat{\theta}_{N-1})$$

where $\hat{\theta}_{N-1}$ is the a posteriori estimate of θ, given y^{N-1},

$$\hat{\theta}_{N-1} \triangleq \theta_1 z_{N-1} + \theta_2 (1 - z_{N-1})$$

and where

$$z_{N-1} \triangleq P[\theta = \theta_1 | y^{N-1}] = \frac{1}{1 + [(1 - z_0)/z_0]\alpha_{N-1}}$$

where

$$\alpha_{N-1} \triangleq \left(\frac{\theta_2}{\theta_1}\right)^{s_{N-1}} \left(\frac{1-\theta_2}{1-\theta_1}\right)^{N-1-s_{N-1}}.$$

Thus γ_N^*, which is generally a function of y^{N-1} and u^{N-2}, is seen to be a function of y_{N-1} and z_{N-1} (or s_{N-1}) only, a reduction in the number of variables from $2N-3$ to just two.

For the optimal control policy for this problem and its relation with optimal control policies of the corresponding stochastic systems, where $\theta = \theta_1$ or θ_2 with probability one, see Chapter 9.

3.5 Control of Random Parameter Systems

Dynamic systems with randomly varying parameters have a destabilizing tendency. This fact can be brought out by considering a simple control problem of a scalar random parameter systems with known probability distributions. This feature is not found in deterministic systems and in systems where "certainty equivalence principle" is used to replace random parameters by their expected values. Information about variability about the mean is completely ignored in this approach, and this is why this fact has not been observed. Complicating the nature of uncertainty with imperfectly known parameters as in adaptive control of dynamics compounds the difficulty to controlling such systems with that of parameter learning. These two can best be described separately before their joint effects are evaluted.

Example 6. To examine the destabilizing effect of random parameters in a simple control problem, consider the next simple example. The problem is

$$\min Ex_N^2 \tag{44}$$

subject to

$$x_{k+1} = a_k x_k + b_k u_k, \qquad k = 0, 1, \ldots, N, \tag{45}$$

where a_k and b_k are jointly Gaussian and independently and identically distributed over time,

$$\begin{pmatrix} a_k \\ b_k \end{pmatrix} \sim N\left[\begin{pmatrix} a \\ b \end{pmatrix}, \boldsymbol{\Sigma}\right], \quad \text{where} \quad \boldsymbol{\Sigma} = \begin{pmatrix} \Sigma_a & \Sigma_{ab} \\ \Sigma_{ba} & \Sigma_b \end{pmatrix}.$$

Control problems with cost that depends only on variables at the end of the (planning) horizon are called terminal control problems. A terminal control problem with quadratic cost is a special case of the class of problems

$$J_t = \sum_{k=t}^{N} (q_k^2 x_k^2 + r_k u_k^2)$$

in which q_s and r_s are both zero except at $k = N$. Later we discuss control problems where cost is incurred throughout the control horizon.

The last control u_{N-1} is chosen to minimize the conditional cost

$$\begin{aligned} \gamma_N &= E_{N-1} x_N^2 \\ &= \alpha x_{N-1}^2 + 2\Theta x_{N-1} u_{N-1} + \beta u_{T-1}^2 \end{aligned} \tag{46}$$

where

$$\alpha = \overline{a_{N-1}^2} = a^2 + \Sigma_a,$$

$$\Theta = \overline{a_{N-1} b_{N-1}} = ab + \Sigma_{ab},$$

and

$$\beta = \bar{b}_{N-1}^2 = b^2 + \Sigma_b.$$

Equation (46) is minimized by

$$u_{N-1} = -\Theta x_{N-1}/\beta,$$

i.e., a feedback control with the feedback gain $-\Theta/\beta$ and

$$\gamma_N^* = \min_{u_{N-1}} \gamma_N = \mu x_{N-1}^2$$

where

$$\mu = \alpha - \Theta^2/\beta. \tag{47}$$

By the nature of the cost expression, μ is nonnegative. The next-to-last control minimizes the conditional cost given x_{N-2},

$$\begin{aligned}\gamma_{N-1} &= E_{N-2}\gamma_N^* \\ &= \mu E_{N-2}x_{N-1}^2.\end{aligned}$$

Except for the time index, and for the scalar factor μ, the minimization problem is exactly analogous to that of minimizing (46). Thus we can immediately deduce that

$$u_{N-2} = -\Theta x_{N-2}/\beta$$

and

$$\begin{aligned}\gamma_{N-1}^* &= \min_{u_{N-2}} \beta_{N-1} \\ &= \mu^2 x_{N-2}^2.\end{aligned}$$

Clearly, the optimal control is given by

$$u_{N-k} = -\Theta x_{N-k}/\beta$$

and $\gamma_{N-k+1}^* = \mu^k x_{N-k}^2$, $k = 1, \ldots, N$. Scalar systems are automatically controllable and observable even when the variable to be controlled is observed only through noise and even when the dynamics contain exogenous additive disturbances. Nonexistence of finite cost expressions *never* arises for deterministic dynamic systems. It is a phenomenon unique to nondeterministic systems.

Quadratic Control Problems

We pointed out the possibility of control cost becoming unbounded when parameter uncertainty exceeds certain thresholds in infinite-horizon problems. This section examines the control of the same random parameter system with a more general control cost. The problem is to minimize $E_0 J_0$ where

$$J_s = \sum_{k=s}^{N} (q_k x_k^2 + r_k u_k^2), \tag{48}$$

subject to the same dynamics as in the previous section. Since u_N does not affect x_N, set it to zero. Then the optimal choice of u_{N-1} is to minimize γ_N

$$\gamma_N = q_{N-1}x_{N-1}^2 + r_{N-1}u_{N-1}^2 + E_{N-1}(q_N x_N^2).$$

The previous example showed that

$$E_{N-1}x_N^2 = \alpha x_{N-1}^2 + 2\Theta x_{N-1}u_{N-1} + \beta u_{N-1}^2.$$

We obtain the optimal control as

$$u_{N-1} = -K_{N-1}x_{N-1}$$

where

$$K_{N-1} = \Theta q_N/(r_{N-1} + q_N\beta)$$

and

$$\gamma_N^* = \min_{u_{N-1}} \gamma_N = \Pi_{N-1}x_{N-1}^2$$

where

$$\Pi_{N-1} = q_{N-1} + q_N[\alpha - \Theta^2 q_N/(r_{N-1} + \beta q_N)].$$

The functional equation of dynamic programming becomes

$$\gamma_{k+1} = q_k x_k^2 + r_k u_k^2 + E_k(\Pi_{k+1}x_{k+1}^2), \qquad k = 0, 1, \ldots, N-2.$$

Noting that

$$\gamma_{k+1} = E_k x_{k+1}^2 = \alpha x_k^2 + 2\Theta x_k u_k + \beta u_k^2,$$

we see immediately that

$$\gamma_{k+1}^* = \min_{u_k} \gamma_{k+1}$$

$$= \Pi_k x_k^2$$

(49) $$\Pi_k = q_k + \alpha\Pi_{k+1} - \frac{\Theta^2\Pi_{k+1}^2}{r_k + \beta\Pi_{k+1}},$$

$$\Pi_N = q_N,$$

and

$$u_k = -K_k x_k$$

where

(50) $$K_k = \frac{\Theta\Pi_{k+1}}{(r_k + \beta\Pi_{k+1})}.$$

To simplify the recursion (49) somewhat, assume that $q_k = q$ and $r_k = r$ for all k from now on.* The variable Π_k should be written as $\Pi_{N,k}$ to make its dependence explicit on N. To lighten notation, this is not done but should be understood since the behavior of $\Pi_{N,k}$ for fixed k as N approaches infinity is the vital question to be examined.

---

* Another simple case is to set $q_k = \theta\delta^k$ and $r_k = r\delta^k$ for some $0 < \delta < 1$. This produces a problem with a discount factor.

When r is zero, the Riccati equation (49) simplifies to

$$\Pi_k = q + \mu \Pi_{k+1}$$

where μ is as defined by (47) of the previous example. This equation can be solved explicitly. $\Pi_{N,N-k}$ become unbounded if and only if $\mu > 1$. This confirms the crucial role of the condition, $\mu > 1$, identified earlier in the terminal control problem.

Steady State Solution

If the Riccati equation (49) has a steady state solution, then Π_∞ is a positive root of the quadratic equation

$$\beta(1-\mu)x^2 + [r(1-\alpha) - q\beta]x - qr = 0.$$

If $qr\beta(1-\mu)$ is positive, i.e., if $(1-\mu) > 0$, since $qr\beta > 0$, then

$$\Pi_\infty = \frac{1}{2[\beta(1-\beta)+\Theta^2]} \times \{-r(1-\alpha) + q\beta + \sqrt{[r(1-\alpha) - q\beta]^2 + 4qr[\Theta^2 + \beta(1-\alpha)]}\}.$$

When $\mu > 1$, no steady state positive solution to (49) exists.

Bounds on Costs

The crucial role of the magnitude of μ is also revealed by examining the rate at which $\Pi_{T,t}$ grows.

When $\Pi_{T,t}$ becomes unbounded when T approaches infinity, the control gain (50) approaches a constant $\mathrm{K}_\infty = \Theta/\beta$, and the dynamics (45) becomes

$$x_{k+1} = (a_k - \mathrm{K}_\infty b_k)x_k.$$

Since

$$\begin{aligned} E_k x_{k+1}^2 &= (a - \mathrm{K}_\infty b)^2 x_k^2 + (\Sigma_a + \mathrm{K}_\infty^2 \Sigma_b - 2\mathrm{K}_\infty \Sigma_{ab})x_k^2 \\ &= \mu x_k^2, \end{aligned}$$

the inequality $\mu > 1$ implies that $(E_k x_{k+1}^2 / x_k^2) > 1$. This means that $\{x_k^2\}$ becomes unbounded as k increases.

Alternately, consider

$$s_k = q + \mu s_{k+1}$$

with the terminal condition $x_{n+1} = 0$. The solution is

$$s_{N,N-k} = \frac{q(1-\mu^k)}{(1-\mu)},$$

If $\mu \neq 1$. With $\mu = 1$, $s_{T,T-t} = (1+t)q$.

Then $\delta_k = \Pi_k - s_k$ is governed by

$$\delta_k = \mu\delta_{k+1} + \frac{r}{r+\beta\Pi_{k+1}}\,\frac{\Theta\Pi_{k+1}}{\beta} > \mu\delta_{k+1}.$$

Solving this with the terminal condition $\delta_{N+1} = 0$ we see that $\delta_{N,k} > 0$ for all $0 \leqslant k < N$, i.e., $\Pi_k > s_k$ for all $0 \leqslant k < N$. Therefore, if $\mu > 1$, $s_{N,N-k}$ grows at the rate μ^k and $\Pi_{N,N-k}$ becomes unbounded.

This phenomenon was first reported by Athans, Ku, and Gershwin (1977) and points to the fundamental difference in qualitative behavior of random parameter dynamic systems from deterministic systems.

3.6 Control of Stationary Processes

We assume that exogenous noises are stationary and have rational spectral density functions. We can then appeal to the powerful Fourier transform methods to simply derive the optimal control or filtering results.

The observed process $\{\mathbf{y}_t\}$ has the autocovariance matrices

$$\mathbf{\Lambda}_k = E(\mathbf{y}_{t+k}\mathbf{y}_t'), \qquad k > 0$$

and

$$\mathbf{\Lambda}_k = \mathbf{\Lambda}_{-k}', \qquad k < 0.$$

Associate with it the z-transform or the autocovariance generating function

$$g(q^{-1}) = \sum_{-\infty}^{\infty} \mathbf{\Lambda}_k q^{-k}.$$

When $\mathbf{y}_t$ has a rational spectral density and $\mathbf{\Lambda}_k$ decays to zero exponentially, the covariance matrix has the representation $\mathbf{CA}^{k-1}\mathbf{M}$ for some matrices $\mathbf{A}, \mathbf{C}$, and $\mathbf{M}$ where $\mathbf{A}$ is asymptotically stable. See Aoki (1987C, Chapter 5) for example. Thus $\|\mathbf{\Lambda}_k\|$ decays to zero exponentially with k, and $g(q^{-1})$ will be analytic in an annulus containing the unit circle.

If one lets $q = e^{-j\omega}$, then

$$\mathbf{g}(e^{j\omega}) = \sum_k \mathbf{\Lambda}_k e^{j\omega k}$$
$$= \mathbf{f}(\omega)$$

defines the spectral density function.

Given a formal power series in q^{-1},

$$\alpha(q^{-1}) = \sum_{-\infty}^{\infty} \alpha_i q^{-i},$$

the projection operator $[\ ]_+$ extracts only the nonnegative power in q^{-1},

$$[\alpha(q^{-1})]_+ = \sum_{0}^{\infty} \alpha_i q^{-i},$$

and $[\ ]_-$ extracts only the nonpositive powers in q^{-1},

$$[\alpha(q^{-1})]_- = \sum_{-\infty}^{0} \alpha_i q^{-i}.$$

A function $\alpha(q^{-1})$ is said to be realizable if $\alpha_j = 0$ for $j < 0$. It is said to be stable if $\alpha(q)$ is analytic in $|q| \leqslant 1$, which implies that

$$\alpha(q) = \sum_{0}^{\infty} \alpha_i q^i$$

converges in $|q| \leqslant 1$.

The function $\alpha(q^{-1})$ is invertible if $\alpha(q^{-1})^{-1}$ is stable. Finally, it is canonical if it is stable and invertible.

With a p-dimensional $\mathbf{y}_t$, the canonical factorization of $\mathbf{g}(q^{-1})$ takes the form

$$\mathbf{g}(q^{-1}) = \boldsymbol{\Psi}(q^{-1})\boldsymbol{\Delta}\boldsymbol{\Psi}'(q),$$

and $\boldsymbol{\Delta}$ is a matrix independent of z.

When y_t is scalar-valued, $\mathrm{g}(q^{-1})$ is scalar-valued. Its canonical factorization decomposes $\mathrm{g}(q^{-1})$ into

$$g(q^{-1}) = \mathrm{h}(q^{-1})\mathrm{h}(q)$$

where h is canonical.

When y_t is, in addition, stationary, one can associate the factorization of the autocovariance generating function with the moving average representation of the data series

$$y_t = \sum_{0}^{\infty} \mathrm{h}_\tau \varepsilon_{t-\tau} \tag{51}$$

where ε_t is uncorrelated with unit variance, or

$$y_t = \mathrm{h}(q^{-1})\varepsilon_t$$

where

$$\mathrm{h}(q^{-1}) = \sum_{0}^{\infty} \mathrm{h}_\tau q^{-\tau}.$$

Since $\mathrm{h}(q^{-1})$ is canonical, the moving average representation can be inverted into an autoregressive form

$$\theta(q^{-1})y_t = \varepsilon_t$$

where

$$\theta(q^{-1}) = \mathrm{h}(q^{-1})^{-1}$$

is stable.

The coefficient h_0 may be zero but if it is not, rewriting (51) as

$$y_t = \sum_0^{\infty} (\mathrm{h}_\tau \mathrm{h}_0^{-1})\zeta_{t-\tau},$$

where we define

$$\zeta_{t-\tau} = \mathrm{h}_0 \varepsilon_{t-\tau},$$

we note that $y_t - y_{t|t-1} = \zeta_t$ is the innovation process of y_t and $\operatorname{cov}\zeta_t = h_0 h_0'$.

Example 7: Scalar Control. Consider a dynamic equation

$$\alpha(q^{-1})x_t = u_{t-1} + \varepsilon_t$$

where

$$\alpha(q^{-1}) = \sum_0^{\infty} \alpha_\tau q^{-\tau}, \qquad \alpha_0 \neq 0,$$

and $\{\varepsilon_t\}$ is a stationary process with the autocovariance generating function $\mathrm{g}(q^{-1})$. The control u_t is to be generated as a linear function of current and past xs:

$$u_t = \beta(q^{-1})x_t \tag{52}$$

where

$$\beta(q^{-1}) = \sum_0^{\theta} \beta_\tau q^{-\tau}.$$

The polynomial $\beta(\cdot)$ is to be chosen to minimize

$$J = E(x_t^2 + \lambda u_t^2).$$

Use of this control role implies that

$$\begin{aligned} x_t &= (\alpha - q^{-1}\beta)^{-1}\varepsilon_t \\ &= \theta\varepsilon_t, \end{aligned}$$

where θ is defined by the preceding equality. Then (52) can be rewritten as

$$u_t = q(\alpha\theta - 1)\varepsilon_t.$$

Since the choice of β can be regarded as the choice of θ and since θ appears linearly in the expression for x_t and u_t, we use $\theta(\cdot)$ in minimizing J. Once θ is determined, then β is obtained by

$$\beta = q(\alpha - \theta^{-1}).$$

By factoring the autocovariance generating function as

$$g_\varepsilon(q^{-1}) = h(q^{-1})h(q)\sigma^2,$$

ε_t can be written as

$$\varepsilon_t = h(q^{-1})\zeta_t$$

in terms of $\{\zeta_t\}$, which is mean zero and serially independent with variance σ^2, i.e., x_t is related to ζ_t by

$$x_t = \delta(q^{-1})\zeta_t$$

where

$$\delta(q^{-1}) = \theta(q^{-1})h(q^{-1}) = \sum_0^\infty \delta_j q^{-j}.$$

This means that

$$Ex_t^2 = \sigma^2 \sum_{j=0}^\infty \delta_j^2.$$

By the Parseval identity, this sum can be evaluated as

$$\begin{aligned}\Sigma\delta_j^2 &= \frac{1}{2\pi j}\oint \delta(z)\delta(1/z)\,dz/z \\ &= \frac{1}{2\pi j}\oint \theta(z)\theta(1/z)g_\varepsilon(z^{-1})\,dz/z.\end{aligned}$$

The expected value of Eu_t^2 is similarly expressible by the integral. Thus

$$J = \frac{1}{2\pi j}\oint [\theta(z)\theta(1/z) + \lambda(1 - \alpha(z)\theta(z))(1 - \alpha(1/z)\theta(1/z)]g(z^{-1})\frac{dz}{z}.$$

The variational argument yields the first-order condition for the optimal choice for θ,

$$0 = \frac{1}{2\pi j}\oint \{\theta(1/z)[1 + \lambda\alpha(1/z)\alpha(z)] - \lambda\alpha(z)\}g_\varepsilon(z^{-1})\frac{dz}{z},$$

or the expression

(53) $$\{\theta(1/z)[1 + \lambda\alpha(1/z)\alpha(z)] - \lambda\alpha(z)\}g_\varepsilon(z^{-1})$$

has no pole in the closed unit disk.

Let

$$1 + \lambda\alpha(1/z)\alpha(z) = d(1/z)\,d(z)$$

be the canonical factorization, i.e., $d(1/z)$ is stable and invertible. The expression (53) can be rewritten as

$$d(1/z)h(1/z)d(z)h(z)\theta(1/z) - \lambda\alpha(z)h(1/z)h(z) \tag{54}$$

which has no pole inside the unit circle. Since the expression $d(z)h(z)$ has zeros outside the unit circle, dividing (54) by $d(z)h(z)$ does not introduce any pole inside the unit circle, i.e.,

$$d(1/z)h(1/z)\theta(1/z) - \lambda\alpha(z)h(1/z)/d(z)$$

has no pole inside the unit circle. Since the first term is analytic in z^{-1}, we must have

$$d(1/z)h(1/z)\theta(1/z) = [\lambda\alpha(z)h(1/z)/d(z)]_1^\infty + D_0h_0\theta_0$$

with $\theta_0 = \alpha_0$, i.e.,

$$\theta(1/z) = \frac{1}{d(1/z)h(1/z)}\left\{\left[\frac{\lambda\alpha(z)h(1/z)}{d(z)}\right]_1^\infty + \frac{D_0h_0}{\alpha_0}\right\},$$

where $[\ ]_1^\infty$ denotes extracting positive powers in z^{-1}.

Minimum Variance Control

This section derives constrained minimum variance control sequences by variational arguments following Burt and Rigby (1982). The system is modeled by

$$y_t = z^{-k}\frac{B(z^{-1})}{A(z^{-1})}u_t + \xi_t \tag{55}$$

where A, B, and C are polynomials in z^{-1}. The factor z^{-k}, $k \geqslant 0$, allows for delays in control u_t affecting y_t and where ξ_t is a white noise sequence.

The control problem is to find a control function $R(z^{-1})$ to generate

$$u_t = -R(z^{-1})\,y_t \tag{56}$$

which minimizes

$$E(y_t^2 + \lambda v_t^2)$$

where v_t is related to u_t by

$$v_t = K(z^{-1})u_t \tag{57}$$

with known $K(z^{-1})$.

This is a convenient way of minimizing the variance of y_t while satisfying some constraint on the variance of a given function of control signal u_t.

Substitute (56) into (55) and solve for y_t as

$$y_t = \frac{A}{A + z^{-h}BR}\xi_t \tag{58}$$

$$= [1 - z^{-k}G(z^{-1})]\xi_t$$

where

$$G(z^{-1}) = \frac{B(z^{-1})R(z^{-1})}{A(z^{-1}) + z^{-k}B(z^{-1})R(z^{-1})}.$$

Then

$$v_t = F(z^{-1})G(z^{-1})\xi_t$$

where

$$F(z^{-1}) = -A(z^{-1})K(z^{-1})/B(z^{-1}).$$

Instead of obtaining R, we obtain G. Once G is known, then R can be given by solving (59) for it as

$$R(z^{-1}) = \frac{AG}{B(1 - z^{-h}G)}, \tag{60}$$

if G is stable and B is of minimum phase, i.e., the zeros of B are of the form $(1 - \beta_i z^{-1})$ with $|\beta_i| < 1$. When B contains the zeros with $|\beta_i| > 1$, then R attempts to cancel this factor out. Since βs must be estimated, inexact cancellation will occur, leading to system instability. To avoid this, G is constrained to contain all factors that are zeros of $z^{-h}B/A$, i.e.,

$$G = H \cdot \prod_{|\beta_i| > 1} (1 - \beta_i z^{-1}), \tag{61}$$

where $H(z^{-1})$ is stable and causal (realizable).

Let Φ be the spectral density function of the process ξ_t

$$\Phi = \sum_{-\infty}^{\infty} \phi_v z^v$$

where

$$\phi_v = E(\xi_t \xi_{t+v}).$$

Define

$$X(z^{-1}) = z^{-k}\Pi(z^{-1})$$

and

$$Y(z^{-1}) = AK\Pi/B.$$

Then

$$y_t = \xi_t - XH\xi_t$$

and

$$v_t = -YH\xi_t.$$

The criterion function to be minimized is a functional of H

$$I(H) = E(y_t^2 + \lambda v_t^2)$$
$$= \frac{1}{2\pi i}\int (P + \lambda v)\, dz/z,$$

where

$$P = (I - XH)(I - \bar{X}\bar{H})\Phi$$
$$= \Phi + X\bar{X}H\bar{H}\Phi - (XH + \bar{X}\bar{H})\Phi$$

and

$$V = Y\bar{Y}H\bar{H}\Phi$$

where

$$H = H(z)$$

and

$$\bar{H} = H(z^{-1}).$$

and similarly for $\bar{X}$ and $\bar{Y}$, i.e.,

$$I(H) = \frac{1}{2\pi j}\int [(X\bar{X} + \lambda Y\bar{Y})H\bar{H}\Phi - (XH + \bar{X}\bar{H})\Phi + \Phi]\, dz/z.$$

We need to find a rational function in z with no pole in the closed unit disk that minimizes $I(H)$.

Let $H(z)$ be the optimal one and denote its Laurent series expansion by

$$H(z) = \sum h_v z^v.$$

Note that $h_v = 0$, $v \leqslant 0$.

Since $H(z)$ is optimal, any variation in the coefficient of z^v, $v > 0$ cannot decrease I, i.e., $f(\varepsilon) = I(H + \varepsilon\eta^v) \geqslant f(0)$ for any $v > 0$. This implies that $(df/d\varepsilon)_0 = 0$ and $(d^2f/d\varepsilon^2)_0 \geqslant 0$.

By straightforward calculation,

(62)
$$(df/d\varepsilon)_0 = \frac{1}{2\pi j}\int [(X\bar{X} + \lambda Y\bar{Y})(Hz^{-v} + \bar{H}z^{v}) - (Xz^{v} + \bar{X}z^{-v})]\Phi dz/z = 0$$

and

(63)
$$(d^2f/d\varepsilon^2)_0 = \frac{1}{2\pi j}\int (X\bar{X} + \lambda Y\bar{Y})dz/z.$$

The integral of (63) is symmetric in z and z^{-1}, and hence the integral is nonnegative.

Let

(64)
$$L = (X\bar{X} + \lambda Y\bar{Y})\bar{H}\Phi - X\Phi.$$

Then (62) is rewritten as

$$(df/d\varepsilon)_0 = \frac{1}{2\pi j}\int (\bar{L}z^{-v-1} + Lz^{v-1})dz = 0.$$

Denote the Laurent series expansion of L by

$$L = \sum_{-\infty}^{\infty} l_v z^v.$$

Then

$$(df/d\varepsilon) = 2l_{-v} = 0, \qquad v \geqslant 0,$$

i.e., $L(0) = 0$ and $L(z)$ has no pole in the closed unit disk.

Factor $X\bar{X} + \lambda Y\bar{Y}$ as $\Lambda\bar{\Lambda}$ because it is symmetric in z and z^{-1}, where $\Lambda(z)$ has all its poles and zeros outside the unit disk. Factor Φ also as $\psi\bar{\psi}$.

Substitute these factored expressions into (64), and write it as

(65)
$$L/\Lambda\psi = \bar{\Lambda}\bar{H}\bar{\psi} - X\Phi/\Lambda$$

where the left-hand side is analytic inside the disk.

The Laurent series for $X\bar{\psi}/\Lambda$ is

(66)
$$\begin{aligned}\frac{X\bar{\psi}}{\Lambda} &= \sum_{v=-\infty}^{\infty} z^v \frac{1}{2\pi j}\int \frac{X(\omega)\Phi(\omega)}{\Lambda(\omega)}\omega^{-v-1}d\omega \\ &= \sum_{1}^{\infty} z^v \frac{1}{2\pi j}\int \frac{X(\omega)\Phi(\omega)}{\Lambda(\omega)}\omega^{-v-1}d\omega \\ &\quad + \sum_{0}^{\infty} z^{-v} \frac{1}{2\pi j}\int \frac{X(\omega)\Phi(\omega)}{\Lambda(\omega)}\omega^{-v-1}d\omega.\end{aligned}$$

The first of the preceding terms has no pole inside the unit disk, while the second term has no pole outside it.

From (65) and (66)

$$L/\Lambda\psi = -\sum_{v=1}^{\infty} z^{v}\frac{1}{2\pi j}\int\frac{X(\omega)\Phi(\omega)}{\Lambda(\omega)}\omega^{-v-1}\,d\omega$$

and

$$\bar{\Lambda}\bar{H}\bar{\psi} = \sum_{0}^{\infty} z^{-v}\frac{1}{2\pi j}\int\frac{X(\omega)\Phi(\omega)}{\Lambda(\omega)}\omega^{v-1}\,d\omega.$$

From the second equality,

$$H(z^{-1}) = \frac{1}{\Lambda(z^{-1})\psi(z^{-1})}\sum_{0}^{\infty} z^{-v}\frac{1}{2\pi j}\int\frac{X(\omega)\Phi(\omega)}{\Lambda(\omega)}\omega^{v-1}\,d\omega. \tag{67}$$

Substitute

$$X(z) = z^{k}\Pi(z)$$

and

$$Y(z) = A(z)K(z\Pi(z)/B(z)$$

into

$$\Lambda(z)\Lambda(z^{-1}) = \frac{\Pi\bar{\Pi}}{B\bar{B}}(B\bar{B} + \lambda A\bar{A}K\bar{K})$$

$$= \frac{\Pi\bar{\Pi}\Gamma\bar{\Gamma}}{B\bar{B}}$$

where

$$\Gamma\bar{\Gamma} = B\bar{B} + \lambda A\bar{A}K\bar{K}$$

where $\Gamma(z)$ is analytic inside the unit disk.

Then

$$\Lambda = \Pi\Gamma/B.$$

When this is substituted into (67), we obtain

$$H(z^{-1}) = \frac{B(z^{-1})}{\Gamma(z^{-1})\psi(z^{-1})\Pi(z^{-1})}\sum_{v=0}^{\infty} z^{-v}\int\frac{B(\omega)\Phi(\omega)}{\Gamma(\omega)}\omega^{k+v-1}\,d\omega,$$

and

$$G(z^{-1}) = \Pi(z^{-1})H(z^{-1})$$

is the solution we are after.

3.7 Open-Loop versus Closed-Loop Control

Controlling nondeterministic systems differs in many important ways from controlling deterministic ones. This section describes one important distinction, namely, the nonequivalence of open-loop and closed-loop controls in nondeterministic systems. Open-loop controls are sequences of controls $u_0, u_1, \ldots,$ where each u_k is a function of the initial condition of the dynamic model system parameters and possibly time but not of the state of the system now or in the past. Open-loop controls seek to minimize the cost functional expressed as functions of $u_0, u_1, \ldots, u_{N-1}$ in the N-stage decision problem after dependence of cost functional J on the state variables is substituted out

$$\min_{u_k, k=1,\ldots,N-1} J(u_0, u_1, \ldots, u_{N-1}).$$

Closed-loop controls, on the other hand, are basically stochastic because u_k at some future time k is expressed as a function of the future state x_k or some future observations on x_k that are stochastic. For deterministic systems, x_k is uniquely defined by the initial state, x_0, of the system and $u_0, u_1, \ldots, u_{k-1}$ which are themselves unique functions of x_0; hence u_k also depends uniquely on x_0, and this distinction is important only in organizing computationally but not in concept.

Bellman (1961) called attention to the difference of describing states of dynamic systems in terms of initial conditions and in terms of current state. This is the essential distinction between open-loop and closed-loop representation of the dynamic systems, which is also emphasized by Dreyfus (1964). We illustrate that open-loop and closed-loop controls of deterministic systems are the same in a simple example. Considere a scalar model

$$x_{k+1} = ax_k + bu_k, \qquad k = 0, 1, \ldots . \tag{68}$$

Its state at any time $t > 0$ can be uniquely specified once we know x_0 and we are given u_0, u_1, u_{t-1} because x_t is uniquely expressible as

$$x_t = a^t x_0 + bu_{t-1} + abu_{t-2} + \ldots + a^{t-1} bu_0.$$

Consider choosing u_0 and u_1 to minimize

$$J = (x_1^2 + x_2^2) + r(u_0^2 + u_1^2).$$

Stack x_2 above x_1 to form a two-dimensional vector $\mathbf{x}$. Do likewise to form u. Then, the cost functional J, now written as

$$J = \mathbf{x}'\mathbf{x} + r\mathbf{u}'\mathbf{u},$$

is to be minimized by u where

$$\mathbf{x} = \phi\mathbf{x}_0 + \mathbf{Hu}, \tag{70}$$

where

$$\mathbf{u}' = (u_1, u_0), \qquad \mathbf{x}' = (x_2, x_1),$$
$$\phi' = (a^2, a),$$

and

$$\mathbf{H} = \begin{bmatrix} b & ab \\ 0 & b \end{bmatrix}.$$

After (70) is substituted into J, J is minimized with respect to $\mathbf{u}$. From the first-order condition, $0 = \partial J/\partial \mathbf{u}$, we obtain

$$(\mathbf{H}'\mathbf{H} + r\mathbf{I})\mathbf{u} = -\mathbf{H}'\boldsymbol{\phi}\mathbf{x}_0,$$

or

$$\mathbf{u} = -(\mathbf{H}'\mathbf{H} + r\mathbf{I})^{-1}\mathbf{H}'\boldsymbol{\phi}\mathbf{x}_0. \tag{71}$$

This vector $\mathbf{u}$ gives the sequence of open-loop controls that minimizes J since $\partial^2 J/\partial \mathbf{u}^2 > 0$. When r is set to zero and if $b \neq 0$, then (71) becomes $\mathbf{u} = -H^{-1}\phi x_0$ indicating that $\mathbf{x} = 0$ because of (70). With zero cost assigned to control, $\mathbf{x} = 0$ achieves the minimal cost $J = 0$. Equation (71) shows that both u_0 and u_1 are proportional to x_0. Since

$$x_1 = r(b^2 + r)ax_0/D$$

with optimal control where

$$D = \det(\mathbf{H}'\mathbf{H} + \mathbf{I}),$$

we also see that the optimal u_1 is proportional to x_1

$$\begin{aligned} u_1 &= -a^2 brx_0/D \\ &= -abx_1/(b^2 + r). \end{aligned}$$

In the deterministic systems, the optimal open-loop control u_k, which is linear in x_0, can be shown to be proportional to x_k in general, $k = 0, 1, \ldots$.

3.8 Time-Consistent and Time-Inconsistent Solutions

In some sequential decision problems involving several decision makers, the decision problems interact in such a way that not all agents' decision problems can be formulated by the Bellman equations as sequential processes.

This phenomenon is known as time inconsistency. This section illustrates this point by an example that is a version of Kydland and Prescott (1971), because the nature of subtleties is best understood in terms of a simple example.

The next example simplifies the one in Buiter (1981).

Example 8. A scalar model is specified by

(72)
$$y_t = \alpha y_{t-1} + \beta x_t + \delta x_{t+1|t} + u_t, \qquad t = 1, 2$$
$$y_0 = 0,$$

and

$$x_3 = 0.$$

This is a stochastic linear quadratic version of the two period model of Kydland and Prescott (1971). The boundary condition $x_3 = 0$ is imposed to have a well-defined solution for the example. Imagine a situation in which the government credibly announces a termination of some program at the third period which corresponds to $x_3 = 0$ in the example and the public believes that $x_{3|2} = 0$. This renders $y_2 = \alpha y_1 + \beta x_2 + u_2$ to be well defined. The cost function is taken to be

$$J = y_1^2 + y_2^2 + r(x_1 - a)^2.$$

The information sets are $I_1 = \{y_0, x_1\}$ and $I_2 = \{y_0, y_1, x_1, x_2\}$. The decision x_t is taken before y_t and u_t are observed.

A time-consistent solution chooses x_2 given y_1 and x_1. Although y_1 is a bygone once period 2 arrives, Eq. (72) shows that x_2 affects y_1 via its expected value $x_{2|1}$ as

$$y_1 = \beta x_1 + \delta x_{2|1} + u_1.$$

The choice of x_2 on the assumption that y_1 is unaffected by x_2 is suboptimal.

Time-Consistent Solution

In this example, the dynamic programming formulation chooses x_2 by minimizing

$$\gamma_2 = E[y_2^2 | I_2] = (\alpha y_1 + \beta x_2)^2 + \sigma_2^2.$$

On the assumption that y_1 and x_1 are given, γ_2 is minimized by

(73)
$$x_2 = -\alpha y_1 / \beta,$$

which yields

$$\gamma_2^* = \sigma_2^2.$$

Dynamic programming then chooses x_1 by minimizing

$$\gamma_1 = E[y_1^2 + r(x_1 - a)^2 + \sigma_2^2 | I_1]$$

or equivalently

$$E[y_1^2 + r(x_1 - a^2) | I_1],$$

on the assumption that x_2 of (73) will be used in period 2, i.e., taking y_1 as

$$y_1 = \beta x_1 + \delta x_2. \tag{74}$$

Then

$$\begin{aligned}\gamma_1 - \sigma_2^2 &= E(y_1^2 + r(x_1 - a)^2 | I_1)\\ &= (\beta x_1 + \delta x_{2|1})^2 + r(x_1 - a)^2\\ &\quad + \delta^2 E[(x_2 - x_{2|1})^2 | I^1] + \sigma_1^2.\end{aligned}$$

To obtain $x_{2|1}$, use (73) and note that

$$x_{2|1} = -(\alpha/\beta) y_{1|1}$$

where from (74)

$$y_{1|1} = \beta x_1 + \delta x_{2|1}.$$

Solving these two equations jointly for $x_{2|1}$,

$$x_{2|1} = -\alpha\beta x_1/(\alpha\delta + \beta), \tag{75}$$

and

$$E((x_2 - x_{2|1})^2 | I_1) = (\alpha/\beta)^2 \sigma_1^2.$$

Thus

$$\gamma_1 - \sigma_2^2 = \beta^4 x_1^2/(\alpha\delta + \gamma)^2 + r(x_1 - a)^2 + (\alpha/\beta)^2 \sigma_1^2,$$

which is minimized by setting x_1 to

$$x_1 = ra(\alpha\delta + \beta)^2/[\beta^4 + r(\alpha\delta + \beta)^2]. \tag{76}$$

The total cost is

$$\gamma_1^* = (1 + \alpha^2/\beta^2)\sigma_1^2 + \sigma_2^2 + ra^2\beta^4/[\beta^4 + r(\alpha\delta + \beta)^2]. \tag{77}$$

Time-Inconsistent Solution

The difference between x_2 and $x_{2|1}$ is proportional to the innovation in the information set which is proportional to u_1. Express x_2 as

$$x_2 = x_{2|1} + \pi u_1$$

and the cost functional as

$$J = [\beta x_1 + \delta x_{2|1} + u_1]^2$$
$$+ [\alpha\beta x_1 + (\alpha\delta + \beta)x_{2|1} + (\alpha + \beta\pi)u_1 + u_2]^2 + r(x_1 - a)^2.$$

The expression $E(J \mid I_1)$ is now minimized with respect to $x_1, x_{2|1}$ and π. Clearly,

$$\pi = -\alpha/\beta.$$

The first-order conditions for x_1 and $x_{2|1}$ are

$$\beta(\beta x_1 + \delta x_{2|1}) + \alpha\beta[\alpha\beta x_1 + (\alpha\delta + \beta)x_{2|1}] + r(x_1 - a) = 0$$

and

(78) $$\delta(\beta x_1 + \delta x_{2|1}) + (\alpha\delta + \beta)[\alpha\beta x_1 + (\alpha\delta + \beta)x_{2|1}] = 0.$$

The expected cost is

$$\sigma_1^2 + \sigma_2^2 + (\beta x_1 + \delta x_{2|1})^2 + \alpha\beta x_1 + [(\alpha\delta + \beta)x_{2|1}]^2 + r(x_1 - a)^2.$$

Equation (78) can be rewritten as

$$[\beta\delta + r\delta(\alpha\delta + \beta)/\beta^2]x_1 + \delta^2 x_{2|1} = ra\delta(\alpha\delta + \beta)/\beta^2$$

and

$$[\beta\delta + \alpha\beta(\alpha\delta + \beta)]x_1 + [\delta^2 + (\alpha\delta + \beta)^2]x_{2|1} = 0$$

which are solved to yield

$$x_1^* = ra[\delta^2 + (\alpha\delta + p)^2]/\{\beta^4 + r[\delta^2 + (\alpha\delta + \beta)^2]\},$$

and

(79) $$x_{2|1}^* = -ra\beta[\delta + \alpha(\alpha\delta + p)]/\{\beta^4 + r[\delta^2 + (\alpha\delta + \beta)^2]\}.$$

The total cost is

(80) $$\sigma_1^2 + \sigma_2^2 + ra^2\beta^4/\{\beta^4 + r[\delta^2 + (\alpha\delta + \beta)^2]\}.$$

Compared with (79), (80) is clearly lower.

For any deterministic model $x_2 = x_{2|1}$, the optimal open-loop decision rule is

$$\overline{x_1} = x_1^*$$

and

$$\overline{x_2} = x_{2|1}^*,$$

i.e., with $\pi = 0$. The term $\alpha^2\sigma_1^2$ is added to (80) as the minimal cost under open-loop rules.

Appendices

Integration by Completing Squares

When densities have sufficient statistics, updating probability densities takes the form of recursion formulas for sufficient statistics. For normal distributions, conditional means and covariances are the sufficient statistics, and their recursion formulas can be obtained directly without explicit integration because conditional means and variances can be directly calculated as shown in this chapter. However, as an exercise we evaluate the integral by completion of squares. Given

$$p(\mathbf{x}) = \text{const} \exp - \tfrac{1}{2}(\mathbf{x} - \mathbf{a})'\mathbf{P}^{-1}(\mathbf{x} - \mathbf{a})$$

and

$$p(\mathbf{y} \mid \mathbf{x}) = \text{const} \exp - \tfrac{1}{2}(\mathbf{y} - \mathbf{Cx})'\mathbf{R}^{-1}(\mathbf{y} - \mathbf{Cx}),$$

the integral $p(\mathbf{y}) = \int p(\mathbf{y} \mid \mathbf{x}) p(\mathbf{x})\, d\mathbf{x}$ can be evaluated by completion of the squares of the exponent expressions

$$\begin{aligned}
&(\mathbf{x} - \mathbf{a})'\mathbf{P}^{-1}(\mathbf{x} - \mathbf{a}) + (\mathbf{y} - \mathbf{Cx})'\mathbf{R}^{-1}(\mathbf{y} - \mathbf{Cx}) \\
&= \mathbf{x}'(\mathbf{P}^{-1} + \mathbf{C}'\mathbf{R}^{-1}\mathbf{C})\mathbf{x} - \mathbf{x}'(\mathbf{P}^{-1}\mathbf{a} + \mathbf{C}'\mathbf{R}^{-1}\mathbf{y}) + \mathbf{a}'\mathbf{P}^{-1}\mathbf{a} + \mathbf{y}'\mathbf{R}^{-1}\mathbf{y} \\
&= (\mathbf{x} - \mathbf{W}^{-1}\mathbf{z})'\mathbf{W}(\mathbf{x} - \mathbf{W}^{-1}\mathbf{z}) + \mathbf{a}'\mathbf{P}^{-1}\mathbf{a} + \mathbf{y}'\mathbf{R}^{-1}\mathbf{y} - \mathbf{z}'\mathbf{W}^{-1}\mathbf{z},
\end{aligned}$$

where

$$\mathbf{W}^{-1} = \mathbf{P}^{-1} + \mathbf{C}'\mathbf{R}^{-1}\mathbf{C}$$

and

$$\mathbf{W}^{-1}\mathbf{z} = \mathbf{P}^{-1}\mathbf{a} + \mathbf{C}'\mathbf{R}^{-1}\mathbf{y}.$$

The last three terms are grouped as

$$(\mathbf{y} - \mathbf{b})'\boldsymbol{\Sigma}^{-1}(\mathbf{y} - \mathbf{b})$$

where

$$\begin{aligned}
\boldsymbol{\Sigma}^{-1} &= \mathbf{R}^{-1} - \mathbf{R}^{-1}\mathbf{CWC}'\mathbf{R}^{-1} \\
&= (\mathbf{R} + \mathbf{CPC}')^{-1},
\end{aligned}$$

and

$$\begin{aligned}
\mathbf{b} &= \boldsymbol{\Sigma}\mathbf{R}^{-1}\mathbf{CWP}^{-1}\mathbf{a} \\
&= \boldsymbol{\Sigma}[\mathbf{R}^{-1} - \mathbf{R}^{-1}\mathbf{CPC}'(\mathbf{R} + \mathbf{CPC}')^{-1}]\mathbf{Ca} \\
&= \boldsymbol{\Sigma}\boldsymbol{\Sigma}^{-1}\mathbf{Ca} \\
&= \mathbf{Ca}.
\end{aligned}$$

Note that

$$\mathbf{a}'(\mathbf{P}^{-1} - \mathbf{P}^{-1}\mathbf{W}\mathbf{P})\mathbf{a} - \mathbf{b}'\boldsymbol{\Sigma}^{-1}\mathbf{b} = \mathbf{a}'(\mathbf{P}^{-1} - \mathbf{P}^{-1}\mathbf{W}\mathbf{P}^{-1} - \mathbf{C}'\boldsymbol{\Sigma}^{-1}\mathbf{C})\mathbf{a} = 0$$

where

$$\mathbf{P}^{-1} - \mathbf{P}^{-1}\mathbf{W}\mathbf{P}^{-1} - \mathbf{C}'\boldsymbol{\Sigma}^{-1}\mathbf{C} = 0,$$

since

$$\mathbf{P}^{-1}\mathbf{W}\mathbf{P}^{-1} = \mathbf{P}^{-1} - \mathbf{C}'(\mathbf{R} + \mathbf{C}\mathbf{P}\mathbf{C}')^{-1}\mathbf{C}.$$

Granger Causality

Denote by $H(x^t)$ the completion of a linear hulls of the subsets $\{x_s, s \leqslant t\}$ with respect to the mean-square norm. The orthogonal projection of x_t onto the subspace $H(x^{t-1})$ is denoted by $x_{t|t-1} = \hat{E}(x_t | H(x^{t-1}))$. This is also written as $E(x_t | x^{t-1})$.

The innovation process e_t^0, where $e_t^0 = x_t - x_{t|t-1}$, is then orthogonal to $H(x^{t-1})$, i.e., $\hat{E}(e_t^0 | H(x^{t-1})) = 0$. This fact is denoted by $e_t^0 \perp H(x^{t-1})$. With any multivariate process $\{(x_t, y_t)\}$, the relation $H(x_t) \subset H(x^t, y^t)$ is obtained. When $\hat{E}[x_t | H(x^{t-1}, y^{t-1})] \in H(x^{t-1})$, $\{y_t\}$ is said not to Granger cause x_t, Granger (1969). This condition is equivalent to the innovation process of x_t with respect to its own past, $e_t^0 = x_t - x_{t|t-1}$, being orthogonal to $H(x^{t-1}, y^{t-1})$ for all t. First, note that $e_t^0 \perp H(x^{t-1})$ implies $e_t^0 \perp H(x^{t-1}, y^{t-1})$ because of the definition of Granger causality. Conversely, suppose that $e_t^0 \perp H(x^{t-1}, y^{t-1})$, i.e., $\hat{E}[x_t | H(x^{t-1}, y^{t-1})] = \hat{E}[x_{t|t-1} | H(x^{t-1}, y^{t-1})]$. Then $x_{t|t-1} \in H(x^{t-1}) \subset H(x^{t-1}, y^{t-1})$ implies that $\hat{E}[x_{t|t-1} | H(x^{t-1}, y^{t-1})] = x_{t|t-1}$. Therefore $\hat{E}[x_t | H(x^{t-1}, y^{t-1})] = \hat{E}[x_t | H(x^{t-1})]$, i.e., $e_t^0 \perp H(x^{t-1})$.

Sims (1972) proposed a condition for $\{y_t\}$ not causing $\{x_t\}$ to be that in the decomposition

$$y_t = \hat{E}(y_t | x^t) + \eta_t,$$

$\eta_t \in H(x)^\perp$ for all t, where $H(x) = \bigcup_t H(x^t)$. Then $H(y^t) \subset H(x^t) \oplus H(x)^\perp$ and hence $H(x^t, y^t) \subset H(x^t) \oplus H(x)^\perp$.

Hosoya (1977) established the equivalence of Granger and Sims conditions that is described next. Denote the revision processes with respect to own past observations by

$$e_t^j = x_{t+j|t} - x_{t+j|t-1}, \qquad j \geqslant 0.$$

We decompose x_{t+p} into a component in $H(x^t)$ and its orthogonal complement as follows:
First,

$$x_{t+p} = e_{t+p}^0 + x_{t+p|t+p-1}$$

where

$$e^0_{t+p} \perp H(x^{t+p-1}).$$

Second,

$$x_{t+p|t+p-1} = x_{t+p|t+p-2} + e^1_{t+p-1}$$

where

$$e^1_{t+p-1} \perp H(x^{t+p-2}).$$

Telescoping the process

$$x_{t+p} = e^0_{t+p} + e^1_{t+p-1} + \ldots + e^{p-1}_{t+1} + x_{t+p|t}$$

where

$$x_{t+p|t} \in H(x^t)$$

and

$$e^{p-1}_{t+1} \perp H(x^t).$$

Since $H(x^t) \subset H(x^{t+1}) \subset \ldots \subset H(x^{t+p-1})$, the term $e^0_{t+p} + \ldots + e^{p-1}_{t+1}$ is orthogonal to $H(x^t)$. Now if

$$e^j_{t+p-j} \perp H(x^t, y^t)$$

for all $p \geqslant j \geqslant 0$, then so is the sum

$$e^0_{t+p} + \ldots + e^{p-1}_{t+1} \qquad \text{for all } p \geqslant 1.$$

Consider e^0_{t+1}. Since $e^0_{t+1} \perp H(x^t)^\perp$, $e^0_{t+1} \perp H(x)^\perp$. Any element g of $H(y^t)$ is decomposed as $g = l + \lambda$ where $l \in H(x^t)$ and $\lambda \in H(x)^\perp$.

From $e^0_{t+1} \perp l$, and $e^0_{t+1} \perp \lambda$ follows $e^0_{t+1} \perp g$.

Since g is arbitrary, $e^0_{t+1} \perp H(y^t)$, i.e., $e^0_{t+1} \perp H(x^t, y^t)$. That is, $\{x^t, y^t\}$ satisfies the Granger condition.

CHAPTER 4

Time Series and Econometric Models: Examples

This chapter collects some simple dynamic models drawn mostly from the literature on econometrics and time series.

4.1 Cointegrated Time Series

***Example 1: A Common Trend in a Stochastic Growth Model* (*King, Plossor, Stock, and Watson* [1987]).** The following is a simple stochastic growth model with exogenous productivity shock postulated to be

$$\ln \lambda_t = \ln \lambda_{t-1} + \mu + \eta_t$$

where

$$\eta_t \sim NID(0, \sigma^2).$$

This factor λ_t appears in the production function $Y_t = \lambda_t K_t^{1-\alpha}$. For simplicity we assume full employment and 100% depreciation so that $K_{t+1} = I_t$.

The social planner's problem is to solve

$$V(K_t) = \max E_t[\theta \ln C_t + \beta V(K_{t+1})]$$

where

$$K_{t+1} = \lambda_t K_t^{1-\alpha} - C_t.$$

Postulate the functional form

$$V(K_t) = a \ln K_t + b_t$$

and substitute it in the functional equation. The optimal level of consumption is given by

$$\ln C_t = \ln[\theta/(\theta + \beta\alpha)] + \ln \lambda_t + (1 - \alpha) \ln K_t.$$

The output is the same as the capital stock in this model,

$$\ln Y_t = \ln I_t = \ln K_{t+1} = \ln \lambda_t + (1 - \alpha) \ln K_t.$$

The state space model is

$$\begin{bmatrix} \ln \lambda_{t+1} \\ \ln k_{t+1} \end{bmatrix} = \begin{bmatrix} 1 & 0 \\ 1 & 1 - \alpha \end{bmatrix} \begin{bmatrix} \ln \lambda_t \\ \ln k_t \end{bmatrix} + \begin{bmatrix} \mu + \eta_t \\ 0 \end{bmatrix}$$

and

$$\begin{bmatrix} \ln c_t \\ \ln Y_t \end{bmatrix} = \begin{bmatrix} 1 \\ 1 \end{bmatrix} \{\ln \lambda_t + (1 - \alpha) \ln k_t\} + \begin{pmatrix} \gamma \\ 0 \end{pmatrix}$$

for some constant γ. Note that the dynamics have "an error-correction" form we later describe. Here, $\ln \lambda_t$ is a common trend factor in these series since it has the unit root. Note that the unit root that is a "common trend" term in them is exogenously imposed by the assumed dynamics for $\ln \lambda_t$.

***Example 2: Quadratic Utility Function* (*Cochrane* [1988]).** The social planner's problem with a quadratic utility function $-\frac{1}{2}E_t \Sigma_{j=0}^{\infty} \beta^j (c_{t+j} - \bar{c})^2$ and wealth A_t that earns a fixed rate of return produces the functional equation

$$V_t(A_t) = \max_{c_t} E_t[-\tfrac{1}{2}(c_t - \bar{c})^2 + \beta V_{t+1}(A_t^+)]$$

where

$$A_t^+ = (1 + r)A_t + w_t - c_t, \qquad \beta = (1 + r)^{-1},$$

where $\{w_t\}$ is a weakly stationary income (wage) stochastic process. Set $\bar{c} = 0$ without loss of generality. Postulate a quadratic form for $V(\cdot)$ as

$$V_t(A) = a_t + b_t A + \frac{d}{2} A^2.$$

Maximization yields the first-order condition for the optimal stationary consumption rule

(1) $$c_t = -[\beta d(1 + r)/(1 - \beta d)]A_t + \gamma_t$$

where

$$\gamma_t = -\beta(b_{t+1|t} + dw_t)/(1 - \beta d).$$

Substitute this into the functional equation and equate the coefficients of the quadratic term in A_t on both sides of the equation

$$1 = (r + 1)/(1 - \beta d)$$

where $\beta(1 + r) = 1$ is used. This determines d to be $-r/\beta$.

The optimal consumption is then equal to

$$c_t = rA_t + \gamma_t,$$

where

$$\gamma_t = \beta r w_t - \beta^2 b_{t+1|t}.$$

Comparison of the coefficients multiplying A_t yields

$$b_t = -rw_t + \beta b_{t+1|t}.$$

Since $0 < \beta < 1$, we solve this equation forward to obtain the solution as

$$b_t = -rE_t \sum_{j=0}^{\infty} \beta^j w_{t+j}.$$

Then

$$\gamma_t = (1 - \beta)E_t \sum_0^{\infty} \beta^j w_{t+j}.$$

The wealth evolves with time according to

$$A_{t+1} = A_t + w_t - (1 - \beta)E_t \sum_{j=0}^{\infty} \beta^j w_{t+j}.$$

Real output (GNP) is the sum of labor income and interest income $y_t = rA_t + w_t$. The optimal consumption level is related to the income by

$$\begin{aligned} c_t &= y_t + (1 - \beta)E_t \sum_0^{\infty} \beta^j w_{t+j} - w_t \\ &= y_t - \beta w_t + (1 - \beta)E_t \sum_1^{\infty} \beta^j w_{t+j}. \end{aligned}$$

Therefore, when $\{w_t\}$ is a weakly stationary stochastic process, $c_t - y_t$ is also a weakly stationary stochastic process. We say c_t and y_t are cointegrated with the cointegrating vector $(1, -1)$. Unlike the model in Example 1, this model of consumption does not allow for growth in consumption and income.

Example 3: Present Value Model. Some intertemporal problems has no mechanisms for linking constraints at different times. Such problems have no state variables as such. Intertemporal aspects come solely from the multi-stages of criterion functions, i.e., a criterion function being determined by variables at several distinct time instants. Consider an example of Nickell (1985)

$$L_t = \sum_{s=t}^{\infty} \delta^t [\lambda(x_{t+s} - y_{t+s})^2 + (x_{t+s} - u_{t+s-1})^2]$$

where y_{t+s} is an exogenously given target, u_{t+s} is the decision variable, and $0 < \delta < 1$ is a discount factor. Direct differentiation of L_t with respect to u_{t+s} generates a difference equation for it as

$$\delta x_{t+s+1} - (1 + \delta + \lambda) x_{t+s} + x_{t+s-1} = -\lambda y_{t+s}, \qquad s \geqslant 0.$$

The dynamics is entirely in the way controls are generated. Write it as

$$[\delta L^{-1} - (1 + \delta + \lambda) + L] x_{t+s} = -\lambda y_{t+s} \tag{2}$$

where the expression in the square bracket is factored as $(1 - \mu L)(1 - \delta\mu L^{-1})$ where μ is a root of $\delta\mu^2 - (1 + \delta + \lambda)\mu + 1 = 0$. Choose one that satisfies $|\delta\mu| < 1$ so that

$$\frac{1}{1 - \delta\mu L^{-1}} = 1 + \delta\mu L^{-1} + (\delta\mu)^2 L^{-2} + \ldots$$

is well defined. Equation 2 is rewritten as

$$\begin{aligned} -(1/\mu)(1 - \mu L) x_t &= -\lambda(1 - \delta\mu L^{-1}) y_t \\ &= -\lambda \sum_{s=0}^{\infty} (\delta\mu)^s y_{t+s} \end{aligned}$$

or

$$x_t = \mu x_{t-1} + \mu\lambda \sum_{s=0}^{\infty} (\delta\mu)^s y_{t+s}. \tag{3}$$

The summation may be thought of as the present value of the streams of target values. Equation (3) is in the form of the present-value model as named by Campbell and Shiller (1988), when deterministic terms are replaced by conditional expectations $\Sigma_{s=0}^{\infty} (\delta\mu)^s E_t(y_{t+s})$.

Example 4: (Hendry et al. [1984]). A one-period minimization of a quadratic cost

$$\Delta \mathbf{y}_t' \mathbf{R} \Delta \mathbf{y}_t + (\mathbf{y}_t^* - \mathbf{y}_t)' \mathbf{Q} (\mathbf{y}_t^* - \mathbf{y}_t),$$

with respect to $\mathbf{y}_t$, where $\mathbf{y}_t^*$ is an exogenously specified equilibrium value for $\mathbf{y}_t$, and $\mathbf{Q}$ and $\mathbf{R}$ are both positive definite, leads to the optimal decision rule

$$\Delta\mathbf{y}_t = \mathbf{D}(\mathbf{y}_t^* - \mathbf{y}_{t-1})$$

where $\mathbf{D} = (\mathbf{Q} + \mathbf{R})^{-1}\mathbf{Q}$. This is a type of decision rule used in some disequilibrium models. If both $\mathbf{R}$ and $\mathbf{Q}$ are diagonal, then so is $\mathbf{D}$. Then $d_j = q_j/(q_i + r_j) < 1$. In this case, decision rule produces a stable difference equation $\mathbf{y}_t = (\mathbf{I} - \mathbf{D})\mathbf{y}_{t-1} + \mathbf{D}\mathbf{y}_t^*$ where $\mathbf{I} - \mathbf{D} = (\mathbf{Q} + \mathbf{R})^{-1}\mathbf{R}$. Now consider a multiperiod loss function

$$\sum_{t=1}^{\infty} [\Delta\mathbf{y}_t'\mathbf{R}\Delta\mathbf{y}_t + (\mathbf{y}_t^* - \mathbf{y}_t)'\mathbf{Q}(\mathbf{y}_t^* - \mathbf{y}_t)]$$

in a stochastic environment. If the decision maker perceives that $E(\mathbf{y}_{t+s}^*|I_t) = \mathbf{y}_t^*$ for all $s \geqslant 0$, this problem essentially reduces to the one-period problem. On the other hand, if the decision maker assumes that $\mathbf{y}_t^*$ is generated by $\Delta\mathbf{y}_t^* = \mathbf{A}\Delta\mathbf{y}_{t-1}^* + \mathbf{v}_t$, then the linear decision rule of the form $\Delta\mathbf{y}_t = \mathbf{D}_1\Delta\mathbf{y}_t^* + \mathbf{D}_2(\mathbf{y}_t^* - \mathbf{y}_{t-1})$ results. This is a type of error-correction mechanism. Error-correction models are discussed later in this chapter. See Nickell (1983) for further detail on the relation of error-correction models with optimization problems.

***Example 5(Pagan* [1985]).** Minimization of a one-period quadratic criterion

$$\min[a(y_t - y_t^*)^2 + b(\Delta y_t - \alpha_t)^2],$$

where y_t^* and α_t are exogenously given, with respect to y_t, yields the optimal decision rule

$$y_t = (a + b)^{-1}(ay_t^* + by_{t-1} + bx_t).$$

This cost function implies that the decision maker is interested in keeping y_t close to a growing target path y_t^* while maintaining the growth rate specified by α_t. To reflect this, we restate the decision rule in terms of Δy_t as

$$\Delta y_t = [a(y_t^* - y_{t-1}) + b\alpha_t]/(a + b).$$

This makes it clear that the decision on Δy_t is formed as a weighted sum of the past deviation $y_t^* - y_{t-1}$ and the desired growth rate α_t. Alternatively we may detrend y_t by α_t and restate the decision rule as

$$y_t - y_{t-1} - \alpha_t = [a/(a + b)](y_t^* - y_{t-1} - \alpha_t),$$

which shows that the decision is proportional to the gap between y_t^* and $y_{t-1} + \alpha_t$, i.e., the detrended gap, or the rule is a partial adjustment rule on the detrended gap. To restate the decision rule to emphasize this point, write the rule as

$$y_t - y_t^* = -\frac{b}{a+b}[(y_{t-1}^* - y_{t-1}) + \Delta y_t^* - \alpha_t].$$

It shows that if Δy_t^* is equal to α_t, this is nothing but a partial adjustment rule, i.e., this is a dynamic partial adjustment rule.

Example 6: Error Correction Model. By imposing a constraint that the coefficients sum to 1, we can transform an autoregressive distributed lag model into an error-correction model. Aoki (1988) discusses the dual relation between error correction models and certain state space models. With the constraint $a_1 + b_0 + b_1 = 1$, the model

$$y_t = a_1 y_{t-1} + b_0 x_t + b_1 x_{t-1} + e_t$$

is transformed into

$$\begin{aligned}\Delta y_t &= (a_1 - 1) y_{t-1} + b_0 \Delta x_t + (b_0 + b_1) x_{t-1} + e_t \\ &= b_0 \Delta x_t - (1 - a_1)(y_{t-1} - x_{t-1}) + e_t\end{aligned}$$

since $b_0 + b_1 = 1 - a_1$.

4.2 Dynamic Multiplier and Average Lag Length

Given a dynamic model either in transfer function or in state space form, we define a sequence $\{H_i\}$, called impulse responses or dynamic multipliers, as a sequence of model responses to a unit impulse i period earlier, $i = 0, 1, 2, \ldots$.

In a representation

$$\mathbf{D}(q^{-1})\mathbf{y}_t = \mathbf{N}(q^{-1})\mathbf{e}_t \tag{4}$$

where

$$\mathbf{D}(z) = \mathbf{I} + \mathbf{D}_1 z + \ldots + \mathbf{D}_n z^n$$

and

$$\mathbf{N}(z) = \mathbf{N}_0 + \mathbf{N}_1 z + \ldots + \mathbf{N}_n z^n,$$

the generating function H(z) of impulse responses is obtained by $\mathbf{y}_t = \mathbf{H}(z)\mathbf{e}_t$ where

$$\mathbf{H}(z) = \mathbf{D}^{-1}(z)\mathbf{N}(z)$$

and

$$\mathbf{H}(z) = \sum_0^\infty \mathbf{H}_i z^i.$$

This is a form of distributed lag models. In a state space representation,

$$\begin{aligned} \mathbf{x}_{t+1} &= \mathbf{A}\mathbf{x}_t + \mathbf{B}\mathbf{e}_t \\ \mathbf{y}_t &= \mathbf{C}\mathbf{x}_t + \mathbf{D}\mathbf{e}_t \\ \mathbf{H}(z^{-1}) &= \mathbf{D} + \mathbf{C}(z\mathbf{I} - \mathbf{A})^{-1}\mathbf{B} \end{aligned} \tag{5}$$

i.e.,

$$\mathbf{H}_i = \mathbf{C}\mathbf{A}^{i-1}\mathbf{B}, \qquad i = 1, 2, \ldots$$

$$\mathbf{H}_0 = \mathbf{D}.$$

The long-run multiplier is defined to be $\mathbf{H}(1) = \Sigma_0^\infty \mathbf{H}_i = \mathbf{D}^{-1}(1)\mathbf{N}(1)$ in the distributed lag model and is equal to $\mathbf{D} + \mathbf{C}(\mathbf{I} - \mathbf{A})^{-1}\mathbf{B}$ in the state space form.

In the distributed lag model (4), the average lag is defined for scalar models by

$$\Psi = \Sigma i H_i / \Sigma H_i = \frac{d}{dL} \ln H(L)|_{L=1}.$$

In terms of $D(L)$ and $N(L)$,

$$\begin{aligned} \frac{d}{dL} \ln H(L) &= \left[-\frac{D'(L)}{D(L)} + \frac{N'(L)}{N(L)} \right]_{L=1} \\ &= -\Sigma_i D_i / D(1) + \Sigma_i N_i / N(1). \end{aligned}$$

The average lag matrix is equal to be $\mathbf{C}(\mathbf{I} - \mathbf{A})^{-2}\mathbf{B}$ in the state space model, which is a multivariate extension of the notion of the average lag.

By rewriting the transfer function $\mathbf{H}(L)$ for Eq. (5) as $\mathbf{H}(L) = \mathbf{H}(1) + (1 - L)\mathbf{H}^*(L)$, we have

$$\begin{aligned} \mathbf{y}_t &= \mathbf{H}(1)\mathbf{x}_t + \mathbf{H}^*(L)\Delta\mathbf{x}_t \\ &= \mathbf{H}(1)[\mathbf{x}_t + \mathbf{H}(1)^{-1}\mathbf{H}^*(L)\Delta\mathbf{x}_t]. \end{aligned} \tag{6}$$

We recognize $\mathbf{H}(1)$ as the long-run multiplier. Since

$$\mathbf{H}^*(1) = -\mathbf{C}(\mathbf{I} - \mathbf{A})^{-2}\mathbf{B},$$

because

$$\begin{aligned} \mathbf{H}(L) - \mathbf{H}(1) &= \mathbf{C}(\mathbf{I} - \mathbf{A}L)^{-1}\mathbf{B}L - \mathbf{C}(\mathbf{I} - \mathbf{A})^{-1}\mathbf{B} \\ &= \mathbf{C}(\mathbf{I} - \mathbf{A}L)^{-1}[(\mathbf{I} - \mathbf{A})L - (\mathbf{I} - \mathbf{A}L)](\mathbf{I} - \mathbf{A})^{-1}\mathbf{B} \\ &= (L - 1)\mathbf{C}(\mathbf{I} - \mathbf{A}L)^{-1}(\mathbf{I} - \mathbf{A})^{-1}\mathbf{B}, \end{aligned}$$

we refer to $-\mathbf{H}(1)^{-1}\mathbf{H}^*(L)$ as the average lag length matrix.

Often, one is interested not in model parameters per se but rather in some parameter combinations that summarize some aspects of model behavior. Long-run multipliers and average lag lengths are two such summaries.

Equation (6) shows that the long-run multiplier can be directly estimated by putting the model in the form of (6). We can similarly transform an autoregressive distributed lag model into an alternative form as follows. Start with

$$\mathbf{D}(L)\mathbf{y}_t = \mathbf{N}(L)\mathbf{x}_t + \mathbf{u}_t$$

where

$$\mathbf{D}(L) = \mathbf{I}_0 - \mathbf{D}_1 L - \ldots - \mathbf{D}_n L^n.$$

Rewrite the polynomial matrices as

$$\mathbf{D}(L) = \mathbf{D}(1) - (1 - L)\mathbf{D}^*(L)$$

and

$$\mathbf{N}(L) = \mathbf{N}(1) - (1 - L)\mathbf{N}^*(L),$$

which is always possible by expanding polynomials not in L but in $L = L - 1 + 1$. If $\mathbf{D}(1)$ is not singular, then the model is restated as

$$\begin{aligned}\mathbf{y}_t = {} & \mathbf{D}(1)^{-1}\mathbf{D}^*(L)\,\Delta y_t + \mathbf{D}(1)^{-1}\mathbf{N}(1)\mathbf{x}_t \\ & - \mathbf{D}(1)^{-1}\mathbf{N}^*(L)\,\Delta\mathbf{x}_t + \mathbf{D}(1)^{-1}\mathbf{u}_t.\end{aligned}$$

In this form, the long-run multiplier matrix $\mathbf{D}(1)^{-1}\mathbf{N}(1)$ is the coefficient matrix that is directly estimated. This is basically the procedure proposed by Bewley (1979).

Because $D'(L) = D^*(L)$ and $N'(L) = N^*(L)$, the average lag matrix defined to be $-D(1)^{-1}D'(1) + N(1)^{-1}N'(1)$ is equal to $-D(1)^{-1}D^*(1) + H^{-1}D(1)^{-1}N^*(1)$ where $\mathbf{H} = D^{-1}(1)N(1)$ is the long-run multiplier matrix where $D(1)^{-1}D^*(1)$ and $D(1)^{-1}N^*(1)$ are directly available as the coefficient matrices of Δy_t and Δx_t, respectively.

***Example** 7.* Given a scalar valued model $y_t = a_1 y_{t-1} + \beta_0 x_t + e_t$, introduce $\Delta y_t = y_t - y_{t-1}$, i.e., $y_{t-1} = y_t - \Delta y_t$, and rewrite as

$$\begin{aligned}y_t &= -\frac{a_1}{1 - a_1}\,\Delta y_t + \frac{\beta_0}{1 - a_1}x_t + \frac{e_t}{1 - a_1} \\ &= -\psi\Delta y_t + \theta x_t + u_t.\end{aligned}$$

In this form, the coefficients of Δy_t and x_t are directly interpretable as the average lag length and the long-run multiplier.

4.3 Partial Adjustment Models and Equilibrium Dynamics Models

Strictly speaking, an equilibrium of a dynamic model is a state from which the model does not move in the absence of any disturbance. For example, in a model described by

$$x_{t+1} = \mathrm{f}(x_t, e_t) \tag{7}$$

where e_t is a mean zero white noise, any x^* satisfying

$$x^* = \mathrm{f}(x^*, 0)$$

is an equilibrium.

In economic models, one often wishes to extend the notion of equilibrium to growth or other dynamic situations. In other words, we often wish to speak of equilibrium dynamics of a sequence $\{x_t^*\}$ generated by

$$x_{t+1}^* = \mathrm{g}(x_t^*) \tag{8}$$

or even with

$$x_{t+1}^* = \mathrm{g}(x_t^*, e_t)$$

where e_t is some exogenous disturbance affecting the equilibrium paths! For (8) to be an "equilibrium" dynamics for the model (7), there is of course, some explicit or implicit understanding that the dynamics of (8) are more gradual than those of (7). Otherwise the notion of "equilibrium" does not make sense. Often, (8) has a mode of an exponential growth and (7) a cyclical fluctuation about such a smooth moving path, or (7) may contain (8) as a submodel in some sense. For example, when (7) and (8) are both linear, dynamic modes of (7) may include those of (8). We make this precise later in connection with the notion of dynamic aggregation.

There is literature on error correction models that addresses just this question. The question is typically phrased in terms of linear models in autoregressive form (also called autoregressive adjustment model)

$$\mathbf{A}(L)\mathbf{x}_t = \boldsymbol{\varepsilon}_t \tag{9}$$

and asks if it contains the equilibrium dynamics

$$\mathbf{B}(L)\mathbf{x}_t^* = \mathbf{e}_t \tag{10}$$

or if (9) is consistent with the dynamic restriction imposed by (10).

Intuitively it is clear that $\mathbf{B}(L)$ should be a factor of $\mathbf{A}(L)$, a right factor to be precise so that $\mathbf{A}(L) = \mathbf{A}_1(L)\mathbf{B}(L)$, because then

$$\mathbf{A}_1(L)\mathbf{B}(L)\mathbf{x}_t = \boldsymbol{\varepsilon}_t$$

implies that

$$\mathbf{A}_1(L)\mathbf{e}_t = \boldsymbol{\varepsilon}_t,$$

which constrains the equilibrium dynamics disturbance $\mathbf{e}_t$.

A simple way to relate (9) with (10) is to work with deviational variables by measuring $\mathbf{x}_t$ from the equilibrium value $\mathbf{x}_t^*$

$$\delta\mathbf{x}_t = \mathbf{x}_t - \mathbf{x}_t^*.$$

Then (9) becomes

$$\mathbf{A}(L)(\delta\mathbf{x}_t + \mathbf{x}_t^*) = \boldsymbol{\varepsilon}_t$$

or

$$\begin{aligned} \mathbf{A}(L)\delta\mathbf{x}_t &= \boldsymbol{\varepsilon}_t - \mathbf{A}(L)\mathbf{x}_t^* \\ &= \boldsymbol{\varepsilon}_t - \mathbf{A}(L)\mathbf{B}(L)^{-1}\mathbf{e}_t. \end{aligned} \tag{11}$$

For $\delta\mathbf{x}_t$ to go to zero in some probability sense as $t \to \infty$, clearly we must require that $\mathbf{A}(L)$ is a stable polynomial matrix and that $\mathbf{A}(L)\mathbf{B}(L)^{-1}$ does not introduce any unstable root so that $\delta\mathbf{x}_t$ becomes weakly stationary, i.e., $\mathbf{A}(L)\mathbf{B}(L)^{-1} = \mathbf{C}(L)$ or $\mathbf{A}(L) = \mathbf{C}(L)\mathbf{B}(L)$ for a polynomial matrix $\mathbf{C}(L)$, i.e., $\mathbf{B}(L)$ is a right factor of $\mathbf{A}(L)$. Furthermore, (11) also constrains e_t by

$$\mathbf{A}(L)\mathbf{B}(L)^{-1}\mathbf{e}_t = \boldsymbol{\varepsilon}_t$$

in equilibrium, or $\mathbf{C}(L)\mathbf{e}_t = \boldsymbol{\varepsilon}_t$, requiring that $\mathbf{C}(L)$ be a stable polynomial. This point is brought out more clearly when (9) is in an error-correction form

$$\mathbf{x}_t = \mathbf{A}(L)(\mathbf{x}_t^* - \mathbf{x}_t) + \boldsymbol{\varepsilon}_t \tag{12}$$

with

$$\mathbf{B}(L)\mathbf{x}_t^* = \mathbf{e}_t,$$

where

$$\mathbf{A}(L) = \mathbf{D}(L)^{-1}\mathbf{N}(L).$$

Then

$$\begin{aligned} [\mathbf{D}(L) + \mathbf{N}(L)](\mathbf{x}_t - \mathbf{x}_t^*) &= -\mathbf{D}(L)(\mathbf{x}_t^* - \boldsymbol{\varepsilon}_t) \\ &= -\mathbf{D}(L)[\mathbf{B}(L)^{-1}\mathbf{e}_t - \boldsymbol{\varepsilon}_t] \\ &= -\mathbf{D}(L)\mathbf{B}(L)^{-1}[\mathbf{e}_t - \mathbf{B}(L)\boldsymbol{\varepsilon}_t]. \end{aligned} \tag{13}$$

If the model (12) is to approach the dynamic equilibrium of (10), i.e., $\mathbf{x}_t - \mathbf{x}_t^* \to 0$ as t goes to infinity in some probability sense, given that $\mathbf{e}_t$ and $\boldsymbol{\varepsilon}_t$ are both mean zero weakly stationary, we must at least require that (13) be a stable dynamic model, i.e., the zeros of $[\mathbf{D}(L) + \mathbf{N}(L)]$ are all outside the unit disk and $\mathbf{D}(L)\mathbf{B}(L)^{-1} = \mathbf{C}(L)$ is a polynomial matrix, i.e., $\mathbf{B}(L)$ is a right factor of $\mathbf{D}(L)$, i.e., $\mathbf{D}(L) = \mathbf{C}(L)\mathbf{B}(L)$, so that no unstable dynamic mode such as a unit root is introduced by $\mathbf{D}(L)\mathbf{B}(L)^{-1}$. This is important since $\mathbf{B}(L)$ will have a unit zero if (10) specifies a random growth condition.

***Example 8: A First-Order ADL Model* (*Hendry et al.* [1978]).** The model is

$$y_t = \alpha_1 y_{t-1} + \beta_1 x_t + \beta_2 x_{t-1} + \varepsilon_t. \tag{14}$$

The equilibrium model is specified by

$$\begin{aligned} y_t^* &= x_t^* \\ x_t^* &= k + e_t. \end{aligned} \tag{15}$$

By rewriting (14) in terms of deviational variables

$$\begin{aligned} \delta y_t = \alpha_1 \delta y_{t-1} &+ \beta_1 \delta x_t + \beta_2 \delta x_{t-1} \\ &+ (-1 + \alpha_1 + \beta_1 + \beta_2)k + (-1 + \beta_1)e_t \\ &+ (\alpha_1 + \beta_2)e_{t-1} + \varepsilon_t, \end{aligned}$$

we immediately derive a restriction on the parameter values

$$1 = \alpha_1 + \beta_1 + \beta_2$$

because otherwise δy_t does not have 0 as the (probability) limit. The equilibrium error process is determined by

$$(1 - \beta_1)e_t - (\alpha_1 + \beta_2)e_{t-1} = \varepsilon_t.$$

For e_t to be a weakly stationary process, another restriction, $|(\alpha_1 + \beta_2)/(1 - \beta_1)| < 1$, must be imposed.

In terms of lag polynomial matrices, (14) is written as

$$(1 - \alpha_1 L, \beta_1 + \beta_2 L)\begin{pmatrix} y_t \\ x_t \end{pmatrix} = \varepsilon_t$$

and (15) as

$$\begin{pmatrix} 1 & -1 \\ 0 & 1 \end{pmatrix}\begin{pmatrix} y \\ x \end{pmatrix}_t^* = \begin{pmatrix} 0 \\ 1 \end{pmatrix} k + \begin{pmatrix} 0 \\ 1 \end{pmatrix} e_t.$$

Note that the matrix $\mathbf{B}$ is a right factor of $\mathbf{A}(L)$:

$$\begin{aligned} \mathbf{A}(L) &= (1 - \alpha_1 L, \beta_1 + \beta_2 L) \\ &= [1 - \alpha_1 L, 1 - \beta_1 - (\alpha_1 + \beta_2)L]\begin{bmatrix} 1 & -1 \\ 0 & 1 \end{bmatrix} \end{aligned}$$

and

$$\mathbf{C}(L) = [1 - \alpha_1 L, 1 - \beta_2 - (\alpha_1 + \beta_2)L],$$

i.e.,

$$\mathbf{A}(L)\begin{pmatrix} \delta y \\ \delta x \end{pmatrix}_t = \varepsilon_t - \mathbf{A}_1(L)e_t,$$

where

$$\mathbf{A}_1(L) = \mathbf{C}(L)\begin{pmatrix} 1 \\ 1 \end{pmatrix}$$

where e_t is being generated by

$$\mathbf{C}(L)e_t = \varepsilon_t$$

in the dynamic equilibrium noise process.

***Example 9:* (*Salmon* [1987]).** Consider a model

$$\Delta y_{1t} + y_{1t} + y_{2t} = e_{1t}$$
$$\Delta y_{1t} + \Delta y_{2t} = e_{2t}$$

where $e_t = (e_{1t}, e_{2t})'$ is mean zero white noise with constant variance.

The second equation shows that $y_{1t} + y_{2t}$ is integrated of order one. The first equation means that y_{1t} is integrated of order two. Precise nature of ys can be obtained by writing the model in autoregressive lag form

(16) $$\mathbf{A}(L)\,y_t = e_t$$

where

$$\mathbf{A}(L) = \begin{bmatrix} 2 - L & 1 \\ 1 - L & 1 - L \end{bmatrix} = \mathbf{A}_0 - \mathbf{A}_1 L$$

where

$$\mathbf{A}_0 = \begin{pmatrix} 2 & 1 \\ 1 & 1 \end{pmatrix}$$

and

$$\mathbf{A}_1 = \begin{pmatrix} 1 & 0 \\ 1 & 1 \end{pmatrix}.$$

Note that $\mathbf{A}(1) \neq 0$ but

$$\det \mathbf{A}(1) = |\mathbf{A}_0 - \mathbf{A}_1| = 0.$$

We can write $\mathbf{A}(L)$ as

$$\mathbf{A}(L) = \mathbf{A}(1) + [\mathbf{A}(L) - \mathbf{A}(1)]$$

where

$$\mathbf{A}(L) - \mathbf{A}(1) = (1 - L)\mathbf{A}_1.$$

Equation (16) is the same as

$$\mathbf{A}(1)\mathbf{y}_t + \mathbf{A}_1 \Delta \mathbf{y}_t = \mathbf{e}_t.$$

We see that $(0, 1)\mathbf{A}(1) = 0$ and we recover $(0, 1)\mathbf{A}_1 \Delta \mathbf{y}_t = \Delta(\mathbf{y}_{1t} + \mathbf{y}_{2t}) = \mathbf{e}_{2t}$. Any vector proportional to $(0, 1)$ is integrating. Inverting (16), we obtain the reduced form

$$\begin{aligned} \mathbf{y}_t &= \mathbf{A}(L)^{-1}\mathbf{e}_t \\ &= \frac{1}{(1-L)^2}\begin{bmatrix} 1 - L & -1 \\ -(1 - L) & 2 - L \end{bmatrix} e_t \end{aligned}$$

or

(17) $$\Delta^2 y_{1t} = \Delta e_{1t} - e_{2t}$$

and

$$\Delta^2 y_{2t} = -\Delta e_{1t} + e_{2t} + \Delta e_{2t}.$$

The model can also be put into state space innovation form. The model in level (undifferenced) variables is

$$\mathbf{A}_o \mathbf{y}_t = \mathbf{A}_1 \mathbf{y}_{t-1} + \mathbf{e}_t$$

or

$$\mathbf{y}_t = \mathbf{A}_0^{-1}\mathbf{A}_1 \mathbf{y}_{t-1} + \mathbf{A}_0^{-1}\mathbf{e}_t.$$

Introduce a state vector $\mathbf{x}_t$ by

$$\mathbf{x}_t = \mathbf{y}_t - \mathbf{A}_0^{-1}\mathbf{e}_t.$$

It evolves over time by

$$\begin{aligned} \mathbf{x}_t &= \mathbf{A}_0^{-1}\mathbf{A}_1 \mathbf{y}_{t-1} \\ &= \mathbf{A}_0^{-1}\mathbf{A}_1(\mathbf{x}_{t-1} + \mathbf{A}_0^{-1}\mathbf{e}_{t-1}) \\ &= \mathbf{F}\mathbf{x}_{t-1} + \mathbf{F}\mathbf{A}_0^{-1}\mathbf{e}_{t-1} \end{aligned}$$

where

$$\mathbf{F} = \mathbf{A}_0^{-1}\mathbf{A}_1 = \begin{pmatrix} 0 & -1 \\ 1 & 2 \end{pmatrix}.$$

Let $\boldsymbol{\varepsilon}_t = \mathbf{A}_0^{-1}\mathbf{e}_t$. Then we obtain a state space innovation model

$$\begin{cases} \mathbf{x}_t = \mathbf{F}\mathbf{x}_{t-1} + \mathbf{F}\varepsilon_{t-1} \\ \mathbf{y}_t = \mathbf{x}_t + \boldsymbol{\varepsilon}_t. \end{cases}$$

Note that $|\lambda\mathbf{I} - \mathbf{F}| = (\lambda - 1)^2$.

Rewrite (16) as

$$y_{1t} = \frac{1}{1-L}e_{1t} - \frac{1}{(1-L)^2}e_{2t}$$

and

$$y_{2t} = -\frac{1}{1-L}e_{1t} + \left[\frac{1}{(1-L)^2} - \frac{1}{1-L}\right]e_{2t}.$$

Then

$$y_{1t} + y_{2t} = -\frac{1}{1-L}e_{2t},$$

which shows that $y_t + y_{2t}$ is a random walk.

4.4 Distributed Lag Models and Error-Correction Models

A linear autoregressive distributed lag model has the form

$$\mathbf{B}(L)\mathbf{y}_t + \boldsymbol{\Gamma}(L)\mathbf{x}_t = \mathbf{u}_t \tag{18}$$

where

$$\mathbf{B}(L) = \sum_{i=0}^{p} \mathbf{B}_i L^i \quad \text{and} \quad \boldsymbol{\Gamma}(L) = \sum_{i=0}^{p} \boldsymbol{\Gamma}_i L^i$$

are matrix lag polynomials and $\mathbf{u}_t$ is a vector of zero mean and weakly stationary unobserved disturbances. When the inverse of $\mathbf{B}(L)$ exists, it is converted into a final form

$$\mathbf{y}_t = \boldsymbol{\Pi}(L)x_t + \mathbf{B}(L)^{-1}\mathbf{u}_t$$

where

$$\boldsymbol{\Pi}(L) = -\mathbf{B}(L)^{-1}\boldsymbol{\Gamma}(L).$$

If $\mathbf{x}_t$ is fixed at $\mathbf{x}$ and $\mathbf{u}_t$ is zero for all t, then

$$\mathbf{y}_t = \mathbf{y} = \mathbf{\Pi}(1)\mathbf{x}$$

where $\mathbf{\Pi}(1)$ is the sum of the lag weights, where

$$\mathbf{\Pi}(1) = -\mathbf{B}(1)^{-1}\mathbf{\Gamma}(1).$$

The idea of a vector of linear equilibrium relations is often stated as

$$\bar{\mathbf{B}}\mathbf{y} + \bar{\mathbf{\Gamma}}\mathbf{x} = 0.$$

To be compatible with the forementioned model then,

$$\mathbf{\Gamma}(1) = -\bar{\mathbf{B}}^{-1}\bar{\mathbf{\Gamma}}$$

or

$$\mathbf{B}(1) = \mathbf{C}\bar{\mathbf{B}}$$

and

$$\mathbf{\Gamma}(1) = \mathbf{C}\bar{\mathbf{\Gamma}}$$

for some nonsingular matrix $\mathbf{C}$. Rewrite $\mathbf{B}(L)$ and $\mathbf{\Gamma}(L)$ not as polynomial matrices in (L) but in $L - 1$ by changing variables from L to $L - 1$, i.e., use $L^k = (L - 1 + 1)^k$ and expand in powers of $(L - 1)$, then

$$\mathbf{B}(L) = \sum_0^p \mathbf{B}^{(j)}(1 - L)^j$$

and

$$\mathbf{\Gamma}(L) = \sum_0^p \mathbf{\Gamma}^{(j)}(1 - L)^j.$$

In particular,

$$\mathbf{B}^{(0)} = \mathbf{B}(1) = \mathbf{C}\bar{\mathbf{B}} \quad \text{and} \quad \mathbf{\Gamma}^{(0)} = \mathbf{\Gamma}(1) = \mathbf{C}\bar{\mathbf{\Gamma}}.$$

For example, $B(L) = L^2 + 3L + 5 = (L - 1 + 1)^2 + 3(L - 1 + 1) + 5 = (L - 1)^2 + 5(L - 1) + 9$. Model (18) now is restated as

$$\mathbf{C}(\bar{\mathbf{B}}\mathbf{y}_t + \bar{\mathbf{\Gamma}}\mathbf{x}_t) + \sum_{j=1}^p (\mathbf{B}^{(j)}\Delta^j\mathbf{y}_t + \mathbf{\Gamma}^{(J)}\Delta^j\mathbf{x}_t) = \mathbf{u}_t.$$

Use the identity

$$\Delta^{j+1}\mathbf{y}_t = \Delta^j(\mathbf{y}_t - \mathbf{y}_{t-1})$$

or

$$\Delta^j\mathbf{y}_t = \Delta^{j+1}\mathbf{y}_t + \Delta^j\mathbf{y}_{t-1} = \dots = \Delta^p\mathbf{y}_t + \Delta^{p-1}\mathbf{y}_{t-1} + \dots + \Delta^j\mathbf{y}_{t-1}$$

to rearrange the second term so that $\Delta^p \mathbf{y}_t$ and $\Delta^p \mathbf{x}_t$ appears only once,

$$\mathbf{C}(\bar{\mathbf{B}}\mathbf{y}_t + \bar{\mathbf{\Gamma}}\mathbf{x}_t) + \left(\sum_{j=0}^{p} \mathbf{B}^{(j)} \right) \Delta^p \mathbf{y}_t + \left(\sum_{j=0}^{p} \mathbf{\Gamma}^{(j)} \right) \Delta^p \mathbf{x}_t$$
$$+ \sum_{j=1}^{p-1} \left[\left(\sum_{k=0}^{j} \mathbf{B}^{(k)} \right) \Delta^j \mathbf{y}_{t-1} + \left(\sum_{k=0}^{j} \mathbf{\Gamma}^{(k)} \right) \Delta^j \mathbf{x}_{t-1} \right] = \mathbf{u}_t .$$

This is a system of error-correction models because in equilibrium the term $\bar{\mathbf{B}}\mathbf{y}_t + \bar{\mathbf{\Gamma}}\mathbf{x}_t$ vanished. When $p = 1$, the summation $\Sigma_{j=1}^{p-1}$ is absent, and

$$\mathbf{C}(\bar{\mathbf{B}}\mathbf{y}_t + \mathbf{\Gamma}\mathbf{x}_t) + (\mathbf{B}^{(0)} + \mathbf{B}^{(1)})\Delta\mathbf{y}_t + (\mathbf{\Gamma}^{(0)} + \mathbf{\Gamma}^{(1)})\Delta\mathbf{x}_t = \mathbf{u}_t .$$

This transformation is used by Davidson (1987) to show that error-correction models are just an alternative representation of autoregressive distributed lag models. As he points out, error-correction models have the advantages of explicitly displaying the target or long-run equilibrium relations together with the adjustment mechanisms in one set of equations. When **C** is diagonal, the model is just a collection of independent equations in error-correction form. A nondiagonal **C**, however, imposes cross-equation restrictions. This algebraic manipulation, however, fails to give us much intuitive grasp of the essence of error-correction models.

Before we turn to an alternative and more geometric view of error-correction models, we define the notion of cointegration in an autoregressive model

$$\mathbf{A}(L)\mathbf{y}_t = \mathbf{e}_t \tag{19}$$

where $\mathbf{A}(L)$ is an $m \times m$ polynomial and $\mathbf{e}_t$ is a mean-zero weakly stationary white noise sequence. A vector $\mathbf{y}_t$ is said to be integreated to order d if at least one of its components must be differenced d times before it becomes stationary.

When $|\mathbf{A}(z)| = 0$ has a unit root, y_t is integrating, since $\mathbf{y}_t$ is nonstationary but $\Delta^d \mathbf{y}_t$ is stationary for some $d > 0$. Since

$$\mathbf{A}(L) = \mathbf{A}(1) + (L - 1)\mathbf{C}(L)$$

for some $\mathbf{C}(L)$,

$$|\mathbf{A}(L)| = |\mathbf{A}(1)| + (L - 1)\mathbf{A}(L).$$

Thus $|\mathbf{A}(z)| = 0$ has a unit root if and only if $\mathbf{A}|(1)| = 0$. We usually discuss only cases with $d = 1$. Rewrite (18) as

$$\Delta(L)\mathbf{y}_t = \Delta(L)\mathbf{A}(L)^{-1}\mathbf{e}_t$$

where $\Delta(L) = \text{diag}[(1 - L)^{d_1}, \ldots (1 - L)^{d_p}]$ where d_i is a nonnegative integer that cancels out all negative powers of $(1 - L)$ in the corresponding row of $\mathbf{A}(L)^{-1}$. Let $\mathbf{\Phi}(L) = \mathbf{\Delta}(L)\mathbf{A}(L)^{-1}$. The matrix Φ has no $1/(1 - L)$ as a factor by construction, i.e., $|\mathbf{\Phi}(1)| \neq 0$. Then because of $\mathbf{A}(L) = \mathbf{\Phi}^{-1}(L)\mathbf{\Delta}(L)$, $|\mathbf{A}(1)| = |\mathbf{\Phi}^{-1}(1)||\mathbf{\Delta}(1)| = 0$. This is a vector version of a univariate model

$$\Delta \mathbf{y}_t = \mathbf{u}_t$$

where

$$\mathbf{u}_t = \mathbf{\Phi}(L)\mathbf{e}_t.$$

This type of model has been examined by Beveridge and Nelson (1981).

Suppose that $\mathbf{y}_t$ has a model in state space representation

$$\mathbf{x}_{k+1} = \mathbf{A}\mathbf{x}_k + \mathbf{B}\mathbf{e}_t$$
$$\mathbf{y}_t = \mathbf{C}\mathbf{x}_k + \mathbf{e}_t.$$

To be concrete, assume that one eigenvalue of $\mathbf{A}$ is equal to one, and all other eigenvalues are strictly less than one in magnitude. (If all eigenvalues of $\mathbf{A}$ are less than one in magnitude then $\{\mathbf{y}_t\}$ is weakly stationary.) Choose a coordinate system in which $\mathbf{A}$ has the form

$$\begin{bmatrix} 1 & \mathbf{X} \\ 0 & \mathbf{F} \end{bmatrix}.$$

Such a coordinate system always exists, i.e., the matrix $\mathbf{A}$, when put into Schur form, has the indicated block triangular form. In the next section, we show how state space representation of error-correction models and the notion of cointegration are closely related from the viewpoint of the aggregating of dynamic systems. The transfer function representation of this model is

(20) $$\hat{\mathbf{y}} = [\mathbf{I} + \mathbf{C}(zI - \mathbf{A})^{-1}\mathbf{B}]\hat{e} = [\mathbf{R}/(z-1) + \mathbf{I} + \mathbf{w}'(z\mathbf{I} - \mathbf{F})^{-1}\zeta]\hat{\mathbf{e}}$$

where

$$\mathbf{C} = [\xi, \boldsymbol{\eta}], \qquad \mathbf{B} = \begin{bmatrix} \gamma \\ \zeta \end{bmatrix}, \qquad \mathbf{R} = \gamma + \boldsymbol{\beta}'(\mathbf{I} - \mathbf{F})^{-1}\zeta,$$

and

$$\mathbf{w}' = \boldsymbol{\eta} - \xi\boldsymbol{\beta}'(I - \mathbf{F})^{-1}.$$

The representation makes clear that $\boldsymbol{\mu}'\mathbf{y}$ is weakly stationary for some vector $\boldsymbol{\mu}$ if and only if $\boldsymbol{\mu}'\mathbf{R}\mathbf{R}'\boldsymbol{\mu} = 0$ and hence $|\mathbf{R}| = 0$. This means that $|\gamma + \boldsymbol{\beta}'(z\mathbf{I} - \mathbf{F})^{-1}\zeta|$ has a unit zero. In other words, the spectral density of $\Delta \boldsymbol{v}$ where $\boldsymbol{v} = \boldsymbol{\mu}'\mathbf{y}$ is zero at zero frequency. When $\boldsymbol{\mu}'\mathbf{y}_t$ is stationary, we call μ a fully cointegrating vector. If $\boldsymbol{\mu}$ is such that $\boldsymbol{\mu}'\mathbf{y}_t$ is not stationary but of a lower order of integration than $\mathbf{y}_t$, then $\boldsymbol{v}$ is called a cointegrating vector.

CHAPTER 5

Estimation

In this chapter, we discuss least squares, Kalman filters, Wiener filters and Bayesian estimators. Not only are the estimation (or identification) problems of interest on their own, but also they are of inherent interest as a part of overall system-optimization problems. This is because some optimal control problems naturally separate into two types of subproblems: one is the construction of the optimal estimators of state vectors or unknown vectors of parameters, and the other is the synthesis of optimal decision rules for known parameter values. When this "separation" holds, the overall optimal control schemes are optimal estimators followed by optimal controllers for related deterministic models. For a much larger class of control problems, however, the overall optimization requirements do not permit such a convenient and simplifying separation of the estimation processes from the control processes. Then, this separation of problems provides initial approximate solutions to the original control problems that may be further improved upon. Such approximation schemes are important in practice.

Unknown parameters occur in both static models and dynamic models. In a static or regression model

$$y_t = \mathrm{f}(x_t, \boldsymbol{\theta}) + \text{noise}, \qquad t = 1, \ldots$$

where y_t is data; x_t is an input, control, or otherwise known (independent) vector; and $\boldsymbol{\theta}$ is a vector of parameters. A linear regression model

$$y_t = \mathbf{x}_t' \boldsymbol{\theta} + \text{noise}, \qquad t = 1 \ldots$$

is perhaps the simplest example of a static model with unknown parameters. Estimation of unknown parameters is usually formulated as minimization of some measure of "error" of estimation or loss functions which is often taken to be a quadratic one. We then speak of the least squares estimators or mean distance estimators where the "distance" is suitably defined. Gallant (1987) has more extensive treatment of asymptotic properties of nonlinear regression models.

Dynamic models may be in state space form

$$\mathbf{s}_{t+1} = \mathbf{f}(\mathbf{s}_t, \boldsymbol{\theta}_1, \mathbf{x}_t) + \text{noise},$$
$$\mathbf{y}_t = \mathbf{h}(\mathbf{s}_t, \boldsymbol{\theta}_2, \boldsymbol{\eta}_t) + \text{noise}, \qquad t = 1, \ldots,$$

or in its special case

$$\mathbf{y}_t = \mathbf{f}(\mathbf{y}_{t-1}, \mathbf{x}_t, \boldsymbol{\theta}) + \text{noise}, \qquad t = 1, \ldots$$

or in vector-valued ARMAX form, which can be converted into regression models with stochastic regressors. Regression models can be estimated in a variety of ways. Books such as Ljung (1987) or Goodwin and Sin (1984) discuss many algorithms. Vector-valued (time-series) data can be used to estimate innovation state space models by a method rather different from those in these books as described in Aoki (1987c). Some examples of special cases are given shortly. In dynamic models, we often wish to estimate, in addition, unobserved state variables that are changing with time such as the state vector $\mathbf{x}_t$ in

$$\mathbf{x}_{t+1} = \mathbf{A}\mathbf{x}_t + \mathbf{B}\xi_t$$
$$\mathbf{y}_t = \mathbf{C}\mathbf{x}_t + \boldsymbol{\eta}_t$$

where **A**, **B** and **C** are assumed unknown. When the model parameters and noise characteristics are known, the vector $\mathbf{x}_t$ can be estimated by Kalman filters. This chapter develops estimation procedures, mostly in the context of constant coefficient linear models.

5.1 Least Squares Method

Parameter Estimates

Many estimation problems can be reduced to finding a vector $\hat{\boldsymbol{\theta}}$ subject to a set of linear constraints $\mathbf{A}\hat{\boldsymbol{\theta}} = \mathbf{b}$ where **A** is an $m \times n$ matrix and **b** is an m-vector, $m > n$. This set of linear constraint equations may arise as the first order condition of some quadratic cost minimization. There are more equations than unknowns, and the system of equation is said to be overdetermined. In such situations, we look for $\hat{\boldsymbol{\theta}}$ that minimizes some error distance function (or error norm), $f(\boldsymbol{\theta}) = \|\mathbf{A}\boldsymbol{\theta} - \mathbf{b}\|_p$, for some positive integer p. The problem with $p = 2$ is known as the least squares problem. Although other choices of p are certainly possible, the function $f(\boldsymbol{\theta})$ ceases to be differentiable for $p = 1$ and $p = \infty$ while it is differentiable for $p = 2$.

The least squares problem has an appealing geometric interpretation. When $\mathbf{b}$ is not in the range space spanned by the column vectors of $\mathbf{A}$, $\mathbf{A}\hat{\boldsymbol{\theta}}$ is the orthogonal projection image of $\mathbf{b}$ on the range space, i.e., $\mathbf{b} - \mathbf{A}\hat{\boldsymbol{\theta}}$ is orthogonal to the range space of $\mathbf{A}$, i.e., $\mathbf{A}'(\mathbf{b} - \mathbf{A}\hat{\boldsymbol{\theta}}) = 0$. (See Aoki [1971, p. 7] or Golub and Van Loan [1983, p. 137].) To simplify exposition, suppose at first that rank $\mathbf{A} = n$. Then $\hat{\boldsymbol{\theta}} = (\mathbf{A}'\mathbf{A})^{-1}\mathbf{A}'\mathbf{b}$ and $\mathbf{A}\hat{\boldsymbol{\theta}} = \mathbf{P}_A\mathbf{b}$ where $\mathbf{P}_A = \mathbf{A}(\mathbf{A}'\mathbf{A})^{-1}\mathbf{A}'$ is the orthogonal projection matrix since $\mathbf{P}_A^2 = \mathbf{P}_A$. The least squares estimate is given by this $\hat{\boldsymbol{\theta}}$ since it minimizes $\|\mathbf{A}\boldsymbol{\theta} - \mathbf{b}\|_2^2 = (\mathbf{A}\boldsymbol{\theta} - \mathbf{b})'(\mathbf{A}\boldsymbol{\theta} - \mathbf{b})$. The first-order condition for minimization is the normal equation

$$\mathbf{A}'\mathbf{A}\boldsymbol{\theta} = \mathbf{A}'\mathbf{b}$$

and the component orthogonal to the range space is the residual $\mathbf{r} = \mathbf{b} - \mathbf{A}\boldsymbol{\theta} = (\mathbf{I} - \mathbf{P}_A)\mathbf{b}$. The second order condition $\mathbf{A}'\mathbf{A} > 0$ is satisfied by assumption.

Note that any orthogonal matrix $\mathbf{Q}'\mathbf{Q} = \mathbf{I}_m$ does not alter the norm

$$\|\mathbf{A}\boldsymbol{\theta} - \mathbf{b}\|_2^2 = \|\mathbf{Q}\mathbf{A}\boldsymbol{\theta} - \mathbf{Q}\mathbf{b}\|_2^2.$$

A suitable choice of $\mathbf{Q}$ can simplify the quadratic expression as discussed next.

Suppose rank $\mathbf{A} = r \leqslant n$. Let the singular value decomposition of $\mathbf{A}$ by $\mathbf{U}\boldsymbol{\Sigma}\mathbf{V}'$, where $\mathbf{U}'\mathbf{U} = \mathbf{I}_m$, $\mathbf{V}'\mathbf{V} = \mathbf{I}_n$, $\boldsymbol{\Sigma} = [\mathrm{diag}(\sigma_1 \ldots \sigma_r, 0 \ldots 0)\mathbf{0}]'$. This decomposition suggests that we choose $\mathbf{Q} = \mathbf{U}'$. See Golub and Van Loan (1983, p. 16) on the singular value decomposition. Then

$$\mathbf{U}'(\mathbf{A}\boldsymbol{\theta} - \mathbf{b}) = \boldsymbol{\Sigma}\mathbf{V}'\boldsymbol{\theta} - \mathbf{U}'\mathbf{b} = \begin{bmatrix} \boldsymbol{\Sigma}_1\mathbf{V}_1'\boldsymbol{\theta} - \mathbf{U}_1'\mathbf{b} \\ -\mathbf{U}_2'\mathbf{b} \end{bmatrix}$$

where

$$\boldsymbol{\Sigma}_1 = \mathrm{diag}(\sigma_1 \ldots \sigma_r), \qquad \mathbf{V}' = \begin{bmatrix} \mathbf{V}_1' \\ \mathbf{V}_2' \end{bmatrix}, \qquad \mathbf{U}' = \begin{bmatrix} \mathbf{U}_1' \\ \mathbf{U}_2' \end{bmatrix}$$

are comfortable partitions of $\mathbf{V}'$ and $\mathbf{U}'$, i.e., $\mathbf{V}_1'$ is $r \times n$ and $\mathbf{U}_1'$ is $r \times m$. We see that

$$\|\mathbf{A}\boldsymbol{\theta} - \mathbf{b}\|_2^2 = \sum_{i=1}^{r} (\sigma_i\alpha_i - \mathbf{u}_i'\mathbf{b})^2 + \sum_{r+1}^{m} (\mathbf{u}_i'\mathbf{b})^2,$$

where $\mathbf{u}_i$ is the ith column vector of $\mathbf{U}_1$, and α_i is the ith component of the vector defined by

$$\alpha = \mathbf{V}_1'\boldsymbol{\theta}.$$

This sum is minimized by setting

$$\alpha_i = \mathbf{u}_i'\mathbf{b}/\sigma_i, \qquad i = 1 \dots r,$$

i.e., the least squares estimate is

$$\hat{\boldsymbol{\theta}} = \mathbf{V}_1\boldsymbol{\alpha} = \sum_{i=1}^{r} (\mathbf{u}_i'\mathbf{b}/\sigma_i)\mathbf{v}_i.$$

Regression Models

Next, consider the parameter estimation problem in a regression model

$$y_t = \mathbf{x}_t'\boldsymbol{\theta} + \text{noise}$$

where $\mathbf{x}_t$ is a nonstochastic n-vector. Define for $T \geqslant n$

$$\mathbf{y}' = (y_1, \dots, y_T) \quad \text{and} \quad \mathbf{X}' = [\mathbf{x}_1, \dots, \mathbf{x}_T].$$

The problem is to minimize

$$\|\mathbf{y} - \mathbf{X}\boldsymbol{\theta}\|_2^2.$$

Then the least squares estimate of $\mathbf{y}$ is

$$\hat{\mathbf{y}} = \mathbf{P}_X\mathbf{y}, \qquad \mathbf{P}_X = \mathbf{X}(\mathbf{X}'\mathbf{X})^{-1}\mathbf{X}', \qquad \text{and} \quad \hat{\boldsymbol{\theta}} = (\mathbf{X}'\mathbf{X})^{-1}\mathbf{X}'\mathbf{y},$$

i.e.,

$$\hat{\boldsymbol{\theta}} = (\Sigma_1^T \mathbf{x}_t\mathbf{x}_t')^{-1}\Sigma_1^T \mathbf{x}_t\mathbf{y}_t.$$

Sometimes a distance function $(\mathbf{y} - \mathbf{X}\boldsymbol{\theta})'\mathbf{W}(\mathbf{y} - \mathbf{X}\boldsymbol{\theta})$ for some $\mathbf{W}' = \mathbf{W} > 0$ may be used instead of $\|\mathbf{y} - \mathbf{X}\boldsymbol{\theta}\|_2^2$. We call this problem a weighted least squares problem. Since

$$\min\|\mathbf{y} - \mathbf{X}\boldsymbol{\theta}\|_2^2 = \|\hat{\mathbf{y}} - \mathbf{X}\boldsymbol{\theta}\|_2^2 = (\mathbf{y} - \mathbf{X}\boldsymbol{\theta})'\mathbf{P}_X(\mathbf{y} - \mathbf{X}\boldsymbol{\theta})$$

where

$$\mathbf{X}\hat{\boldsymbol{\theta}} = \mathbf{P}_X\mathbf{X}\boldsymbol{\theta}$$

and

$$\mathbf{P}_X^2 = \mathbf{P}_X,$$

the least squares problem is that of the weighted one

$$\min \| \mathbf{y} - \mathbf{X}\boldsymbol{\theta} \|_2^2 = \min \| \mathbf{y} - \mathbf{X}\boldsymbol{\theta} \|_{\mathbf{W}}^2$$

with $\mathbf{W} = \mathbf{P}_X^2 = \mathbf{P}_X$. We continue our discussion of parameter estimates in regression models in a later section and in Chapter 6.

Suppose that we have another $(T \times q)$ matrix $\mathbf{Z}$ where $q \geqslant n$. Then the weighted least squares problem with $\mathbf{P}_Z = \mathbf{Z}(\mathbf{Z}'\mathbf{Z})^{-1}\mathbf{Z}'$ as the weight yields the instrumental variable estimator with $\mathbf{Z}$ as the instrument,

$$\hat{\boldsymbol{\theta}} = (\mathbf{X}'\mathbf{P}_Z\mathbf{X})^{-1}\mathbf{X}'\mathbf{P}_Z\mathbf{y}.$$

Given a quadratic cost below the least squares problem's tie to nonlinear programming is most obvious,

$$J_k(\boldsymbol{\theta}) = \sum_{t=0}^{k} \| \xi\boldsymbol{\theta} - \mathbf{y}_t \|_2^2.$$

Sometimes the weighted sum $\| \xi\boldsymbol{\theta} - \mathbf{y}_t \|_{\mathbf{V}_t}^2$ for some symmetric positive definite matrix $\mathbf{V}_t$ is used, where $\|\mathbf{x}\|_{\mathbf{V}}^2 = \mathbf{x}'\mathbf{V}\mathbf{x}$.

This quadratic expression is expanded into

$$\begin{aligned} J_k(\boldsymbol{\theta}) &= J_k(\boldsymbol{\theta}_k^*) + \langle \operatorname{grad} J_k(\hat{\boldsymbol{\theta}}_k^*), \boldsymbol{\theta} - \hat{\boldsymbol{\theta}}_k \rangle \\ &\quad + \tfrac{1}{2}\langle \boldsymbol{\theta} - \hat{\boldsymbol{\theta}}_k, H_k(\boldsymbol{\theta} - \hat{\boldsymbol{\theta}}_k^*) \rangle \end{aligned} \tag{1}$$

where H_k is the Hessian of $J_k(\boldsymbol{\theta})$, which is a constant positive definite matrix in the least squares problem. The optimal $\boldsymbol{\theta}_k^*$ minimizes $J_k(\boldsymbol{\theta})$, i.e.,

$$0 = \langle \operatorname{grad} J_k(\boldsymbol{\theta}_k^*), \boldsymbol{\theta} - \boldsymbol{\theta}_k^* \rangle \quad \text{for arbitrary } \boldsymbol{\theta}$$

is the first-order condition for optimality, which is both necessary and sufficient. This is the normal equation.

State Vector Estimation

Estimation of state vectors in dynamic models may be phrased as a least squares problem. Instead of estimating a fixed-parameter vector $\boldsymbol{\theta}$, we may wish to estimate a time-varying vector $\mathbf{x}_t$ that is a state vector of a dynamic model

$$\mathbf{x}_{t+1} = \mathbf{f}(\mathbf{x}_t),$$

and

$$\mathbf{y}_t = \mathbf{h}(\mathbf{x}_t, \boldsymbol{\eta}_t),$$

where $\boldsymbol{\eta}_t$ is a noise vector, or its linear version

$$\mathbf{x}_{t+1} = \mathbf{A}_t \mathbf{x}_t,$$

and

$$\mathbf{y}_t = \mathbf{C}_t \mathbf{x}_t + \boldsymbol{\eta}_t.$$

A related problem is that of estimating a fixed vector $\mathbf{x}_0$, say, from an ever-expanding set of observations $\mathbf{y}^t$. The next example illustrates how an initial-condition vector may be estimated using a quadratic distance measure.

Example 1: Estimation of the Initial Condition

Suppose that a linear state space model

$$\mathbf{x}_{t+1} = \mathbf{A}\mathbf{x}_t$$

is observed through

$$\mathbf{y}_t = \mathbf{C}\mathbf{x}_t + \boldsymbol{\eta}_t$$

from time $t = 0$ to $t = T - 1$. The quadratic loss function to be minimized is

$$J_T(\mathbf{x}_0) = \sum_{t=0}^{T-1} \|\mathbf{H}_t \mathbf{x}_0 - \mathbf{y}_t\|^2$$

where $\mathbf{H}_t = \mathbf{C}\mathbf{A}^t$, $t = 0, \ldots, T - 1$.

We are interested in assessing the effect of the next measurement on the estimate and the effect of initial reading as the number of data points increases. Stack the T data vector as

$$\mathbf{y} = \begin{bmatrix} \mathbf{y}_0 \\ \vdots \\ \mathbf{y}_{T-1} \end{bmatrix} = \mathbf{O}_{\mathrm{T}} \mathbf{x}_0 + \text{noise terms}$$

where

$$\mathbf{O}'_{\mathrm{T}} = [\mathbf{C}' \mathbf{A}' \mathbf{C}' \ldots (\mathbf{A}')^{T-1} \mathbf{C}'].$$

The least squares estimate of $\mathbf{x}_0$ is

$$\hat{\mathbf{x}}_0 = (\mathbf{O}'_{\mathrm{T}} \mathbf{O}_{\mathrm{T}})^{-1} \mathbf{O}'_{\mathrm{T}} \mathbf{y}$$

if rank $\mathbf{O}_{\mathrm{T}} = \dim \mathbf{x}_0$. The rank condition is the observability condition of the model.

With one additional datum, the least squares estimate, denoted by $\hat{\mathbf{x}}_0(T + 1)$ is

$$\begin{aligned} \hat{\mathbf{x}}_0(T+1) &= \boldsymbol{\Omega}_{T+1}^{-1}[\mathbf{O}'_{\mathrm{T}} \mathbf{y} + (\mathbf{A}')^T \mathbf{C}' \mathbf{y}_T] \\ &= \hat{\mathbf{x}}_0(T) + \boldsymbol{\Omega}_{T+1}^{-1} (\mathbf{A}')^T \mathbf{C}' [\mathbf{y}_T - \mathbf{C}\mathbf{A}^T \hat{\mathbf{x}}_0(T)] \end{aligned} \tag{2}$$

where $\hat{\mathbf{x}}_0(T) = \mathbf{\Omega}_T^{-1}\mathbf{O}_T'\mathbf{y}$ and $\mathbf{\Omega}_T = \mathbf{O}_T'\mathbf{O}_T$. Here we use the fact that $\mathbf{O}_T'\mathbf{y} = \mathbf{\Omega}_T\hat{\mathbf{x}}_0(T)$ and that $\mathbf{\Omega}_{T+1} = \mathbf{\Omega}_T + (\mathbf{A}')^T\mathbf{C}'\mathbf{C}\mathbf{A}^T$. The second term gives the correction due to one more piece of information. Equation (2) shows that the discrepancy between the predicted and the actual value is incorporated with weight $\mathbf{\Omega}_{T+1}^{-1}(\mathbf{A}')^T\mathbf{C}'$. When $\mathbf{A}$ is asymptotically stable, $\mathbf{\Omega}_T$ converges to the observability gramian introduced in Chapter 1.

$$\mathbf{\Omega} = \mathbf{C}'\mathbf{C} + \mathbf{A}'\mathbf{\Omega}\mathbf{A},$$

which is finite, and $\mathbf{A}^T \to 0$. The correction term therefore goes to zero as $T \to \infty$.

Next we assess the effect of the initial data point by rewriting the stacked data vector as a partitioned vector

$$\begin{bmatrix} \mathbf{y} \\ \\ \mathbf{y}_T \end{bmatrix} = \begin{bmatrix} \mathbf{y}_0 \\ \\ \mathbf{w} \end{bmatrix} = \begin{bmatrix} \mathbf{C} \\ \\ \mathbf{O}_T\mathbf{A} \end{bmatrix} \mathbf{x}_0 + \text{noise},$$

where $\mathbf{w}$ stacks y_1 through y_T. Then the least squares estimate of $\mathbf{x}_0$ is given by

$$\hat{\mathbf{x}}_0(T+1) = \mathbf{\Omega}_{T+1}^{-1}(\mathbf{C}'\mathbf{y}_0 + \mathbf{A}'\mathbf{O}_T'\mathbf{w}).$$

Denote by $\bar{\mathbf{x}}_0(T) = (\mathbf{A}'\mathbf{\Omega}_T\mathbf{A})^{-1}\mathbf{A}'\mathbf{O}_T'\mathbf{w}$ the least squares estimate of $\mathbf{x}_0$ using T latest data points, $y_1, \ldots, y_T$ rather than $y_0, y_1, \ldots, y_T$. Using this estimate and the fact that $\mathbf{A}'\mathbf{O}_T'\mathbf{w} = (\mathbf{A}'\mathbf{\Omega}_T\mathbf{A})\bar{\mathbf{x}}_0(T)$ and $\mathbf{\Omega}_{T+1} = \mathbf{C}'\mathbf{C} + \mathbf{A}'\mathbf{\Omega}_T\mathbf{A}$, we can rewrite $\hat{\mathbf{x}}_0(T+1)$ as

$$\hat{\mathbf{x}}_0(T+1) = \bar{\mathbf{x}}_0(T) + \mathbf{\Omega}_{T+1}^{-1}\mathbf{C}'[\mathbf{y}_0 - \mathbf{C}\bar{\mathbf{x}}_0(T)],$$

which shows the effect of including $\mathbf{y}_0$, which is ignored or dropped in calculating $\bar{\mathbf{x}}_0(T)$. This equation shows that unless $\mathbf{\Omega}_T^{-1} \to 0$ as $T \to \infty$, the effect of initial noise $\boldsymbol{\eta}_0$ persists. When $(\mathbf{A}, \mathbf{C})$ is observable, either $\mathbf{A}$ is asymptotically stable and $\mathbf{\Omega}_T$ is finite, or $\mathbf{\Omega}_T$ diverges. Thus when $(\mathbf{A}, \mathbf{C})$ is observable and $\mathbf{A}$ is asymptotically stable, then the effect of $\mathbf{y}_0$ persists as T goes to infinity.

Now consider a weighted least squares problem

$$J_k(\mathbf{x}) \triangleq \sum_{i=0}^{k} \|\mathbf{H}_i\mathbf{x} - \mathbf{y}_i\|_{\mathbf{V}_i^2} \tag{3}$$

where $\mathbf{V}_i$ is a $p \times p$ symmetric positive (semi-)definite matrix and acts as a weight for different $\mathbf{y}_i$. This implies that we have some idea of the relative magnitudes of the random noises involved, otherwise $\mathbf{V}$ may be taken to be the identity matrix. J_k is the criterion function of the estimation problem. The optimal $\mathbf{x}$ will be determined sequentially so that one need not resolve the least-squares problems from the beginning when new observation becomes available.

From (3), one readily sees that

$$J_{k+1}(\mathbf{x}) = J_k(\mathbf{x}) + \| H_{k+1}\mathbf{x} - \mathbf{y}_{k+1} \|_{\mathbf{V}^2_{k+1}}. \tag{4}$$

Denote by * the optimal of (3), that is

$$J_k(\mathbf{x}_k^*) \leqslant J_k(\mathbf{x}) \tag{5}$$

for any $\mathbf{x}$ in the Euclidean n space. From (5) and (3), by considering $\mathbf{x} = \mathbf{x}_k^* + \Delta\mathbf{x}$, the first-order condition analogous to (1) is

$$\Delta\mathbf{x}' \sum_{i=0}^{k} \mathbf{H}_i'\mathbf{V}_i(\mathbf{H}_i\mathbf{x}_k^* - \mathbf{y}_i) = 0 \tag{6}$$

for any $\Delta\mathbf{x}$ in the Euclidean n space.

5.2 Extended Least Squares Method

We follow Solo (1981) to discuss the behavior of the least squares estimator of a regression model

$$y_t = \mathbf{x}_t'\boldsymbol{\theta} + \varepsilon_t$$

where $\mathbf{x}_t$ is a k-vector of nonstochastic regressors and ε_t a disturbance sequence that is not necessarily Gaussian. The least square estimator of $\boldsymbol{\theta}$ minimizes

$$L(\boldsymbol{\theta}) = \sum_{s=1}^{t} (y_s - \mathbf{x}_s'\boldsymbol{\theta})^2,$$

i.e.,

$$\boldsymbol{\theta}_t = \mathbf{V}_t^{-1}\Sigma_1^t \mathbf{x}_s y_s, \qquad t \geqslant p \tag{7}$$

where

$$\mathbf{V}_t = \sum_{s=1}^{t} \mathbf{x}_s\mathbf{x}_s'.$$

Assume that $\mathbf{V}_p > 0$. Then $\mathbf{V}_n \geqslant \mathbf{V}_p > 0$ for all $n \geqslant p$. Equation (7) can be put in a recursive form

$$\begin{aligned} \boldsymbol{\theta}_t &= \mathbf{V}_t^{-1}\Sigma_1^{t-1}\mathbf{x}_t y_t + V_t^{-1}\mathbf{x}_t y_t \\ &= \mathbf{V}_t^{-1}\mathbf{V}_{t-1}\mathbf{V}_{t-1}^{-1}\Sigma_1^{t-1}\mathbf{x}_s y_s + V_t^{-1}\mathbf{x}_t y_s \\ &= \mathbf{V}_t^{-1}\mathbf{V}_{t-1}\boldsymbol{\theta}_{t-1} + \mathbf{V}_t^{-1}\mathbf{x}_t y_t \\ &= \mathbf{V}_t^{-1}(\mathbf{V}_{t-1} + \mathbf{x}_t\mathbf{x}_t')\boldsymbol{\theta}_{t-1} + \mathbf{V}_t^{-1}(y_t - \mathbf{x}_t'\boldsymbol{\theta}_{t-1})\mathbf{x}_t \\ &= \boldsymbol{\theta}_{t-1} + \mathbf{V}_t^{-1}(y_t - \mathbf{x}_t'\boldsymbol{\theta}_{t-1})\mathbf{x}_t \end{aligned} \tag{8}$$

since

$$\mathbf{V}_t = \mathbf{V}_{t-1} + \mathbf{x}_t \mathbf{x}_t'.$$

Here the last term is a correction term in a Newton–Raphson-type algorithm because

$$\begin{aligned} -(\tfrac{1}{2})(\partial L/\partial \boldsymbol{\theta})_{\boldsymbol{\theta}_{t-1}} &= [\Sigma_{s=1}^{t-1}(y_s - \mathbf{x}_s'\boldsymbol{\theta})\mathbf{x}_s + (y_t - \mathbf{x}_t'\boldsymbol{\theta})\mathbf{x}_t]_{\boldsymbol{\theta}_{t-1}} \\ &= (y_t - \mathbf{x}_t'\boldsymbol{\theta}_{t-1})\mathbf{x}_t \end{aligned}$$

and

$$\frac{\partial^2 L}{\partial \boldsymbol{\theta}^2} = \Sigma \mathbf{x}_s \mathbf{x}_s' = \mathbf{V}_t > 0.$$

In (8), $y_t - \mathbf{x}_t'\boldsymbol{\theta}_{t-1}$ is the prediction error ε_t.

When the method is extended to an ARMAX model, $\mathbf{x}_t$ is not longer exogenous but depends on $\boldsymbol{\theta}$, and the model becomes

$$y_t + a_1 y_{t-1} = b_1 u_{t-1} + b_2 u_{t-2} + \varepsilon_t,$$

for a univariate y_t, and it is written in a "regression" model form

$$y_t = \mathbf{x}_t'\boldsymbol{\theta} + \varepsilon_t,$$

where

$$\begin{aligned} \mathbf{x}_t' &= [-y_{t-1} \quad u_{t-1} \quad u_{t-2}], \\ \boldsymbol{\theta}' &= [a_1 \quad b_1 \quad b_2]. \end{aligned}$$

Now y_t certainly depends on $\boldsymbol{\theta}$, which in turn implies that $\mathbf{x}_t$ depends on $\boldsymbol{\theta}$ as well.

There are two ways to extend (8). One is to use

(9) $$\boldsymbol{\theta}_t = \boldsymbol{\theta}_{t-1} + \mathbf{V}_t^{-1}\boldsymbol{\phi}_t e_t$$

where the prediction error is

$$e_t = y_t - \boldsymbol{\theta}_{t-1}'\boldsymbol{\phi}_t$$

and $\boldsymbol{\phi}_t = \mathbf{x}_t(\boldsymbol{\theta}_{t-1})$, which contains the prediction errors $e_{t-1}, e_{t-2}, \ldots$ among other things. The other replaces prediction error in the definition of $\boldsymbol{\theta}_t$ by the residuals

(10) $$\boldsymbol{\theta}_t = \boldsymbol{\theta}_{t-1} + \mathbf{V}_t^{-1}\boldsymbol{\psi}_t \eta_t$$

where the residual is

$$\eta_t = y_t - \boldsymbol{\theta}_t'\boldsymbol{\psi}_t$$

and $\boldsymbol{\psi}_t = \mathbf{x}_t(\boldsymbol{\theta}_t)$ which contains residuals $\eta_t, \eta_{t-2} \ldots$ among other things. The recursion (9) is called RML_1 (recursive maximum likelihood of type 1) estimate and (10) AML (approximate maximum likelihood) estimate. The difference of these two schemes lies in the fact that in (9) the prediction error e_t is calculated by the last estimate $\boldsymbol{\theta}_{t-1}$ while in (10) it is calculated by the updated (latest) estimate $\boldsymbol{\theta}_t$, not $\boldsymbol{\theta}_{t-1}$. The almost sure (a.s.) convergence properties of AML are established by Solo (1979). See Chapter 6 for further topics on convergence.

For a univariate model

$$y_t + a_1 y_{t-1} + \ldots + a_p y_{t-p} = b_1 u_{t-1} + \ldots + b_q u_{t-q} \\ + \varepsilon_t + c_1 \varepsilon_{t-1} + \ldots + c_r \varepsilon_{t-r},$$

collect the unknown parameters as

$$\boldsymbol{\theta}' = (a_1 \ldots a_p, b_1 \ldots b_q, c_1 \ldots c_r)$$

and define

$$\mathbf{x}_t' = (-y_{t-1} \ldots -y_{t-p}, u_{t-1} \ldots u_{t-q}, \bar{\varepsilon}_{t-1} \ldots \bar{\varepsilon}_{t-r})$$

where

$$\bar{\varepsilon}_t = y_t - \mathbf{x}_t' \boldsymbol{\theta}_t.$$

The method of extended least squares updates the estimate by

$$\boldsymbol{\theta}_t = \boldsymbol{\theta}_{t-1} + \gamma_t \mathbf{V}_t^{-1} \mathbf{x}_t (y_t - \hat{y}_t)$$
$$\hat{y}_t = \mathbf{x}_t' \boldsymbol{\theta}_{t-1}$$
$$\mathbf{V}_t = \mathbf{V}_{t-1} + \gamma_t (\mathbf{x}_t \mathbf{x}_t' - \mathbf{V}_{t-1})$$

for some positive sequence $\gamma_t \downarrow 0$. Note that $\mathbf{x}_t$ contains entries that are constructed from past estimates.

Example 2: Estimating the* C *Matrix

Suppose the dynamic matrix $\mathbf{A}$ is known in a model

$$\mathbf{x}_t = \mathbf{A}\mathbf{x}_{t-1} + \boldsymbol{\xi}_t$$
$$\mathbf{y}_t = \mathbf{C}\mathbf{x}_t + \boldsymbol{\eta}_t,$$

but the matrix $\mathbf{C}$ is unknown, where $\boldsymbol{\xi}$ and $\boldsymbol{\eta}$ are mean-zero weakly stationary serially uncorrelated noises. The future observations are forecasted from the data by the orthgonal projection $\hat{E}(\mathbf{y}_{t+1}^+ \mid \mathbf{y}_t^-)$ where $\mathbf{y}_{t+1}^+$ stacks $\mathbf{y}_{t+1}, \ldots, \mathbf{y}_{t+J}$ and $\mathbf{y}_t^-$ stacks $y_t, \ldots, y_{t-K}$, for some J and K. We have

$$\hat{E}(\mathbf{y}_{t+1}^+ \mid \mathbf{y}_t^-) = \mathbf{H}\mathbf{R}^{-1}\mathbf{y}_t^-$$

where $\mathbf{H}$ is the Hankel matrix whose (i,j)th submatrix is $\mathbf{\Lambda}_{i+j-1}$ where $\Lambda_k = E(\mathbf{y}_{t+k}\mathbf{y}_t') = \mathbf{CA}^{k-1}\mathbf{M}$, $k \geqslant 1$, where $\mathbf{M} = E\mathbf{x}_{t+1}\mathbf{y}_t'$. The first submatrix row of $\mathbf{H}$, $\mathbf{H}_1.$ is $[\Lambda_1 \Lambda_2 \ldots \Lambda_{K+1}]$, and the first submatrix column of $\mathbf{H}$ is $[\mathbf{\Lambda}_1' \mathbf{\Lambda}_2' \ldots \mathbf{\Lambda}_J']'$, for example. Note that

$$\mathbf{H}_1. = \mathbf{C\Omega}$$

where

$$\mathbf{\Omega} = [\mathbf{M} \quad \mathbf{AM} \ldots \mathbf{A}^K\mathbf{M}].$$

This set of algebraic equations is solved as

$$\mathbf{C} = \mathbf{H}_1.\mathbf{\Omega}'(\mathbf{\Omega\Omega}')^{-1}.$$

Alternatively, when rewriting the equation as

$$\operatorname{vec}\mathbf{H}_1. = (\mathbf{\Omega}' \otimes \mathbf{I}) \operatorname{vec}\mathbf{C},$$

use the relation in Appendix of Chapter 1 to solve it as

$$\begin{aligned} \operatorname{vec}\mathbf{C} &= [(\mathbf{\Omega} \otimes \mathbf{I})(\mathbf{\Omega}' \otimes \mathbf{I})]^{-1}(\mathbf{\Omega} \otimes \mathbf{I}) \operatorname{vec}\mathbf{H}_1. \\ &= [(\mathbf{\Omega\Omega}')^{-1}\mathbf{\Omega} \otimes \mathbf{I}] \operatorname{vec}\mathbf{H}_1., \end{aligned}$$

which is the same as the forementioned. Next note that the Hankel matrix can be factored as $\mathbf{H} = \mathbf{O\Omega}$. Let $\mathbf{U\Sigma V}'$ be the singular value decomposition. Choose a coordinate system in which $\mathbf{O} = \mathbf{U\Sigma}^{1/2}$ and $\mathbf{\Omega} = \mathbf{\Sigma}^{1/2}\mathbf{V}'$. (Such a coordinate exists as shown in Aoki [1987c, Chapter 5].) Then $\mathbf{\Omega}^+ = \mathbf{V\Sigma}^{-1/2}$, and the solution $\mathbf{C} = \mathbf{H}_1.\mathbf{\Omega}^+ = \mathbf{H}_1.\mathbf{V\Sigma}^{-1/2}$ coincides with the previous solution since $\mathbf{\Omega}'(\mathbf{\Omega\Omega}')^{-1} = \mathbf{V\Sigma}^{-1/2}$. This estimate of $\mathbf{C}$ is the same as the one obtained from writing the observation equation as

$$\operatorname{vec}\mathbf{Y} = (\mathbf{X}' \otimes \mathbf{I}_p) \operatorname{vec}\mathbf{C} + \operatorname{vec}\mathbf{W},$$

and note that the covariance matrix of $\operatorname{vec}\mathbf{W}$ is block triangular since ηs are serially uncorrelated

$$E \operatorname{vec}\mathbf{W} \operatorname{vec}\mathbf{W}' = \mathbf{I}_J \otimes \mathbf{R}.$$

In practice, $\mathbf{X}$ is not available. Since $E\hat{\mathbf{X}}\mathbf{X}' = E\hat{\mathbf{X}}\hat{\mathbf{X}}'$, using $\hat{\mathbf{X}}$ as an instrument of $\mathbf{X}$, $\operatorname{vec}\mathbf{C}$ is estimated by

$$\begin{aligned} \operatorname{vec}\hat{\mathbf{C}} &= [(\hat{\mathbf{X}} \otimes \mathbf{I}_p)(\mathbf{I}_J \otimes \mathbf{I}_p)]^{-1}(\hat{\mathbf{X}} \otimes \mathbf{I}_p)(\mathbf{I}_J \otimes \mathbf{R}^{-1}) \operatorname{vec}\mathbf{Y} \\ &= [(\hat{\mathbf{X}}\hat{\mathbf{X}}')^{-1}\hat{\mathbf{X}} \otimes \mathbf{I}_p] \operatorname{vec}\mathbf{Y}, \end{aligned}$$

because $\mathbf{X} = \hat{\mathbf{X}} + \boldsymbol{\Xi}$ and $E\boldsymbol{\Xi}'\hat{\mathbf{X}} = 0$. We note that $\boldsymbol{\Omega}$ serves as an instrument. As far as estimating $\mathbf{C}$, the method in Aoki (1987c) is the same as the generalized least squares estimate.

Example 3: Estimating Noise Covariance Matrices. Continuing our discussion of a state space model with a known $\mathbf{A}$ matrix, use the relation

$$\mathbf{H}_{\cdot 1} = \mathbf{OM}$$

to estimate M by

$$\begin{aligned}\mathbf{M} &= (\mathbf{O}'\mathbf{O})^{-1}\mathbf{O}'\mathbf{H}_{\cdot 1} \\ &= \boldsymbol{\Sigma}^{-1/2}\mathbf{U}'\mathbf{H}_{\cdot 1}\end{aligned}$$

since

$$\mathbf{O} = \mathbf{U}\boldsymbol{\Sigma}^{-1/2}.$$

From definition,

$$\begin{aligned}\mathbf{M} &= E\mathbf{x}_{t+1}\mathbf{y}_t' \\ &= \mathbf{C}(\mathbf{Ax}_t + \boldsymbol{\xi}_t)(\mathbf{Cx}_t + \boldsymbol{\eta}_t)' \\ &= \mathbf{A\Pi C}' + \mathbf{N}\end{aligned}$$

where

$$\boldsymbol{\Pi} = \operatorname{cov} \mathbf{x}_t \qquad \text{and} \qquad \mathbf{N} = E\boldsymbol{\xi}_t\boldsymbol{\eta}_t'.$$

From the dynamics

$$\boldsymbol{\Pi} = \mathbf{A\Pi A}' + \mathbf{Q}$$

where

$$\mathbf{Q} = \operatorname{cov} \boldsymbol{\xi}_t$$

and

$$\begin{aligned}\boldsymbol{\Lambda}_0 &= E\mathbf{y}_t\mathbf{y}_t' \\ &= \mathbf{C\Pi C}' + \mathbf{R}\end{aligned}$$

where

$$\mathbf{R} = \operatorname{cov} \boldsymbol{\eta}_t.$$

The covariance matrix

$$\operatorname{cov}\begin{pmatrix}\boldsymbol{\xi} \\ \boldsymbol{\eta}\end{pmatrix}_t = \begin{pmatrix}\mathbf{Q} & \mathbf{N} \\ \mathbf{N}' & \mathbf{R}\end{pmatrix}$$

must be positive (semi-)definite. This can be ensured by factorizing the covariance matrix as

$$\begin{pmatrix}\mathbf{Q} & \mathbf{N} \\ \mathbf{N}' & \mathbf{R}\end{pmatrix} = \begin{pmatrix}\mathbf{B} \\ \mathbf{I}\end{pmatrix}\boldsymbol{\Delta}(\mathbf{B}' \quad \mathbf{I}),$$

i.e., solving for $\mathbf{B}$ and $\mathbf{\Delta}$ such that

$$\mathbf{Q} = \mathbf{B}\mathbf{\Delta}\mathbf{B}', \qquad \mathbf{N} = \mathbf{B}\mathbf{\Delta}, \qquad \text{and} \quad \mathbf{R} = \mathbf{\Delta}.$$

Thus,

$$\mathbf{\Delta} = \mathbf{\Lambda}_0 - \mathbf{C}\mathbf{\Pi}\mathbf{C}'$$

$$\mathbf{B} = (\mathbf{M} - \mathbf{A}\mathbf{\Pi}\mathbf{C}')(\mathbf{\Lambda}_0 - \mathbf{C}\mathbf{\Pi}\mathbf{C}')^{-1},$$

and

$$\mathbf{Q} = (\mathbf{M} - \mathbf{A}\mathbf{\Pi}\mathbf{C}')(\mathbf{\Lambda}_0 - \mathbf{C}\mathbf{\Pi}\mathbf{C}')^{-1}(\mathbf{M} - \mathbf{A}\mathbf{\Pi}\mathbf{C}')'$$

where the matrix $\mathbf{\Pi}$ is the solution of

$$\mathbf{\Pi} = \mathbf{A}\mathbf{\Pi}\mathbf{A}' + (\mathbf{M} - \mathbf{A}\mathbf{\Pi}\mathbf{C}')(\mathbf{\Lambda}_0 - \mathbf{C}\mathbf{\Pi}\mathbf{C}')^{-1}(\mathbf{M} - \mathbf{A}\mathbf{\Pi}\mathbf{C}')'.$$

This is a Riccati equation. The matrix $\mathbf{A}$ is known and $\mathbf{C}$ and $\mathbf{M}$ are estimated as explained. On the assumption that $\mathbf{\Lambda}_0 - \mathbf{C}\mathbf{\Pi}\mathbf{C}' > 0$ and that the state vector is minimal, the smallest positive definite solution of this equation (in the partial ordering of positive semidefinite matrices) exists. See Aoki (1987c, Chapter 7).

5.3 Wide-Sense Kalman Filters

Kalman filters are used both to forecast and to obtain estimates of state vectors of some dynamic models given a set of data over some fixed period. The latter applications are called smoothing of data. In both types of applications, model dynamics must be known and at least means and second moments of noise statistics are needed. Outputs of Kalman filters are the same as that of maximum likelihood estimates when noises are Gaussian and the same as the (weighted) least squares with non-Gaussian noises. Kalman filters provide computationally efficient ways for incorporating new or additional information in revising the estimates. When model parameters or noise statistics are unknown, they must be estimated before Kalman filters can be constructed. One way is to use a smoothing algorithm as in Clark (1987). Another method that is probably computationally more efficient and more robust is given in Aoki (1987c).

Some applications of Kalman filters are improving published preliminary estimates of monthly statistics, such as retail sales, as in Conrad and Corrado (1979); filling in missing observations in dynamic models, as in Palm and Nijman (1984); and utilizing predictions by a monthly model to improve forecasts by a quarterly model, as in Kalchbrenner and Tinsley (1977) and Corrado and Greene (1984).

We have already derived Kalman filters with uncorrelated dynamic and observation noises in an example in Chapter 3. Here we give the derivation when the noises are mutually correlated.

Given a system

$$\mathbf{x}_{t+1} = \mathbf{A}\mathbf{x}_t + \boldsymbol{\xi}_t$$

$$\mathbf{y}_t = \mathbf{C}\mathbf{x}_t + \boldsymbol{\eta}_t$$

where $\boldsymbol{\xi}_t$ and $\boldsymbol{\eta}_t$ are mean zero and serially uncorrelated with covariance

$$\operatorname{cov}\begin{pmatrix} \boldsymbol{\xi}_t \\ \boldsymbol{\eta}_t \end{pmatrix} = \begin{bmatrix} \mathbf{Q} & \mathbf{N} \\ \mathbf{N}' & \mathbf{S} \end{bmatrix},$$

this section discusses the estimate of $\mathbf{x}_t$ by its orthogonal projection on the subspace spanned by $\mathbf{y}_{t-1}, \mathbf{y}_{t-2}, \ldots$, denoted by $\mathbf{x}_{t|t-1}$.

The vector

$$\mathbf{e}_t = \mathbf{y}_t - \mathbf{y}_{t|t-1}$$

is uncorrelated with $\mathbf{y}_{t|t-1}$ by construction. Since

$$\begin{aligned} \mathbf{y}_{t+1|t} &= \hat{E}(\mathbf{y}_{t+1} \mid \mathbf{y}^{t-1}, \mathbf{y}_t) \\ &= \hat{E}(\mathbf{y}_{t+1} \mid \mathbf{y}^{t-1}, \mathbf{y}_t - \mathbf{y}_{t|t-1}) \\ &= \hat{E}(\mathbf{y}_{t+1} \mid \mathbf{y}^{t-1}, \mathbf{e}_t), \end{aligned}$$

$\mathbf{e}_{t+1} = \mathbf{y}_{t+1} - \mathbf{y}_{t+1|t}$ is uncorrelated with $\mathbf{y}^{t-1}$ and $\mathbf{e}_t$. Proceeding this way, we see that $\mathbf{e}_t$ is uncorrelated with $\mathbf{e}_s$, $s < t$. The dynamic equation implies that

$$\begin{aligned} \mathbf{x}_{t+1|t-1} &= \mathbf{A}\mathbf{x}_{t|t-1} + \boldsymbol{\xi}_{t|t-1} \\ &= \mathbf{A}\mathbf{x}_{t|t-1}. \end{aligned}$$

Next note that

$$\begin{aligned} \mathbf{x}_{t+1|t} &= \mathbf{x}_{t+1|t-1} + \hat{E}(\mathbf{x}_{t+1} \mid \mathbf{e}_t) \\ &= \mathbf{x}_{t+1|t-1} + \mathbf{K}_t\mathbf{e}_t \end{aligned} \tag{11}$$

where $\mathbf{K}_t = \boldsymbol{\Theta}_t\boldsymbol{\Delta}_t^{-1}$,

$$\boldsymbol{\Theta}_t = E\mathbf{x}_{t+1}\mathbf{e}_t'$$

and

$$\boldsymbol{\Delta}_t = E\mathbf{e}_t\mathbf{e}_t'.$$

The observation equation is used to express $\mathbf{e}_t$ as

$$\mathbf{e}_t = \mathbf{y}_t - \mathbf{y}_{t|t-1} = \mathbf{C}(\mathbf{x}_t - \mathbf{x}_{t|t-1}) + \boldsymbol{\eta}_t,$$

because $\boldsymbol{\eta}_{t|t-1} = 0$. Note that

$$\begin{aligned}\boldsymbol{\Theta}_t &= E\mathbf{x}_{t+1}\mathbf{e}_t' \\ &= E\{(\mathbf{A}\mathbf{x}_t + \xi_t)[\mathbf{C}(\mathbf{x}_t - \mathbf{x}_{t|t-1}) + \boldsymbol{\eta}_t]'\} \\ &= \mathbf{A}E[\mathbf{x}_t(\mathbf{x}_t - \mathbf{x}_{t|t-1})'\mathbf{C}' + E\xi_t\boldsymbol{\eta}_t', \\ &= \mathbf{A}\boldsymbol{\Xi}_t\mathbf{C}' + \mathbf{N},\end{aligned}$$

where

$$\boldsymbol{\Xi}_t = E(\mathbf{x}_t - \mathbf{x}_{t|t-1})(\mathbf{x}_t - \mathbf{x}_{t|t-1})'.$$

Here we use the facts that ξ_t is uncorrelated with $\mathbf{x}_t - \mathbf{x}_{t|t-1}$ (since the latter depends only on ξ_s and $\boldsymbol{\eta}_s$, $s < t$), that $\boldsymbol{\eta}_t$ is uncorrelated with $\mathbf{x}_t$ for the same reason, and that $\mathbf{x}_t - \mathbf{x}_{t|t-1}$ is uncorrelated with $\mathbf{x}_{t|t-1}$. Note also

$$\begin{aligned}\boldsymbol{\Delta}_t &= E\mathbf{e}_t\mathbf{e}_t' \\ &= \mathbf{C}\boldsymbol{\Xi}_t\mathbf{C}' + \mathbf{S}. \qquad (12)\end{aligned}$$

Therefore, the optimal Kalman filter gain is

$$\mathbf{K}_t = (\mathbf{A}\boldsymbol{\Xi}_t\mathbf{C}' + \mathbf{N})(\mathbf{C}\boldsymbol{\Xi}_t\mathbf{C}' + \mathbf{S})^{-1}. \tag{13}$$

Subtract (11) from the model dynamic equation to see that the one-step-ahead forecast error propagates with time by

$$\begin{aligned}\mathbf{x}_{t+1} - \mathbf{x}_{t+1|t} &= \mathbf{A}(\mathbf{x}_t - \mathbf{x}_{t|t-1}) + \xi_t - \boldsymbol{\Theta}_t\boldsymbol{\Delta}_t^{-1}\mathbf{e}_t \\ &= (\mathbf{A} - \mathbf{K}_t\mathbf{C})(\mathbf{x}_t - \mathbf{x}_{t|t-1}) + \xi_t - \mathbf{K}_t\boldsymbol{\eta}_t.\end{aligned}$$

Its covariance matrix at $t + 1$ is related to that at t by

$$\boldsymbol{\Xi}_{t+1} = (\mathbf{A} - \mathbf{K}_t\mathbf{C})\boldsymbol{\Xi}_t(\mathbf{A} - \mathbf{K}_t\mathbf{C})' + \mathbf{Q} + \mathbf{K}_t\mathbf{S}\mathbf{K}_t' - \mathbf{N}\mathbf{K}_t' - \mathbf{K}_t\mathbf{N}'. \tag{14}$$

The linear terms in $\mathbf{K}_t$ are grouped together using (12) as $\boldsymbol{\Theta}_t\boldsymbol{\Delta}_t^{-1}(\mathbf{C}\boldsymbol{\Xi}\mathbf{A}' + \mathbf{N}') = -\boldsymbol{\Theta}_t\boldsymbol{\Delta}_t'$ and its transpose. The quadratic term in $\boldsymbol{\Theta}_t$ is just

$$\mathbf{K}_t(\mathbf{C}\boldsymbol{\Xi}_t\mathbf{C}' + \mathbf{S})\mathbf{K}_t' = \boldsymbol{\Theta}_t\boldsymbol{\Delta}_t^{-1}\boldsymbol{\Theta}_t$$

in view of (13). Thus, the updating equation for the error covariance is simply

$$\boldsymbol{\Xi}_{t+1} = \mathbf{A}\boldsymbol{\Xi}_t\mathbf{A}' + \mathbf{Q} - \boldsymbol{\Theta}_t\boldsymbol{\Delta}_t^{-1}\boldsymbol{\Theta}_t' \tag{15}$$

where

$$\boldsymbol{\Theta}_t = \mathbf{A}\boldsymbol{\Xi}_t\mathbf{C}' + \mathbf{N}.$$

Equations (14) and (15) constitute the optimal Kalman filter.

Exercise

Let $\{\mathbf{y}_t\}$ be a mean-zero weakly stationary process. Let $\mathbf{y}_t^+$ denote a vector that stacks $\mathbf{y}_t, \mathbf{y}_{t+1}, \ldots, \mathbf{y}_{t+J}$ in that order and $\mathbf{y}_{t-1}^-$ denote a vector that stacks $\mathbf{y}_{t-1}, \mathbf{y}_{t-2}, \ldots, \mathbf{y}_{t-K}$ in that order for some positive J and K. Then $\hat{E}(\mathbf{y}_t^+ \mid \mathbf{y}_{t-1}^-) = \mathbf{H}\mathbf{R}^{-1}\mathbf{y}_{t-1}^-$ where $\mathbf{H}$ is a Hankel matrix whose (i,j)th submatrix is $E(\mathbf{y}_{t+i+j-1}\mathbf{y}_t') = \mathbf{\Lambda}_{i+j-1}$ and $\mathbf{R} = E(\mathbf{y}_{t-1}^-\mathbf{y}_{t-1}^{-\prime})$.

a. Now suppose that an unobservable dynamic factor (state space) model is posited as

$$\mathbf{z}_{t+1} = \mathbf{A}\mathbf{z}_t + \boldsymbol{\xi}_t$$

$$\mathbf{y}_t = \mathbf{C}\mathbf{z}_t + \boldsymbol{\eta}_t$$

where $\{\boldsymbol{\xi}, \boldsymbol{\eta}\}$ is a mean-zero serially uncorrelated process. Show that

$$\hat{E}(\mathbf{z}_t \mid \mathbf{y}_{t-1}^-) = \mathbf{\Omega}\mathbf{R}^{-1}\mathbf{y}_{t-1}^-$$

where

$$\mathbf{\Omega} = [\mathbf{D}, \mathbf{A}\mathbf{D}, \ldots, \mathbf{A}^{K-1}\mathbf{D}]$$

where

$$\mathbf{D} = E(\mathbf{z}_t\mathbf{y}_{t-1}').$$

b. Noting that $\mathbf{D} = E(\mathbf{z}_{t+1}\mathbf{y}_t')$ and that $\mathbf{z}_{t+1} = \mathbf{A}(\mathbf{z}_{t|t-1} + \boldsymbol{\zeta}_t) + \boldsymbol{\xi}_t$ where $\boldsymbol{\zeta}_t$ is uncorrelated with $\mathbf{y}_{t-1}^-$, show that $\mathbf{D} = \mathbf{A}(\mathbf{\Pi} + \mathbf{Z})\mathbf{C}' + \mathbf{N}$ where $\mathbf{\Pi} = E(\mathbf{z}_{t|t-1}\mathbf{z}_{t|t-1}') = \mathbf{\Omega}\mathbf{R}^{-1}\mathbf{\Omega}$, $\mathbf{Z} = E(\boldsymbol{\zeta}_t\boldsymbol{\zeta}_t')$ and $\mathbf{N} = E\boldsymbol{\xi}_t\boldsymbol{\eta}_t'$.

c. Derive from part (a)

$$\mathbf{z}_{t+1|t} = \mathbf{A}\mathbf{z}_{t|t-1} + \mathbf{B}\mathbf{e}_t$$

$$\mathbf{y}_t = \mathbf{C}\mathbf{z}_{t|t-1} + \mathbf{e}_t$$

where $\mathbf{e}_t = \mathbf{y}_t - \mathbf{y}_{t|t-1}$, $\mathbf{B} = E(\mathbf{z}_{t+1}\mathbf{e}_t')\mathbf{\Delta}^{-1}$, $\mathbf{\Delta} = \operatorname{cov}\mathbf{e}_t = \mathbf{C}\mathbf{Z}\mathbf{C}' + \mathbf{\Sigma}_\eta$, and $\mathbf{\Sigma}_\eta = \operatorname{cov}\boldsymbol{\eta}_t$.

d. Show that $\mathbf{D} = \mathbf{A}\mathbf{\Pi}\mathbf{C}' + \mathbf{B}\mathbf{\Delta}$ and that $\mathbf{\Pi} = \mathbf{A}\mathbf{\Pi}\mathbf{A}' + \mathbf{B}\mathbf{\Delta}\mathbf{B}'$ where $\mathbf{\Pi} = \mathbf{\Omega}\mathbf{R}^{-1}\mathbf{\Omega}'$ and that $\mathbf{\Lambda}_0 = E\mathbf{y}_t\mathbf{y}_t' = \mathbf{C}\mathbf{\Pi}\mathbf{C}' + \mathbf{\Delta}$ where $\mathbf{\Delta} = \mathbf{C}\mathbf{Z}\mathbf{C}' + \mathbf{\Sigma}_\eta$.

e. Show that $E(\mathbf{y}_{t+k}\mathbf{y}_t') = \mathbf{\Lambda}_k = \mathbf{C}\mathbf{A}^{k-1}\mathbf{D}$, $k \geqslant 1$. Therefore the Hankel matrix $\mathbf{H}$ at the start of the exercise can be factored as $\mathbf{H} = \mathbf{O}\mathbf{\Omega}$ where $\mathbf{O}' = [\mathbf{C}' \quad \mathbf{A}'\mathbf{C}' \ldots (\mathbf{A}')^{J-1}\mathbf{C}']$.

f. Show that $\operatorname{cov}\boldsymbol{\zeta}_t = \mathbf{Z}_t$ is governed by a Riccati equation

$$\mathbf{\Sigma}_\xi + \mathbf{A}\mathbf{Z}\mathbf{A}' = \mathbf{A} + (\mathbf{N} + \mathbf{A}\mathbf{Z}\mathbf{C}')(\mathbf{\Sigma}_\eta + \mathbf{C}\mathbf{Z}\mathbf{C}')^{-1}(\mathbf{N} + \mathbf{A}\mathbf{Z}\mathbf{C}')'$$

where $\mathbf{\Sigma}_\xi = \operatorname{cov}\boldsymbol{\xi}$, $\mathbf{\Sigma}_\eta = \operatorname{cov}\boldsymbol{\eta}$, and $\mathbf{N} = E\boldsymbol{\xi}\boldsymbol{\eta}'$.

Variational Argument

Variational argument gives the optimal Kalman filter gain simply as follows. First, recognize that $\mathbf{y}_t - \mathbf{y}_{t|t-1}$ is the "innovation" containing new information, i.e., information not correlated with the previous forecast, $\mathbf{y}_{t|t-1}$. The forecast $\mathbf{x}_{t+1|t}$ is related to the previous forecast $\mathbf{x}_{t+1|t-1} = \mathbf{A}\mathbf{x}_{t|t-1}$ by

$$\mathbf{x}_{t+1|t} = \mathbf{A}\mathbf{x}_{t|t-1} + \mathbf{K}_t\mathbf{e}_t$$

where

(16)
$$\begin{aligned}\mathbf{e}_t &= \mathbf{y}_t - \mathbf{y}_{t|t-1}\\ &= \mathbf{C}(\mathbf{x}_t - \mathbf{x}_{t|t-1}) + \boldsymbol{\eta}_t.\end{aligned}$$

The estimation error is propagated over time by

$$\begin{aligned}\mathbf{x}_{t+1} - \mathbf{x}_{t+1|t} &= \mathbf{A}(\mathbf{x}_t - \mathbf{x}_{t|t-1}) + \xi_t - \mathbf{K}_t\mathbf{e}_t\\ &= (\mathbf{A} - \mathbf{K}_t\mathbf{C})(\mathbf{x}_t - \mathbf{x}_{t|t-1}) + \xi_t - \mathbf{K}_t\boldsymbol{\eta}_t.\end{aligned}$$

This equation is valid for any $\mathbf{K}_t$. The (conditional) covariance matrix is

$$\boldsymbol{\Xi}_{t+1} = (\mathbf{A} - \mathbf{K}_t\mathbf{C})\boldsymbol{\Xi}_t(\mathbf{A} - \mathbf{K}_t\mathbf{C})' + \mathbf{K}_t\mathbf{S}\mathbf{K}_t' - \mathbf{N}\mathbf{K}_t' - \mathbf{K}_t\mathbf{N}'$$

where $\boldsymbol{\Xi}_t = \operatorname{cov}(\mathbf{x}_t - \mathbf{x}_{t|t-1})$ because $\mathbf{x}_t - \mathbf{x}_{t|t-1}$ is uncorrelated with ξ_t and $\boldsymbol{\eta}_t$.

The optimal gain $\mathbf{K}_t$ minimized $\boldsymbol{\Xi}_{t+1}$ given $\boldsymbol{\Xi}_t$, i.e., $\delta\boldsymbol{\Xi}_{t+1} = 0$ in response to variation $\delta\mathbf{K}_t$ from the optimal gain. The first variation then satisfies

$$0 = -(\mathbf{A} - \mathbf{K}_t\mathbf{C})\boldsymbol{\Xi}_t\mathbf{C}'\delta\mathbf{K}_t' + \mathbf{K}_t\mathbf{S}\delta\mathbf{K}_t' - \mathbf{N}\delta\mathbf{K}_t'.$$

Since $\delta\mathbf{K}_t$ is arbitrary, the optimal gain is given by

$$\mathbf{K}_t = (\mathbf{A}\boldsymbol{\Xi}_t\mathbf{C}' + \mathbf{N})(\mathbf{S} + \mathbf{C}\boldsymbol{\Xi}_t\mathbf{C}')^{-1} = \boldsymbol{\Theta}_t\boldsymbol{\Delta}_t^{-1},$$

since (16) implies $\operatorname{cov}\mathbf{e}_t = \boldsymbol{\Delta}_t = \mathbf{C}\boldsymbol{\Xi}_t\mathbf{C}' + \mathbf{S}_t$. This is exactly the expression (12). From the dynamic equation,

(17)
$$\boldsymbol{\Pi}_{t+1} = \mathbf{A}\boldsymbol{\Pi}_t\mathbf{A}' + \mathbf{Q}$$

where

$$\boldsymbol{\Pi}_t = \operatorname{cov}\mathbf{x}_t$$

since ξ_t and $\mathbf{x}_t$ are uncorrelated. The Kalman filter equation gives

(18)
$$\boldsymbol{\Sigma}_{t+1} = \mathbf{A}\boldsymbol{\Sigma}_t\mathbf{A}' + \boldsymbol{\Theta}_t\boldsymbol{\Delta}_t^{-1}\boldsymbol{\Theta}_t' + \mathbf{Q}.$$

Note that

$$\boldsymbol{\Xi}_t = \operatorname{cov}(\mathbf{x}_t - \mathbf{x}_{t|t-1}) = \boldsymbol{\Pi}_t - \boldsymbol{\Sigma}_t \geqslant 0.$$

The difference of (17) and (18) is exactly (15).

In some applications, $\mathbf{y}_t$ is available in estimating $\mathbf{x}_t$ in addition to $\mathbf{y}^{t-1}$. Then $\mathbf{x}_{t|t}$ rather than $\mathbf{x}_{t|t-1}$ may be more appropriate. The relation between $\mathbf{x}_{t|t-1}$ and $\mathbf{x}_{t|t}$ is straightforward:

$$
\begin{aligned}
\mathbf{x}_{t|t} &= E(\mathbf{x}_t \mid \mathbf{y}^{t-1}, \mathbf{y}_t) \\
&= E(\mathbf{x}_t \mid \mathbf{y}^{t-1}, \mathbf{y}_t - \mathbf{y}_{t|t-1}) \\
&= E(\mathbf{x}_t \mid \mathbf{y}^{t-1}) + E(\mathbf{x}_t \mid \mathbf{y}_t - \mathbf{y}_{t|t-1}) \\
&= \mathbf{x}_{t|t-1} + \mathbf{L}_t \mathbf{e}_t
\end{aligned} \tag{19}
$$

where

$$
\begin{aligned}
\mathbf{L}_t &= (E\mathbf{x}_t\mathbf{e}_t')(E\mathbf{e}_t\mathbf{e}_t')^{-1}, \\
E\mathbf{x}_t\mathbf{e}_t' &= E(\mathbf{x}_t - \mathbf{x}_{t|t-1})(\mathbf{x}_t - \mathbf{x}_{t|t-1})'\mathbf{C}' \\
&= \mathbf{\Xi}_t\mathbf{C}',
\end{aligned}
$$

and

$$E\mathbf{e}_t\mathbf{e}_t' = \mathbf{C}\mathbf{\Xi}_t\mathbf{C}' + \mathbf{R}.$$

Thus

$$
\begin{aligned}
\mathbf{x}_{t+1|t+1} &= \mathbf{x}_{t+1|t} + \mathbf{L}_{t+1}\mathbf{e}_{t+1} \\
&= \mathbf{A}\mathbf{x}_{t|t} + \mathbf{L}_{t+1}(\mathbf{y}_{t+1} - \mathbf{C}\mathbf{A}\mathbf{x}_{t|t})
\end{aligned}
$$

where

$$\mathbf{L}_{t+1} = \mathbf{\Xi}_{t+1}\mathbf{C}'(\mathbf{C}\mathbf{\Xi}_{t+1}\mathbf{C}' + \mathbf{R})^{-1}.$$

Compared with (13), we note that the structure of updating remains the same. Only the expression for the gain is different. Let $\mathbf{\Pi}_t = \operatorname{cov}(\mathbf{x}_t - \mathbf{x}_{t|t-1})$ and $\mathbf{\Xi}_t = \operatorname{cov}(\mathbf{x}_t - \mathbf{x}_{t|t-1})$ as before. Then use (19) to obtain the relation between the two covariance matrices as

$$
\begin{aligned}
\mathbf{\Pi}_t &= \mathbf{\Xi}_t - \mathbf{L}_t\mathbf{C}\mathbf{\Xi}_t - \mathbf{\Xi}_t\mathbf{C}'\mathbf{L}_t' + \mathbf{L}_t\mathbf{\Delta}\mathbf{L}_t' \\
&= (\mathbf{I} - \mathbf{L}_t\mathbf{C})\mathbf{\Xi}_t(\mathbf{I} - \mathbf{L}_t\mathbf{C})' + \mathbf{L}_t\mathbf{R}\mathbf{L}_t'
\end{aligned}
$$

or

$$\mathbf{\Pi}_t = \mathbf{\Xi}_t - \mathbf{\Xi}_t\mathbf{C}'\mathbf{\Delta}_t^{-1}\mathbf{C}\mathbf{\Xi}_t. \tag{20}$$

From (20), by the Woodbury identity,

$$\mathbf{\Pi}_t^{-1} = \mathbf{\Xi}_t^{-1} + \mathbf{C}'\mathbf{R}^{-1}\mathbf{C}.$$

Multiply (20) from the right by $\mathbf{C}'\mathbf{R}^{-1}$ and group terms as follows to see that

$$
\begin{aligned}
\mathbf{\Pi}_t\mathbf{C}'\mathbf{R}^{-1} &= \mathbf{\Xi}_t\mathbf{C}'\mathbf{R}^{-1} + \mathbf{\Xi}_t\mathbf{C}'\mathbf{\Delta}_t^{-1}\mathbf{C}\mathbf{\Xi}_t\mathbf{C}'\mathbf{R}^{-1} \\
&= \mathbf{\Xi}_t\mathbf{C}'(\mathbf{R}^{-1} - \mathbf{\Delta}_t^{-1}\mathbf{C}\mathbf{\Xi}_t\mathbf{C}'\mathbf{R}^{-1}) \\
&= \mathbf{\Xi}_t\mathbf{C}'(\mathbf{I} - \mathbf{\Delta}_t^{-1}\mathbf{C}\mathbf{\Xi}_t\mathbf{C}')\mathbf{R}^{-1}
\end{aligned}
$$

$$= \boldsymbol{\Xi}_t \mathbf{C}' \boldsymbol{\Delta}_t^{-1} (\boldsymbol{\Delta}_t - \mathbf{C}\boldsymbol{\Xi}_t \mathbf{C}') \mathbf{R}^{-1}$$
$$= \boldsymbol{\Xi}_t \mathbf{C}' \boldsymbol{\Delta}_t^{-1}$$
$$= \mathbf{L}_t,$$

i.e., $\mathbf{L}_t$ has an alternative expression

$$\text{(21)} \qquad \begin{aligned} \mathbf{L}_t &= \boldsymbol{\Xi}_t \mathbf{C}' \boldsymbol{\Delta}_t^{-1} \\ &= \boldsymbol{\Pi}_t \mathbf{C}' \mathbf{R}^{-1}. \end{aligned}$$

Innovation Model

A state space model

$$\text{(22)} \qquad \begin{aligned} \mathbf{x}_{t+1|t} &= \mathbf{A}\mathbf{x}_{t|t-1} + \mathbf{B}\mathbf{e}_t \\ \mathbf{y}_t &= \mathbf{C}\mathbf{x}_{t|t-1} + \mathbf{e}_t, \end{aligned}$$

where $\mathbf{e}_t$ is serially uncorrelated, is called the innovation model, because $\mathbf{e}_t = \mathbf{y}_t - \mathbf{y}_{t|t-1}$. In the innovation model

$$\operatorname{cov}\begin{pmatrix} \mathbf{B}\mathbf{e}_t \\ \mathbf{e}_t \end{pmatrix} = \begin{pmatrix} \mathbf{B} \\ \mathbf{I} \end{pmatrix} \boldsymbol{\Delta}_t (\mathbf{B}' \quad \mathbf{I}).$$

Compare this with

$$\operatorname{cov}\begin{pmatrix} \boldsymbol{\xi}_t \\ \boldsymbol{\eta}_t \end{pmatrix} = \begin{pmatrix} \mathbf{Q} & \mathbf{N} \\ \mathbf{N}' & \mathbf{S} \end{pmatrix}$$

where

$$\boldsymbol{\Delta}_t = \mathbf{C}\boldsymbol{\Xi}_t \mathbf{C}' + \mathbf{S}.$$

Clearly, in the innovation model, $\boldsymbol{\Xi}_t = 0$ so that $\boldsymbol{\Delta}_t = \mathbf{S}$. Similarly, $\boldsymbol{\Theta}_t = \mathbf{N} = \mathbf{B}\boldsymbol{\Delta}_t$ and $\mathbf{Q} = \boldsymbol{\Theta}_t \boldsymbol{\Delta}_t^{-1} \boldsymbol{\Theta}_t' = \mathbf{B}\boldsymbol{\Delta}_t \mathbf{B}'$. In the innovation model then, $\mathbf{x}_{t|t-1}$ itself serves as the state vector $\mathbf{x}_t$. We have

$$\begin{aligned} \boldsymbol{\Lambda}_t &= E\mathbf{y}_t \mathbf{y}_t' \\ &= \mathbf{C}\boldsymbol{\Pi}_t \mathbf{C}' + \mathbf{S} \end{aligned}$$

and

$$\begin{aligned} \boldsymbol{\Lambda}_{t,s} &= E\mathbf{y}_t \mathbf{y}_s' \\ &= \mathbf{C}\mathbf{A}^{t-s-1} \mathbf{M}_s \end{aligned}$$

where

$$\mathbf{M}_s = \mathbf{A}\boldsymbol{\Pi}_s \mathbf{C}' + \mathbf{N}, \qquad t > s.$$

If $\{y_t\}$ is a weakly stationary process,

$$\Lambda_0 = \mathbf{C\Pi C}' + \mathbf{S}$$

$$\Lambda_k = \mathbf{CA}^{k-1}\mathbf{M}$$

where

$$\mathbf{M} = \mathbf{A\Pi C}' + \mathbf{N}.$$

The dynamics show

$$\mathbf{\Pi} = \mathbf{A\Pi A}' + \mathbf{Q}.$$

Then recognizing that $\mathbf{\Theta}_t$ becomes $\mathbf{N} = \mathbf{M} - \mathbf{A\Pi C}'$ and $\mathbf{\Delta}_t\mathbf{S} = \Lambda_0 - \mathbf{C\Pi C}'$, the Kalman filter gain is a constant matrix

$$\mathbf{K} = (\mathbf{M} - \mathbf{A\Pi C}')(\Lambda_0 - \mathbf{C\Pi C}')^{-1}$$

where

$$\mathbf{\Pi} = \mathbf{A\Pi A}' + (\mathbf{M} - \mathbf{A\Pi C}')(\Lambda_0 - \mathbf{C\Pi C}')^{-1}(\mathbf{M} - \mathbf{A\Pi C}')'$$

because $\mathbf{Q} = \mathbf{\Theta}_t\mathbf{\Delta}_t^{-1}\mathbf{\Theta}_t$.

Equation (22) shows that the innovation model is invertible as

$$\mathbf{e}_t = \mathbf{y}_t - \mathbf{C}\mathbf{x}_{t|t-1}$$

$$\mathbf{x}_{t+1|t} = (\mathbf{A} - \mathbf{BC})\mathbf{x}_{t|t-1} + \mathbf{B}\mathbf{y}_t.$$

This shows that $\mathbf{e}^t$ and $\mathbf{y}^t$ span the same subspace in the Hilbert space of random variables with finite variances.

An Alternative Derivation of Kalman Filters

Given $\mathbf{y}_t = \mathbf{C}\mathbf{x}_t + \mathbf{e}_t$ where $\mathbf{e}_t$ is orthogonal to $\mathbf{y}_{t-1}^-$, stack the data vector $\mathbf{y}_{t-1}, \mathbf{y}_{t-2}, \ldots$ as $\mathbf{y}_{t-1}^-$. The orthogonal projection of $\mathbf{y}_t$ onto the subspace spanned by data is

$$\hat{E}(\mathbf{y}_t \mid \mathbf{y}_{t-1}^-) = \mathbf{C}\hat{E}(\mathbf{x}_t \mid \mathbf{y}_{t-1}^-)$$

$$= \mathbf{C}E(\mathbf{x}_t\mathbf{y}_{t-1}^{-\prime})\mathbf{R}_{t-1}^{-1}\mathbf{y}_{t-1}^-$$

where

$$\mathbf{R}_{t-1} = \operatorname{cov}(\mathbf{y}_{t-1}^-).$$

From the left-hand side,

$$\hat{E}(\mathbf{y}_t \mid \mathbf{y}_{t-1}^-) = \mathbf{H}\mathbf{R}_{t-1}^{-1}\mathbf{y}_{t-1}^-$$

where

$$\mathbf{H} = E(\mathbf{y}_t \mathbf{y}_{t-1}^{-\prime}) = [\mathbf{\Lambda}_1 \quad \mathbf{\Lambda}_2 \ldots].$$

When the covariance matrix has the factored form*

$$\mathbf{\Lambda}_k = E(\mathbf{y}_t \mathbf{y}_{t-k}') = \mathbf{C}\mathbf{A}^{k-1}\mathbf{M}, \qquad k \geqslant 1,$$

we can write H as $\mathbf{C}\mathbf{\Omega}_t$ where

$$\mathbf{\Omega}_t = [\mathbf{M} \quad \mathbf{A}\mathbf{M} \quad \mathbf{A}^2\mathbf{M} \ldots].$$

Denote $\hat{E}(\mathbf{x}_t \mid \mathbf{y}_{t-1}^-)$ by $\mathbf{x}_{t|t-1}$. To see how it propagates with time, advance the time index by 1. Then

$$\mathbf{x}_{t+1|t} = \mathbf{C}\hat{E}(\mathbf{x}_{t+1} \mid \mathbf{y}_t^-) = \mathbf{C}\mathbf{\Omega}_t R_t^{-1} \mathbf{y}_t^-,$$

where, by definition,

$$\mathbf{y}_t^- = \begin{bmatrix} \mathbf{y}_t \\ \mathbf{y}_{t-1}^- \end{bmatrix}.$$

Note that

$$\mathbf{\Omega}_t = [\mathbf{M} \quad \mathbf{A}\mathbf{\Omega}_{t-1}] \tag{23}$$

and

$$\mathbf{R}_t = \begin{bmatrix} \mathbf{\Lambda}_0 & \mathbf{C}\mathbf{\Omega}_{t-1} \\ \mathbf{\Omega}_{t-1}'\mathbf{C}' & \mathbf{R}_{t-1} \end{bmatrix}, \tag{24}$$

where $\mathbf{\Lambda}_0 = \operatorname{cov} \mathbf{y}_t$. Write the inverse of $\mathbf{R}_t$ as

$$\mathbf{R}_t^{-1} = \begin{bmatrix} \mathbf{R}^{11} & -\mathbf{R}^{12} \\ -\mathbf{R}^{21} & \mathbf{R}^{22} \end{bmatrix}$$

where we define

$$\mathbf{\Sigma}_t = \mathbf{\Omega}_{t-1} \mathbf{R}_{t-1}^{-1} \mathbf{\Omega}_{t-1}'$$

(25) $\mathbf{R}^{11} = (\mathbf{\Lambda}_0 - \mathbf{C}\mathbf{\Sigma}_t\mathbf{C}')^{-1}, \qquad \mathbf{R}^{12} = \mathbf{R}^{11}\mathbf{C}\mathbf{\Omega}_{t-1}\mathbf{R}_{t-1}^{-1}$

$\mathbf{R}^{21} = (\mathbf{R}^{12})' \qquad$ and $\qquad \mathbf{R}^{22} = \mathbf{R}_{t-1}^{-1} + \mathbf{R}_{t-1}^{-1}\mathbf{\Omega}_{t-1}'\mathbf{C}'\mathbf{R}^{11}\mathbf{C}\mathbf{\Omega}_{t-1}\mathbf{R}_{t-1}^{-1}.$

---

* When $\mathbf{y}_t$ has a rational spectral density, the covariance matrices indeed have the indicated structure for some finite dimensional matrices.

Arranging terms as shown,

$$\begin{aligned}\mathbf{x}_{t+1|t} &= \mathbf{M}(\mathbf{R}^{11}\mathbf{y}_t - \mathbf{R}^{12}\mathbf{y}_{t-1}^-) + \mathbf{A}\boldsymbol{\Omega}_{t-1}(-\mathbf{R}^{21}\mathbf{y}_t + \mathbf{R}^{22}\mathbf{y}_{t-1}^-)\\ &= (\mathbf{A}\boldsymbol{\Omega}_{t-1}\mathbf{R}^{22} - \mathbf{M}\mathbf{R}^{12})\mathbf{y}_{t-1}^- + (\mathbf{M}\mathbf{R}^{11} - \mathbf{A}\boldsymbol{\Omega}_{t-1}\mathbf{R}^{21})\mathbf{y}_t,\end{aligned}$$

we recognize the two terms on the right as

(26) $$\mathbf{x}_{t+1|t} = \mathbf{A}\mathbf{x}_{t|t-1} + \mathbf{K}_t(\mathbf{y}_t - \mathbf{C}\mathbf{x}_{t|t-1})$$

where

$$\mathbf{K}_t = (\mathbf{M} - \mathbf{A}\boldsymbol{\Sigma}_t\mathbf{C}')\mathbf{R}^{11}.$$

From the definition, $\boldsymbol{\Sigma}_{t+1} = \boldsymbol{\Omega}_t\mathbf{R}_t^{-1}\boldsymbol{\Omega}_t'$. After substituting (23) and (24) and making use of (25), we see that it evolves with time according to

$$\boldsymbol{\Sigma}_{t+1} = \mathbf{A}\boldsymbol{\Sigma}_t\mathbf{A}' + \mathbf{K}_t\boldsymbol{\Delta}\mathbf{K}_t,$$

where

(27) $$\boldsymbol{\Delta} = \mathbf{R}^{11}.$$

In view of (26) and (27), $\boldsymbol{\Sigma}_t$ can be interpreted as the covariance of $\mathbf{x}_{t|t-1}$ and $\boldsymbol{\Delta} = \operatorname{cov}(\mathbf{y}_t - \mathbf{C}\mathbf{x}_{t|t-1})$ since $\mathbf{x}_{t|t-1}$ and $\mathbf{y}_t - \mathbf{C}\mathbf{x}_{t|t-1}$ are uncorrelated. From $\mathbf{y}_t - \mathbf{y}_{t|t-1} = \mathbf{C}(\mathbf{x}_t - \mathbf{x}_{t|t-1}) + \mathbf{e}_t$,

$$\operatorname{cov}(\mathbf{y}_t \mid \mathbf{y}_{t-1}^-) = \mathbf{C}\operatorname{cov}(\mathbf{x}_t \mid \mathbf{y}_{t-1}^-)\mathbf{C}' + \operatorname{cov}\mathbf{e}_t.$$

If the equation for the state vector is given by

$$\mathbf{x}_{t+1} = \mathbf{A}\mathbf{x}_t + \boldsymbol{\xi}_t$$

and the observation equation is

$$\mathbf{y}_t = \mathbf{C}\mathbf{x}_t + \boldsymbol{\eta}_t$$

where $\boldsymbol{\xi}_t$ and $\boldsymbol{\eta}_t$ are serially uncorrelated with

$$\operatorname{cov}\begin{pmatrix}\boldsymbol{\xi}_t\\ \boldsymbol{\eta}_t\end{pmatrix} = \begin{bmatrix}\mathbf{Q} & \mathbf{N}\\ \mathbf{N}' & \mathbf{S}\end{bmatrix},$$

then $\operatorname{cov}\mathbf{x}_t = \boldsymbol{\Pi}_t$ evolves according to

$$\boldsymbol{\Pi}_{t+1} = \mathbf{A}\boldsymbol{\Pi}_t\mathbf{A}' + \mathbf{Q}$$

and the one-step prediction error covariance $\boldsymbol{\Xi}_t = \operatorname{cov}(\mathbf{x}_t - \mathbf{x}_{t|t-1})$ evolves over time by

$$\boldsymbol{\Xi}_{t+1} = \mathbf{A}\boldsymbol{\Xi}_t\mathbf{A}' + \mathbf{Q} - \mathbf{K}_t\boldsymbol{\Delta}\mathbf{K}_t'.$$

See Aoki (1987c, Section 7.2) for detail.

***Example 4:* (*Goodwin and Sin*).** In Chapter 3, we discussed problems caused by the presence of unit zeros in forecasting. Goodwin and Sin (1984, p. 298) discuss the unit zero from a state space point of view. Briefly put, the zero prevents the process from being invertible. Then the information set spanned by $y_t, y_{t-1}, \ldots$ is not the same as the information set spanned by its innovation sequences $\varepsilon_t, \varepsilon_{t-1}, \ldots$ because y_t is expressible in terms of $\varepsilon_t, \varepsilon_{t-1}, \ldots$, but ε_t is not expressible in terms of $y_t, y_{t-1}, \ldots$ when the spectral density of $\{y_t\}$ has a zero on the unit circle.

Its state space model is a trivial model

$$x_t = e_{t-1}$$

and

$$y_t = -x_t + e_t,$$

i.e., $A = 0$, $B = 1$, $C = -1$ in the standard model representation.

Its Kalman filter is then of the form

$$\hat{x}_{t+1} = K_t(y_t + \hat{x}_t) \tag{28}$$

where K_t is time-verifying filter gain. The one-step-ahead forecasting error is governed by

$$\begin{aligned}\tilde{x}_{t+1} &= x_{t+1} - \hat{x}_{t+1}\\ &= e_t - K_t(y_t + \hat{x}_t)\\ &= e_t - K_t(-x_t + e_t + \hat{x}_t)\\ &= (1 - K_t)e_t + K_t\tilde{x}_t.\end{aligned}$$

Noting that e_t is uncorrelated with $\tilde{x}_t$, we derive its ereror variance as

$$\pi_{t+1} = K_t\pi_t + (1 - K_t)^2,$$

where $\pi_t = \operatorname{var}\tilde{x}_t$. Normalize the variances by $\Sigma_t = \pi_t/\sigma^2$. Then the recursion is

$$\Sigma_{t+1} = K_t\Sigma_t + (1 - K_t)^2. \tag{29}$$

Minimization with respect to K_t yields the optimal filter gain

$$K_t = 1/(\Sigma_t + 1).$$

Substitute this back in (29) and obtain

$$\Sigma_{t+1} = \Sigma_t/(1 + \Sigma_t). \tag{30}$$

As t goes to infinity, (30) shows that $\Sigma_t^{-1} \to \infty$ or $\Sigma_t \to 0$ and $K_t \to 1$. In the limit, then, the Kalman filter takes the form

$$\hat{x}_{t+1} = \hat{x}_t + y_t. \tag{31}$$

Rewriting this recursion as $\Sigma_{t+1}^{-1} = 1 + \Sigma_t^{-1}$, the explicit expression $\Sigma_t = 1/(t + \Sigma_0^{-1})$ is obtained. Setting $\Sigma_0 = 1$ for convenience, the close-form expression for the one-step forecasting error is

$$\tilde{x}_t = \frac{1}{t+1}\tilde{x}_0 + \frac{1}{t+1}\Sigma_{\tau=0}^{t} e_\tau .$$

This is another way of seeing that var $\tilde{x}_t \to 0$ as $t \to \infty$ regardless of $\hat{x}_0$, i.e., regardless of the initial condition $\hat{x}_0$ employed. On the other hand, if the limit, i.e., steady state, Kalman filter (31) is used from the start, then the estimate is given by

$$\begin{aligned}\hat{x}_t &= y_{t-1} + y_{t-2} + \ldots y_0 + \hat{x}_0 \\ &= \hat{x}_0 + e_{t-1} - e_{-1}\end{aligned}$$

or

$$\tilde{x}_t = e_{-1} - \hat{x}_0 ,$$

and its variance does not approach zero! If $e_{-1} - \hat{x}_0$ is called the initial error $\tilde{x}_0$ (since $x_0 = e_{-1}$), then var $\tilde{x}_t$ = var $\tilde{x}_0$. In this sense, the constant coefficient Kalman filter forever retains the initial error. This is one feature of the unit zero. In the sense that the past error is forever retained, this feature is analogous to that of unit roots.

***Example 5: Combining Rival Forecasts* (*Corrado and Greene* [1984]).** Corrado and Greene discussed ways for combining forecasts from a quarterly model and a monthly model (appropriately aggregated to a quarterly frequency). Each model is of state space form

$$\mathbf{A}_0 \mathbf{y}_t = \mathbf{A}_1 \mathbf{y}_{t-1} + \mathbf{B}\mathbf{x}_t + \mathbf{q}_t ,$$

$\mathbf{A}_0$ nonsingular, where $\mathbf{x}_t$ is the instrument vector and $\mathbf{q}_t$ is an exogenous disturbance. The one-period-ahead forecast is

$$\mathbf{y}_{t|t-1} = \mathbf{H}\mathbf{A}_1 \mathbf{y}_{t-1} + \mathbf{H}\mathbf{B}\mathbf{x}_t \qquad \text{where} \qquad \mathbf{H} = \mathbf{A}_0^{-1}$$

and hence

$$\mathbf{y}_t = \mathbf{y}_{t|t-1} + \mathbf{H}\mathbf{q}_t . \tag{33}$$

This forecast can be improved by incorporating the discrepancy between the two forecasts

$$\mathbf{z}_t = \mathbf{y}_{t|t-1}^{\mathrm{M}} - \mathbf{y}_{t|t-1}^{\mathrm{Q}} \tag{34}$$

where superscript M refers to the monthly model and Q to the quarterly model, because $\mathbf{z}_t$ contains information on $\mathbf{q}_t$. A combined forecast is

$$\hat{\mathbf{y}}_t = \mathbf{y}^{Q}_{t|t-1} + \mathbf{H}\hat{\mathbf{q}}_t \tag{35}$$

where

$$\hat{\mathbf{q}}_t = \hat{E}(\mathbf{q}_t \,|\, \mathbf{z}_t) = \boldsymbol{\Sigma}_{qz}\boldsymbol{\Sigma}_{zz}^{-1}\mathbf{z}_t,$$

which is the regression model of $\mathbf{q}_t$ on $\mathbf{z}_t$. This is best in the sense of minimizing the forecast error covariance

$$\operatorname{cov}(\mathbf{y}_t - \hat{\mathbf{y}}_t) = \mathbf{H}(\boldsymbol{\Sigma}_{qq} - \boldsymbol{\Sigma}_{qz}\boldsymbol{\Sigma}_{zz}^{-1}\boldsymbol{\Sigma}_{zq})\mathbf{H}' \leqslant \mathbf{H}\boldsymbol{\Sigma}_{qq}\mathbf{H}'$$

which is smaller than the forecast error of the quarterly model. Alternatively, rewrite (34) as

$$\hat{\mathbf{y}}_t = \mathbf{y}^{Q}_{t|t-1} + \mathbf{H}\boldsymbol{\Sigma}_{qq}\boldsymbol{\Sigma}_{zz}^{-1}(\mathbf{y}^{M}_{t|t-1} - \mathbf{y}^{Q}_{t|t-1}) = (\mathbf{I} - \mathbf{W})\mathbf{y}^{Q}_{t|t-1} + \mathbf{W}\mathbf{y}^{M}_{t|t-1}$$
$$\mathbf{W} = \mathbf{H}\boldsymbol{\Sigma}_{qz}\boldsymbol{\Sigma}_{zz}^{-1}$$

to see that $\hat{\mathbf{y}}_t$ is a pooled forecast. In practice, monthly models deal with a subset of economic variables of those tracked by quarterly models. Then

$$\mathbf{z}_t = \mathbf{y}^{M}_t - \mathbf{R}\mathbf{y}^{Q}_t$$

where $\mathbf{R}$ is $p \times m$, $p < m$ is likely to hold rather than (34). The pooled model still provides better forecasts.

Exercise

The optimal Kalman filter for the process

$$\mathbf{z}_{k+1} = \mathbf{A}\mathbf{z}_k + \boldsymbol{\xi}_k$$
$$\mathbf{y}_k = \mathbf{C}\mathbf{z}_k + \boldsymbol{\eta}_k$$

where

$$\operatorname{cov}\begin{pmatrix} \boldsymbol{\xi}_k \\ \boldsymbol{\eta}_k \end{pmatrix} = \begin{pmatrix} \mathbf{Q} & \mathbf{O} \\ \mathbf{O} & \mathbf{R} \end{pmatrix}$$

is given by

$$\mathbf{s}_{k+1} = \mathbf{A}\mathbf{s}_k + \mathbf{K}(\mathbf{y}_k - \mathbf{C}\mathbf{s}_k)$$

where

$$\mathbf{K} = \mathbf{A}\boldsymbol{\Pi}\mathbf{C}'(\mathbf{R} + \mathbf{C}\boldsymbol{\Pi}\mathbf{C}')^{-1}$$

and where

$$\boldsymbol{\Pi} = \mathbf{A}\boldsymbol{\Pi}\mathbf{A}' + \mathbf{Q} - \mathbf{K}'(\mathbf{R} + \mathbf{C}\boldsymbol{\Pi}\mathbf{C}')^{-1}\mathbf{K}.$$

The expression

$$\mathbf{G}(z) = \mathbf{I} + \mathbf{C}(z\mathbf{I} - \mathbf{A})^{-1}\mathbf{K}$$

is known as the return difference. It is the transfer function from the filter correction signal (which is the innovation vector if **K** is set optimally) to the output of the filter. This is easiest to see when the covariance matrix is factored as

$$\operatorname{cov}\begin{pmatrix}\boldsymbol{\xi}_k \\ \boldsymbol{\eta}_k\end{pmatrix} = \begin{pmatrix}\mathbf{B} \\ \mathbf{I}\end{pmatrix}\boldsymbol{\Delta}(\mathbf{B}'\mathbf{I})$$

because then

$$\mathbf{s}_{k+1} = \mathbf{A}\mathbf{s}_k + \mathbf{K}\mathbf{e}_k$$

$$\mathbf{y}_k = \mathbf{C}\mathbf{s}_k + \mathbf{e}_k$$

has $\mathbf{G}(z)$ as its transfer matrix.

Show that $S_y(z)$ is then expressible as

$$\begin{aligned}\mathbf{S}_y(z) &= \boldsymbol{\Delta} + \mathbf{C}(z\mathbf{I} - \mathbf{A})^{-1}\mathbf{B}\boldsymbol{\Delta}\mathbf{B}'(z^{-1}\mathbf{I} - \mathbf{A}')^{-1}\mathbf{C}' \\ &= \mathbf{G}(z)(\boldsymbol{\Delta} + \mathbf{C}\boldsymbol{\Pi}\mathbf{C}')\mathbf{G}'(1/z).\end{aligned}$$

Show that the transfer function of the filter is such that

$$\rho_c(\lambda) = |\lambda\mathbf{I} - \mathbf{A} + \mathbf{K}\mathbf{C}| = \rho_o(\lambda)|\mathbf{G}(z)|$$

where

$$\rho_o(\lambda) = |\lambda\mathbf{I} - \mathbf{A}|.$$

5.4 Sensitivity and Error Analysis of Kalman Filters

It is important to have some measures of the variations of filter outputs when some of the underlying assumptions are not true, since the system parameters or noise statistics will not be generally known exactly in real problems. Such inaccuracy may arise as a result of numerically evaluating model parameters (round-off errors and/or error in quadrature). For example, the linear system may be merely an approximate expression of nonlinear dynamic equations obtained by linearizing them about some nominal trajectories. Then, model parameters are evaluated, perhaps numerically, by taking certain partial derivatives. Another reason for such analysis is that, for problems with complex dynamic structure, it is of interest to examine the effect of a simplified approximate expression for model matrices on the accuracy of the estimation. As for noise statistics, we usually have only their rough estimates. Therefore, it is important to evaluate the effects of inaccuracies in model parameters and noise covariances on the estimates, i.e., on the error covariance matrices of the outputs of Kalman filters. It is also important to know the effects of nonoptimal filter gains on the way error-covariance matrices propagate. We are interested in nonoptimal gains (1) to study the sensitivity of the estimates and of the error-covariance matrices with respect to the filter gain and (2) to study the effects of the simplified suboptimal method of gain computation on the filter performance since the gain computation is the most time-consuming operation in generating the estimates. The error-covariance matrix of the optimal estimate is calculated by

(36)
$$\begin{aligned}\mathbf{\Gamma}_t &\triangleq E[\mathbf{x}_t - \mathbf{x}_t^*)(\mathbf{x}_t - \mathbf{x}_t^*)' | \mathbf{y}^t] \\ &= (\mathbf{I} - \mathbf{K}_t\mathbf{H}_t)\mathbf{M}_t(\mathbf{I} - \mathbf{K}_t\mathbf{H}_t)' + \mathbf{K}_t\mathbf{R}_t\mathbf{K}_t'\end{aligned}$$

or equivalently by

(37)
$$\mathbf{\Gamma}_t = (\mathbf{I} - \mathbf{K}_t\mathbf{H}_t)\mathbf{M}_t$$

where

(38)
$$\mathbf{M}_t = \mathbf{A}_t\mathbf{\Gamma}_t\mathbf{A}_t' + \mathbf{Q}_t.$$

Gain Variation

Let us first consider the effects of $\mathbf{\Gamma}$ on the gain changes from its optimal values $\mathbf{K}_t$ by $\delta\mathbf{K}_t$. Denoting by $\delta\mathbf{\Gamma}_t$ the deviation of the error-covariance matrix from its optimal form $\mathbf{\Gamma}_t$ and dropping the subscripts, (36) leads to

$$\begin{aligned}\delta\mathbf{\Gamma} &= [\mathbf{I}(\mathbf{K} + \delta\mathbf{K})\mathbf{H}]\mathbf{M}[\mathbf{I} - (\mathbf{K} + \delta\mathbf{K})]' + (\mathbf{K} + \delta\mathbf{K})\mathbf{R}(\mathbf{K} + \delta\mathbf{K})' \\ &= \delta\mathbf{K}[-\mathbf{H}\mathbf{M}(\mathbf{I} - \mathbf{K}\mathbf{H})' + \mathbf{R}\mathbf{K}'] + [(\mathbf{I} - \mathbf{K}\mathbf{H})\mathbf{M}\mathbf{H}' + \mathbf{K}\mathbf{R}]\delta\mathbf{K}'\end{aligned}$$

where the second-order terms are neglected.

Since the expressions in the square brackets define the optimal gain, we have

$$\delta\mathbf{\Gamma} = 0.$$

The alternative expression for optimal error-covariance (37) gives

$$\delta\mathbf{\Gamma} = -\delta\mathbf{K}\mathbf{H}\mathbf{M}.$$

Therefore, when numerically evaluating $\mathbf{\Gamma}$, the expression (37) is more sensitive than (36) to small variation in $\mathbf{K}$.

The Variation of the Transition Matrix

We now investigate the effects of changes in $\mathbf{A}_t$ on the accuracy of computing $\mathbf{M}_{t+1}$ from $\mathbf{\Gamma}_t$. The noises are taken to be Gaussian random variables. See Toda and Patel (1980) for a more comprehensive analysis.

Denoting the small variation of $\mathbf{A}_t$ by $\delta\mathbf{A}_t$ and dropping subscripts, $\delta\mathbf{M} = \delta\mathbf{A}\mathbf{\Gamma}\mathbf{A}' + \mathbf{A}\mathbf{\Gamma}\delta\mathbf{A}'$ from (38).

Since $\delta\mathbf{M}$ will be small compared with $\mathbf{M}$, write $\mathbf{M} + \delta\mathbf{M} = \mathbf{M} + \varepsilon\mathbf{N}$, where ε is a small positive constant.

Since $\mathbf{M}$ is symmetric, by appropriate linear transformation on $\mathbf{x}$, $\mathbf{M}$ can be made diagonal, i.e., the components of the estimation error $\mathbf{x} - \hat{\mathbf{x}}$ after the linear transformation can be taken to be uncorrelated, hence independent. The variances of these independent errors are the eigenvalues of $\mathbf{M}$. Therefore, the change in the eigenvalues of $\mathbf{M}$ due to a small change in $\mathbf{A}$ may be regarded approximately as the changes in the variances of the components of the estimation error $\mathbf{x} - \hat{\mathbf{x}}$. (This is only approximately true since $\delta\mathbf{M}$ will not generally be diagonal even if $\mathbf{M}$ is.)

We will now investigate the difference between the eigenvalues of $\mathbf{M}$ and of $\mathbf{M} + \delta\mathbf{M}$. Denote by λ_t the ith eigenvalue of $\mathbf{M}$ with its normalized eigenvector denoted by e_t. We define $\tilde{\lambda}_t$ and $\tilde{e}_t$ as the corresponding quantities for $\mathbf{M} + \varepsilon\mathbf{N}$. Writing

$$\lambda = \lambda_t + \varepsilon\lambda_{t1} + o(\varepsilon)$$

$$\tilde{e}_t = e_t + \varepsilon e_{t1} + o(\varepsilon),$$

the relation $(\mathbf{M} + \varepsilon\mathbf{N})\tilde{e} = \tilde{\lambda}\tilde{e}$ yields, to the order ε,

$$\begin{aligned}
\varepsilon\lambda_{t1} &= \varepsilon_t'\varepsilon\mathbf{N}e_t \\
&= e_t'(\delta\mathbf{A}\mathbf{\Gamma}\mathbf{A}' + \mathbf{A}\mathbf{\Gamma}\delta\mathbf{A}')e_t \\
&= 2e_t'\delta\mathbf{A}\mathbf{\Gamma}\mathbf{A}'e_t \\
&= 2e_t'\delta\mathbf{A}\mathbf{A}^{-1}\mathbf{A}\mathbf{\Gamma}\mathbf{A}'e_t \\
&= 2e_t'(\delta\mathbf{A}\mathbf{A}^{-1})(\lambda_t e_t - Qe_t).
\end{aligned}$$

Therefore,

$$|\varepsilon\lambda_{t1}| \leqslant 2|\lambda_t|\,\|\delta\mathbf{A}_t\mathbf{A}_t^{-1}\| + 2\|\mathbf{Q}_t\|$$

or

$$\frac{|\varepsilon\lambda_{1i}|}{|\lambda_t|} \leqslant 2\left(\|\delta\mathbf{A}_t\mathbf{A}_t^{-1}\| + \frac{\|\mathbf{Q}_t\|}{\lambda_t}\right).$$

If a major contribution to $\mathbf{M}$ comes from $\mathbf{A}\mathbf{\Gamma}\mathbf{A}'$ and not from $\mathbf{Q}$, then $\|Q_t\|/\lambda_t \ll 1$, and one has approximately

$$2\sum_{0}^{N-1} \|\delta\mathbf{A}_t\mathbf{A}_t^{-1}\|$$

or

$$2N\|\delta\mathbf{A}\mathbf{A}^{-1}\|$$

if $\mathbf{A}$ is a constant matrix.

Therefore, as a rule of thumb, one must have

$$\|\delta \mathbf{A}\mathbf{A}^{-1}\| \ll \frac{1}{2N}$$

in such applications where N total number of estimates are generated.

Imprecise Noise Covariance Matrices

Since the statistics of the random noises are known only very roughly, the effects of large variations of $\mathbf{Q}$ and $\mathbf{R}$, rather than their small variations on $\mathbf{\Gamma}$, need be investigated. Such investigations must generally be done numerically in designing filters.

One may take the min–max point of view in evaluating the effect of different $\mathbf{Q}$s and $\mathbf{R}$s on $\mathbf{\Gamma}$, using techniques similar to those in Aoki (1965) where the effects of an unknown gain (distribution) matrix on the performance index have been discussed.

5.5 Suboptimal Filters by Observers

Given the dynamics of the model and its measurement equation

$$\begin{aligned} \mathbf{z}_{t+1} &= \mathbf{A}_t \mathbf{z}_t + \mathbf{B}_t \mathbf{x}_t + \boldsymbol{\xi}_t \\ \mathbf{y}_t &= \mathbf{C}_t \mathbf{z}_t + \boldsymbol{\eta}_t, \end{aligned} \tag{39}$$

we consider a computationally easy, albeit suboptimal, way to estimate the state vector $\mathbf{z}_t$. The estimation schemes we discuss are of interest for linear models with a large number of variables since they are computationally less demanding than estimates by the Kalman filter and are reasonably good when observation noises are "small."

When new information $\mathbf{y}_t - \mathbf{C}\mathbf{x}_{t|t-1}$ contained in $\mathbf{y}_t$ is not incorporated with the optimal weight but by a suboptimal one, a suboptimal Kalman filter results. Another suboptimal state vector estimator, known as an observer, can be constructed if additional current state vector measurement is available as

$$\mathbf{s}_t = \mathbf{T}_t \mathbf{z}_t + \boldsymbol{\varepsilon}_t. \tag{40}$$

Together with the original measurement equation, an augmented set of observations

$$\begin{pmatrix} \mathbf{y}_t \\ \mathbf{s}_t \end{pmatrix} = \begin{pmatrix} \mathbf{C}_t \\ \mathbf{T}_t \end{pmatrix} \mathbf{z}_t + \begin{pmatrix} \boldsymbol{\varepsilon}_t \\ \boldsymbol{\eta}_t \end{pmatrix}$$

is available. In particular, if the matrix multiplying $\mathbf{z}_t$ is invertible or has full column rank, then the observer estimates can be shown to have the same form as the suboptimal Kalman filter estimates. If $(\mathbf{C}_t'\mathbf{T}_t')$ is not invertible, these two estimates generally have different structures. Since both are suboptimal the choice must be made on individual applications. This section develops the observer and its equivalence with the Kalman filter following Aoki and Huddle (1967) and Aoki (1976, Chapter 10).

Let $\dim s_t = l$. We take $l = \dim \mathbf{z}_t - \dim \mathbf{y}_t$ to be specific. Suppose the vector $\mathbf{s}_t$ is governed by the dynamic equation

$$\mathbf{s}_{t+1} = \mathbf{F}_t\mathbf{s}_t + \mathbf{G}_t\mathbf{x}_t + \mathbf{D}_t\mathbf{y}_t$$

where $\mathbf{F}_t$ is an $l \times l$ matrix.

For (40) to be consistent with (39), it is necessary that the matrices $\mathbf{F}_t$, $\mathbf{G}_t$, and $\mathbf{D}_t$ in (40) satisfy the relations in (41) obtained from

$$\mathbf{T}_{t+1}(\mathbf{A}_t\mathbf{z}_t + \mathbf{B}_t\mathbf{x}_t + \xi_t) + \boldsymbol{\varepsilon}_{t+1} = \mathbf{F}_t(\mathbf{T}_t\mathbf{z}_t + \boldsymbol{\varepsilon}_t) + \mathbf{G}_t\mathbf{x}_t + \mathbf{D}_t(\mathbf{C}_t\mathbf{z}_t + \boldsymbol{\eta}_t)$$

for all $\mathbf{z}_t$ and $\mathbf{x}_t$, or

$$\begin{aligned} \mathbf{T}_{t+1}\mathbf{A}_t &= \mathbf{F}_t\mathbf{T}_t + \mathbf{D}_t\mathbf{C}_t \\ \mathbf{T}_{t+1}\mathbf{B}_t &= \mathbf{G}_t, \end{aligned} \tag{41}$$

and the aggregation error propagates with time according to the dynamic equation

$$\boldsymbol{\varepsilon}_{t+1} = \mathbf{F}_t\boldsymbol{\varepsilon}_t + \mathbf{D}_t\boldsymbol{\eta}_t - \mathbf{T}_{t+1}\boldsymbol{\xi}_t. \tag{42}$$

The equations (41) and (42) are also sufficient for (39) and (40) to be aggregation. To see this, we have from (39) and (41)

$$\begin{aligned} \mathbf{s}_{t+1} - \mathbf{T}_{t+1}\mathbf{z}_{t+1} &= \mathbf{F}_t\mathbf{s}_t + \mathbf{G}_t\mathbf{x}_t + \mathbf{D}_t(\mathbf{C}_t\mathbf{z}_t + \boldsymbol{\eta}_t) - \mathbf{T}_{t+1}(A_t\mathbf{z}_t + B_tx_t + \xi_t) \\ &= \mathbf{F}_t(\mathbf{s}_t - \mathbf{T}_t\mathbf{z}_t) + \mathbf{D}_t\boldsymbol{\eta}_t - \mathbf{T}_{t+1}\boldsymbol{\xi}_t \end{aligned}$$

or

$$\boldsymbol{\varepsilon}_{t+1} = \mathbf{F}_t\boldsymbol{\varepsilon}_t + \mathbf{D}_t\boldsymbol{\eta}_t - \mathbf{T}_{t+1}\boldsymbol{\xi}_t. \tag{43}$$

We now consider choosing $(\mathbf{T}_t\}$ to minimize the error of estimating $\mathbf{z}_t$. Write the first equation of (41) as

$$\mathbf{T}_{t+1}\mathbf{Z}_t = (\mathbf{D}_t \quad \mathbf{F}_t)\begin{pmatrix} \mathbf{C}_t \\ \mathbf{T}_t \end{pmatrix}. \tag{44}$$

We impose a condition that $\mathbf{T}_t$ is chosen to make $\binom{\mathbf{C}_t}{\mathbf{T}_t}$ nonsingular. Let

$$\begin{pmatrix} \mathbf{C}_t \\ \mathbf{T}_t \end{pmatrix}^{-1} = (\mathbf{V}_t \quad \mathbf{P}_t) \tag{45}$$

where $\mathbf{P}_t$ is an $n \times l$ matrix, and $\mathbf{V}_t$ is an $n \times m$ matrix, $l = n - m$. From (44) and (45), the matrices $\mathbf{D}$ and $\mathbf{F}$ in (40) are expressible as

$$\begin{aligned} \mathbf{F}_t &= \mathbf{T}_{t+1} A_t \mathbf{P}_t \\ \mathbf{D}_t &= \mathbf{T}_{t+1} A_t \mathbf{V}_t \end{aligned} \tag{46}$$

since

$$(\mathbf{D}_t \quad \mathbf{F}_t) = \mathbf{T}_{t+1} A_t (\mathbf{V}_t \quad \mathbf{P}_t).$$

In other words, a choice of $\mathbf{T}_{t+1}$ determines the matrices in the dynamic equation for the aggregated state vector (40) uniquely, since $\mathbf{G}_t$ is determined also uniquely by $\mathbf{T}_{t+1}$.

Let

$$\mathbf{E}_t = \operatorname{cov} \boldsymbol{\varepsilon}_t.$$

From (43), it propagates with time according to

$$\mathbf{E}_{t+1} = \mathbf{T}_{t+1} \boldsymbol{\Omega}_t \mathbf{T}'_{t+1} \tag{47}$$

where

$$\boldsymbol{\Omega}_t = \mathbf{A}_t \mathbf{P}_t \mathbf{E}_t \mathbf{P}'_t \mathbf{A}'_t + \mathbf{A}_t \mathbf{V}_t \mathbf{R}_t \mathbf{V}'_t \mathbf{A}'_t + \mathbf{Q}_t \tag{48}$$

and where we assume

$$\begin{aligned} E\boldsymbol{\eta}_t \boldsymbol{\eta}'_\tau &= \mathbf{R}_t \delta_{t\tau} \\ E\boldsymbol{\xi}_t \boldsymbol{\xi}'_\tau &= \mathbf{Q}_t \delta_{t\tau}, \\ E\boldsymbol{\xi}_t \boldsymbol{\eta}'_\tau &= \mathbf{0} \qquad \text{for all } t, \tau. \end{aligned}$$

Augmenting the measurement vector $\mathbf{y}_t$ by the aggregated state vector $\mathbf{s}_t$, we obtain an equation that can be used to estimate $\mathbf{z}_t$,

$$\begin{pmatrix} \mathbf{y}_t \\ \mathbf{s}_t \end{pmatrix} = \begin{pmatrix} \mathbf{C}_t \\ \mathbf{T}_t \end{pmatrix} \mathbf{z}_t + \begin{pmatrix} \boldsymbol{\eta}_t \\ \boldsymbol{\varepsilon}_t \end{pmatrix}, \tag{49}$$

or we estimate $\mathbf{z}_t$ by

$$\hat{\mathbf{z}}_t = \begin{pmatrix} \mathbf{C}_t \\ \mathbf{T}_t \end{pmatrix}^{-1} \begin{pmatrix} \mathbf{y}_t \\ \mathbf{s}_t \end{pmatrix}.$$

Aggregation

From (45), the estimate can be written as

$$\hat{\mathbf{z}}_t = (\mathbf{V}_t \quad \mathbf{P}_t)\begin{pmatrix} \mathbf{y}_t \\ \mathbf{s}_t \end{pmatrix}. \tag{50}$$

Advance t by 2 in (50) and use the dynamic relation (40) to write

$$\begin{aligned}
\hat{\mathbf{z}}_{t+1} &= \mathbf{V}_{t+1}\mathbf{y}_{t+1} + \mathbf{P}_{t+1}\mathbf{s}_{t+1} \\
&= \mathbf{V}_{t+1}\mathbf{y}_{t+1} + \mathbf{P}_{t+1}(\mathbf{F}_t\mathbf{s}_t + \mathbf{G}_t x_t + \mathbf{D}_t\mathbf{y}_t) \\
&= \mathbf{V}_{t+1}\mathbf{y}_{t+1} + \mathbf{P}_{t+1}\mathbf{T}_{t+1}\mathbf{A}_t(\mathbf{V}_t\mathbf{D}_t)\begin{pmatrix} \mathbf{y}_t \\ \mathbf{s}_t \end{pmatrix} + \mathbf{P}_{t+1}\mathbf{G}_t x_t \\
&= \mathbf{V}_{t+1}\mathbf{y}_{t+1} + \mathbf{P}_{t+1}(\mathbf{T}_{t+1}\mathbf{A}_t\hat{\mathbf{z}}_t + \mathbf{G}_t\mathbf{x}_t) \\
&= \mathbf{V}_{t+1}\mathbf{y}_{t+1} + \mathbf{P}_{t+1}\mathbf{T}_{t+1}(\mathbf{A}_t\hat{\mathbf{z}}_t + \mathbf{B}_t\mathbf{x}_t)
\end{aligned}$$

where (46) is used. By definition,

$$\mathbf{P}_{t+1}\mathbf{T}_{t+1} = \mathbf{I} - \mathbf{V}_{t+1}\mathbf{C}_{t+1}.$$

Thus, the forementioned becomes

$$\begin{aligned}
\hat{\mathbf{z}}_{t+1} &= \mathbf{V}_{t+1}\mathbf{y}_{t+1} + (\mathbf{I} - \mathbf{V}_{t+1}\mathbf{C}_{t+1})(\mathbf{A}_t\hat{\mathbf{z}}_t + \mathbf{B}_t\mathbf{x}_t) \\
&= \mathbf{A}_t\hat{\mathbf{z}}_t + \mathbf{B}_t\mathbf{x}_t + \mathbf{V}_{t+1}[\mathbf{y}_{t+1} - \mathbf{C}_{t+1}(\mathbf{A}_t\hat{\mathbf{z}}_t + \mathbf{B}_t\mathbf{x}_t)].
\end{aligned}$$

This is seen as a Kalman filter with the filter gain $\mathbf{V}_{t+1}$. The condition that rank $\mathbf{P}_t = n - p$, where $p = \dim y_t$ and rank $V_t = p$, is the same that $\binom{\mathbf{C}_t}{\mathbf{T}_t}$ is invertible. The estimation error is expressible from (49) as

$$\begin{aligned}
\hat{\mathbf{z}}_t &= \hat{\mathbf{z}}_t - \mathbf{z}_t \\
&= -(\mathbf{V}_t \quad \mathbf{P}_t)\begin{pmatrix} \boldsymbol{\eta}_t \\ \boldsymbol{\varepsilon}_t \end{pmatrix}.
\end{aligned}$$

Let $\mathbf{Z}_t = \operatorname{cov}(\hat{z}_t)$. Then the estimation error covariance is related to the covariance matrices of the observation noise and the aggregation error and to the aggregation matrix $\mathbf{T}_{t+1}$ by

$$\mathbf{Z}_t = (\mathbf{V}_t \quad \mathbf{P}_t)\begin{pmatrix} \mathbf{R}_t & 0 \\ 0 & \mathbf{E}_t \end{pmatrix}(\mathbf{V}_t \quad \mathbf{P}_t)' \tag{51}$$

since we see from (43) that $\boldsymbol{\varepsilon}_t$ is uncorrelated with $\boldsymbol{\eta}_t$.

We now introduce some simplifying assumptions without loss of generality in order to exhibit an optimal aggregation matrix explicitly. First, we may assume that $\mathbf{C}_t$ is of the form

$$\mathbf{C}_t = (\mathbf{I}_m \quad 0).$$

To see this, we note that we can partition $\mathbf{C}_t$ as $\mathbf{C}_t = (\mathbf{C}_t^1 \quad \mathbf{C}_t^2)$, where $\mathbf{C}_t^1$ is a nonsingular $m \times m$ matrix, by renumbering the components of the state vector if necessary. Then by transforming $\mathbf{z}_t$ by a nonsingular matrix to a new state vector $\mathbf{q}_t$ by

$$\mathbf{z}_t = \begin{bmatrix} (\mathbf{C}_t^1)^{-1} & -(\mathbf{C}_t^1)^{-1}\mathbf{C}_t^2 \\ 0 & \mathbf{I}_t \end{bmatrix} \mathbf{q}_t,$$

we can write the observation equation as

$$\begin{aligned} \mathbf{y}_t &= \mathbf{C}_t \mathbf{z}_t + \boldsymbol{\eta}_t \\ &= (\mathbf{I}_m \quad 0)\mathbf{q}_t + \boldsymbol{\eta}_t. \end{aligned}$$

Second, we may assume $\mathbf{T}_t$ to be of the form $\mathbf{T}_t = (\mathbf{K}_t \quad \mathbf{I}_t)$. This is because $\mathbf{C}_t$ and $\mathbf{T}_t$ must form a nonsingular matrix by choice, which necessitates that $\mathbf{T}_t^2$ in $\mathbf{T}_t = (\mathbf{T}_t^1 \quad \mathbf{T}_t^2)$ be nonsingular. Again by a nonsingular transformation of the state vector, we can take $\mathbf{K}_t = \mathbf{T}_t^1(\mathbf{T}_t^2)^{-1}$. In this form, the choice of the aggregation matrix amounts to that of the matrix $\mathbf{K}_t$. Thus

$$\begin{pmatrix} \mathbf{C}_t \\ \mathbf{T}_t \end{pmatrix}^{-1} = \begin{pmatrix} \mathbf{I}_m & 0 \\ -\mathbf{K}_t & \mathbf{I}_t \end{pmatrix}$$

or, from (45),

$$\mathbf{P}_t = \begin{pmatrix} 0 \\ \mathbf{I}_t \end{pmatrix} \quad \text{and} \quad \mathbf{V}_t = \begin{pmatrix} \mathbf{I}_m \\ -\mathbf{K}_t \end{pmatrix}.$$

In the coordinate system in which $\mathbf{C}_t$ and $\mathbf{T}_t$ take the assumed simplified form, (50) shows that $\hat{\mathbf{w}}_{1t} = \mathbf{y}_t$ and $\hat{\mathbf{w}}_{2t} = -\mathbf{K}_t\mathbf{y}_t + \mathbf{s}_t$ where $\mathbf{w}_t$ is the state vector in this new coordinate system.

Using these special structures of $\mathbf{P}_t$ and $\mathbf{V}_t$, from (51),

(52)
$$\begin{aligned} &\mathbf{Z}_{11,t+1} = \mathbf{R}_{t+1}, \qquad \mathbf{Z}_{12,t+1} = \mathbf{R}_{t+1}\mathbf{K}'_{t+1} \\ &\mathbf{Z}_{22,t+1} = -\mathbf{K}_{t+1}\mathbf{R}_t\mathbf{E}_{t+1} + \mathbf{K}_{t+1}\mathbf{R}_{t+1}\mathbf{K}'_{t+1}. \end{aligned}$$

Partitioning $\boldsymbol{\Omega}_t$ of (48) conformably, we see from (47) that

$$\mathbf{E}_{t+1} = \mathbf{K}_{t+1}\boldsymbol{\Omega}_{11t}\mathbf{K}'_{t+1} + \mathbf{K}_{t+1}\boldsymbol{\Omega}_{12t} + \boldsymbol{\Omega}_{21t}\mathbf{K}'_{t+1} + \boldsymbol{\Omega}_{22t}.$$

Take the trace of $\mathbf{Z}_{t+1}$ in (52). It is a quadratic function of $\mathbf{K}_{t+1}$,

$$\operatorname{tr}\mathbf{Z}_{t+1} = \operatorname{tr}\mathbf{R}_{t+1} + \operatorname{tr}\{\mathbf{K}_{t+1}(\mathbf{\Omega}_{11t} + \mathbf{R}_{t+1})\mathbf{K}'_{t+1} + \mathbf{K}_{t+1}\mathbf{\Omega}_{12t} + \mathbf{\Omega}_{21t}\mathbf{K}'_{t+1} + \mathbf{Q}_{22t}\}.$$

The best choice of the aggregation matrix $\mathbf{T}_{t+1}$ is therefore determined by

$$\min_{K_{t+1}} \operatorname{tr}\mathbf{Z}_{t+1}$$

or

$$\mathbf{K}^*_{t+1} = -\mathbf{\Omega}_{21t}(\mathbf{\Omega}_{11t} + \mathbf{R}_{t+1})^{-1}.$$

This determines the optimal $\mathbf{T}_{t+1}$. The dynamics of this optimal aggregation model are therefore given by (40), where from (46),

$$\mathbf{F}_t = \mathbf{A}_{22t} + \mathbf{K}_{t+1}\mathbf{A}_{12t}$$

$$\mathbf{D}_t = \mathbf{A}_{21t} - \mathbf{A}_{22}\mathbf{K}_t + \mathbf{K}_{t+1}(\mathbf{A}_{11t} - \mathbf{A}_{12t}\mathbf{K}_t).$$

The aggregation model (40) is known as the observer in engineering literature (Luenberger [1966]). The estimate of $\mathbf{z}_t$ given by (50), $\hat{\mathbf{z}}_t = \mathbf{V}_t\mathbf{y}_t + \mathbf{P}_t\mathbf{s}_t$, is not a truly optimal estimate in the sense of the conditional expectation of $\mathbf{z}_t$ given $\mathbf{y}_0, \ldots, \mathbf{y}_t$ generated via the Kalman filter, since for example, the estimate uses only the current values of $\mathbf{y}_t$ and $\mathbf{s}_t$, not the past values. However, it is much simpler computationally and it is known that $\hat{\mathbf{z}}_t$ obtained via the aggregation model is exact when observation noise is absent.

5.6 Wiener Filters

Whittle (1963) solved the problem of prediction and regulation of a stationary stochastic process using the Wiener filter. His contributions were somewhat neglected when Kalman filters became very popular. Much later the economic profession rediscovered Whittle's contribution mostly in connection with solutions of rational expectations models, as in Whiteman (1983) or Sargent (1987), because in economic concept, one is often interested in stationary state solutions and willing to forget the effects of transient phenomena due to nonzero initial conditions. Many intertemporal optimization problems have complicated solution structures. In such cases, one may wish to know simpler benchmark solution forms when the state of weakly stationary is imposed on the solution. For these reasons, we describe how the optimal filtering and prediction problem is solved when data is available on a semi-infinite horizon, and we discuss Wiener filters as solutions of the Wiener-Hopf equation.

Consider estimating an nth element of a scalar-valued sequence $\{s_k\}$, s_n, by a linear filter

$$\hat{s}_n = \sum_{i=0}^{\infty} h_i u_{n-i}$$

using only the current and past input sequences, $\{u_k\}$, which are assumed to be observed without error.* The error of estimation is

$$e_n = s_n - \hat{s}_n.$$

The weight hs are chosen to minimize the expected value of e_n^2,

$$Ee_n^2 = E(s_n - \hat{s}_n)^2.$$

The set of first-order conditions

$$\partial Ee_n^2/\partial h_i = 0, \qquad i = 0, 1, 2, \ldots$$

shows that optimal hs are such that the estimation errors are uncorrelated with us

(53) $$E[(s_n - \hat{s}_n)u_{n-i}] = 0, \qquad i = 0, 1, 2, \ldots .$$

Equation (53) produces a set of equations for covariances

$$R_{su}(i) = \sum_{j=0}^{\infty} h_j R_{uu}(i-j), \qquad i \geqslant 0.$$

This set of equations is called the Wiener-Hopf equation. If we ignore the restriction of $i \geqslant 0$, this equation can be converted into the z-transform relations which is

$$S_{su}(z) = H(z)S_{uu}(z)$$

from which we can solve for

$$H(z) = S_{su}(z)/S_{uu}(z),$$

as the transfer function of the optimal Wiener filter. In particular then, if u is a white noise sequence, then $S_{uu}(z)$ is merely a constant σ_u^2, and

$$H(z) = S_{su}(z)/\sigma_u^2$$

would be the optimal filter. Because S_{su} is a two-sided z-transform, $H(z)$ is not a causal filter, however. When u is a white noise sequence,

(54) $$H(z) = \frac{1}{\sigma_u^2}[S_{su}(z)]_+$$

* Extension to vector-valued processes is trivial.

where $[\cdot]_+$ retains only the causal part, i.e. positive powers in z^{-1} is the optimal Wiener filter. (See Whittle [1963].) When u is not a white noise, its spectral density decomposition

$$S_{uu}(z) = \sigma^2 B(z) B'(1/z)$$

is used to generate a white noise sequence by passing the u-sequence through a filter with transfer function $1/B(z)$ because

$$S_{\varepsilon\varepsilon}(z) = \frac{1}{B(z)} \sigma^2 B(z) B'(1/z) \frac{1}{B'(1/z)} = \sigma^2 = \text{const},$$

to which the filter (54) is optimal. Recall the discussions in Section 2.7.

Since $S_{su}(z) = S_{s\varepsilon} B(z^{-1})$, we have, analogous to (54)

$$H(z) = \frac{1}{\sigma^2} [S_{s\varepsilon}(z)]_+ = \frac{1}{\sigma^2 B(z)} \left[\frac{S_{su}(z)}{B(z^{-1})} \right]_+ \tag{55}$$

and

$$S_{\varepsilon u}(z) = \frac{1}{B(z)} S_{uu}.$$

Note from (53) that $S_{su}(z) = F(z) S_{uu}(z)$ is the filter from u to $\hat{s}$, i.e.,

$$F(z) = S_{su}(z)/S_{uu}(z)$$

without the causality restriction. When causality is imposed, we obtain

$$\begin{aligned} H(z) &= [S_{su}(z)/S_{uu}(z)]_+ \\ &= \frac{1}{\sigma^2 B(z)} [S_{su}(z)/B(1/z)]_+ \end{aligned}$$

as in (55). By the Parseval identity, the filter error $e_k = s_k - \hat{s}_k$ has the variance

$$\begin{aligned} E(e_k^2) &= E(s_k - \hat{s}_k)^2 \\ &= E s_k^2 - E s_k \hat{s}_k \\ &= \frac{1}{2\pi} \int_{-\pi}^{\pi} S_{ss}(e^{i\omega})\, d\omega - \frac{1}{\pi} \int_{-\pi}^{\pi} S_{s\hat{s}}(e^{i\omega})\, d\omega \end{aligned}$$

where

$$S_{s\hat{s}} = S_{su} H(1/z).$$

Prediction and Filtering with Semi-Infinite Data, and the Wiener-Hopf Equation

The following developments follow Whittle (1963, Chapter 6). Suppose that a scalar purely nondeterministic process $\{y_t\}$ is an ARMA process with no zero on the unit circle and consider predicting y_{t+v} by

$$\hat{y}_{t+v} = \sum_0^\infty \gamma_j y_{t-j}$$

or alternatively as

$$\hat{y}_{t+v} = \sum_0^\infty \phi_j \varepsilon_{t-j}$$

where $\{\varepsilon_t\}$ is the innovation process associated with $\{y_t\}$. Their transforms are denoted by $\hat{y}(q^{-1}) = \Sigma_{j=0}^{s} yq^{-1}$ and $\hat{\varepsilon}(q^{-1}) = \Sigma_{j=0}^{\infty} \Sigma_j q^{-j}$. We impose $\Sigma_0^\infty \phi_j^2 < \infty$. Thus, y_t can be represented as $y_t = \Sigma_0^\infty b_j \varepsilon_{t-j}$. Construct polynomials in q^{-1} by

$$B(q^{-1}) = \sum_0^\infty b_j q^{-j}$$

$$\gamma(q^{-1}) = \sum_0^\infty \gamma_j q^{-j}$$

and

$$\phi(q^{-1}) = \sum_0^\infty \phi_j q^{-j}.$$

Then

$$\begin{aligned} \phi(q^{-1})\hat{\varepsilon}(q^{-1}) &= \gamma(q^{-1})\hat{y}(q^{-1}) \\ &= \gamma(q^{-1})B(q^{-1})\hat{\varepsilon}(q^{-1}) \qquad (56) \\ \phi(q^{-1}) &= \gamma(q^{-1})B(q^{-1}). \end{aligned}$$

The prediction error variance σ_v^2 can be calculated as

$$\begin{aligned} \sigma_v^2 &= E(\hat{y}_{t+v} - y_{t+v})^2 \\ &= E\left(\sum_{j=0}^\infty \phi_j \varepsilon_{t-j} - \sum_{j=0}^\infty b_j \varepsilon_{t+v-j}\right)^2 \\ &= \sigma^2\left[\sum_0^{v-1} b_j^2 + \sum_0^\infty (\phi_j - b_{j+v})^2\right], \end{aligned}$$

which is minimized by setting $\phi_j = b_{j-v}, j \geqslant 0$. It is more convenient to use past ys rather than innovations in the prediction formula. We can use (56) to deduce the required γs:

$$\gamma(z) = \phi(z)/B(z) = \sum_0^\infty b_{j+v} z^{-j}/B(z)$$
$$= [B(z)z^n]_+/B(z)$$

where $[\ ]_+$ is a projection operator that extracts only the nonnegative powers in z^{-1}, i.e., $b_v + b_{v+1}z^{-1} + \ldots$. This is the optimal prediction filter. It is a special case of the Wiener-Hopf equation we shall derive shortly.

To set the stage, we next turn to the prediction of one series from another, i.e., estimating z_t from y_t in a bivariate stationary process

$$\hat{z}_t = \sum_0^\infty \gamma_j y_{t-j}.$$

If we set $z_t = y_{t+v}$, the problem is reduced to the one we just discussed. The spectral density function of the y-process admits the canonical factorization

$$g_{yy}(q^{-1}) = \sigma^2 B(q^{-1})B(q),$$

i.e., y_t has the innovation representation

$$y_t = \sum_0^\infty b_j \varepsilon_{t-j}.$$

In terms of the innovations, suppose that

$$\hat{y}_t = \sum_0^\infty \phi_j \varepsilon_{t-j}.$$

Then the estimation error variance is expressible by

$$E(y_t - \hat{y}_t)^2 = \text{var}(y) - 2\sum \phi_j c_j + \sum \phi_j^2$$

where

$$c_j = \text{cov}(y_t, \varepsilon_{t-j}).$$

Denote the cross-covariance generating function by

$$g_{z\varepsilon}(q^{-1}) = \sum_0^\infty c_j q^{-j}.$$

Then the coefficients of the optimal estimator satisfy the set of first-order conditions

$$\phi_j = c_j, \qquad j \geqslant 0,$$

i.e.,

(57)
$$\phi(q^{-1}) = \sum \phi_j q^{-j}$$
$$= [g_{y\varepsilon}(q^{-1})]_+ .$$

To express (57) in terms of y, note that the spectral densities $g_{z\varepsilon}$ and g_{zy} are related by $g_{yy} = B(q)g_{z\varepsilon}(q^{-1})$, and (57) is equal to

$$\phi(q^{-1}) = \left[\frac{g_{zy}(q^{-1})}{B(q)}\right]_+ .$$

Then

$$\gamma(q^{-1}) = \frac{1}{B(q^{-1})}\left[\frac{g_{zy}(q^{-1})}{B(q)}\right]_+$$

follows.

Now consider predicting a random vector $\mathbf{z}$ from a vector-valued observation process $\{\mathbf{y}_t\}$. Let the optimal predictor be given by

$$\hat{\mathbf{z}}_t = \sum_0^\infty \boldsymbol{\Gamma}_j \mathbf{y}_{t-j} = \boldsymbol{\Gamma}(q^{-1})\mathbf{y}_t .$$

The covariance generating matrix is

$$\mathbf{g}_{zy}(q^{-1}) = \sum_0^\infty \operatorname{cov}(\mathbf{z}, \mathbf{y}_{t-s})q^{-s}.$$

Since the prediction error $\mathbf{z} - \hat{\mathbf{z}}_t$ is orthogonal to $\mathbf{y}_{t-s}$, $s \geqslant 0$, i.e., $E(\mathbf{z}\mathbf{y}'_{t-s}) - \Sigma\boldsymbol{\Gamma}_j E(y_{t-j}y'_{t-s}) = 0$, $s \geqslant 0$, and the covariance-generating matrices are such that $\mathbf{g}_{zy}(q^{-1}) - \boldsymbol{\Gamma}(q^{-1})\mathbf{g}_{yy}^{-1}(q)$ is some polynomial in q without a constant term. Let the canonical factorization of $\mathbf{g}_{yy}$ be

$$\mathbf{g}_{yy}(q^{-1}) = \mathbf{B}(q^{-1})\boldsymbol{\Delta}\mathbf{B}'(q)$$

where

$$\mathbf{B}(q^{-1}) = \sum_0^\infty \mathbf{B}_j q^{-1}, \qquad \mathbf{B}_0 = \mathbf{I}$$

and $\boldsymbol{\Delta}$ is a matrix independent of q^{-1}. This of course means that

$$\mathbf{y}_t = \sum \mathbf{B}_j \boldsymbol{\varepsilon}_{t-j}$$

where $\{\boldsymbol{\varepsilon}_t\}$ is the innovation process associated with $\{\mathbf{y}_t\}$, and if we write

$$\hat{\mathbf{z}}_t = \sum \boldsymbol{\theta}_j \boldsymbol{\varepsilon}_{t-j},$$

then

$$\boldsymbol{\theta}_j = \operatorname{cov}(\mathbf{z}, \boldsymbol{\varepsilon}_{t-1})\boldsymbol{\Delta}^{-1},$$

i.e.,

$$\boldsymbol{\theta}(q^{-1}) = [\mathbf{g}_{z\varepsilon}(q^{-1})]_+ \boldsymbol{\Delta}^{-1}$$

where $[\cdot]_+$ retains only the nonnegative powers in q^{-1}.

Noting that

$$\theta(q^{-1}) = \mathbf{\Gamma}(q^{-1})\mathbf{B}(q^{-1}),$$

$$\mathbf{\Gamma}(q^{-1}) = \theta(q^{-1})\mathbf{B}(q^{-1})^{-1}.$$

Since

$$\mathbf{g}_{zy}(q^{-1}) = \mathbf{g}_{z\varepsilon}(q^{-1})\mathbf{B}'(q),$$

the optimal filter is given by

$$\mathbf{\Gamma}(q^{-1}) = [\mathbf{g}_{zy}(q^{-1})\mathbf{B}'(q)^{-1}]_+\mathbf{\Delta}^{-1}\mathbf{B}(q^{-1})^{-1}.$$

5.7 Likelihood Function

Schweppe (1965) used outputs from Kalman filters to calculate the likelihood functions of observed data sequences. The procedure applies to nonstationary data as well (Goodrich and Caines [1975]). Write the joint density function as a product of successive conditional probability densities

$$p(\mathbf{y}_0 \ldots \mathbf{y}_T \,|\, \mathbf{x}_0) = \left[\prod_{t=0}^{T-1} p(\mathbf{y}_{t+1} \,|\, \mathbf{y}^t, \mathbf{x}_0)\right] \cdot p(\mathbf{y}_0 \,|\, \mathbf{x}_0).$$

For a reasonably sized T, and asymptotically stable dynamics, the preceding may be approximated by

$$(2\pi)^{-(T+1)p/2}|\mathbf{\Delta}|^{-(T+1)/2} \exp -\tfrac{1}{2}\Sigma_0^T(\mathbf{y}_t - \mathbf{y}_{t|t-1})'\mathbf{\Delta}^{-1}(\mathbf{y}_t - \mathbf{y}_{t|t-1})$$

where $\mathbf{\Delta} = \operatorname{cov}(\mathbf{y}_t - \mathbf{y}_{t|t-1})$ is taken to be weakly stationary and $p = \dim \mathbf{y}_t$.

In nonstationary situations, $\mathbf{\Delta}_T$ needs be used. In that case, the logarithm of the likelihood function is proportional to

$$L = -\Sigma \ln \mathbf{\Delta}_t - \tfrac{1}{2}\Sigma \mathbf{e}_t'\mathbf{\Delta}_t^{-1}\mathbf{e}_t$$

where $\mathbf{e}_t = \mathbf{y}_t - \mathbf{y}_{t|t-1}$ and $\mathbf{\Delta}_t = \mathbf{R} + \mathbf{C}\mathbf{\Sigma}_{t|t-1}\mathbf{C}'$ where $\mathbf{\Sigma}_{t|t-1}$ is calculated as in the Kalman filter gain matrix by

$$\mathbf{\Sigma}_{t+1|t} = \mathbf{A}\mathbf{\Sigma}_{t|t}\mathbf{A}' + \mathbf{Q}$$

and

$$\mathbf{\Sigma}_{t|t} = \mathbf{\Sigma}_{t|t-1} - \mathbf{\Sigma}_{t|t-1}\mathbf{C}'(\mathbf{C}'\mathbf{\Delta}_t\mathbf{C} + \mathbf{R})^{-1}\mathbf{C}\mathbf{\Sigma}_{t|t-1}.$$

Note that the last equation can be equivalently put as

$$\mathbf{\Sigma}_{t|t} = (\mathbf{I} - \mathbf{K}_t\mathbf{C})\mathbf{\Sigma}_{t|t-1}$$

where

$$\mathbf{K}_t = \mathbf{\Sigma}_{t|t-1}\mathbf{C}'(\mathbf{C}'\mathbf{\Delta}_t\mathbf{C} + \mathbf{R})^{-1}$$

is the Kalman filter gain. However, the former is less sensitive numerically than the latter.

5.8 Smoothing Filters

Take the model

$$\mathbf{x}_{t+1} = \mathbf{A}\mathbf{x}_t + \boldsymbol{\xi}_t$$
$$\mathbf{y}_t = \mathbf{C}\mathbf{x}_t + \boldsymbol{\eta}_t$$

to which Kalman filters are applied. When $\boldsymbol{\xi}$ and $\boldsymbol{\eta}$ are Gaussian random vectors, the maximum likelihood estimates of $\mathbf{x}_t$ given the entire data $\mathbf{y}^T$ can be extracted recursively, again by a simple application of the orthogonal projection method. This is known as the fixed-interval smoother since the sample interval T remains fixed. First,

$$\begin{aligned}\mathbf{x}_{t|t+1} &= \hat{E}(\mathbf{x}_t \,|\, \mathbf{y}^{t+1}) \\ &= \hat{E}(\mathbf{x}_t \,|\, \mathbf{y}^t, \mathbf{y}_{t+1} - \mathbf{y}_{t+1|t}) \\ &= \mathbf{x}_{t|t} + \mathbf{K}_{t|t+1}\mathbf{e}_{t+1}\end{aligned}$$

where

$$\mathbf{e}_{t+1} = \mathbf{y}_{t+1} - \mathbf{y}_{t+1|t},$$
$$\mathbf{K}_{t|t+1} = E(\mathbf{x}_t\mathbf{e}'_{t+1})\boldsymbol{\Delta}_{t+1}^{-1},$$

and

$$\boldsymbol{\Delta}_{t+1} = \mathbf{C}\boldsymbol{\Sigma}_{t+1|t}\mathbf{C}' + \mathbf{R}.$$

Recalling that $\mathbf{x}_{t|t}$ is uncorrelated with $\mathbf{e}_{t+1}$, we can evaluate the factor in $\mathbf{K}_{t|t+1}$ as

$$\begin{aligned}E(\mathbf{x}_t\mathbf{e}'_{t+1}) &= E(\mathbf{x}_t - \mathbf{x}_{t|t})(\mathbf{x}_{t+1} - \mathbf{A}\mathbf{x}_{t+1|t})'\mathbf{C}' \\ &= \boldsymbol{\Sigma}_{t|t}\mathbf{A}'\mathbf{C}'.\end{aligned}$$

Since $\mathbf{C}'\boldsymbol{\Delta}_{t+1}^{-1}$ is equal to $\boldsymbol{\Sigma}_{t+1|t}^{-1}\mathbf{K}_{t+1}$ because the Kalman filter gain is $\mathbf{K}_{t+1} = \boldsymbol{\Sigma}_{t+1|t}\mathbf{C}'\boldsymbol{\Delta}_{t+1}^{-1}$, we can write the basic smoother relation as

$$\begin{aligned}\mathbf{x}_{t|t+1} &= \mathbf{x}_{t|t} + \mathbf{N}_{t|t+1}\mathbf{K}_{t+1}\mathbf{e}_{t+1} \\ &= \mathbf{x}_{t|t} + \mathbf{N}_{t|t+1}(\mathbf{x}_{t+1|t+1} - \mathbf{x}_{t+1|t})\end{aligned}$$

where

$$\mathbf{N}_{t|t+1} = \boldsymbol{\Sigma}_{t|t}\mathbf{A}'\boldsymbol{\Sigma}_{t+1|t}^{-1}.$$

By induction it is established that

$$\mathbf{x}_{t|T} = \mathbf{x}_{t|t} + \mathbf{N}_{t|t+1}(\mathbf{x}_{t+1|T} - \mathbf{x}_{t+1|t}).$$

(See Meditch [1969, pp. 216–220].) This is used backward in time. For example, starting from $x_{T|T}$ which is available on the output of a Kalman filter, we generate

$$\mathbf{x}_{T-1|T} = \mathbf{x}_{T-1|T-1} + \mathbf{N}_{T-1|T}(\mathbf{x}_{T|T} - \mathbf{x}_{T|T-1}),$$
$$\mathbf{x}_{T-2|T} = \mathbf{x}_{T-2|T-2} + \mathbf{N}_{T-2|T-1}(\mathbf{x}_{T-1|T} - \mathbf{x}_{T-1|T-2}), \quad \text{etc.}$$

5.9 Backward Dynamics

Box and Jenkins (1970) used the notion of backcasting to obtain the smoothed estimate of initial conditions from data. Smoothing calculates $\hat{E}(\mathbf{x}_0 \mid \mathbf{y}^T)$ recursively from the estimated model to calculate the same thing. A more general procedure is to associate the forward and backward dynamic models to any data series since the notion of Markovian state space models can be formulated for time running in both forward and backward directions. These two models can be used to advantage in estimating unobserved variables because the error covariance associated with forecasts and backcasts have confidence ellipsoids with major and minor axes oriented differently, and the intersections of the ellipsoids can provide much more precise estimates of latent variables than from forward or backward error covariance matrices alone.

State space models in which time runs forward can be constructed by the algorithm in Aoki (1987c, Chapter 9). This section develops a similar state space model for the backcast purpose. We follow Pavon (1980) in the development that follows.

Let $\mathbf{\Pi} = E\mathbf{x}_t\mathbf{x}_t'$. From the dynamic equation, this matrix obeys

$$\mathbf{\Pi} = \mathbf{A\Pi A}' + \mathbf{BB}'.$$

For later use, we rewrite it when $\mathbf{A}^{-1}$ exists as follows:

$$\mathbf{\Pi}^{-1} - \mathbf{A}'\mathbf{\Pi}^{-1}\mathbf{A} = \mathbf{A}'\mathbf{\Pi}^{-1}\mathbf{BB}'(\mathbf{A}')^{-1}\mathbf{\Pi}^{-1} \tag{58}$$

and

$$\begin{aligned}\mathbf{B}'\mathbf{\Pi}^{-1}\mathbf{A} &= \mathbf{B}'(\mathbf{A}')^{-1}\mathbf{\Pi}^{-1} - \mathbf{B}'\mathbf{\Pi}^{-1}\mathbf{BB}'(\mathbf{A}')^{-1}\mathbf{\Pi}^{-1} \\ &= (\mathbf{I} - \mathbf{B}'\mathbf{\Pi}^{-1}\mathbf{B})\mathbf{B}'(\mathbf{A}')^{-1}\mathbf{\Pi}^{-1}.\end{aligned} \tag{59}$$

Given a forward state space model

$$\mathbf{x}_{k+1} = \mathbf{Ax}_k + \mathbf{Bw}_k$$
$$\mathbf{y}_t = \mathbf{Cx}_k + \mathbf{Dw}_k$$

where

$$\operatorname{cov} \mathbf{w}_k = \mathbf{I},$$

write

$$\mathbf{x}_k = \hat{E}(\mathbf{x}_k \mid \mathbf{X}_{k+1}) + \mathbf{x}_k - \hat{E}(\mathbf{x}_k \mid \mathbf{X}_{k+1})$$

where

$$\mathbf{X}_{k+1} = \mathrm{Sp}\,(\mathbf{x}_{k+1})$$

is the minimal dimensional subspace (the minimal splitting subspace). See Lindquist, Picci, and Ruckebusch (1979) or Caines (1988) for further explanation of the splitting subspaces. Thus,

$$\begin{aligned}\mathbf{x}_{k|k+1} &= E(\mathbf{x}_k\mathbf{x}'_{k+1})\mathbf{\Pi}^{-1}\mathbf{x}_{k+1}\\ &= E(\mathbf{x}_k\mathbf{x}'_k\mathbf{A}')\mathbf{\Pi}^{-1}\mathbf{x}_{k+1}\\ &= \mathbf{\Pi A'\Pi}^{-1}\mathbf{x}_{k+1} = \mathbf{\Pi A'\Pi}^{-1}\mathbf{x}_k + \mathbf{\Pi A'\Pi}^{-1}\mathbf{Bw}_k\end{aligned}$$

since $\mathbf{x} \perp \mathbf{w}_k$.

Note that

$$\begin{aligned}\mathbf{x}_k - \mathbf{x}_{k|k+1} &= (\mathbf{I} - \mathbf{\Pi A'\Pi}^{-1}\mathbf{A})\mathbf{x}_k - \mathbf{\Pi A'\Pi}^{-1}\mathbf{Bw}_k\\ &= \mathbf{\Pi}(\mathbf{\Pi}^{-1} - \mathbf{A'\Pi}^{-1}\mathbf{A})\mathbf{x}_k - \mathbf{\Pi A'\Pi}^{-1}\mathbf{Bw}_k\\ &= \mathbf{\Pi A'\Pi}^{-1}\mathbf{B}[\mathbf{B}'(\mathbf{A}')^{-1}\mathbf{\Pi}^{-1}\mathbf{x}_k - \mathbf{w}_k]\end{aligned}$$

or

$$\mathbf{\Pi}^{-1}(\mathbf{x}_k - \mathbf{x}_{k|k+1}) = -\mathbf{A'\Pi}^{-1}\mathbf{B}[\mathbf{w}_k - \mathbf{B}'(\mathbf{A}')^{-1}\mathbf{\Pi}^{-1}\mathbf{x}_k], \tag{60}$$

where (58) is used.

From (59) an alternative expression

$$\mathbf{A'\Pi}^{-1}\mathbf{B} = \mathbf{\Pi}^{-1}\mathbf{A}^{-1}\mathbf{B}(\mathbf{I} - \mathbf{B'\Pi}^{-1}\mathbf{B}) \tag{61}$$

is derived for later use.

Define

$$\bar{\mathbf{x}}_k = \mathbf{\Pi}^{-1}\mathbf{x}_{k+1}$$

and

$$\begin{aligned}\bar{\mathbf{w}}_k &= (\mathbf{I} - \mathbf{B'\Pi}^{-1}\mathbf{B})^{1/2}[\mathbf{w}_k - \mathbf{B}'(\mathbf{A}')^{-1}\mathbf{\Pi}^{-1}\mathbf{x}_k]\\ &= (\mathbf{I} - \mathbf{B'\Pi}^{-1}\mathbf{B})^{1/2}[\mathbf{w}_k - \mathbf{B}'(\mathbf{A}')^{-1}\bar{\mathbf{x}}_{k-1}].\end{aligned}$$

We now have derived the dynamic equation for the backward realization

$$\bar{\mathbf{x}}_{k-1} = \mathbf{A}'\bar{\mathbf{x}}_k + \bar{\mathbf{B}}\bar{\mathbf{w}}_k \tag{62}$$

where

$$\bar{\mathbf{B}} = -\mathbf{\Pi}^{-1}\mathbf{A'B}(\mathbf{I} - \mathbf{B'\Pi}^{-1}\mathbf{B})^{1/2} \tag{63}$$

where $\bar{\mathbf{w}}_k$ is normalized to have the identity covariance

$$\mathrm{cov}\,\bar{\mathbf{w}}_k = \mathbf{I}_p. \tag{64}$$

Note the dynamic matrix is $\mathbf{A}'$ rather than $\mathbf{A}$. This is one indication that the forward and backward models are "dual" in some sense. Another evidence supporting this point appears in (63) below.

To verify (64), observe from (59) that

$$\mathbf{X} = (\mathbf{I} - \mathbf{X})\mathbf{Y}$$

where

$$\mathbf{X} = \mathbf{B}'\mathbf{\Pi}^{-1}\mathbf{B},$$
$$\mathbf{Y} = \mathbf{B}'(\mathbf{A}')^{-1}\mathbf{\Pi}^{-1}\mathbf{A}^{-1}\mathbf{B},$$

and

$$\operatorname{cov} \bar{\mathbf{w}}_k = (\mathbf{I} - \mathbf{X})^{1/2}(\mathbf{I} - \mathbf{X})^{-1}(\mathbf{I} - \mathbf{X})^{1/2} = \mathbf{I}_p.$$

The observation equation for the backward realization is obtained by calculating

$$\mathbf{y}_t = \mathbf{y}_{t|t+1} + \mathbf{y}_t - \mathbf{y}_{t|t+1}$$

where

$$\begin{aligned} \mathbf{y}_{t|t+1} &= \hat{E}(\mathbf{y}_t \mid \mathbf{X}_{t+1}) \\ &= E(\mathbf{y}_t\mathbf{x}'_{t+1})\mathbf{\Pi}^{-1}\mathbf{x}_{t+1} \\ &= (\mathbf{C}\mathbf{\Pi}\mathbf{A}' + \mathbf{D}\mathbf{B}')\mathbf{\Pi}^{-1}\mathbf{x}_{t+1} \\ &= \mathbf{M}'\bar{\mathbf{x}}_t \end{aligned} \tag{65}$$

where the matrix $\mathbf{M}$ is defined by the above equality. This is the same matrix $\mathbf{M}$ that appears in the sum decomposition of the spectral density matrix, and that is estimated in conducting the forward state space models.

Also,

$$\mathbf{y}_t - \mathbf{y}_{t|t+1} = \mathbf{C}(\mathbf{x}_t - \mathbf{C}_{t|t+1}) + \mathbf{D}(\mathbf{w}_t - \mathbf{w}_{t|t+1})$$

where from (3) and (4)

$$\mathbf{C}(\mathbf{x}_t - \mathbf{x}_{t|t+1}) = \mathbf{C}\mathbf{A}^{-1}\mathbf{B}(\mathbf{I} - \mathbf{B}'\mathbf{\Pi}^{-1}\mathbf{B})^{1/2}\bar{\mathbf{w}}_t \tag{66}$$

and

$$\begin{aligned} \mathbf{D}(\mathbf{w}_t - \mathbf{w}_{t|t+1}) &= \mathbf{D}(\mathbf{w}_t - \mathbf{B}'\mathbf{\Pi}^{-1}\mathbf{x}_{t+1}) \\ &= \mathbf{D}[(\mathbf{I} - \mathbf{B}'\mathbf{\Pi}^{-1}\mathbf{B})\mathbf{w}_t - \mathbf{B}'\mathbf{\Pi}^{-1}\mathbf{A}\mathbf{x}_t] \\ &= \mathbf{D}(\mathbf{I} - \mathbf{B}'\mathbf{\Pi}^{-1}\mathbf{B})^{1/2}\bar{\mathbf{w}}_t. \end{aligned} \tag{67}$$

The last equality follows from (59) and the definition of $\bar{\mathbf{x}}_t$.

Combine (65), (66), and (67) to derive

$$\mathbf{y}_t = \mathbf{M}'\bar{\mathbf{x}}_t + \bar{\mathbf{D}}\bar{\mathbf{w}}_t \tag{68}$$

where

$$\bar{\mathbf{D}} = (\mathbf{D} - \mathbf{C}\mathbf{A}^{-1}\mathbf{B})(\mathbf{I} - \mathbf{B}'\mathbf{\Pi}^{-1}\mathbf{B})^{1/2}.$$

Equations (62) and (68) constitute the backward state space model.

Note that $\operatorname{cov} \bar{\mathbf{x}}_k = \mathbf{\Pi}^{-1}$.

Such backward dynamics should be important in dynamic models with rational expectations, since the future "determines" the present. See Grandmont and Laroque (1986) for an indication of this.

5.10 Kalman Filters for Linearized Nonlinear Systems

In this section, we shall construct Kalman filters for nonlinear systems by linearizing the dynamic and the observation equations. It should be noted, however, that the procedure of linearizing the dynamic and observation equations first and constructing an optimal linear filter for the linearized system does not necessarily yield a better approximation than that of obtaining the exact or approximate conditional probability densities for the nonlinear system. Outputs of such linearized filters may also fail to converge.

Construction of Filter

In order to construct an approximate filter for a nonlinear system, we linearize a nonlinear plant and observation equations

$$x(i+1) = \mathrm{F}(x(i), \xi(i), i)$$
$$y(i) = \mathrm{G}(x(i), \eta(i), i)$$

by a Taylor series expansion about a nominal sequence of state vectors $\bar{x}$s and about $\xi(i) = \eta(i) = 0$, retaining only terms up to the first order. The subscripts are now used to denote components of vectors. The time index is carried as the argument

$$x(i+1) = \mathrm{F}(\bar{x}(i), 0, i) + A(i)(x(i) - \bar{x}(i)) + v(i)$$
$$y(i) = \mathrm{G}(\bar{x}(i), 0, i) + H(i)(x(i) - \bar{x}(i)) + w(i)$$

where $\bar{x}(i)$ is the nominal $x(i)$ and where $A(i)$ is the Jacobian matrix of F with respect to x. Its (j,k)th element is given by

$$A_{jk}(i) \triangleq \left(\frac{\partial F}{\partial x}\right)_{jk}$$

which is the partial derivative of the jth component of F by the kth component of x, and the partial derivative is evaluated at $x(i) = \bar{x}(i)$, $\xi(i) = \eta(i) = 0$, and i. Similarly, we compute

$$H(i) \triangleq \left[\left(\frac{\partial G}{\partial x}\right)_{jk}\right]_{(\bar{x}(i),0,i)}$$

$$v_j(i) = \sum_k \frac{\partial \mathrm{F}_j(\bar{x}(i),0,i)}{\partial \xi_k}\xi_k(i)$$

$$w_j(i) = \sum_k \frac{\partial \mathrm{G}_j(\bar{x}(i),0,i)}{\partial \eta_k}\eta_k(i)$$

where $v_j(i)$ is the jth component of the vector $v(i)$. Assume

$$\mathrm{E}(v(i)) = \mathrm{E}(w(i)) = 0$$

$$\mathrm{E}(v(i)v'(j)) = V(i)\delta_{ij}$$

$$\mathrm{E}(w(i)w'(j)) = W(i)\delta_{ij}$$

$$\mathrm{E}(v(i)w'(j)) = 0.$$

Define

$$\alpha(i) = x(i) - \bar{x}(i)$$

$$\beta(i+1) = x(i+1) - \mathrm{F}(\overline{x(i)},0,i) \tag{69}$$

$$\gamma(i) = y(i) - \mathrm{G}(\overline{x(i)},0,i).$$

Then

$$\beta(i+1) = A(i)\alpha(i) + v(i)$$

$$\gamma(i) = H(i)\alpha(i) + w(i)$$

are the linearized dynamic and observation equations. Using symbols defined as

$$\alpha^*(i) = \mathrm{E}(\alpha(i) \mid \gamma(0),\ldots,\gamma(i))$$

$$\hat{\beta}(i+1) = \mathrm{E}(\beta(i+1) \mid \gamma(0),\ldots,\gamma(i))$$

$$\beta^*(i+1) = \mathrm{E}(\beta(i+1) \mid \gamma(0),\ldots,\gamma(i+1))$$

$$\hat{\beta}(i+1) = A(i)\alpha^*(i),$$

The Kalman filter for the system of (69) is governed by

$$\beta^*(i+1) = \hat{\beta}(i+1) + K(i+1)[\gamma(i+1) - H(i+1)\hat{\beta}(i+1].$$

Then from (69) the approximately optimal estimate is given by

$$x^*(i+1) = \mathrm{F}(\overline{x(i)}, 0, i) + \beta^*(i+1)$$

and

$$\alpha^*(i) = x^*(i) - \bar{x}(i).$$

The expressions for error-covariance matrices can be similarly obtained.

If the nominal $\bar{x}$s are chosen to satisfy the nonlinear dynamic equation with no noise, then $\alpha(i)$ and $\beta(i)$ coincide. If F and G are linearized about $x^*(i)$, instead of about the nominal $\bar{x}$, then

$$\alpha^*(i) = 0$$

since

$$\alpha(i) = x(i) - x^*(i).$$

Hence

$$\hat{\beta}(i+1) = 0$$

and

$$\beta^*(i+1) = K(i+1)\gamma(i+1).$$

Define

$$\hat{x}(i+1) = \mathrm{F}(x^*(i), 0, i).$$

Then

$$x^*(i+1) = \hat{x}(i+1) + K(i+1)[y(i+1) - \mathrm{G}(\hat{x}(i+1), 0, i)]$$

is the Kalman filter for the linearized system, where the gain is computed recursively from

$$\begin{aligned}\Gamma(i+1) = {} & [I - K(i+1)H(i+1)]M(i+1)[I - K(i+1)H(i+1)]' \\ & + K(i+1)W(i+1)K'(i+1)\end{aligned}$$

where

$$\begin{aligned}M(i+1) &= A(i)\Gamma(i)A'(i) + V(i) \\ K(i+1) &= M(i+1)H'(i+1) \\ & \quad \times [H(i+1)M(i+1)H'(i+1) + W(i+1)]^{-1}.\end{aligned}$$

5.11 Nonlinear Dynamic Model

We now discuss the least squares estimation problems of a nonlinear system. Consider the model dynamic and the observation equation given by

(70) $$x_{k+1} = f_k(x_k)$$

(71) $$y_k = h_k(x_k) + \eta_k, \qquad k = 0, 1, \ldots .$$

If (70) can be solved for x_k uniquely in terms of x_{k+1} as $x_k = g_k(x_{k+1})$, where g_k is the inverse of f_k, then x_0, x_1, x_{k-1} can be expressed in terms of x_k as before and a criterion of estimators, J_k, defined analogous to (3), can be regarded as a function of x_k. Then

$$J_{k+1}(x_{k+1}) = J_k(g_k(x_{k+1})) + (h_{k+1}(x_{k+1}) - y_{k+1})' \times V_{k+1}(h_{k+1}(x_{k+1}) - y_{k+1}). \tag{72}$$

The estimate x^*_{k+1}, then, will be expressed as

$$x^*_{k+1} = f_k(x^*_k) + \Delta x_{k+1}.$$

The optimal correction term is Δx_{k+1} which minimizes J_{k+1} with respect to Δx_{k+1}. Equation (72) is, however, no longer quadratic in Δx_{k+1}.

As mentioned, we may linearize (70) and (71). Then we can apply the technique of Section 5.8.

We next derive approximately optimal least-squares estimates recursively by the method of invariant imbedding.

Approximate Solution

The method is given for one-dimensional systems in order to present the basic procedure most clearly. It is a discrete-time version of the method proposed by Bellman (1964).

Consider a model

$$x_{t+1} = x_t + f_t(x_t), \qquad t = 0, 1, \ldots \tag{73}$$

where f_t is assumed differentiable, and also consider the observation equation

$$y_t = h_t(x_t) + \eta_t, \qquad t = 0, 1, \ldots \tag{74}$$

where ηs are noises in observation, and where h_t is assumed twice differentiable.

Denote by x^*_t the best least squares estimate of x_t at time t. Namely x^*_t is the best estimate of the current state variable x_t of the system.

We will look for the recursion formula for x^*_t in the form of

$$x^*_{t+1} = x^*_t + g_t(x^*_t). \tag{75}$$

Define the criterion function of estimation by

$$J(x, N) = \sum_0^N v_t[y_t - h_t(x_t)]^2, \qquad v_t \leqslant 0 \tag{76}$$

with the understanding that x is the state variable at time N, i.e., $x_N = x$.

The v_ts are weights of the observations, which implies that some ideas of the relative magnitudes of the variance of η_t are available. Otherwise, one may take $v_t = 1$, $t = 0, 1, \ldots, N$.

The optimal x_N^*, therefore, satisfies

$$\text{(77)} \qquad \mathrm{J}_x(x_N^*, N) = 0$$

where J_x is the partial derivative of J with respect to x. Similarly,

$$\text{(78)} \qquad J_x(x_{N+1}^*, N+1) = 0.$$

From (73), (77), and (78),

$$\text{(79)} \qquad \begin{aligned} 0 &= \mathrm{J}_x(x_{N+1}^*, N+1) \\ &= \mathrm{J}_x(x_N^*, N+1) + \mathrm{J}_{xx}(x_N^*, N+1)g_N(x_N^*) + \cdots. \end{aligned}$$

Therefore, we obtain the expression for the best correction term as

$$\text{(80)} \qquad g_N(x_N^*) \doteqdot -\mathrm{J}_x(x_N^*, N+1)/\mathrm{J}_{xx}(x_N^*, N+1).$$

To compute J_x and J_{xx} in (80), note from (73) and (76) that

$$\text{(81)} \qquad \begin{aligned} &J(x, N) + v_{N+1}[y_{N+1} - h_{N+1}(x + f_N(x))]^2 \\ &\quad = J(x + f_N(x), N+1) \\ &\quad = J(x, N+1) + J_x(x, N+1)f_N(x) + \tfrac{1}{2}J_{xx}(x, N+1)f_N^2(x) + \cdots \end{aligned}$$

where the last line is obtained by the Taylor series.

Differentiating (81) with respect to x,

$$\text{(82)} \qquad \begin{aligned} J_x(x, N) = [1 + f_N'(x)]\{&J_x(x, N+1) + J_{xx}(x, N+1)f_N(x) \\ &+ 2v_{N+1}[y_{N+1} - h_{N+1}(x + f_N(x))]h_{N+1}'(x + f_N(x))\} \\ &+ (\text{terms in } J_{xxx} \text{ and higher}). \end{aligned}$$

Substituting x_N^* for x in (82) and noting (77),

$$\text{(83)} \qquad \begin{aligned} \mathrm{J}_x(x_N^*, N+1) \doteqdot &-\mathrm{J}_{xx}(x_N^*, N+1)f_N(x_N^*) \\ &- 2v_{N+1}[y_{N+1} - h_{N+1}(\hat{x}_{N+1})]h_{N+1}'(\hat{x}_{N+1}) \end{aligned}$$

where

$$\text{(84)} \qquad \hat{x}_{N+1} \triangleq x_N^* + f_N(x_N^*)$$

is the best estimate of x_{N+1} at time N. Since $\mathrm{J}(x, N)$ is nearly quadratic in the neighborhood of x_N^*, terms in J_{xxx} and higher derivatives are neglected in obtaining (83) from (82).

From (80) and (83), therefore

$$x^*_{N+1} = \hat{x}_{N+1} + K_{N+1}[y_{N+1} - h_{N+1}(\hat{x}_{N+1})] \tag{85}$$

where

$$K_{N+1} = \frac{2v_{N+1}h'_{N+1}(\hat{x}_{N+1})}{\mathrm{J}_{xx}(x^*_N, N+1)}. \tag{86}$$

Note that (83) has the form identical with the one-dimensional version of the recursion formula of the Kalman filter previously obtained for linear dynamic and observation equations. There $\hat{x}_{N+1}$ is $A_N x^*_N$ since the dynamic equation is linear.

CHAPTER 6

Convergence Questions

Questions on convergence of one kind or another frequently arise in sequential decision problems, especially under imperfect information. Certain decision rules are often proposed because of ease of implementation or perhaps because they make only modest demands on information required to implement the rules. Do repeated uses of such rules lead to convergence to optimal rules under some circumstances? Any policy-improving iteration scheme or (suboptimal) estimation scheme for learning unknown parameter values in decision problems must be evaluated for convergence behavior. What are the conditions under which Bayesian parameter adaptive schemes converge to true parameters? This chapter answers these questions and collects some tools to answer comparable ones.

Besides the questions related to the convergence (such as that of learning models to rational expectations or that of suboptimal algorithms in estimation or decision rules to optimal ones), stability questions may also be thought of as a special case of convergence problems. "Does the dynamic process converge to its equilibrium state, and if so under what conditions?" is one of the basic questions one should ask when dealing with dynamic processes. The stochastic Lyapunov function approach to stability has been introduced in Chapter 2. Stochastic Lyapunov functions are also used in establishing convergence of parameter estimation schemes as we will discuss in this chapter.

A small number of basic procedures or facts are used to answer these questions on convergence in this chapter. Some algorithms in learning or estimation are natural extensions of nonlinear programming algorithms to stochastic context, such as various versions of stochastic gradient methods, Newton-like methods, stochastic Lyapunov functions, and stochastic approximation methods due to Robbins and Monro or Kiefer and Wolfowitz. Martingale convergence theorems, ergodic theorems of Markov processes, and certain ordinary differential equations are also used for examining convergence questions. This chapter touches on some of these methods to indicate their potential applications. Since each method has a long history and often merits a book on its own, we are not systematic but rather brief in their discussion.

6.1 Regression Models

Deterministic Regressors

We start with a model

$$y_\tau = \boldsymbol{\theta}'\mathbf{x}_\tau + \varepsilon_\tau, \qquad \tau = 1, 2, \ldots$$

where x_t is nonstochastic. Then stacking ys and εs, write the model by defining $\mathbf{X}_t' = [\mathbf{x}_1 \ldots \mathbf{x}_t]$,

$$\mathbf{y}^t = \mathbf{X}_t\boldsymbol{\theta} + \boldsymbol{\varepsilon}^t$$

where

$$\mathbf{y}^t = \begin{bmatrix} y_1 \\ \vdots \\ y_t \end{bmatrix} \quad \text{and} \quad \boldsymbol{\varepsilon}^t = \begin{bmatrix} \varepsilon_1 \\ \vdots \\ \varepsilon_t \end{bmatrix}.$$

The least squares estimator for $\boldsymbol{\theta}$ is

$$\begin{aligned} \hat{\boldsymbol{\theta}}_t &= \mathbf{V}_t^{-1}\mathbf{X}_t'\mathbf{y}^t \\ &= \boldsymbol{\theta} + \mathbf{V}_t^{-1}\mathbf{X}_t'\boldsymbol{\varepsilon}^t \end{aligned} \tag{1}$$

where

$$\mathbf{V}_t = \Sigma_1^t \mathbf{x}_\tau \mathbf{x}_\tau'$$

and

$$\mathbf{X}_t'\boldsymbol{\varepsilon}^t = \Sigma_1^t \mathbf{x}_\tau \boldsymbol{\varepsilon}_\tau .$$

Equation (1) can be put into a recursive form

$$\hat{\boldsymbol{\theta}}_t = \hat{\boldsymbol{\theta}}_{t-1} + \mathbf{V}_t^{-1}\mathbf{x}_t\mathbf{e}_t$$

where

$$\mathbf{e}_t = \mathbf{y}_t - \mathbf{x}_t'\hat{\boldsymbol{\theta}}_{t-1}.$$

The estimation error variance matrix is

$$\operatorname{var}(\hat{\boldsymbol{\theta}}_t - \boldsymbol{\theta}) = \sigma^2\mathbf{V}_t^{-1}$$

when

$$\operatorname{var}(\boldsymbol{\varepsilon}^t) = \sigma^2\mathbf{I}_t.$$

Thus $\hat{\boldsymbol{\theta}}_t$ converges to $\boldsymbol{\theta}$ in the mean square, and hence in probability by the Chebychev inequality if the smallest eigenvalue of $\mathbf{V}_t$, $\lambda_{\min}(\mathbf{V}_t)$, goes to infinity as $t \to \infty$, or $\liminf \lambda_{\min}(\mathbf{V}_t) > 0$.

Stochastic Regressors

When xs are random variables independent of εs, the same sufficient conditions apply for almost sure convergence. This condition or something analogous is known as the "persistent" excitation condition in the system literature, e.g., Åström (1965) or Aoki and Staley (1970) for some early works and Moore (1983) for a more recent contribution. See Lai and Wei (1982) which weakened this persistent excitation condition.

When x_t is scalar, we can use Kronecker's lemma (Lukacs [1975, p. 96]) to see that the error term $(\Sigma_t x_t^2)^{-1}\Sigma_\tau x_\tau\varepsilon_\tau$ converges to zero a.s. if $V_\tau^{-1} \to 0$ (the persistent excitation condition), and if $\Sigma_\tau V_\tau^{-1}x_\tau\varepsilon_\tau$, converges a.s. where $V_t = \Sigma_1^t x_\tau^2$. The variable

$$y_t = \Sigma_{\tau=1}^t V_\tau^{-1}x_\tau\varepsilon_\tau$$

is a martingale where ε_t are i.i.d. mean zero, since $\sup_t Ey_t^2 \leqslant \Sigma_1^\infty x_\tau^2/V_\tau^2 < \infty$. To see this, note

$$\begin{aligned}
\sum_1^\infty x_\tau^2/V_\tau^2 &= \sum_2^\infty (V_\tau - V_{\tau-1})/V_\tau^2 \\
&\leqslant \sum_2^\theta (V_\tau - V_{\tau-1})/V_\tau V_{\tau-1} \\
&= \sum_2^\infty (V_{\tau+1}^{-1} - V_\tau^{-1}) \\
&= V_1^{-1}.
\end{aligned}$$

With a vector valued $\boldsymbol{\theta}$, Kronecker's lemma is not valid. The a.s. convergence of $\hat{\boldsymbol{\theta}}_t$ was established by Lai and Robbins (1979). Solo (1986, p. 256) has the proof when εs are independent Gaussian random variables. Solo writes $\hat{\boldsymbol{\theta}}_t$ in a recursive form shown below (1). Assume that $\mathbf{x}_t$ is predictable. The estimation error, $\tilde{\boldsymbol{\theta}}_t = \tilde{\boldsymbol{\theta}}_t - \boldsymbol{\theta}$, obeys

(2) $$\tilde{\boldsymbol{\theta}}_t = \tilde{\boldsymbol{\theta}}_{t-1} + \mathbf{V}_t^{-1}\mathbf{x}_t e_t.$$

Since e_t is a linear combination of $\boldsymbol{\varepsilon}^t$, $\{e_t\}$ is a sequence of uncorrelated, i.e., independent, random variables with

$$E(e_t^2 \mid \mathscr{F}_{t-1}) = \sigma^2(1 + \mathbf{x}_t'\mathbf{V}_{t-1}^{-1}\mathbf{x}_t).$$

Let $T_t = \boldsymbol{\alpha}'\tilde{\boldsymbol{\theta}}_t$ for an arbitrary $\boldsymbol{\alpha}'\boldsymbol{\alpha} = 1$. Then (2) implies that

$$E(T_t^2 \mid \mathscr{F}_{t-1}) = T_{t-1}^2 + \beta_t.$$

where

$$\beta_t = \sigma^2(1 + \mathbf{x}_t'(\mathbf{X}'\mathbf{X})_{t-1}^{-1}\mathbf{x}_t)\mathbf{x}_t'(\mathbf{X}'\mathbf{X})_t^{-2}\mathbf{x}_t.$$

Then T_t converges a.s. by the martingale convergence theorem of Chapter 2 if $\Sigma\beta_t < \infty$. Since $T_t \to 0$ in probability if $\lambda_{\min}(\mathbf{X}'\mathbf{X})_t \to \infty$, then $T_t \to 0$ a.s.

6.2 Stochastic Gradient Method

Consider a system

$$y_t = \mathbf{x}_t'\boldsymbol{\theta} + \eta_t$$

where $\boldsymbol{\theta}$ is a parameter vector and $\mathbf{x}_t$ is a function of past observations and controls. For example, estimation by the extended least squares method of the coefficients in

$$y_{t+1} = \sum_{i=0}^{n} (a_i y_{t-i} + b_i u_{t-i}) + \eta_{t+1}$$

is put as just mentioned by defining a parameter vector by

$$\boldsymbol{\theta}' = (a_0 \ldots a_n, b_0 \ldots b_n)$$

and the regressor vector by

$$\mathbf{x}_t' = (y_{t-1} \ldots y_{t-n-1}, u_{t-1}, \ldots u_{t-n-1}), \qquad t - 1 \geqslant n.$$

Define

$$J_t(\theta) = \sum_{\tau=0}^{t} (y_t - \mathbf{x}_t'\boldsymbol{\theta})^2.$$

The gradient of this with respect to $\boldsymbol{\theta}$ is

$$\partial J_t / \partial \boldsymbol{\theta} = -2\Sigma_\tau (y_\tau - \mathbf{x}_\tau'\boldsymbol{\theta})\mathbf{x}_\tau,$$

and one stochastic gradient–type algorithm estimates θ iteratively by

$$\hat{\boldsymbol{\theta}}_{t+1} = \hat{\boldsymbol{\theta}}_t - \frac{\mu}{\gamma_t}\left(\frac{\partial J_t}{\partial \boldsymbol{\theta}}\right)_{\hat{\boldsymbol{\theta}}_t},$$

for some γ_t, where $\mu/\gamma_t \to 0$ is some step-size-adjusting sequence. Convergence of this algorithm can be established using the stochastic Lyapunov function and also by the method due to Ljung (1977), who studied the stability of a certain associated ordinary differential equation. His method is briefly explained later in this chapter.

Example 1. Consider an iterative scheme for estimating a parameter vector $\boldsymbol{\theta}$

$$\hat{\boldsymbol{\theta}}_{t+1} = \hat{\boldsymbol{\theta}}_t + \frac{\mathbf{x}_t}{\gamma_t}(y_{t+1} - \mathbf{x}_t'\hat{\boldsymbol{\theta}}_t)$$

where

(3) $$\gamma_t = 1 + \Sigma_0^t \mathbf{x}_\tau'\mathbf{x}_\tau.$$

The observation process is specified by

(4) $$y_{t+1} = \boldsymbol{\theta}'\mathbf{x}_t + \eta_{t+1},$$

where $E(\eta_{t+1}) = 0$, $E(\eta_{t+1}) = \sigma^2$ and the noises are serially uncorrelated, and $\mathbf{x}_t$ is a vector of past observation and past decisions that satisfy $\mathbf{x}_t'\hat{\boldsymbol{\theta}}_t = 0$. Errors of estimation $\tilde{\boldsymbol{\theta}}_t = \hat{\boldsymbol{\theta}}_t - \boldsymbol{\theta}$ are equal to

$$\tilde{\boldsymbol{\theta}}_{t+1} = \tilde{\boldsymbol{\theta}}_t + \frac{\mathbf{x}_t}{\gamma_t}(y_{t+1} - \mathbf{x}_t'\hat{\boldsymbol{\theta}}_t)$$

$$= \tilde{\boldsymbol{\theta}}_t + \frac{\mathbf{x}_t}{\gamma_t} y_{t+1}.$$

Define

$$J_t = \tilde{\boldsymbol{\theta}}_t'\tilde{\boldsymbol{\theta}}_t.$$

Let $\mathscr{F}_t$ be the σ-field generated by past and current observation and controls. This function is certainly positive definite, and it will serve as the stochastic Lyapunov function.

To see this, calculate

$$E(J_{t+1}\,|\,\mathscr{F}_t) = J_t + \frac{2}{\gamma_t}\mathbf{x}_t'\tilde{\boldsymbol{\theta}}_t y_{t+1|t} + \frac{\mathbf{x}_t'\mathbf{x}_t}{\gamma_t^2}E(y_{t+1}^2\,|\,\mathscr{F}_t), \tag{5}$$

where $y_{t+1|t} = E(y_{t+1}\,|\,\mathscr{F}_t)$.

Noting that

$$\boldsymbol{\theta}'\mathbf{x}_t = \hat{\boldsymbol{\theta}}_t'\mathbf{x}_t - \tilde{\boldsymbol{\theta}}_t'\mathbf{x}_t = -\tilde{\boldsymbol{\theta}}_t'\mathbf{x}_t,$$

(4) becomes

$$y_{t+1} = -\tilde{\boldsymbol{\theta}}_t'\mathbf{x}_t + \eta_{t+1},$$

i.e.,

$$y_{t+1|t} = -\tilde{\boldsymbol{\theta}}_t'\mathbf{x}_t = \boldsymbol{\theta}'\mathbf{x}_t,$$

and

$$\begin{aligned} E(y_{t+1}^2\,|\,\mathscr{F}_t) &= (\boldsymbol{\theta}'\mathbf{x}_t)^2 + \sigma^2 \\ &= (y_{t+1|t})^2 + \sigma^2. \end{aligned}$$

Substituting these back into (5), we write it as

$$\begin{aligned} E(V_{t+1}\,|\,\mathscr{F}_t) &= J_t - \frac{2}{\gamma_t}(y_{t+1|t})^2 \\ &\quad + \frac{\mathbf{x}_t'\mathbf{x}_t}{\gamma_t^2}(y_{t+1|t})^2 + \frac{\mathbf{x}_t'\mathbf{x}_t}{\gamma_t^2}\sigma^2 \\ &\leqslant J_t - \frac{2}{\gamma_t}(y_{t+1|t})^2 + \frac{\mathbf{x}_t'\mathbf{x}_t}{\gamma_t^2}\sigma^2 \end{aligned} \tag{6}$$

where (3) is used.

Equation (6) is in a form to which Solo's extension of Neveu's theorem applies if

$$\Sigma_t \mathbf{x}_t'\mathbf{x}_t/\gamma_t^2 < \infty.$$

To see this, note that

$$\mathbf{x}_t'\mathbf{x}_t = \gamma_t - \gamma_{t-1},$$

and because $\gamma_{t-1} < \gamma_t$, we have

$$\begin{aligned} \Sigma_1^T \mathbf{x}_t'\mathbf{x}_t/\gamma_t^2 &= \Sigma(\gamma_t - \gamma_{t-1})/\gamma_t^2 \\ &\leqslant \Sigma(\gamma_t - \gamma_{t-1})/\gamma_t\gamma_{t-1} \\ &= \Sigma_1^T(1/\gamma_{t-1} - 1/\gamma_t) \leqslant 1/\gamma_0 = 1. \end{aligned}$$

Therefore J_t converges a.s. and

$$\Sigma_t (y_{t+1|t})^2/\gamma_t < \infty \text{ a.s.}$$

Since γ_t is monotonically increasing, and $\gamma_T \to \infty$,* Kronecker's lemma applies and

$$\frac{1}{\gamma_T}\Sigma_1^T (y_{t+1|t})^2 \to 0 \quad \text{as} \quad T \to \infty. \tag{7}$$

To relate this to the behavior of $\Sigma_1^T y_t^2/T$, note that

$$\Sigma_2^{T+1} y_t^2 = \Sigma_1^T [(y_{t+1|t})^2 + \eta_{t+1}^2 + 2\eta_{t+1} y_{t+1|t}]$$

where

$$\Sigma_1^T \eta_{t+1} y_{t+1|t} \leqslant \alpha_T [\Sigma_1^T (y_{t+1|t})^2 (\Sigma_1^T \eta_{t+1}^2)^{1/2}]$$

by the Schwarz inequality for some $|\alpha_T| \leqslant 1$. Under some technical conditions, it can be shown that γ_T/T is bounded a.s.** Then (7) is replaceable by

$$\frac{1}{T}\Sigma (y_{t+1|t})^2 \to 0, \quad \text{a.s. as} \quad T \to \infty,$$

and

$$\frac{1}{T}\Sigma_1^{T-1} y_t^2 \to \sigma^2$$

follows from

$$\lim_{T\to\infty} \frac{1}{T}\Sigma_0^{T-1} \eta_t^2 = \sigma^2 \quad \text{a.s.}$$

---

* If γ_T converges to a finite limit, then $\Sigma_0^T \mathbf{x}_t' \mathbf{x}_t < \infty$, and $\mathbf{x}_t' \mathbf{x}_t \to 0$. Then $y_t - \eta_{t+1} \to 0$. But this can happen only on a set of measure zero.

** Boundedness a.s. of the sequence $\{\gamma_T\}$ can be established if (4) is a minimum-phase system, because x_t can be regarded as the output of the inverse system with y_{t+1} and η_{t+1} as inputs. Since the inverse system is asymptotically stable, there are constants C_1 and C_2 such that

$$\Sigma_1^T x_t^2/T \leqslant C_1 \Sigma_1^T y_{t+1}^2/T + C_2 \Sigma_1^T \eta_{t+1}^2/T \qquad \text{for all } T > 0.$$

Since we know that $\Sigma_1^T \eta_{t+1}^2/T \to \sigma^2$ at $T \to \infty$, this inequality becomes

$$\gamma_T/T \leqslant C_3 \Sigma_1^T y_{t+1}^2/T + C_4$$

for some C_3 and C_4. Substitute $y_{t+1} = y_{t+1|t} + \eta_{t+1}$ to obtain

$$\gamma_T/T \leqslant C_5 \Sigma_1^T y_{t+1|t}^2/T + C_5$$

or

$$\Sigma_1^T y_{t+1|t}/\gamma_T \geqslant (\gamma_T - C_5 T)/C_6 \gamma_T$$

for some $C_6 > 0$. If γ_T/T were not bounded a.s., then along some subsequence of $\{T\}$, $T_k \to \infty$, the right side converges to a strictly positive number that contradicts (7).

6.3 Stochastic Approximation Method

A variant of the stochastic gradient method is the prototype stochastic approximation method proposed by Robbins and Munro (1951). The Robbins-Monro scheme solves an equation $M(\theta)$ for its zero $M(\theta) = E\ Q(\theta, \eta) = 0$ where η_t is a (disturbance) stochastic process and $Q(\theta, \eta_t)$ is assumed to be observed. Such an equation often arises as the first-order condition of some intertemporal optimization problem. Their scheme is to find the root of $M(\theta) = 0$ by

$$\theta_t = \theta_{t-1} + \alpha_t Q(\theta_{t-1}, \eta_t)$$

where $\{\alpha_t\}$ is typically chosen to satisfy

$$\sum_1^\infty \alpha_t = \infty, \qquad \alpha_t > 0, \qquad \text{and} \qquad \lim \alpha_t = 0.$$

To see that this is a variant of the gradient method, let $V(\theta) = E[e(\theta)^2]$. Then

$$\frac{dV}{d\theta} = E\left(e(\theta)\frac{de}{d\theta}\right) = -E\left[e(\theta)\frac{d\hat{y}_t}{d\theta}\right],$$

and the Robbins-Munro scheme becomes

$$\theta_t = \theta_{t-1} + \alpha_t \frac{d\hat{y}_t}{d\theta} e_t.$$

Kiefer and Wolfowitz (1952) proposed another method when gradient vectors cannot be evaluated analytically but must be estimated.

A stochastic regression model

$$y_t = \beta_1 + \beta_2 x_t + \varepsilon_t, \tag{8}$$

where ε_t is i.i.d., mean zero, and finite variance, has been discussed by Lai and Robbins (1979) and Lai and Wei (1982) as an adaptive design (control) problem. The problem is to set x_t to render y_t zero in the absence of exogenous disturbances. Later the model is generalized to

$$y_t = M(x_t) + \varepsilon_t$$

where $M'(\theta) = \beta$. In one version, β is assumed known and in another version only the sign of β is assumed known. If β_1 and β_2 are known, then this is easily done. When the parameter values are not known, one may analyze the problem from a Bayesian point of view or, alternatively, one may obtain their least squares estimate and set x_t using the estimated values. This type of reasoning produces a gradual adjustment rule for x_t

(9) $$x_{t+1} = \bar{x}_t - c\bar{\varepsilon}_t$$

for some constant c, where "$-$" denotes arithmetic average. The Bayesian approach is discussed later.

To see how (9) arises, rewrite the model as

$$y_t = \beta(x_t - \theta) + \varepsilon_t,$$

where θ is the desired setting for the control. The least squares estimate θ is

$$\begin{aligned}\hat{\theta}_t &= \bar{x}_t - \beta^{-1}\bar{y}_t \\ &= \theta - \beta^{-1}\bar{\varepsilon}_t,\end{aligned}$$

and setting x_{t+1} to $\hat{\theta}_t$ produces the adjustment rule

$$x_{t+1} = \theta - \beta^{-1}\bar{\varepsilon}_t.$$

Approximating the unknown θ by $\bar{x}_t$, we obtain (9) for some constant c. Since the adjustment scheme (9) is the same as

$$x_{t+1} = x_t - \frac{c}{t}\varepsilon_t,$$

any sequence $\{c_t\}$ may be used in (9)

$$x_{t+1} = \bar{x}_t - c_t\bar{\varepsilon}_t$$

or equivalently

(10) $$x_{t+1} = x_t - \frac{c_t}{t}\varepsilon_t.$$

When $c_t/t \to 0$ and $\Sigma\, c_t/t = \infty$ in (10), this scheme is a stochastic approximation scheme of Robbins–Monro (1951) type.

A stochastic approximation scheme for price adjustment was discussed in Aoki (1974). Aoki compared a stochastic approximation price-adjustment rule by a marketer in an organized market for a single commodity. This pricing rule is compared with three Bayesian pricing rule, two of which have a one-period critierion function and a third a multiperiod criterion function. He showed that the two one-period (myopic) Bayesian pricing rules asymptotically approach the stochastic approximation pricing rule, and the multiperiod pricing rule similarly converges to the stochastic approximation rule except for a stock adjustment term.

This section next examines the convergence behavior of (8) with the adjustment rule (9) and shows that the least squares estimates for β_1 and β_2 are biased. Subtract the arithmetic average of (8), $\bar{y}_{t-1} = \beta_1 + \beta_2 \bar{x}_{t-1} + \bar{\varepsilon}_{t-1}$, from (8)

$$y_t - \bar{y}_{t-1} = \beta_2(x_t - \bar{x}_{t-1}) + \varepsilon_t - \bar{\varepsilon}_{t-1}.$$

The least squares estimate of β_2 is

$$\hat{\beta}_{2t} = \beta_2 + \frac{\Sigma(x_\tau - \bar{x}_{\tau-1})(\varepsilon_\tau - \varepsilon_{\tau-1})}{\Sigma(x_\tau - \bar{x}_{\tau-1})^2}.$$

From (9), $x_\tau - \bar{x}_{\tau-1} = -c\bar{\varepsilon}_{\tau-1}$ is substituted into the preceding to yield

$$\hat{\beta}_{2t} = \beta_2 + \frac{1}{c} - \frac{1}{c}\frac{\Sigma\,\varepsilon_\tau \bar{\varepsilon}_{\tau-1}}{\Sigma\,\bar{\varepsilon}^2_{\tau-1}}.$$

From Fact 2 of Chapter 2, we know that

$$\Sigma^t \bar{\varepsilon}^2_{\tau-1} \sim O(\ln t)$$

and the martingale transform $\Sigma\,\varepsilon_\tau \bar{\varepsilon}_{\tau-1}$ is such that

$$\overline{\lim_n} \frac{\Sigma^t \varepsilon_\tau \bar{\varepsilon}_{\tau-1}}{d_t^{1/2}(\ln d_t)^{1/2+\delta}} = 0$$

where $d_t = \sigma^2 \Sigma\,\bar{\varepsilon}^2_{\tau-1} \sim O(\ln t)$.

Therefore,

$$\hat{\beta}_{2t} \to \beta_2 + \frac{1}{c} \text{ a.s.,}$$

and

$$\begin{aligned}\hat{\beta}_{1t} &= \bar{y}_t - \hat{\beta}_2 \bar{x}_t \\ &= \beta_1 + (\beta_2 - \hat{\beta}_2)\bar{x}_t + \bar{\varepsilon}_t.\end{aligned}$$

Now $x_{t+1} = -c\,\Sigma_1^t \varepsilon_\tau/\tau$ converges a.s. (Fact 3, Chapter 2). Then from $\bar{x}_t = x_{t+1} + c\bar{\varepsilon}_t$, $\bar{x}_t \to -c\,\Sigma_1^\infty \varepsilon_\tau/\tau$ and $\hat{\beta}_{1t} \to \beta_1 + \Sigma_1^\infty \varepsilon_\tau/\tau$ a.s. Both $\hat{\beta}_{1t}$ and $\hat{\beta}_{2t}$ have biases.

***Example 2* (*Lai and Wei* [*1982*]).** Consider a model

$$\begin{aligned}y_t &= ay_{t-1} + bx_t + \varepsilon_t \\ &= (y_{t-1}, x_t)\begin{pmatrix} a \\ b \end{pmatrix} + \varepsilon_t, \qquad y_0 = 0,\end{aligned}$$

where ε_t are i.i.d. random variables with mean zero and variance $\sigma_1^2 > 0$ and the sequence of exogenous variables (inputs) are i.i.d. mean-zero random variables independent of εs with $Ex_t^2 = \sigma_2^2 > 0$.

By the strong law of large numbers,

$$\Sigma^t x_\tau^2 \sim t\sigma_2^2 \text{ a.s.}$$

The expression $\Sigma_1^t y_{t-1} x_\tau$ is a martingale transform and by the usual truncation method, the strong law of large numbers for martingales is applicable to produce

$$\begin{aligned}\Sigma_1^t y_{\tau-1} x_\tau &\sim o[(\Sigma_1^t y_{\tau-1}^2)^{1/2} (\ln \Sigma_1^t y_{\tau-1}^2)^{1/2+\delta}] \\ &\sim o[(\Sigma_1^t y_{\tau-1}^2)^{1/2} \ln \Sigma_1^t y_{\tau-1}^2]\end{aligned}$$

for any $\delta > 0$.

Then the determinant of

$$\mathbf{Z}_t'\mathbf{Z}_t = \Sigma_\tau \begin{bmatrix} y_{\tau-1} \\ x_\tau \end{bmatrix} [y_{\tau-1} \quad x_\tau]$$

is given by

$$\det \mathbf{Z}_t'\mathbf{Z}_t = (\Sigma y_{\tau-1}^2)(\Sigma x_\tau^2) - (\Sigma y_{\tau-1} x_\tau)^2,$$

and its trace is

$$\operatorname{tr} \mathbf{Z}_t'\mathbf{Z}_t = \Sigma y_{\tau-1}^2 + \Sigma x_\tau^2.$$

The first term dominates the second and we see that

$$\lambda_{\max}(\mathbf{Z}_t'\mathbf{Z}_t) \sim \Sigma y_{\tau-1}^2$$

and

$$\lambda_{\min}(\mathbf{Z}_t'\mathbf{Z}_t) \sim t\sigma_2^2.$$

In the extreme case of $a = 1$ and $b = 0$, $y_t = \Sigma_1^t \varepsilon_\tau$, and $\Sigma y_{\tau-1}^2$ is of the order $t^2 \ln \ln t$ because by the law of iterated logarithm,

$$\overline{\lim} \frac{y_t}{\sqrt{2\sigma_1^2 t \ln \ln t}} = 1 \text{ a.s.}$$

6.4 Bayesian Solution

One way to approach the problem is from a Bayesian point of view and formulate a minimization problem

$$\int_{-\infty}^{\infty} E_{\alpha,\beta} \sum_{t=1}^{T} (y_t - y^*)^2 \, d\pi(\alpha, \beta),$$

where $\pi(\alpha, \beta)$ is a prior distribution of α and β, where the "cost" of deviating from the optimal input is

$$y_t - y^* = \beta(x_t - \theta) + \varepsilon_t.$$

A simpler version of this problem is described shortly, and the cost of control is shown to be of the same order as that in the least squares approach. If β is known, the least squares estimate of θ is

$$\hat{\theta}_t = \bar{x}_t - \beta^{-1}(\bar{y}_t - y^*) = \theta - \bar{\varepsilon}_t/\beta,$$

where an overbar denotes the arithmetic average, e.g., $\bar{x}_t = (x_t + \dots + x_t)/t$ is the least squares estimate of θ. Then $E(\hat{\theta}_t - \theta)^2 = \sigma^2/\beta^2 t$ and $\sqrt{t}(\hat{\theta}_t - \theta)$ converges in distribution to $N(0, \sigma^2/\beta^2)$, and the "cost" of control due to deviation from θ is

$$\Sigma_1^T(x_t - \theta)^2 = (x_1 - \theta)^2 + \Sigma_1^{T-1}\bar{\varepsilon}_t^2/\beta^2 \sim \ln T/\beta^2$$

by Fact 2 in Chapter 2.

Consider

$$y_t = y_t^* + \beta(x_t - \theta) + \varepsilon_t$$

where ε_t is i.i.d. mean zero with variance σ^2 and where $\beta \neq 0, y^*$, and σ^2 are known. The problem is to minimize

$$\int_{-\infty}^{\infty} E_\theta[\Sigma_1^T(y_t - y^*)^2]\, d\pi(\theta).$$

We note that x_t depends on x^{t-1} and y^{t-1}. Since

$$E_\theta(x_t - \theta)\varepsilon_t = E\{E_\theta[(x_t - \theta)\varepsilon_t \mid x^{t-1}, y^{t-1}]\} = 0,$$

we have

$$E_\theta \Sigma_1^T(y_t - y^*)^2 = \beta^2 \Sigma_1^T E_\theta(x_t - \theta)^2 + T\sigma^2.$$

For this reason we consider an equivalent problem of minimizing

$$\Sigma_1^T E(x_t - \theta)^2.$$

The best Bayesian estimate is

$$\theta_t^* = E[\theta \mid x^t, y^t] = \left(\sigma_0^2\hat{\theta}_t + \frac{\sigma^2}{t\beta^2}\theta_0\right)\Big/(\sigma_0^2 + \sigma^2/t\beta^2)$$

where $\theta_0 = E\theta$. By direct calculation

$$\Sigma_1^T(x_t - \theta)^2 = (\theta - \theta_0)^2 + \Sigma_1^{T-1}\left[\frac{\sigma^2/t\beta^2}{\sigma_0^2 + \sigma^2/t\beta^2}(\theta - \theta_0) + \frac{\sigma_0^2}{\sigma_0^2 + \sigma^2/t\beta^2}\frac{\bar{\varepsilon}_t}{\beta}\right]^2$$

$$= \Sigma_1^T\left(\frac{\sigma_0^2}{+ \sigma^2/t\beta^2}\frac{\bar{\varepsilon}_t}{\beta}\right)^2 + 0(1).$$

Since $\Sigma_1^T \bar{\varepsilon}_t^2 \sim \ln T$, and the minimal cost under the optimal Bayesian control is under the assumption of normally and independently distributed (n.i.d.),

$$\int E_\theta \Sigma_1^T (x_t - \theta)^2 \, d\pi(\theta) \approx (\sigma^2/\beta^2) \ln T,$$

we see that the cost is the same as the optimal control problem when θ is known.

6.5 Learning Models

Models in which expectations are being revised by some rules, not necessarily optimally, as new observations become available are learning models. Some are Bayesian parameter learning or parameter adaptive models that use the Bayes rule for updating conditional probability distributions, while others employ more mechanical rules for expectation revisions as in adaptive models.

This section examines some learning models and their convergence behavior. The first type of learning models was examined by de Canino (1979) and later modified or extended by Bray (1982) and Evans (1983). In this class of models, expectations are hypothesized to be linear functions of a set of variables. This class of learning models is motivated or related to the method of the undetermined coefficient for solving rational expectations models. Expectations are revised by changing the weights assigned to the individual variables in the set by iterative schemes suggested by the rational expectations solutions.

Expectation Revisions

The method of undetermined coefficient for solving rational expectations models must first hypothesize a linear combination of some variables as the solution form then determines the values of the coefficients. (McCallum calls the variables that enter the solution state variables, but this is *not* a standard use of *state* in the systems literature.) To illustrate the method, consider the so-called Cagan model of inflation in McCallum (1983):

(11) $$m_t - p_t = \gamma - \alpha(p_{t+1|t} - p_t) + u_t, \qquad \alpha > 0$$

(12) $$m_t = \mu_0 + \mu_1 m_{t-1} + e_t, \qquad |\mu_1| < 1$$

where u_t and e_t are independent mean-zero white noise processes. The information set contains current and past values of both ps and ms. The notation is $p_{t|t-1} = E(p_t | p_{t-1}, \ldots, m_{t-1} \ldots)$. The solution

(13) $$p_t = \pi_0 + \pi_1 m_t + \pi_2 u_t$$

is hypothesized. The coefficients π_0, π_1, and π_2 are determined as follows.

Advance time by one unit and take the expectation E_t to write $p_{t+1|t}$ as

$$(14) \qquad \begin{aligned} p_{t+1|t} &= \pi_0 + \pi_1 m_{t+1|t} \\ &= \pi_0 + \pi_1(\mu_0 + \mu_1 m_t) \end{aligned}$$

where (12) and the fact that $u_{t+1|t} = 0$ are used. Now substitute (13) and (14) back into (11) and obtain

$$m_t = \gamma + \pi_0 - \alpha\pi_1\mu_0 + \pi_1[1 + \alpha(1 - \mu_1)]m_t + [1 + (1 + \alpha)\pi_2]u_t.$$

Equating coefficients of this identity, obtain

$$0 = \gamma + \pi_0 - \alpha\pi_1\mu_0,$$

$$1 = \pi_1[1 + \alpha(1 - \mu_1)],$$

and

$$0 = 1 + (1 + \alpha)\pi_2.$$

In this example, the preceding can be solved for

$$\pi_0 = -\gamma + \alpha\mu_0/[1 + \alpha(1 - \mu_1)],$$

$$\pi_1 = 1/[1 + \alpha(1 - \mu_1)],$$

and

$$\pi_2 = -1/(1 + \alpha).$$

Of course, there is no guarantee that the hypothesized (13) is the only solution form. As McCallum shows, a term $\pi_3\psi^t$ can be added to p_t without changing the values of π_0, π_1, and π_2 if $\psi = (1 + \alpha)/\alpha$.

Now consider an iterative scheme to generate $\{p_t^n\}$ and the sequences of coefficients, $\{\pi_i^n\}$, $i = 0, 1, 2$, by

$$\begin{aligned} p_t^{n+1} &= \pi_0^{n+1} + \pi_1^{n+1} m_t + \pi_2^{n+1} u_t \\ &= m_t + \alpha(p_{t+1|t}^n - p_t^n) - u_t - \gamma \end{aligned}$$

after rewriting (11) as

$$p_t = m_t + \alpha(p_{t+1|t} - p_t) - u_t - \gamma.$$

The coefficients are generated by

$$\pi_0^{n+1} = -\gamma + \alpha\mu_0\pi_1^n$$

$$\pi_1^{n+1} = 1 + \alpha\pi_1^n(\mu_1 - 1),$$

and

$$\pi_2^{n+1} = -1 - \alpha\pi_2^n.$$

These equations reduce to the algebraic equation if the limits exist as n goes to infinity. The limits exist when $|\alpha| < 1$ and $|\alpha(1 - \mu_1)| < 1$. These iterative schemes are not unique. For example, (11) can be regrouped as

$$(1 + \alpha)p_t = -\gamma + m_t - u_t + \alpha p_{t+1|t}.$$

This yields another iterative or learning scheme

$$p_t^{n+1} = -\frac{\gamma}{1+\alpha} + \frac{1}{1+\alpha}(m_t - u_t) + \frac{\alpha}{1+\alpha} p_{t+1|t}^n$$

or

$$\pi_0^{n+1} + \pi_1^{n+1} m_t + \pi_2^{n+1} u_t = -\frac{\gamma}{(1+\alpha)} + \frac{(m_t - u_t)}{(1+\alpha)}$$
$$+ \frac{\alpha}{1+\alpha}[\pi_0^n + \pi_1^n(\mu_0 + \mu_1 m_t)],$$

leading to a set of revision relations

$$\pi_0^{n+1} = -\frac{\gamma}{(1+\alpha)} + \alpha(1+\alpha)^{-1}[\pi_0^n + \mu_0 \pi_1^n],$$

$$\pi_1^{n+1} = \frac{1}{1+\alpha} + \frac{\alpha\mu_1\pi_1^n}{1+\alpha},$$

and

$$\pi_2^{n+1} = -\frac{1}{a+\alpha}.$$

Here, the learning process converges if $|\alpha/(1+\alpha)| < 1$, since $|\mu_1| < 1$. Note that the conditions for convergence are also not unique. See Chapter 8 also.

Example 3. This example is from Fourgeaud, Gourieroux, and Pradel (1984). The demand is assumed to be linear in the market price

$$d_t = -\alpha_1 P_t + \beta_1 z_{1t},$$

and the supply is proportional to the anticipated price $\hat{P}_t$ formed at time $t - 1$,

$$s_t = \alpha_2 \hat{P}_t + \beta_2 z_{2t},$$

where one period is the time span required for the production of the agricultural commodity being modeled and z_1 and z_2 are some exogeneous variables. The market-clearing condition yields the dynamic equation for the market price

(15) $$P_t = -a\hat{P}_t + u_t$$

where

$$a = \alpha_2/\alpha_1 > 0,$$

and

$$u_t = (\beta_1 z_{1t} - \beta_2 z_{2t})/\alpha_1.$$

See Aoki (1976, Sec. 12.1) for analysis for the dynamics of this market under the adaptive expectation

$$\hat{P}_t = \hat{P}_{t-1} + \lambda_t(P_{t-1} - \hat{P}_{t-1}).$$

The rational expectation model posits $\hat{P}_t = E(P_t \mid \Omega_{t-1})$ where Ω_t is the information set available to the producer of the commodity at time t. Then (15) is replaced by a rational expectation model

$$P_t = -aP_{t|t-1} + u_t. \tag{16}$$

Since

$$P_{t|t-1} = \frac{1}{1+a} u_{t|t-1},$$

the market price is determined by

$$\begin{aligned} P_t &= u_t - \frac{a}{1+a} u_{t|t-1} \\ &= \frac{1}{1+a} u_{t|t-1} + \varepsilon_t^0 \end{aligned}$$

where

$$\varepsilon_t^0 = u_t - u_{t|t-1}$$

is the martingale difference describing the expectation error in forecasting the exogenous variable u_t.

Suppose now that the producer uses a price model

$$P_t = \alpha x_{t-1} + e_t$$

where α is unknown, with some exogenous "information" variable x_{t-1} that is related to the exogenous disturbance u_t in some way to be specified later, and form $\hat{P}_t$ is

$$\hat{P}_t = \hat{\alpha}_t x_{t-1}$$

where $\hat{\alpha}_t$ is the least squares estimate

$$\hat{\alpha}_t = \sum_{s=1}^{t-1} x_{s-1} P_s \Big/ \sum_{s=1}^{t-1} x_{s-1}^2. \tag{17}$$

The information set is

$$\Omega_t = \{x_t, x_{t-1}, \ldots, P_{t-1}, P_{t-2}, \ldots\}.$$

The estimate of α is revised at time $t + 1$ to incorporate x_t and P_t. From (17), the revision rule is given by

$$\begin{aligned} \hat{\alpha}_{t+1} &= \sum_{s=1}^{t} x_{s-1} P_s \bigg/ \sum_{s=1}^{t} x_{s-1}^2 \\ &= (S_{t-2}\hat{\alpha}_t + x_{t-1}P_t)/S_{t-1} \end{aligned} \tag{18}$$

where

$$S_t = \sum_{s=1}^{t} x_s^2.$$

When the producer forms his expectation thus, the market price evolves according to

$$P_t = -a\hat{\alpha}_t x_{t-1} + u_t. \tag{19}$$

Substitute this into (18) and rewrite it as

$$\hat{\alpha}_{t+1} = [(S_{t-2} - a x_{t-1}^2)\hat{\alpha}_t + x_{t-1}u_t]/S_{t-1}.$$

Assume that the information variable is such that $c x_{t-1} = u_{t|t-1}$. Then the information variable is such that

$$u_t = c x_{t-1} + v_t$$

where

$$v_t = u_t - u_{t|t-1}.$$

Then (18) becomes

$$\begin{aligned} \hat{\alpha}_{t+1} &= \frac{(S_{t-2} - a x_{t-1}^2)}{S_{t-1}}\hat{\alpha}_t + \frac{c x_{t-1}^2}{S_{t-1}} + \frac{x_{t-1}v_t}{S_{t-1}} \\ &= \left[1 - (1 + a)\frac{S_{t-1} - S_{t-2}}{S_{t-1}}\right]\hat{\alpha}_t + \frac{c(S_{t-1} - S_{t-2})}{S_{t-1}} + \frac{x_{t-1}v_1}{S_{t-1}}. \end{aligned} \tag{20}$$

Noting that the information set at time t contains x^{t-1} we have, for sufficiently large t and assuming that $S_t \to \infty$,

$$\begin{aligned} E(\hat{\alpha}_{t+1} \mid \Omega_t) &= \left[1 - (1 + a)\frac{S_{t-1} - S_{t-2}}{S_{t-1}}\right]\hat{\alpha}_t + \frac{c x_{t-1}^2}{S_{t-1}} \\ &\leqslant \hat{\alpha}_t + c x_{t-1}^2/S_{t-1} \end{aligned}$$

since $E(v_t | \Omega_t) = 0$.* By assumption $\hat{\alpha}_t > 0$. We apply the positive supermartingale convergence theorem and conclude that $\hat{\alpha}_t$ converges to a finite positive limit a.s. if $\Sigma x_{t-1}^2 / S_{t-1}$ is bounded. The limit is $c/(1 + a)$.

Assuming this convergence for the moment, we note that (16) becomes

$$\hat{P}_t = \frac{c}{1+a} x_{t-1}$$

in the limit, i.e., in the model in which α has been learned.

From (19), we may then say that, corresponding to the converged state of the learning, the market price is given by

$$P_t^* = -\frac{ac}{1+a} x_{t-1} + u_t,$$

or since the conditional expectation is

$$\begin{aligned} P_{t|t-1}^* &= -\frac{ac}{1+a} x_{t-1} + u_{t|t-1} \\ &= -\frac{ac}{1+a} x_{t-1} + c x_{t-1} \\ &= \frac{c x_{t-1}}{1+a}, \end{aligned}$$

the market price dynamics is given by

$$P_t^* = -aP_{t|t-1}^* + u_t. \tag{21}$$

This is the same as (16), i.e., the price movements produced in the rational expectations model.

6.6 Convergence of Parameter Estimates

Take the transpose of a vector-valued ARMA model

$$\mathbf{y}_t + \mathbf{A}_1 \mathbf{y}_{t-1} + \ldots + \mathbf{A}_p \mathbf{y}_{t-p} = \mathbf{B}_0 \mathbf{u}_t + \ldots + \mathbf{B}_q \mathbf{u}_{t-q} + \mathbf{e}_t$$

and write it in the form of a regression model

$$\mathbf{y}_{t-1}' = \boldsymbol{\phi}_t \boldsymbol{\theta} + \mathbf{e}_t'$$

---

* So long as this inequality holds, S_t need not diverge.

where

$$\boldsymbol{\phi}_{t-1} = [-\mathbf{y}'_{t-1} \ldots -\mathbf{y}'_{t-p} \quad \mathbf{u}'_t \ldots \mathbf{u}'_{t-q}]$$

and

$$\boldsymbol{\theta}' = (\mathbf{A}_1 \mathbf{A}_2 \ldots \mathbf{A}_p \mathbf{B}_0 \ldots \mathbf{B}_q).$$

Using the "vec" notation, we convert it into an observation equation of a state space model

$$\mathbf{y}_t = \mathbf{C}_t \mathbf{x} + \mathbf{e}_t \tag{22}$$

where

$$\mathbf{C}_t = \mathbf{I} \otimes \boldsymbol{\phi}_t$$

and

$$\mathbf{x} = \operatorname{vec} \boldsymbol{\theta}.$$

The fact that $\mathbf{x}$ throughout the observation process remains the same is modeled by an identity

$$\mathbf{x}_{t+1} = \mathbf{x}_t. \tag{23}$$

Treating $\mathbf{x}$ as a random variable with its prior distribution $N(\hat{\mathbf{x}}_0, \mathbf{P}_0)$, the state space models (22) and (23) are in a form in which $\mathbf{x}_t$ can be estimated by a Kalman filter. The parameter estimates are sequentially generated by

$$\hat{\mathbf{x}}_t = \hat{\mathbf{x}}_{t-1} + \mathbf{K}_t(\mathbf{y}_t - \mathbf{C}_t \hat{\mathbf{x}}_{t-1})$$

where

$$\hat{\mathbf{x}}_t = E(\mathbf{x} \mid \mathbf{y}^t).$$

The optimal $\mathbf{K}_t$ has been shown to be equal to

$$\mathbf{K}_t = \boldsymbol{\Xi}_{t-1} \mathbf{C}'_t (\mathbf{R} + \mathbf{C}_t \boldsymbol{\Xi}_{t-1} \mathbf{C}'_t)^{-1} = \boldsymbol{\Xi}_t \mathbf{C}'_t \mathbf{R}^{-1},$$

where

$$\boldsymbol{\Xi}_t^{-1} = \boldsymbol{\Xi}_{t-1}^{-1} + \mathbf{C}'_t \mathbf{R}^{-1} \mathbf{C}_t, \tag{24}$$

where $\mathbf{R} = \operatorname{cov} \mathbf{e}_t$.

The dynamics for the estimates are updated by

$$\hat{\mathbf{x}}_t = (\mathbf{I} - \mathbf{K}_t \mathbf{C}_t) \hat{\mathbf{x}}_{t-1} + \mathbf{K}_t \mathbf{y}_t$$

where

$$\mathbf{I} - \mathbf{K}_t \mathbf{C}_t = \mathbf{I} - \boldsymbol{\Xi}_{t-1} \mathbf{C}'_t (\mathbf{R} + \mathbf{C}_t \boldsymbol{\Xi}_{t-1} \mathbf{C}'_t)^{-1} \mathbf{C}_t = \boldsymbol{\Xi}_t \boldsymbol{\Xi}_{t-1}^{-1}.$$

Iterate this relation back to zero to obtain

$$\hat{\mathbf{x}}_t = \boldsymbol{\Xi}_t \boldsymbol{\Xi}_0^{-1} \hat{\mathbf{x}}_0 + \boldsymbol{\Xi}_t \Sigma_1^t \mathbf{C}'_s \mathbf{R}^{-1} \mathbf{y}_s.$$

The effects of the $\hat{\mathbf{x}}_0$ vanish if $\boldsymbol{\Xi}_t \to 0$ as t goes to infinity. Iterating (24), we obtain

$$\boldsymbol{\Xi}_t^{-1} = \boldsymbol{\Xi}_0^{-1} + \sum_{s=1}^{t} \mathbf{C}_s'\mathbf{R}^{-1}\mathbf{C}_s. \tag{25}$$

Equation (25) shows that $\boldsymbol{\Xi}_t \to 0$ if and only if $\Sigma_{s=1}^{t}\, \mathbf{C}_s'\mathbf{R}^{-1}\mathbf{C}_s$ diverges as t goes to infinity. Since $\mathbf{R}$ is positive definite, there are constraints $\varepsilon_1 > 0$ and ε_2 such that

$$\boldsymbol{\varepsilon}_1 \mathbf{I} \leqslant \sum_{s=1}^{t} \mathbf{C}_s'\mathbf{C}_s \leqslant \boldsymbol{\varepsilon}_2 \mathbf{I}.$$

Note that

$$\sum_{s=1}^{t} \mathbf{C}_s'\mathbf{C}_s = \sum_{s=1}^{t} (\mathbf{I} \otimes \boldsymbol{\phi}_s'\boldsymbol{\phi}_s).$$

For every vector $\boldsymbol{\alpha}$,

$$\boldsymbol{\alpha}'\left(\sum_{s=1}^{t} \mathbf{C}_s'\mathbf{C}_s\right)\boldsymbol{\alpha} = \sum_{s=1}^{t} \boldsymbol{\alpha}'(\mathbf{I} \otimes \boldsymbol{\phi}_s'\boldsymbol{\phi}_s)\boldsymbol{\alpha} = \sum_{s=1}^{t} (\mathbf{a}'\boldsymbol{\phi}_s)^2$$

where $\boldsymbol{\alpha} = \mathbf{e} \otimes \mathbf{a}'$.

Therefore $\{\omega: \boldsymbol{\Xi}_t \to 0\} = \{\omega: \Sigma_{s=1}^{\infty}\, (\mathbf{a}'\boldsymbol{\phi}_s)^2$ diverges for every $\mathbf{a}\}$. In the weighted least squares with weight $\mathbf{R}^{-1}$, $\boldsymbol{\Xi}_0^{-1}$ in (25) is zero. To interpret this condition, take a simple univariate model $y_t = bu_{t-1} + e_t$. Then, setting a to be one in $\Sigma_{s=1}^{\infty}\, (\mathbf{a}'\boldsymbol{\phi}_s)^2$, this condition causes $\Sigma_1^{\infty}\, u_{t-1}^2$ to diverge, and it is weaker than the persistent excitation condition $\lim\, (1/N)\Sigma_1^N u_t^2 > 0$ of Åström and Bohlin (1966). (See also Lai and Wei [1982] for similar conditions.) Equation (25) can be used to assess the effects of different prior information on the model parameters. Suppose that for some $\boldsymbol{\Gamma} > 0$,

$$\boldsymbol{\Theta}_0^{-1} = \boldsymbol{\Xi}_0^{-1} + \boldsymbol{\Gamma} > \boldsymbol{\Xi}_0^{-1},$$

or by the Woodbury identity,

$$\boldsymbol{\Theta}_0 = \boldsymbol{\Xi}_0 - \boldsymbol{\Xi}_0(\boldsymbol{\Gamma}^{-1} + \boldsymbol{\Xi}_0)^{-1}\boldsymbol{\Xi}_0 \leqslant \boldsymbol{\Xi}_0.$$

Then (25) generates $\boldsymbol{\Theta}_t$ where $\boldsymbol{\Theta}_t^{-1} = \boldsymbol{\Xi}_t^{-1} + \boldsymbol{\Gamma} \geqslant \boldsymbol{\Xi}_t^{-1}$ for all $t \geqslant 0$, i.e., $\boldsymbol{\Theta}_t \leqslant \boldsymbol{\Xi}_t$ for all $t \leqslant 0$. This shows that $\boldsymbol{\Theta}_0 \leqslant \boldsymbol{\Xi}_0$ implies that $\boldsymbol{\Theta}_t \leqslant \boldsymbol{\Xi}_t$ for all $t \geqslant 0$. Next, drop the assumption on $\boldsymbol{\Gamma}$. Still they are related by

$$\boldsymbol{\Theta}_t = \boldsymbol{\Xi}_t - \boldsymbol{\Xi}_t\boldsymbol{\Gamma}(\mathbf{I} + \boldsymbol{\Xi}_t\boldsymbol{\Gamma})^{-1}\boldsymbol{\Xi}_t.$$

Thus, if $\boldsymbol{\Xi}_t \to 0$ as t goes to infinity, then so does $\boldsymbol{\Theta}_t \to 0$. Then $\hat{\mathbf{x}}_t$ converges to $\hat{\mathbf{x}}_\infty$ independent of the mean of the a priori distribution function. Since $\mathbf{x}_t$ is the mean of the normal conditional probability distribution function of

$\mathbf{x} = \operatorname{vec}\boldsymbol{\theta}$ given $\mathbf{y}^t$, $\hat{\mathbf{x}}_t = E(\mathbf{x} \mid \mathbf{y}^t)$, the martingale convergence theorem can be applied to discuss the behavior of $\hat{\mathbf{x}}_t$ as t goes to infinity. If $\mathbf{x} \in L^p$ for all real numbers $p \in (1, \infty)$, then $\hat{\mathbf{x}}_t$ converges in L^p to $E(\mathbf{x} \mid \mathbf{y}^\infty)$ where $\mathbf{y}^\infty$ stands for σ-field $\vee_t \mathscr{B}_t$ generated by the σ-fields over y^t. See Neveu (1975, p. 96). For the almost sure convergence of an integrable $\hat{\mathbf{x}}$, see Neveu (1975, p. 104).

Strong Consistency of Least Squares Estimates

Let $\{\varepsilon_t\}$ be a martingale difference sequence such that $E_t\varepsilon_{t+1}^2 = \sigma^2$ and define

$$z_t = \Sigma_1^t c_s \varepsilon_s .$$

Then $\{z_t\}$ is a square-integrable martingale and $\{z_t^2\}$ is a positive integrable submartingale since by Schwarz inequality,

$$z_t^2 = (E_t z_{t+1})^2 \leqslant E_t z_{t+1}^2 .$$

The process $\{z_t^2\}$ has the Doob decomposition

$$z_t^2 = M_t + A_t$$

where

$$\begin{aligned} A_{t+1} - A_t &= E_t z_{t+1}^2 - z_t^2 \\ &= E_t(z_{t+1} - z_t)^2 \\ &= E_t(c_{t+1}\varepsilon_{t+1})^2 \\ &= c_{t+1}^2\sigma^2, \\ A_0 &= 0. \end{aligned}$$

Thus $A_t = \sigma^2 \Sigma_1^t c_s^2$. We know from Neveu (1975, pp. 149, 151) that (1) if $A_\infty < \infty$, then z_t converges in L^2; (2) the martingale z_t converges a.s. to a finite limit in the event $\{A_\infty < \infty\}$, i.e., assumption (1) is satisfied; and (3) in the event $\{A_\infty = \infty\}$,

$$z_t = o[f(A_t)] \text{ a.s.}$$

for every increasing function $f: R_+ \to R_+$ with a finite integral $\int_0^\infty [1 + f(t)]^2 \, dt < \infty$. $f(t) = t^{(1+\varepsilon)/2}$ and $[t(\ln^+ t)^{1+\varepsilon}]^{1/2}$, $\varepsilon > 0$ are examples of such f.

For a least square problem

$$y_t = x_t\theta + \varepsilon_t$$

where θ is a scalar and x_t is nonrandom, the least squares estimate is equal to

$$\begin{aligned} \theta_t &= \Sigma_1^t x_s y_s / \Sigma_1^t x_s^2 \\ &= \theta + \Sigma_1^t x_s \varepsilon_s / \Sigma_1^t x_s^2 . \end{aligned}$$

Now take

$$z_t = \Sigma_1^t x_s \varepsilon_s.$$

Then

$$A_t = \sigma^2 \Sigma_1^t x_s^2.$$

Thus in the event $\{A_\infty = \infty\}$, i.e. $\{\Sigma_0^\infty x_s^2 = \infty\}$, then

$$\frac{z_T}{\sqrt{A_T \ln A_T^{1+\varepsilon}}} \to 0 \text{ a.s.}$$

Since

$$\frac{z_T}{A_T} = \frac{z_T}{\sqrt{A_T \ln A_t^{1+\varepsilon}}} \sqrt{\frac{\ln A_T^{1+\varepsilon}}{A_T}}$$

and

$$\ln A_T^{1+\varepsilon}/A_T \to 0 \text{ a.s. } A_T \to \infty,$$

$$z_T/A_T \to 0 \text{ a.s.}$$

For vector-valued parameter cases, Lai, Robbins, and Wei (1979) establish strong consistency of least squares estimate of $y_t = \mathbf{x}_t' \boldsymbol{\beta}_t + \varepsilon_t$, $t = 1, 2, \ldots$ for εs such that (i) $\Sigma_1^\infty c_t \varepsilon_t$ converges a.s. for all real sequences $\{c_t\}$ such that $\Sigma_1^\infty c_t^2 < \infty$ and (ii) for nonrandom xs such that $V_T = \Sigma_{t=1}^T x_t x_t'$ is nonsingular for sufficiently large T and $V_t^{-1} \to 0$ as T goes to infinity. Lai, Robbins, and Wei (1979) also show that for $\{\varepsilon_t\}$, Gaussian with covariance function r_k satisfying $\Sigma |r_k| < \infty$, $V^{-1} \to 0$ implies that $\theta_t \to \theta$ a.s. They also show that if $\{\varepsilon_t\}$ is a uniformly bounded and strongly mixing sequence with $\Sigma_1^\infty k\phi_k < \infty$ where ϕ_k is the mixing coefficient, then $\theta_k \to \theta$ a.s.

Exercise

Given $z_t = az_{t-1} + be_{t-1}$, $|a| < 1$ where e_t is a martingale difference sequence with $E(e_t^2 \mid \mathscr{F}_{t-1}) = \sigma^2 = \text{const}$ and $E(|e_t|^\alpha \mid \mathscr{F}_{t-1}) < \infty$ a.s. for some $\alpha > 2$,

a) Show that $T^{-1}\Sigma_{t=1}^T e_t z_t \to 0$ a.s.

b) Show that $T^{-1/2}\Sigma e_t z_{t-1}$ converges a.s. to a mean zero finite variance normal random variable.

6.7 Ljung's Method

Ljung (1977) proposed an intuitively very appealing method that associates certain deterministic differential equations with iterative adjustment schemes of stochastic gradient and Newton-Raphson variety. Consider a typical rule

$$\hat{\boldsymbol{\theta}}_t = \hat{\boldsymbol{\theta}}_{t-1} + \gamma_t \mathbf{R}_t^{-1} \boldsymbol{\psi}_t \varepsilon_t$$

in which ψ_t denotes a gradient search direction, γ_t is a gain, and $\mathbf{R}_t$ is a Newton step

$$\mathbf{R}_t = \mathbf{R}_{t-1} + \gamma_t(\psi_t\psi_t' - \mathbf{R}_{t-1}).$$

To analyze asymptotic behavior of this algorithm, note that on the average, the updating direction would be $f(\boldsymbol{\theta}) = E(\psi_t\varepsilon_t)$, and the algorithm would behave relative to a new time scale τ according to

$$d\boldsymbol{\theta}/d\tau \simeq \mathbf{R}^{-1}(\tau)\mathbf{f}[\boldsymbol{\theta}(\tau)]$$

and

$$\frac{d\mathbf{R}(\tau)}{d\tau} \cong \mathbf{G}[\boldsymbol{\theta}(\tau)] - \mathbf{R}(\tau)$$

where

$$\mathbf{G}(\boldsymbol{\theta}) = E(\psi_t\psi_t').$$

Goodwin and Sin (1984, p. 335) provide heuristic reasonings why these equations may determine the asymptotic behavior. To justify the method rigorously and find technical conditions under which the method can be used, see Kushner (1984, Chapter 5). For example,

$$\hat{\theta}^{\varepsilon}_{t+1} = \hat{\theta}^{\varepsilon}_{t} + \varepsilon G(\hat{\theta}^{\varepsilon}_{t}, \xi^{\varepsilon}_{t})$$

has the limit differential equation

$$\dot{\theta} = \int P^{\theta}(d\xi)G(x, \xi) = \bar{G}(x)$$

and the limit of the projection process that projects back onto it under a set of technical conditions that are outside the scope of this book to discuss. Marcet and Sargent (1987) have applied this procedure to a number of examples including Bray (1983), Bray and Savin (1986), and Fourgeaud, Gourieroux, and Pradel (1986).

Ljung (1977) and Tsypkin et al. (1981) assumed that V_t^{-1}/t converges a.s. to a positive definite matrix in establishing the convergence of the AML method for ARMAX models by stability analysis of an associated ordinary differential equation. This assumption is stronger than that used by Lai and Wei (1986).

Positive Real Condition

In model reference adaptive control schemes, the difference in the outputs of an unknown true system with parameter vector $\boldsymbol{\theta}_0$ and its model with parameter $\boldsymbol{\theta}$ is used to adjust the model parameter vectors. Let

$$y_t = \boldsymbol{\theta}_0'\psi_t + \omega_t$$

where $\boldsymbol{\psi}_t = (-y_{t-1}, \ldots -y_{t-n}, u_{t-2} \ldots u_{t-m})$ and where $\eta_t = \boldsymbol{\theta}'\boldsymbol{\phi}_t$ and where $\boldsymbol{\phi}_t = (-y_{t-1}, \ldots, -y_{t-n}, u_{t-m})$ since the same exogenous signal is applied to the true system and its model in the usual translation of ARMAX models into stochastic regression format. Denote the difference in outputs by

$$v_t = y_t - \eta_t.$$

Since

$$\begin{aligned} v_t &= \boldsymbol{\theta}_0'\boldsymbol{\psi}_t - \boldsymbol{\theta}'\boldsymbol{\phi}_t + w_t \\ &= \boldsymbol{\psi}_t'\boldsymbol{\theta}_0 - \boldsymbol{\phi}_t'\boldsymbol{\theta}_0 + \boldsymbol{\phi}_t'(\boldsymbol{\theta}_0 - \boldsymbol{\theta}) + \omega_t \\ &= (\boldsymbol{\psi}_t - \boldsymbol{\phi}_t)'\boldsymbol{\theta}_0 + \boldsymbol{\phi}_t'(\boldsymbol{\theta}_0 - \boldsymbol{\theta}) + \omega_t \end{aligned}$$

where from definition

$$\boldsymbol{\psi}_t - \boldsymbol{\phi}_t = (-v_{t-1} \ldots -v_{t-n}, 0 \ldots 0)$$

and

$$(\boldsymbol{\psi}_t - \boldsymbol{\phi}_t)'\boldsymbol{\theta}_0 = [1 - A(q^{-1})]v_t(\boldsymbol{\theta}),$$

we obtain

(26) $$A(q^{-1})v_t(\boldsymbol{\theta}) = \boldsymbol{\phi}_t(\boldsymbol{\theta})'(\boldsymbol{\theta}_0 - \boldsymbol{\theta}) + \omega_t.$$

Parameter values are adjusted by first generating an error signal by

(27) $$\varepsilon_t(\boldsymbol{\theta}) = D(q^{-1})v_t(\boldsymbol{\theta})$$

where

$$D(q^{-1}) = 1 + d_1 q^{-1} + \ldots + d_n q^{-n}$$

is constrained as described below and by

(28) $$\begin{cases} \hat{\boldsymbol{\theta}}_t = \boldsymbol{\theta}_{t-1} + (1/t)\mathbf{R}_t^{-1}\boldsymbol{\phi}_t\varepsilon_t(\theta) \\ \mathbf{R}_t = \mathbf{R}_{t-1} + (1/t)(\boldsymbol{\phi}_t\boldsymbol{\phi}_t' - \mathbf{R}_{t-1}). \end{cases}$$

From (26) and (27), ε_t is governed by

$$\varepsilon_t(\theta) = \frac{D(q^{-1})}{A(q^{-1})}\boldsymbol{\phi}_t'(\boldsymbol{\theta})(\boldsymbol{\theta}_0 - \boldsymbol{\theta}) + \frac{D(q^{-1})}{A(q^{-1})}\omega_t.$$

Define a transfer function by

$$H(q^{-1}) = \frac{D(q^{-1})}{A(q^{-1})}.$$

The transfer function $D(q^{-1})$ in (27) is constrained to produce $H(q^{-1})$, which is the transfer function of an asymptotically stable dynamic relation. This condition on $H(q^{-1})$ is the positive realness condition on $H(q^{-1})$ used initially by Landau (1976) and then by Ljung (1977). Imposing thus the asymptotic stability on $H(q^{-1})$ ensures that ε_t is a weakly stationary stochastic process when ω_t is a weakly stationary stochastic process. Since the updating direction $\phi_t\varepsilon_t(\theta)$ in (28) is equal to

$$\boldsymbol{\phi}_t\varepsilon_t(\boldsymbol{\theta}) = \boldsymbol{\phi}_t\tilde{\boldsymbol{\theta}}_t'(\boldsymbol{\theta}_0\boldsymbol{\theta}_0) + \boldsymbol{\phi}_t\omega_t$$

where

$$\tilde{\boldsymbol{\phi}}_t(\boldsymbol{\theta}) = H(q^{-1})\boldsymbol{\phi}_t(\boldsymbol{\theta}),$$

the asymptotic "average" will be "equal" to

$$E[\boldsymbol{\phi}_t\varepsilon_t(\boldsymbol{\theta})] = \tilde{\mathbf{G}}(\boldsymbol{\theta}) = (\boldsymbol{\theta} - \boldsymbol{\theta}_0)$$

where

$$\tilde{\mathbf{G}}(\boldsymbol{\theta}) = E(\boldsymbol{\phi}_t\tilde{\boldsymbol{\phi}}_t')$$

since ϕ_t and ω_t will be uncorrelated. This is the heuristic reasoning behind the ordinary differential equations

$$d\boldsymbol{\theta}/d\tau = \mathbf{R}(\boldsymbol{\theta})^{-1}f(\boldsymbol{\theta})$$

and

$$d\mathbf{R}/d\tau = \mathbf{G}(\theta) - \mathbf{R}(\theta).$$

Analogous reasoning applied to the extended least squares scheme produces $H(q^{-1}) = 1/C(q^{-1})$ where $C(q^{-1}) = 1 + c_1q^{-1} + \ldots + c_nq^{-n}$ is in the true data-generating scheme $A(q^{-1})y_t = \beta(q^{-1})u_t$. The positive realness of $H(q^{-1})$ in this case is the same as

$$\text{Re}\,1/C(e^{i\theta}) > \tfrac{1}{2} \Leftrightarrow |1 - C(e^{i\theta})| < 1$$

for all $\theta \in [0, 2\pi]$.

Appendix: Representation for Residuals

Solo (1981) obtained an explicit representation of the sequence of the least squares residuals in terms of the original disturbances in the regression model. The model is

$$y_t = \boldsymbol{\theta}'\mathbf{x}_t + \varepsilon_t,$$

and the least squares estimate (l.s.e.) is

(A1) $$\hat{\boldsymbol{\theta}}_t = \mathbf{V}_t^{-1}\Sigma_1^t\mathbf{x}_\tau y_\tau = \boldsymbol{\theta} + \mathbf{V}_t^{-1}\Sigma_1^t\mathbf{x}_\tau\varepsilon_\tau,$$

where

$$\mathbf{V}_t = \Sigma_1^t\mathbf{x}_\tau\mathbf{x}_\tau'.$$

The sequence of residuals is defined by

(A2) $$e_t = y_t - \hat{\boldsymbol{\theta}}_{t-1}'\mathbf{x}_t.$$

It is a linear combination of $\varepsilon_t, \varepsilon_{t-1} \ldots$, as can be seen from (A1) and (A2)

$$\text{(A3)}\qquad \begin{aligned} e_t &= \boldsymbol{\theta}'\mathbf{x}_t + \varepsilon_t - \hat{\boldsymbol{\theta}}'_{t-1}\mathbf{x}_t \\ &= \varepsilon_t + (\boldsymbol{\theta} - \hat{\boldsymbol{\theta}}_{t-1})'\mathbf{x}_t \\ &= \varepsilon_t - \mathbf{x}_t' V_{t-1}^{-1}\Sigma_1^t \mathbf{x}_t \varepsilon_\tau \\ &= \Sigma_1^t a_{t,\tau}\varepsilon_\tau \end{aligned}$$

where

$$a_{t,t} = 1, \qquad a_{t,\tau} = -\mathbf{x}_t'\mathbf{V}_{t-1}^{-1}\mathbf{x}_\tau .$$

By direct calculation, the sequence of residuals is seen to be uncorrelated if εs are. For $t > s$,

$$\begin{aligned} E(e_t e_s') &= \sigma^2 \Sigma_1^s a_{t,\tau} a_{\tau,s}' \\ &= \sigma^2 a_{t,s} + \Sigma_1^{s-1} a_{t,\tau} a_{\tau,s}' \\ &= \sigma^2 - \mathbf{x}_t'\mathbf{V}_{t-1}^{-1}\mathbf{x}_s + \mathbf{x}_t'\mathbf{V}_{t-1}^{-1}\Sigma_1^{s-1}(\mathbf{x}_\tau\mathbf{x}_\tau')\mathbf{V}_{s-1}^{-1}\mathbf{x}_s \\ &= 0. \end{aligned}$$

The variance is

$$E(e_t e_t') = \sigma^2(1 + \mathbf{x}_t'\mathbf{V}_{t-1}^{-1}\mathbf{x}_t).$$

The l.s.e. satisfies the recursion

$$\begin{aligned} \hat{\boldsymbol{\theta}}_t &= \mathbf{V}_t^{-1}\Sigma_1^{t-1}\mathbf{x}_\tau y_\tau \\ &= \mathbf{V}_t^{-1}\mathbf{V}_{t-1}\mathbf{V}_{t-1}^{-1}\Sigma_1^{t-1}\mathbf{x}_\tau y_\tau + \mathbf{V}_t^{-1}\mathbf{x}_t y_t \\ &= \mathbf{V}_t^{-1}\mathbf{V}_{t-1}\hat{\boldsymbol{\theta}}_{t-1} + \mathbf{V}_t^{-1}\mathbf{x}_t(y_t - \mathbf{x}_t'\hat{\boldsymbol{\theta}}_{t-1}) + \mathbf{V}_t^{-1}\mathbf{x}_t\gamma_t\mathbf{x}_t'\hat{\boldsymbol{\theta}}_{t-1} \\ &= \mathbf{V}_t^{-1}(\mathbf{V}_{t-1} + \mathbf{x}_t\mathbf{x}_t')\hat{\boldsymbol{\theta}}_{t-1} + \mathbf{V}_t^{-1}\mathbf{x}_t e_t \\ &= \hat{\boldsymbol{\theta}}_{t-1} + \mathbf{V}_t^{-1}\mathbf{x}_t e_t . \end{aligned}$$

For any vector $\alpha \neq 0$, let $\boldsymbol{\alpha}'\hat{\boldsymbol{\theta}}_t = s_t$. Then

$$s_t = s_{t-1} + c_t e_t$$

where

$$\text{(A4)}\qquad c_t = \boldsymbol{\alpha}'\mathbf{V}_t^{-1}\mathbf{x}_t .$$

From (A3), given a sequence of constants $\{c_t\}$,

$$\begin{aligned} \sum_{t=1}^{T} c_t e_t &= \sum_{t=1}^{T} \Sigma_1^t c_t a_{t,\tau}\varepsilon_\tau \\ &= \sum_{t=1}^{T} b_{t,\tau}\varepsilon_\tau \end{aligned}$$

where

$$b_{t,\tau} = \sum_{t=\tau}^{T} c_t a_{t,\tau}.$$

Solo (1981) treats a stationary sequence of $\{\varepsilon_t\}$ with autovariance γ_l and the spectrum $f(\omega)$ that is bounded by a constant K a.s. He shows that $\Sigma_1^t \varepsilon_\tau$ converges a.s. if $\Sigma_1^\infty c_s^2(1 + \mathbf{x}_s'\mathbf{V}_{s-1}^{-1}\mathbf{x}_s)(\ln s)^2 < \infty$. When εs are orthogonal, this is the condition of Rademacher-Menchoff fundamental convergence theorem (Stout [1974, p. 20]).

CHAPTER 7

Adaptive Control Systems and Optimal Bayesian Control Policies

In Chapter 3, we developed our method for deriving optimal control policies with complete information, i.e., either the dynamics are known and deterministic or the transition probability distributions are known. In this chapter, we follow the same line of attack to discuss optimal control problems under imperfect information, i.e., for a class of dynamic systems where some of the key items of information on the optimal control problems are not known exactly. More specifically, we discuss parameter adaptive control or decision problems where "imperfectness" of information is modeled as uncertain values of some parameter vectors in model or noise process specification. By assumption there is at least one unknown parameter in the problem description, i.e., we asume that at least one unknown parameter is contained in the collection of dynamic transition equation, the observation equation, various probability densities of the noises, initial conditions, and/or in the descriptions of the inputs. The unknown parameter is assumed to be a member of a known compact parameter space. Later in the chapter we shall suitably augment state vectors to endow them with the Markovian properties. Although this augmentation renders only the subset of the state vector components observable, we can still employ the method of Chapter 3 to this partially observed Markov processes. We generalize the procedure introduced in Chapter 3 of deriving optimal decision rules by using recursively generated conditional probabilities.

The main recursion equations for optimal control policies for stochastic and adaptive systems are identical for these two classes of systems. The slight differences are in the auxiliary recursion equations for conditional probability densities for these two classes. The only quantities that are not immediately available and must be generated recursively as the conditional probability densities that are $p(\mathbf{x}_t \mid \mathbf{y}^t, \mathbf{u}^{t-1})$ in the case of purely stochastic systems and are $p(\mathbf{x}_t, \boldsymbol{\theta} \mid \mathbf{y}^t, \mathbf{u}^{t-1})$ in the parameter adaptive systems where $\boldsymbol{\theta}$ is a vector of all unknown parameters in problem descriptions or probability distributions. Other probability densities needed in computing γs are immediately available from the transition and observation equations and from the assumed probability distribution functions of noises and/or random system parameters.

7.1 Problem Statement

A model is now assumed to be described by the dynamic equation

(1) $$\mathbf{x}_{t+1} = \mathbf{f}_t(\mathbf{x}_t, \mathbf{u}_t, \xi_t, \boldsymbol{\alpha}), \qquad t = 0, 1, \ldots, N-1$$

and is observed by $\mathbf{y}_t$,

(2) $$\mathbf{y}_t = \mathbf{h}_t(\mathbf{x}_t, \boldsymbol{\eta}_t, \boldsymbol{\beta}),$$

where $\mathbf{x}_t, \mathbf{u}_t, \xi_t$, and $\boldsymbol{\eta}_t$ are the same vectors as before and where $\boldsymbol{\alpha}$ and $\boldsymbol{\beta}$ are parameter vectors of Eqs. (1) and (2),* where $\boldsymbol{\alpha} \in \Theta_\alpha$ and $\boldsymbol{\beta} \in \Theta_\beta$, and where Θ_α and Θ_β are parameter spaces assumed given. The random vectors ξ and $\boldsymbol{\eta}$ are assumed to be such that their joint probability density $p(\xi^t, \boldsymbol{\eta}^t \mid \boldsymbol{\theta}_1, \boldsymbol{\theta}_2)$ is given, $\boldsymbol{\theta}_1 \in \Theta_1$, and $\boldsymbol{\theta}_2 \in \Theta_2$, where Θ_1 and Θ_2 are known parameter spaces. When each of these parameter spaces contains a single element, all the parameter values are therefore known, and the problem will reduce to a purely stochastic or deterministic one. Therefore, at least one parameter space is assumed to contain more than a single element in this chapter. The method of this chapter generalizes that in Chapter 3 and applies equally well to situations where the probability distribution functions of the model parameters α and/or β contain unknown parameters $\boldsymbol{\theta}_\alpha$ and $\boldsymbol{\theta}_\beta$, $\boldsymbol{\theta}_\alpha \in \Theta_\alpha$, $\boldsymbol{\theta}_\beta \in \Theta_\beta$, with Θ_α and Θ_β known with unknown mean. Since both types of systems can be treated by the method of this chapter, no careful distinction

---

* These parameters could be time-varying. For the simplicity of exposition, they are assumed to be time-invariant in this chapter.

will be made in deriving their optimal control policies. A priori probability density functions are assumed to be given for each and all unknown parameters as well as for the initial state vector $\mathbf{x}_0$. The distribution function for the initial state vector $\mathbf{x}_0$ may contain an unknown parameter $\boldsymbol{\theta}_3$, while $\boldsymbol{\theta}_3 \in \Theta_3$ is assumed known. Known deterministic inputs and/or disturbances to the systems are assumed to be incorporated in (1) and (2). If inputs to the systems are assumed to be stochastic, then the probability density function of such an input may also contain a parameter vector that may be unknown, $\boldsymbol{\mu} \in \Theta_\mu$, where Θ_μ is assumed given. The symbol $\mathbf{d}_t(\boldsymbol{\mu})$, $t = 0, 1, \ldots, N$, is used to denote a sequence of such random inputs. The actual inputs to the systems are denoted by $\mathbf{z}_t$ and are assumed to be related to $\mathbf{d}_t$ by

$$\mathbf{z}_t = \mathbf{K}_t(\mathbf{d}_t, \zeta_t)$$

where ζ_t are random noise with known probability densities. The random variables $\mathbf{z}_t$ are assumed to be observed. Therefore, the observation data at any time k consists of $\mathbf{z}^k$, $\mathbf{y}^k$, and $\mathbf{u}^{k-1}$. The criterion function J is taken to be essentially the same as before

$$J = \sum_{k=1}^{N} W_t(\mathbf{x}_t, \mathbf{d}_t, \mathbf{u}_{t-1}).$$

It is desired to find an optimal closed-loop control policy that minimizes the conditional future cost function(al). Control variables $\mathbf{u}$ to be considered in minimizing the conditional cost depend only on the initially available information plus past and current observations and past controls and do not depend explicitly on unknown parameter values. As in Chapter 3, optimal control policies are nonrandomized, i.e., each control $\mathbf{u}_t$ is specified to be a definite function of observed data only. For simpler notation, we drop $\mathbf{d}_t$ from the arguments in W_t.

The basic idea for the optimal control problem in this general formulation is well conveyed by examining a class of problems where $\boldsymbol{\theta}_1$ and $\boldsymbol{\theta}_2$, where $\boldsymbol{\theta}_1 \in \Theta_1$ and $\boldsymbol{\theta}_2 \in \Theta_2$ are the only parameter vectors assumed unknown. Problems with unknown $\boldsymbol{\alpha}$ and/or $\boldsymbol{\beta}$ or a control problem where the distribution for the initial state vector $\mathbf{x}_0$ contains an unknown parameter can be treated quite analogously. Only examples are discussed in order to avoid repetitions of general formulations.

We consider only nonrandomized control policies. The arguments exactly parallel those in Chapter 3. We just record formulas for the tth stage.

General Case

Proceeding recursively, it is now easy to see that an optimal control function at time t, u_t, is determined by

$$\min_{\mathbf{u}_t} E(W_{t+1} + W_{t+2}^* + \dots + W_N^* \mid \mathbf{y}^t)$$
$$= \min_{u_t} \int \left[\lambda_{t+1} + \int \gamma_{t+2}^* p(\mathbf{y}_{t+1} \mid \mathbf{u}^t, \mathbf{y}^t)\, d\mathbf{y}_{t+1} \right]$$

where

(3) $$\gamma_{t+1} \triangleq \lambda_{t+1} + \int \gamma_{t+2}^* p(\mathbf{y}_{t+1} \mid \mathbf{u}^t, \mathbf{y}^t)\, d\mathbf{y}_t$$

and where

(4) $$\lambda_{t+1} \triangleq \int W_{t+1}(\mathbf{x}_{t+1}, \mathbf{u}_t) p(\mathbf{x}_{t+1} \mid \mathbf{x}_t, \mathbf{u}_t, \boldsymbol{\theta}_1) p(\mathbf{x}_t, \boldsymbol{\theta}_1 \mid \mathbf{y}^t)\, d(\mathbf{x}_t, \mathbf{x}_{t+1}, \boldsymbol{\theta}_1),$$

where $\mathbf{u}_t^*$ minimizes γ_{t+1}:

(5) $$\min_{\mathbf{u}_t \in U_t} \gamma_{t+1} = \gamma_{t+1}^*, \qquad t = 0, \dots, N-1.$$

Thus, by computing λs and γ^*s recursively, the optimal control policy is obtained. The density functions $p(\mathbf{x}_t, \boldsymbol{\theta}_1 \mid \mathbf{y}^t)$ and $p(\mathbf{y}_{t+1} \mid \mathbf{u}^t, \mathbf{y}^t)$, which are needed in computing λs and γ^*s, are derived next.

Remarks similar to those made in Chapter 3 can be made here about the possible simplifications of the recursion formula and of the control policy implementations when sufficient statistics exist.

Recursion Formula for Conditional Probability Densities

We have shown that, if $p(\mathbf{x}_t, \boldsymbol{\theta}_1 \mid \mathbf{y}^t)$ and $p(\mathbf{y}_{t+1} \mid \mathbf{y}^t, \mathbf{u}^t)$, $t = 0, \dots, N-1$, are available, then Eqs. (3), (4), and (5) allow us to derive an optimal control policy as a nonrandomized sequence of control vectors $\mathbf{u}_0^*, \dots, \mathbf{u}_{N-1}^*$. We now generate these conditional probability density functions recursively.

Since it is easier to obtain $p(\mathbf{x}_t, \boldsymbol{\theta}_1, \boldsymbol{\theta}_2 \mid \mathbf{y}^t, \mathbf{u}^{t-1})$ recursively, this conditional density is generated instead of $p(\mathbf{x}_t, \boldsymbol{\theta}_1 \mid \mathbf{y}^t, \mathbf{u}^{t-1})$. These two densities are related by

$$p(\mathbf{x}_t, \boldsymbol{\theta}_1 \mid \mathbf{y}^t, \mathbf{u}^{t-1}) = \int p(\mathbf{x}_t, \boldsymbol{\theta}_1, \boldsymbol{\theta}_2 \mid \mathbf{y}^t, \mathbf{u}^{t-1})\, d\boldsymbol{\theta}_2.$$

From now on we drop $\mathbf{u}^{t-1}$ from the conditioning variables for simpler notations.

We will now derive $p(\boldsymbol{\theta}_1, \boldsymbol{\theta}_2, \mathbf{x}_{t+1} \mid \mathbf{y}^{t+1})$, assuming that $p(\boldsymbol{\theta}_1, \boldsymbol{\theta}_2, \mathbf{x}_t \mid \mathbf{y}^t)$ is available. For this purpose, consider $p(\boldsymbol{\theta}_1, \boldsymbol{\theta}_2, \mathbf{x}_t, \mathbf{x}_{t+1}, \mathbf{y}_{t+1} \mid \mathbf{y}^t)$. It can be written by applying the chain rule

$$
\begin{aligned}
(6) \quad & p(\boldsymbol{\theta}_1, \boldsymbol{\theta}_2, \mathbf{x}_t, \mathbf{x}_{t+1}, \mathbf{y}_{t+1} \mid \mathbf{y}^t) \\
&= p(\boldsymbol{\theta}_1, \boldsymbol{\theta}_2, \mathbf{x}_t \mid \mathbf{y}^t) p(\mathbf{x}_{t+1} \mid \boldsymbol{\theta}_1, \boldsymbol{\theta}_2, \mathbf{x}_t, \mathbf{y}^t) p(\mathbf{y}_{t+1} \mid \boldsymbol{\theta}_1, \boldsymbol{\theta}_2, \mathbf{x}_t, \mathbf{x}_{t+1}, \mathbf{y}^t) \\
&= p(\boldsymbol{\theta}_1, \boldsymbol{\theta}_2, \mathbf{x}_t \mid \mathbf{y}^t) p(\mathbf{x}_{t+1} \mid \boldsymbol{\theta}_1, \mathbf{x}_t, \mathbf{u}^t) p(\mathbf{y}_{t+1} \mid \boldsymbol{\theta}_2, \mathbf{x}_{t+1})
\end{aligned}
$$

where the independence assumptions on ξs and $\boldsymbol{\eta}$s are used to obtain the expression on the last line.

In (6), the first term of the right-hand side is assumed to be given, and the other two terms are computed from the model dynamics, the observation equation, and the assumed density functions for $p(\xi_t \mid \boldsymbol{\theta}_1)$ and $p(\boldsymbol{\eta}_{t+1} \mid \boldsymbol{\theta}_2)$.

Now integrate the left-hand side of (6) with respect to $\mathbf{x}_t$ to obtain

$$
\begin{aligned}
(7) \quad \int p(\boldsymbol{\theta}_1, \boldsymbol{\theta}_2, \mathbf{x}_t, \mathbf{x}_{t+1}, \mathbf{y}_{t+1} \mid \mathbf{y}^t)\, d\mathbf{x}_t &= p(\boldsymbol{\theta}_1, \boldsymbol{\theta}_2, \mathbf{x}_{t+1}, \mathbf{y}_{t+1} \mid \mathbf{y}^t) \\
&= p(\mathbf{y}_{t+1} \mid \mathbf{y}^t) p(\boldsymbol{\theta}_1, \boldsymbol{\theta}_2, \mathbf{x}_{t+1} \mid \mathbf{y}^{t+1}).
\end{aligned}
$$

Hence, from (6) and (7), the desired recursion equation is obtained as

$$
\begin{aligned}
(8) \quad p(\boldsymbol{\theta}_1, \boldsymbol{\theta}_2, \mathbf{x}_{t+1} \mid \mathbf{y}^{t+1}) &= \int p(\boldsymbol{\theta}_1, \boldsymbol{\theta}_2, \mathbf{x}_t \mid \mathbf{y}^t) p(\mathbf{x}_{t+1} \mid \boldsymbol{\theta}_1, \mathbf{x}_t, \mathbf{u}_t)\, d\mathbf{x}_t \\
&\quad \times \frac{p(\mathbf{y}_{t+1} \mid \boldsymbol{\theta}_2, \mathbf{x}_{t+1})}{\int [\text{numerator}]\, d(\mathbf{x}_{t+1}, \boldsymbol{\theta}_1, \boldsymbol{\theta}_2)}
\end{aligned}
$$

where the denominator gives $p(\mathbf{y}_{t+1} \mid \mathbf{y}^t)$. The probability density $p(\boldsymbol{\theta}_1, \mathbf{x}_t \mid \mathbf{y}^t)$ is obtained by integrating (8) with respect to $\boldsymbol{\theta}_2$. By the repeated applications of (8), $p(\boldsymbol{\theta}_1, \boldsymbol{\theta}_2, \mathbf{x}_t \mid \mathbf{y}^t)$, $t = 0, 1, \ldots, N-1$, are computed recursively, starting from

$$
\begin{aligned}
(9) \quad p(\boldsymbol{\theta}_1, \boldsymbol{\theta}_2, \mathbf{x}_0 \mid \mathbf{y}^0) &= \frac{p(\boldsymbol{\theta}_1, \boldsymbol{\theta}_2, \mathbf{x}_0, \mathbf{y}_0)}{p(\mathbf{y}_0)} \\
&= \frac{p_0(\boldsymbol{\theta}_1, \boldsymbol{\theta}_2, \mathbf{x}_0) p(\mathbf{y}_0 \mid \boldsymbol{\theta}_2, \mathbf{x}_0)}{\int p_0(\boldsymbol{\theta}_1, \boldsymbol{\theta}_2, \mathbf{x}_0) p(\mathbf{y}_0 \mid \boldsymbol{\theta}_2, \mathbf{x}_0)\, d(\boldsymbol{\theta}_1, \boldsymbol{\theta}_2, \mathbf{x}_0)}
\end{aligned}
$$

where $p_0(\boldsymbol{\theta}_1, \boldsymbol{\theta}_2, \mathbf{x}_0)$ is the given a priori joint density of $\boldsymbol{\theta}_1, \boldsymbol{\theta}_2$, and $\mathbf{x}_0$.

Note that

$$p(\boldsymbol{\theta}_1, \mathbf{x}_t \mid \mathbf{y}^t) = p(\boldsymbol{\theta}_1 \mid \mathbf{y}^t) p(\mathbf{x}_t \mid \boldsymbol{\theta}_1, \mathbf{y}^t)$$

where $p(\boldsymbol{\theta}_1 \mid \mathbf{y}^t)$ is the tth a posteriori probability density of $\boldsymbol{\theta}_1$ given $\mathbf{y}^t$, i.e., given all past and current observations on the systems state vectors $\mathbf{y}_0, \ldots, \mathbf{y}_t$.

If the state vectors happen to be perfectly observable, i.e., if the observation equation is noise free and can be solved for $\mathbf{x}$ uniquely, then instead of $p(\boldsymbol{\theta}_1, \mathbf{x}_t \mid \mathbf{y}^t)$, we use simply $p(\boldsymbol{\theta}_1 \mid \mathbf{x}^t)$ in the preceding operation.

We can derive a similar recursive relationship for $p(\mathbf{x}_t \mid \mathbf{y}^t)$ or $p(\xi_t, \mathbf{x}_t \mid \mathbf{y}^t)$ or, more generally, for $p(\xi_t, \boldsymbol{\eta}_t, \mathbf{x}_t \mid \mathbf{y}^t)$, $t = 0, \ldots, N-1$. In some problems, the latter is more convenient in computing $E(J \mid \mathbf{y}^t)$. For example, the conditional density $p(\mathbf{x}_{t+1} \mid \mathbf{x}_t, \mathbf{u}_t, \xi_t)$ may be simpler to manipulate than that of $p(\mathbf{x}_{t+1} \mid \mathbf{x}_t, \mathbf{u}_t, \boldsymbol{\theta}_1)$. If this is the case, write $E(W_t \mid \mathbf{y}^{t-1})$ as

$$
\begin{aligned}
E(W_t \mid \mathbf{y}^{t-1}) &= \int W_t(\mathbf{x}_t, \mathbf{u}_{t-1}) p(\mathbf{x}_t \mid \mathbf{y}^{t-1})\, d\mathbf{x}_t \\
&= \int W_t(\mathbf{x}_t, \mathbf{u}_{t-1}) p(\mathbf{x}_t \mid \mathbf{x}_{t-1}, \mathbf{u}_{t-1}, \xi_{t-1}) \\
&\quad \times p(\mathbf{x}_{t-1}, \xi_{t-1} \mid \mathbf{y}^{t-1})\, d(\mathbf{x}_t, \mathbf{x}_{t-1}, \xi_{t-1})
\end{aligned}
\tag{10}
$$

were $t = 1, \ldots, N$ and where $\mathbf{u}$s are taken to be nonrandomized. Now to obtain $p(\mathbf{x}_s, \xi_s \mid \mathbf{y}^s)$, $0 \leqslant s \leqslant N-1$, recursively, write

$$
\begin{aligned}
&p(\boldsymbol{\theta}_1, \boldsymbol{\theta}_2, \xi_t, \mathbf{x}_t, \xi_{t+1}, \mathbf{x}_{t+1}, \mathbf{y}_{t+1} \mid \mathbf{y}^t) \\
&\quad = p(\boldsymbol{\theta}_1, \boldsymbol{\theta}_2, \xi_t, \mathbf{x}_t \mid \mathbf{y}^t) p(\xi_{t+1} \mid \boldsymbol{\theta}_1, \xi_t, \mathbf{x}_t, \mathbf{y}^t) \\
&\qquad \times p(\mathbf{x}_{t+1} \mid \boldsymbol{\theta}_1, \xi_t, \mathbf{x}_t, \xi_{t+1}, \mathbf{y}^t) \\
&\qquad \times p(\mathbf{y}_{t+1} \mid \boldsymbol{\theta}_2, \xi_t, \mathbf{x}_t, \xi_{t+1}, \mathbf{x}_{t+1}, \mathbf{y}^t) \\
&\quad = p(\boldsymbol{\theta}_1, \boldsymbol{\theta}_2, \xi_t, \mathbf{x}_t \mid \mathbf{y}^t) p(\xi_{t+1} \mid \boldsymbol{\theta}_1) p(\mathbf{x}_{t+1} \mid \xi_t, \mathbf{x}_t, \mathbf{u}_t) p(\mathbf{y}_{t+1} \mid \mathbf{x}_{t+1}, \boldsymbol{\theta}_2)
\end{aligned}
\tag{11}
$$

where the independence assumption of ξs and $\boldsymbol{\eta}$s for any $\boldsymbol{\theta}_1 \in \Theta_1$ and $\boldsymbol{\theta}_2 \in \Theta_2$ is used. From the dynamics (1)

$$
p(\mathbf{x}_{t+1} \mid \mathbf{x}_t, \xi_t, \mathbf{u}_t) = \delta(\mathbf{x}_{t+1} - f_t(\mathbf{x}_t, \xi_t, \mathbf{u}_t)).
$$

The left-hand side of (11), after $\mathbf{x}_t$ and ξ_t are integrated out, can be written as

$$
p(\boldsymbol{\theta}_1, \boldsymbol{\theta}_2, \xi_{t+1}, \mathbf{x}_{t+1}, \mathbf{y}_{t+1} \mid \mathbf{y}^t) = p(\mathbf{y}_{t+1} \mid \mathbf{y}^t) p(\boldsymbol{\theta}_1, \boldsymbol{\theta}_2, \xi_{t+1}, \mathbf{x}_{t+1} \mid \mathbf{y}^{t+1}).
$$

Thus,

$$
\begin{aligned}
p(\boldsymbol{\theta}_1, \boldsymbol{\theta}_2, \xi_{t+1}, \mathbf{x}_{t+1} \mid \mathbf{y}^{t+1}) &= \int p(\boldsymbol{\theta}_1, \boldsymbol{\theta}_2, \xi_t, \mathbf{x}_t \mid \mathbf{y}^t) p(\xi_{t+1} \mid \boldsymbol{\theta}_1) \delta(\mathbf{x}_{t+1} - \mathbf{f}_t) \\
&\quad \times \frac{p(\mathbf{y}_{t+1} \mid \mathbf{x}_{t+1}, \boldsymbol{\theta}_2)\, d(\mathbf{x}_t, \xi_t)}{\int [\text{numerator}]\, d(\boldsymbol{\theta}_1, \boldsymbol{\theta}_2, \xi_{t+1}, \mathbf{x}_{t+1})}.
\end{aligned}
\tag{12}
$$

Similarly we can compute $p(\xi_t, \boldsymbol{\eta}_t, \mathbf{x}_t \mid \mathbf{y}^t)$ recursively to obtain the optimal control policies when $\boldsymbol{\theta}_1$ and $\boldsymbol{\theta}_2$ are unknown. Sometimes this form is more convenient to use in deriving optimal control policies. To obtain the recursion relation for $p(\xi_t, \boldsymbol{\eta}_t, \mathbf{x}_t \mid \mathbf{y}^t)$, write

$$
\begin{aligned}
&p(\boldsymbol{\theta}_1, \boldsymbol{\theta}_2, \mathbf{x}_t, \xi_t, \boldsymbol{\eta}_t, \mathbf{x}_{t+1}, \xi_{t+1}, \boldsymbol{\eta}_{t+1}, \mathbf{y}_{t+1} \mid \mathbf{y}^t) \\
&\quad = p(\boldsymbol{\theta}_1, \boldsymbol{\theta}_2, \mathbf{x}_t, \xi_t, \boldsymbol{\eta}_t \mid \mathbf{y}^t) p(\mathbf{x}_{t+1} \mid \mathbf{x}_t, \xi_t, \mathbf{u}_t) \\
&\qquad \times p(\xi_{t+1} \mid \boldsymbol{\theta}_1) p(\boldsymbol{\eta}_{t+1} \mid \boldsymbol{\theta}_2) p(\mathbf{y}_{t+1} \mid \mathbf{x}_{t+1}, \boldsymbol{\eta}_{t+1})
\end{aligned}
\tag{13}
$$

where the conditional independence assumption on ξs and ηs have been used. From (1),

$$p(\mathbf{x}_{t+1} \mid \mathbf{x}_t, \mathbf{u}_t, \xi_t) = \delta(\mathbf{x}_{t+1} - \mathbf{f}_t).$$

From (2),

$$p(\mathbf{y}_{t+1} \mid \mathbf{x}_{t+1}, \boldsymbol{\eta}_{t+1}) = \delta(\mathbf{y}_{t+1} - \mathbf{h}_{t+1}).$$

Integrating (13) with respect to $\mathbf{x}_t$, ξ_t, and $\boldsymbol{\eta}_t$,

$$
\begin{aligned}
&\int p(\boldsymbol{\theta}_1, \boldsymbol{\theta}_2, \mathbf{x}_t, \xi_t, \boldsymbol{\eta}_t, \mathbf{x}_{t+1}, \xi_{t+1}, \boldsymbol{\eta}_{t+1}, \mathbf{y}_{t+1} \mid \mathbf{y}^t)\, d(\mathbf{x}_t, \xi_t, \boldsymbol{\eta}_t) \\
&\quad = p(\boldsymbol{\theta}_1, \boldsymbol{\theta}_2, \mathbf{x}_{t+1}, \xi_{t+1}, \boldsymbol{\eta}_{t+1}, \mathbf{y}_{t+1} \mid \mathbf{y}^t) \\
&\quad = p(\boldsymbol{\theta}_1, \boldsymbol{\theta}_2, \mathbf{x}_{t+1}, \xi_{t+1}, \boldsymbol{\eta}_{t+1} \mid \mathbf{y}^{t+1}) p(\mathbf{y}_{t+1} \mid \mathbf{y}^t).
\end{aligned}
$$

Hence

$$
\begin{aligned}
&\mathrm{p}(\boldsymbol{\theta}_1, \boldsymbol{\theta}_2, \mathbf{x}_{t+1}, \xi_{t+1}, \boldsymbol{\eta}_{t+1} \mid \mathbf{y}^{t+1}) \\
&\quad = \int \mathrm{p}(\boldsymbol{\theta}_1, \boldsymbol{\theta}_2, \mathbf{x}_t, \xi_t, \boldsymbol{\eta}_t \mid \mathbf{y}^t) \delta(\mathbf{x}_{t+1} - \mathbf{f}_t) \delta(\mathbf{y}_{t+1} - \mathbf{h}_{t+1}) \\
&\qquad \times \frac{p(\xi_{t+1} \mid \boldsymbol{\theta}_1) p(\boldsymbol{\eta}_{t+1} \mid \boldsymbol{\theta}_2)\, d(\mathbf{x}_t, \xi_t, \boldsymbol{\eta}_t)}{\int [\text{numerator}]\, d(\boldsymbol{\theta}_1, \boldsymbol{\theta}_2, \mathbf{x}_{t+1}, \xi_{t+1}, \boldsymbol{\eta}_{t+1})}.
\end{aligned}
\tag{14}
$$

Integrating both sides of (14) with respect to $\boldsymbol{\theta}_1$ and $\boldsymbol{\theta}_2$, the desired recursion equation results. Equations (8) and (14) have been obtained assuming that ξ_t, $t = 0, 1, \ldots$, and $\boldsymbol{\eta}_t$, $t = 0, 1, \ldots$, are all independent for each $\boldsymbol{\theta}_1$ and $\boldsymbol{\theta}_2$.

If this conditional independence assumption is weakened and if it is assumed that ξ_t, $t = 0, 1, \ldots$, is dependent for each $\boldsymbol{\theta}_1$, and that the ξ process and the $\boldsymbol{\eta}$ process are still independent, then by considering $\mathrm{p}(\boldsymbol{\theta}_1, \boldsymbol{\theta}_2, \xi_t, \mathbf{x}_t, \xi_{t+1}, \mathbf{y}_{t+1} \mid \mathbf{y}^t)$, for example, instead of Eq. (11), the recursion equation for $\mathrm{p}(\boldsymbol{\theta}_1, \boldsymbol{\theta}_2, \xi_t, \mathbf{x}_t \mid \mathbf{y}^t)$ is obtained as

$$
\begin{aligned}
&p(\boldsymbol{\theta}_1, \boldsymbol{\theta}_2, \xi_{t+1}, \mathbf{x}_{t+1} \mid \mathbf{y}^{t+1}) \\
&\quad = \int p(\boldsymbol{\theta}_1, \boldsymbol{\theta}_2, \xi_t, \mathbf{x}_t \mid \mathbf{y}^t) p(\xi_{t+1} \mid \boldsymbol{\theta}_1, \xi^t) \delta(\mathbf{x}_{t+1} - \mathbf{f}_{t+1}) \\
&\qquad \times \frac{p(\mathbf{y}_{t+1} \mid \mathbf{x}_{t+1}, \boldsymbol{\theta}_2)\, d(\mathbf{x}_t, \xi_t)}{\int [\text{numerator}]\, d(\boldsymbol{\theta}_1, \boldsymbol{\theta}_2, \xi_{t+1}, \mathbf{x}_{t+1})}.
\end{aligned}
\tag{15}
$$

Obviously, the most convenient form should be used for a problem under consideration. The indicated integrations in these recursion equations should not be carried out by themselves when they involve delta functions. Perform the integration in evaluating $E(W_t \mid y^{t-1})$. It is not usually necessary to compute the denominators in expressions such as (8), (9), and (11) since they are normalization constants.

7.2 Systems with Unknown Parameters

In this section, we shall derive optimal control policies for several examples with unknown parameters, for example the random noises in the dynamic equation or in the observation equation have probability density functions with unknown parameters. Inputs to the systems are assumed to be deterministic and known:

Equation 1: Systems with Unknown Initial Condition
In the model,

$$x_{t+1} = ax_t + bu_t, \qquad u_t \in (-\infty, \infty)$$

$$y_t = x_t + \eta_t, \qquad 0 \leqslant i \leqslant N-1$$

where $\eta_0, \ldots, \eta_{N-1}$ are all independent Gaussian random variables with known mean and variances. They are also assumed to be independent of x_0. Assume that the mean of x_0 is unknown, i.e., assume

$$\mathrm{L}(x_0) = \mathrm{N}(\theta, \sigma^2)$$

where $\mathrm{L}(\cdot)$ denotes the distribution function, where $N(\cdot, \cdot)$ stands for a normal distribution, and where θ is a random variable, an a priori distribution of which is given by an independent Gaussian distribution

$$L(\theta) = N(\hat{\theta}, \hat{\sigma}^2).$$

With the cost index

$$J = \sum_{1}^{N} x_t^2,$$

the optimal control policy is given by

$$u_t^* = -a\mu_t/b, \qquad 0 \leqslant t \leqslant N-1$$

where μ_t is the conditional mean of x_t, given y^t. The Kalman filter gives the formula for updating μ_t. The expression for μ_0 is the only difference that results from the assumption of unknown θ. The initial value μ_0 is now given by computing

$$p(x_0 \mid y_0) = \frac{\int p_0(\theta) p_0(x_0 \mid \theta) p(y_0 \mid x_0)\, d\theta}{\int [\text{numerator}]\, dx_0}$$

$$= \text{const} \exp\left(- \frac{(x_0 - \mu_0)^2}{2\sigma_0^2} \right)$$

where

$$\mu_0 = \frac{y_0/r_0^2 + \hat{\theta}/(\sigma^2 + \hat{\sigma}^2)}{1/r_0^2 + 1/(\sigma^2 + \hat{\sigma}^2)}$$

and

$$1/\sigma_0^2 = 1/r_0^2 + 1/(\sigma^2 + \hat{\sigma}^2).$$

Thus, the unknown mean θ of x_0 has the effect of replacing σ^2, the variance of x_0, by $\sigma^2 + \hat{\sigma}^2$.

Example 2: Min–Max Control against Unknown Noise Probability. Consider a system with perfect observation,

$$x_1 = a x_0 + u_0 + \xi_0$$

$$y_0 = x_0 \text{ given}$$

where it is assumed that a is known and that ξ_0 is a random variable with

$$\xi_0 = \begin{cases} \theta_1 & \text{with probability } p \\ \theta_2 & \text{with probability } 1 - p \end{cases}$$

where θ_1 and θ_2 are given, $\theta_1 > \theta_2$. The probability p is unknown.

The criterion function is taken to be

$$J = x_1^2 = (a x_0 + u_0 + \xi_0)^2.$$

The expected value of J is given as

$$EJ = p(a x_0 + u_0 + \theta_1)^2 + (1 - p)(a x_0 + u_0 + \theta_2)^2.$$

Therefore, the control given by

$$u_0^* = -[a x_0 + p\theta_1 + (1 - p)\theta_2]$$

minimizes EJ:

$$\gamma_1^* \triangleq \min_{u_0} EJ = p(1 - p)(\theta_1 - \theta_2)^2.$$

Note that γ_1^* is maximized when $p = \frac{1}{2}$. When p is known, the control u_0 is called the optimal Bayes control for the problem. If p is not given, u_0^* cannot be obtained. Let us look for the control which makes J independent of θ_1 or θ_2. Namely, consider $\hat{u}_0$ given by

$$\hat{u}_0 = -\left(ax_0 + \frac{\theta_1 + \theta_2}{2}\right).$$

Then

$$J(1, \hat{u}_0) = J(0, \hat{u}_0) = \left(\frac{\theta_1 - \theta_2}{2}\right)^2.$$

Thus, if $\hat{u}_0$ is employed, x_1^2 is the same regardless of p values.

Such a control policy is called an equalizer control policy (Sworder [1964]). The value of J is seen to be equal to γ_1^* when $p = \frac{1}{2}$.

In other words, the control $\hat{u}_0$ minimizes the criterion function for the worst possible case $p = \frac{1}{2}$. Therefore $\hat{u}_0$ may be called the min–max control since it minimizes the maximal possible EJ value. Comparing $\hat{u}_0$ and u_0^*, $\hat{u}_0$ is seen to be the optimal Bayes control for $p = \frac{1}{2}$.

For this example, an equalizer control policy is a min–max control policy, which is equal to the optimal Bayes control policy for the worst possible a priori distribution function for the unknown parameter θ.

It is known that the preceding statements are true generally when the unknown parameter θ can take on only a finite number of possible values. When θ can take an infinite number of values, similar but weaker statements are known to be true. See Ferguson (1967) and Sworder (1964) for details.

As before, an optimal control policy minimizes the conditional expected value of J.

Example 3: System with Noisy Observation Having Unknown Bias

Consider a one-dimensional linear control system described by

$$x_{t+1} = \mathrm{a}x_t + \mathrm{b}u_t + \xi_t, \qquad u_t \in (-\infty, \infty), \qquad 0 \leqslant t \leqslant N-1$$

and observed by

$$y_t = x_t + \eta_t, \qquad 0 \leqslant t \leqslant N-1$$

where a and b are known constant model parameters and where ξs are independently and identically distributed with the distribution function given by

$$\mathscr{L}(\xi_t) = N(0, \sigma_0^2).$$

The random variables η are assumed to be independent of ξs. The random variables η are also assumed to be independently and identically distributed with

$$\mathscr{L}(\eta_t) = N(\theta, \sigma_1^2)$$

for each $\theta \in (-\infty, \infty)$ where θ is the unknown parameter of the distribution.

Its a priori distribution function is assumed to be given by

$$\mathscr{L}_0(\theta) = N(\mu, \sigma_2^2).$$

The initial state x_0 of the model is assumed to be a random variable, independent of all other random variables and assumed to be distributed according to

$$\mathscr{L}(x_0) = N(\alpha, \sigma_3^2)$$

where α and σ_3 are know. The cost function is taken to be

$$J = \sum_1^N \mathrm{q}_t x_t^2, \qquad q_t > 0.$$

The optimal control policy is now derived for the preceding system.

In the previous two examples, the recursion equations for conditional probability density functions involve a one-dimensional normal distribution function. In this example, we will see that, because of noisy observations, two-dimensional normal distribution functions must be computed in the recursive generations of the conditional density functions. To obtain u_{N-1}^*, compute first

$$\begin{aligned}
\lambda_N &\triangleq E(q_N x_N^2 \mid y^{N-1}) \\
&= \int q_N x_N^2 p(x_N \mid y^{N-1})\, dx_N \\
&= \int q_N x_N^2 p(x_N \mid x_{N-1}, u_{N-1}, \xi_{N-1}) p(x_{N-1}, \xi_{N-1} \mid y^{N-1})\, d(x_N, x_{N-1}, \xi_{N-1}) \qquad (16) \\
&= \int q_N(\mathrm{a}x_{N-1} + \mathrm{b}u_{N-1} + \xi_{N-1})^2 p(x_{N-1}, \xi_{N-1} \mid y^{N-1})\, d(x_{N-1}, \xi_{N-1}).
\end{aligned}$$

In Eq. (16),

$$p(x_{N-1}, \xi_{N-1} \mid y^{N-1}) = p(\xi_{N-1}) p(x_{N-1} \mid y^{N-1}) \tag{17}$$

since $\boldsymbol{\xi}_t$ is independent of $\boldsymbol{\xi}^{t-1}$ and of η^t by assumption. Note that

$$p(x_t \mid y^t) = \int p(x_t \mid \eta_t, y^t) p(\eta_t \mid y^t)\, d\eta_t, \qquad 0 \leqslant t \leqslant N-1. \tag{18}$$

Substituting Eqs. (17) and (18) into Eq. (22),

$$\lambda_N = \int q_N[(ax_{N-1} + bu_{N-1})^2 + \sigma_0^2]p(x_{N-1} \mid \eta_{N-1}, y^{N-1}) \times p(\eta_{N-1} \mid y^{N-1})\, d(x_{N-1}, \eta_{N-1})$$

$$= \int q_N\{[a(y_{N-1} - \eta_{N-1}) + bu_{N-1}]^2 + \sigma_0^2\} p(\eta_{N-1} \mid y^{N-1})\, d\eta_{N-1}.$$

Defining the conditional mean and variance of η_t by

$$\hat{\eta}_t = E(\eta_t \mid y^t), \qquad 0 \leqslant t \leqslant N-1$$

and

(19) $$\Gamma_t^2 = \operatorname{var}(\eta_t \mid y^t),$$

we have

(20) $$\gamma_N = \lambda_N = q_N\{[a(y_{N-1} - \hat{\eta}_{N-1}) + bu_{N-1}]^2 + \sigma_0^2 + a^2\Gamma_{N-1}^2\}.$$

Since $\hat{\eta}_{N-1}$ and Γ_{N-1} are independent of u_{N-1}, by minimizing Eq. (20) with respect to u_{N-1}, the optimal control variable at time $(N-1)$ is given as

$$u_{N-1}^* = -a(y_{N-1} - \hat{\eta}_{N-1})/b$$

and

(21) $$\gamma_N^* = q_N(\sigma_0^2 + a^2\Gamma_{N-1}^2).$$

Therefore, if μ_{N-1} and Γ_{N-1} are known, so are u_{N-1}^* and γ_N^*.

We also see from Eq. (21) that, since γ_N^* is independent of y_{N-1}, each control stage can be optimized separately, and the optimal control policy consists of a sequence of one stage optimal policy

$$u_t^* = -a(y_t - \hat{\eta}_t)/b, \qquad 0 \leqslant t \leqslant N-1.$$

In order to compute μs and Γs, we show first that the conditional probability densities of θ and η_t are jointly normally distributed, i.e.,

(22) $$p(\theta, \eta_t \mid y^t) = \text{const} \exp\left[-\tfrac{1}{2}(\theta - \hat{\theta}_t, \eta_t - \hat{\eta}_t) M_t^{-1} \begin{pmatrix} \theta - \hat{\theta}_t \\ \eta_t - \hat{\eta}_t \end{pmatrix}\right]$$

where

(23) $$M_t = \operatorname{cov}(\theta, \eta_t \mid y^t) = \begin{pmatrix} M_{11}^t & M_{12}^t \\ M_{21}^t & M_{22}^t \end{pmatrix}$$

is a constant covariance matrix and where

$$\hat{\theta}_t = E(\theta \mid y^t) \qquad M_{11}^t = \operatorname{var}(\theta \mid y^t),$$
$$\hat{\eta}_t = E(\eta_t \mid y^t) \qquad M_{22}^t = \operatorname{var}(\eta_t \mid y^t),$$

and

$$M_{12}^t = M_{21}^t = E[(\theta - \hat{\theta}_t)(\eta_t - \hat{\eta}_t) \mid y^t].$$

Then with the notation defined by (19),

$$\Gamma_t = M_{22}^t.$$

To verify (22)–(24) for $t = 0$, consider

(25) $$p(\theta, \eta_0 \mid y^0) = \frac{p(\theta, \eta_0, y_0)}{p(y_0)} = \frac{p_0(\theta)p(\eta_0 \mid \theta)p(y_0 \mid \theta, \eta_0)}{p(y_0)}$$

where

(26) $$p_0(\theta) = \text{const} \exp\left(-\frac{(\theta - \mu)^2}{2\sigma_2^2}\right),$$

(27) $$p(\eta_0 \mid \theta) = \text{const} \exp\left(-\frac{(\eta_0 - \theta)^2}{2\sigma_1^2}\right)$$

and

(28) $$p(y_0 \mid \theta, \eta_0) = \text{const} \exp\left(-\frac{(y_0 - \alpha - \eta_0)^2}{2\sigma_3^2}\right).$$

From (25)–(28), Eq. (22) is seen to hold for $t = 0$ with

$$\hat{\theta}_0 = \frac{\mu/\sigma_2^2 + (y_0 - \alpha)/(\sigma_1^2 + \sigma_3^2),}{1/\sigma_2^2 + 1/(\sigma_1^2 + \sigma_3^2)}$$
$$\hat{\eta}_0 = \frac{\mu/(\sigma_1^2 + \sigma_2^2) + (y_0 - \alpha)/\sigma_3^2}{1/(\sigma_1^2 + \sigma_2^2) + 1/\sigma_3^2},$$
$$M_{11}^0 = (\sigma_1^2 + \sigma_3^2)\sigma_2^2/(\sigma_1^2 + \sigma_2^2 + \sigma_3^2),$$
$$M_{12}^0 = M_{21}^0 = \sigma_2^2\sigma_3^2/(\sigma_1^2 + \sigma_2^2 + \sigma_3^2),$$
$$M_{22}^0 = (\sigma_1^2 + \sigma_2^2)\sigma_3^2/(\sigma_1^2 + \sigma_2^2 + \sigma_3^2).$$

Note that M_0 is a constant matrix.

Thus (25) is verified for $t = 0$. Next, assume that (25) is true for some $t > 0$. We shall show that the equation is true for $t + 1$, thus completing the mathematical induction on (25). To go from t to $(t + 1)$, consider $p(\theta, \eta_t, \eta_{t+1} \mid y^t)$. By the chain rule, this conditional density can be written as

$$p(\theta, \eta_t, \eta_{t+1}, y_{t+1} \mid y^t) = p(\theta, \eta_t \mid y^t) p(\eta_{t+1} \mid \theta, \eta_t, y^t) \times p(y_{t+1} \mid \theta, \eta_t, \eta_{t+1}, y^t) \tag{29}$$

where the second factor reduces to $p(\eta_{t+1} \mid \theta)$ since ηs are conditionally independent of all other ηs and ξs by assumption. From the model specification, y satisfies the difference equation

$$y_{t+1} = \mathrm{a} y_t + \mathrm{b} u_t + \xi_t - \mathrm{a}\eta_t + \eta_{t+1}.$$

The third factor of (29) is given, therefore, by

$$p(y_{t+1} \mid \theta, \eta_t, \eta_{t+1}, y^t) = \text{const} \exp\left(\frac{(y_{t+1} - \mathrm{a} y_t - \mathrm{b} u_t + \mathrm{a}\eta_t - \eta_{t+1})^2}{2\sigma_0^2} \right).$$

By integrating both sides of (29) with respect to η_t,

$$\begin{aligned} p(\theta, \eta_{t+1}, y_{t+1} \mid y^t) &= p(y_{t+1} \mid y^t) p(\theta, \eta_{t+1} \mid y^{t+1}) \\ &= \int p(\theta, \eta_t \mid y^t) p(\eta_{t+1} \mid \theta) p(y_{t+1} \mid \theta, \eta_t, \eta_{t+1}, y^t)\, d\eta_t. \end{aligned}$$

Therefore,

$$p(\theta, \eta_{t+1} \mid y^{t+1}) = \frac{\int p(\theta, \eta_t \mid y^t) p(\eta_{t+1} \mid \theta) p(y_{t+1} \mid \theta, \eta_t, \eta_{t+1}, y^t)\, d\eta_t}{\int p(\theta, \eta_t \mid y^t) p(\eta_{t+1} \mid \theta) p(y_{t+1} \mid \theta, \eta_t, \eta_{t+1}, y^t)\, d(\theta, \eta_t, \eta_{t+1})}.$$

After carrying out the tedious integration in (30), we establish (22) for $t + 1$ with

$$\hat{\theta}_{t+1} = \frac{[(C_t + D_t)/\sigma_t^2 + B_t D_t - C_t^2]\hat{\theta}_t + (B_t + C_t) Z_t/\sigma_1^2}{\Delta_t^2},$$

$$\hat{\eta}_{t+1} = \frac{(C_t + D_t)\hat{\theta}_t/\sigma_1^2 + [(B_t + C_t)/\sigma_1^2 + B_t D_t - C_t^2] Z_t}{\Delta_t^2},$$

$$M_{11}^{t+1} = (1/\sigma_1^2 + B_t)/\Delta_t^2,$$

$$M_{12}^{t+1} = (1/\sigma_1^2 - C_t)/\Delta_t^2,$$

$$M_{22}^{t+1} = (1/\sigma_1^2 + D_t)/\Delta_t^2,$$

$$Z_t = y_{t+1} - \mathrm{a}(y_t - \hat{\eta}_t) - \mathrm{b} u_t$$

where

$$\Delta_t^2 = (1/\sigma_1^2 + B_t)(1/\sigma_1^2 + D_t) - (1/\sigma_1^2 - C_t)^2,$$

$$B_t = \frac{v_{22}^t/\sigma_0^2}{v_{22}^t + \mathrm{a}^2/\sigma_0^2},$$

$$C_t = \frac{v_{12}^t \mathrm{a}/\sigma_0^2}{v_{22}^t + \mathrm{a}^2/\sigma_0^2},$$

$$D_t = v_{11}^t - \frac{(v_{12}^t)^2}{v_{22}^t + \mathrm{a}^2/\sigma_0^2},$$

and

$$M_t^{-1} = \begin{pmatrix} v_{11}^t & v_{12}^t \\ v_{21}^t & v_{22}^t \end{pmatrix}.$$

Therefore, M_t in (22) all turn out to be constants that can be precalculated (i.e., calculated off-line) The only on-line computation in generating u_t^* is that of updating η_t and θ_t where Z_t depends on y_t, y_{t+1}, and u_t.

Example 4: System and Unknown Noise Variance. The technique similar to that used in the previous example can be used to treat control problems with unknown noise variances. As an illustration, we derive an optimal control policy for the system of Example 2, assuming now that the variances of the observation noises are unknown and that the mean is known.

We only give a summary of the steps involved. Another approach to treat unknown noise variances is to use $C\eta_t$ instead of η_t with $\mathscr{L}(\eta_t) = N(0, 1)$ and assume C to be the unknown parameter.

We now assume that

$$\mathscr{L}(\eta_t) = N(0, \Sigma)$$

where Σ is the variance which is assumed to be distributed according to

$$p_0(\Sigma) = z_{0.1}\delta(\Sigma - \Sigma_1) + z_{0.2}\delta(\Sigma - \Sigma_2)$$

where

$$z_{0.1} + z_{0.2} = 1.$$

Namely, the unknown variance of the observation noise is assumed to be either Σ_1 or Σ_2 with the a priori probability given by $z_{0.1}$ and $z_{0.2}$ respectively.

Other assumptions on the distribution functions of ξs and x_0 are the same as before. The probability of Σ is now taken to be independent of x_0. With the criterion function as before, each control stage can be optimized separately. To obtain u_t^*, we compute

$$\lambda_{t+1} = q_{t+1} \int x_{t+1}^2 p(x_{t+1} \mid y^t)\, dx_{t+1}$$
$$= q_{t+1} \int [(ax_t + bu_t)^2 + \sigma_0^2] p(x_t \mid y^t)\, dx_t.$$

Defining

(31) $E(x_t \mid y^t) = \hat{x}_t$ and $\operatorname{var}(x_t \mid y^t) = \hat{\Gamma}_t, \qquad 0 \leqslant t \leqslant N-1,$

λ_{t+1} can be expressed as

$$\lambda_{t+1} = q_{t+1}[(a\hat{x}_t + bu_t)^2 + a^2\hat{\Gamma}_t + \sigma_0^2].$$

Under the assumption, to be verified later, that $\hat{x}_t$ and $\hat{\Gamma}_t$ are independent of u_t, we obtain

$$u_t^* = -aq_{t+1}\hat{x}_t/b, \qquad 0 \leqslant t \leqslant -1$$

as the optimal and control variable at time t.

In order to compute $\hat{x}_t$, consider the joint probability density function $p(x_t, \Sigma \mid y^t)$.

It is easy to see that it satisfies the recursion equation

$$p(x_{t+1}, \Sigma \mid y^{t+1}) = \frac{\int p(x_t, \Sigma \mid y^t) p(x_{t+1} \mid x_t, u_t) p(y_{t+1} \mid x_{t+1}, \Sigma)\, dx_t}{\int [\text{numerator}]\, d(x_{t+1}, \Sigma)}.$$

It is also easy to show inductively that

(32) $$p(x_t, \Sigma \mid y^t) = [z_{t,1}\delta(\Sigma - \Sigma_1) + z_{t,2}\delta(\Sigma - \Sigma_2)] N[\mu_t(\Sigma), \Gamma_t(\Sigma)].$$

The second factor in (32) is the Gaussian probability density function with mean $\mu_t(\Sigma)$ and variance $\Gamma_t(\Sigma)$, where

$$\mu_{t+1}(\Sigma) = \{[a\mu_t(\Sigma) + bu_t]/[a\Gamma_t(\Sigma) + \sigma_0^2]$$
$$+ y_{t+1}/\Sigma\}/\{1/[a^2\Gamma_t(\Sigma) + \sigma_0^2] + 1/\Sigma\},$$
$$\mu_0(\Sigma) = (a/\sigma_3^2 + y_0/\Sigma)/(1/\sigma_3^2 + 1/\Sigma),$$
$$1/\Gamma_{t+1}(\Sigma) = 1/[a^2\Gamma_t(\Sigma) + \sigma_0^2] + 1/\Sigma,$$
$$1/\Gamma_0(\Sigma) = 1/\sigma_3^2 + 1/\Sigma,$$
$$z_{t+1,1} = z_{t,1}\omega_{t,1}/(z_{t,1}\omega_{t,1} + z_{t,2}\omega_{t,2}),$$
$$z_{t+1,2} = z_{t,2}\omega_{t,2}/(z_{t,1}\omega_{t,1} + z_{t,2}\omega_{t,2}) \qquad (0 \leqslant t \leqslant N-1),$$

and

$$\omega_{t,s} = \frac{\Gamma_t^{1/2}(\Sigma_s)\sigma_0\Sigma_s^{1/2}}{(\Sigma_s + a^2\Gamma_t(\Sigma_s) + \sigma_0^2)^{1/2}} \exp\left(- \frac{(y_{t+1} - a\mu_t(\Sigma_s) - bu_t)^2}{2(\Sigma_s + a^2\Gamma_t(\Sigma_s) + \sigma_0^2)} \right) \qquad s = 1, 2.$$

Then from (31) and (32)

$$\hat{x}_t = z_{t,1}\mu_t(\Sigma_1) + z_{t,2}\mu_t(\Sigma_2)$$
$$\tilde{\Gamma}_t = z_{t,1}\Gamma_t(\Sigma_1) + z_{t,2}\Gamma_t(\Sigma_2) + z_{t,1}z_{t,2}[\mu_t(\Sigma_1) - \mu_t(\Sigma_2)]^2.$$

The assumption that $\hat{x}_t$ and $\hat{\Gamma}_t$ are independent of u_t is thus seen to be satisfied.

Example 5: System with Unknown Gain. Let us consider a stochastic control system

$$x_{t+1} = \mathrm{a}x_t + bu_t + \xi_t, \qquad u_t \in (-\infty, \infty)$$
$$x_0 \text{ given}$$
$$y_t = x_t, \qquad 0 \leqslant t \leqslant N-1$$

where a is a known constant but where b is now assumed to be a random variable, independent of ξs with finite mean and variance. The disturbance ξs are assumed to be independently and identically distributed random variables with

$$E(\xi_t) = 0$$
$$\mathrm{var}(\xi_t) = \Sigma_0^2, \qquad 0 \leqslant t \leqslant N-1.$$

Suppose we replace the model with a (conditionally) deterministic one

(33) $$x_N = ax_{N-1} + b_{N-1}u_{N-1}$$

where

$$b_{N-1} \triangleq E(b \mid x^{N-1}).$$

The optimal u_{N-1} for this problem is

$$u_{N-1} = -ax_{N-1}/b_{N-1}.$$

With this control, the conditional expected value of J, i.e., the contribution to EJ from the last control stage, is given by

$$E(x_N^2 \mid x^{N-1}, u^{N-1}) = E\left[\left(ax_{N-1} - \frac{b}{b_{N-1}}ax_{N-1} + \xi_{N-1}\right)^2 \middle| \, x^{N-1}\right]$$
$$= \frac{\sigma_{N-1}^2(ax_{N-1})^2}{b_{N-1}^2} + \Sigma_0^2$$

where

$$\sigma_{N-1}^2 = \operatorname{var}(b \mid x^{N-1}).$$

Let us try another control variable

$$u_{N-1} = -\frac{b_{N-1}}{b_{N-1}^2 + \sigma_{N-1}^2}(ax_{N-1}). \tag{34}$$

With this control, the conditional cost is

$$E(x_N^2 \mid x^{N-1}, u^{N-1}) = E\left[\left(ax_{N-1} - \frac{bb_{N-1}}{b_{N-1}^2 + \sigma_{N-1}^2}ax_{N-1} + \xi_{N-1}\right)^2 \middle| x^{N-1}\right]$$

$$= \frac{\sigma_{N-1}^2}{b_{N-1}^2 + \sigma_{N-1}^2}(ax_{N-1})^2 + \Sigma_0^2. \tag{35}$$

Comparing these two costs, we see the optimal control for the deterministic one (33) is not optimal since the control variable (34) produces a lower conditional cost (35) and is better.

Let us now compute the cost under the assumption that the random variable b is Gaussian, independent of ξs and of x_0, and has the unknown mean θ,

$$L(b) = N(\theta, \Sigma_1^2)$$

and that θ itself is a random variable with a priori density function given by

$$L(\theta) = N(\theta_0, \Sigma_2^2)$$

where θ_0, Σ_1, and Σ_2 are assumed known.

We first show that

$$p(b \mid x^t) = \text{const} \exp\left(-\frac{(b - b_t)^2}{2\sigma_t^2}\right), \qquad t = 0, 1, \ldots. \tag{36}$$

Since

$$p(b \mid x_0) = p_0(b) = \int p(b \mid \theta) p_0(\theta)\, d\theta$$

$$= \text{const} \exp\left(-\frac{(b - \theta_0)^2}{2\sigma_0^2}\right),$$

(33) is certainly true for $t = 0$ with

$$b_0 = \theta_0$$

$$1/\sigma_0^2 = 1/\Sigma_1^2 + 1/\Sigma_2^2.$$

From the recursion relation

$$p(b \mid x^{t+1}) = \frac{p(b \mid x^t) p(x_{t+1} \mid b, x_t, u_t)}{\int p(b \mid x^t) p(x_{t+1} \mid x_t, b, u_t)\, db}$$

where

$$p(x_{t+1} \mid x_t, b, u_t) = \text{const} \exp\left(- \frac{(x_{t+1} - ax_t - bu_t)^2}{2\Sigma_0^2} \right),$$

one gets

$$p(b \mid x^{t+1}) = \text{const} \exp\left(- \frac{(b - b_{t+1})^2}{2\sigma_{t+1}^2} \right)$$

where

(37a) $$b_{t+1} = \frac{b_t/\sigma_t^2 + u_t(x_{t+1} - ax_t)/\Sigma_0^2}{1/\sigma^2 + u_t^2/\Sigma_0^2}$$

where

(37b) $$1/\sigma_{t+1}^2 = 1/\sigma_t^2 + u_t^2/\Sigma_0^2$$

where Σ_0^2 is the variance of the dynamic noise, thus verifying (36) for all $t = 0, 1, \ldots$. The statistics (b_t, σ_t^2) are the sufficient statistics for b.

From (37), therefore,

$$b_{N-1} = \frac{b_{N-2}/\sigma_{N-2}^2 + u_{N-2}(x_{N-1} - ax_{N-2})/\Sigma_0^2}{1/\sigma_{N-2}^2 + u_{N-2}^2/\Sigma_0^2}$$

where b_{N-2} is a function of x^{N-2}.

Since the criterion function depends only on x_N, $\lambda_t = 0$, $t = 1, \ldots, N-1$, and we have

$$\begin{aligned} \gamma_{N-1}^* &= \int \gamma_N^* p(x_{N-1} \mid x^{N-2})\, dx_{N-1} \\ &= \int \gamma_N^* p(x_{N-1} \mid x_{N-2}, b, u_{N-2}) p(b \mid x^{N-2})\, d(x_{N-1}, b). \end{aligned}$$

Notice that x_{N-1}^* appears not only explicitly in γ_N^* but also through b_{N-1} in carrying out the integration with respect to x_{N-1}.

The integration

(38) $$\int \gamma_N^* p(x_{N-1} \mid x_{N-2}, b, u_{N-2})\, dx_{N-1}$$

is of the form

(39) $$I = \int_{-\infty}^{\infty} \frac{x^2}{(ax+b)^2 + c^2} \exp\left(- \frac{(x-\mu)^2}{2} \right) dx \qquad a, b, c \text{ real}$$

after a suitable change of variables or, more generally,

$$\int \frac{Q_1(x)}{Q_2(x)} \exp\left(-\frac{x^2}{2}\right) dx$$

where Q_1 and Q_2 are quadratic polynomials in x.

The integration of Eq. (39) cannot be carried explicitly to give an analytically closed expression.

Therefore, it is impossible to obtain expressions for $u_0^*, \ldots, u_{N-2}^*$ explicitly analytically in a closed form. They must be obtained either numerically or by approximation.

Here is an example of a common type of difficulty we face in obtaining optimal control policies. The system of this example is rather simple, yet the assumption of the unknown mean of the random gain made it impossible to obtain the optimal control policy explicitly.

There are so many such examples, even when the systems are linear with quadratic performance indices, that they do not admit analytic solutions for optimal control policies.

We note that the integrand γ_N^* in (38) is bounded from above by

$$\gamma_N^* < (ax_{N-1})^2 + \Sigma_0^2.$$

If this upper bound is used in (38), then we have an approximation of γ_{N-1}^* given by

$$\gamma_{N-1}^* \leqslant \Sigma_0^2 + \int (ax_{N-1})^2 p(x_{N-1} \mid x_{N-2}, b, u_{N-2}) p(b \mid x^{N-2})\, d(x_{N-1}, b).$$

This approximation is equivalent to one-stage optimization where each control stage is optimized separately.

This approximation yields a suboptimal control policy where

$$u_t = -\frac{b_t}{b_t^2 + \sigma_t^2}(ax_t), \qquad 0 \leqslant t \leqslant N-1.$$

Even though (35) cannot be carried out analytically, some insight can be obtained into the structure of optimal control policies.

From (33) and (34) we note that

$$b_{t+1} \to b \qquad \text{and} \qquad \sigma_{t+1} \to 0 \qquad \text{as} \qquad |u_t| \to \infty.$$

Therefore, the coefficient multiplying x_{N-1}^2 in the expression for γ_N^* approaches zero:

$$\frac{a^2 \sigma_{N-1}^2}{b_{N-1}^2 + \sigma_{N-1}^2} \to 0 \qquad \text{as} \qquad |u_{N-2}| \to \infty.$$

Substituting $ax_{N-2} + bu_{N-2} + \xi_{N-2}$ for x_{N-1} in γ_N^*, we have

$$\gamma_N^* \to \Sigma_0^2(1 + a^2) \qquad \text{as} \qquad |u_{N-2}| \to \infty.$$

If b is known, then the minimum of the expected control cost for the last two stages using an open-loop control policy is equal to $\Sigma_0^2(1 + a^2)$, i.e., the contribution from ξ_{N-1} and $a\xi_{N-2}$ terms.

Let us now examine the effects of $|u_{N-3}| \to \infty$. Then

$$x_{N-1} \to a\xi_{N-3} + \xi_{N-2}$$

and

$$x_N \to ax_{N-1} + bu_{N-1} + \xi_{N-1}.$$

By employing the controls given by

$$u_{N-1} = -ab_{N-1}x_{N-1}/(b_{N-1}^2 + \sigma_{N-1}^2)$$

and

$$u_{N-2} = -ab_{N-2}x_{N-2}/(b_{N-2}^2 + \sigma_{N-2}^2)$$

and noting that $b_{N-2} \to b$ and $\sigma_{N-2} \to 0$, we have

$$x_N \to \xi_{N-1}.$$

Therefore,

$$E(x_N^2 \mid x^{N-3}) \to \Sigma_3^2.$$

Thus, one optimal policy is to let $|u_{N-3}| \to \infty$.

These considerations indicate that this example is singular, i.e., the control cosst does not attain its minimum for control policies using finite control variables. To be more meaningful, the criterion function must be modified to include the cost of control, for example, from $J = x_N^2$ to

$$J = x_N^2 + \lambda \sum_0^{N-1} u_t^2.$$

With this modified criterion function, we can easily derive an optimal control policy when b is known.

It is given by

$$u_{N-t}^* = -a^{2t-1}x_{N-t}b\bigg/\left(\lambda^2 + b\sum_{\tau=1}^{t} a^{2(\tau-1)}\right).$$

When b is assumed to be the random variable as originally assumed in this example, we encounter the same difficulties in deriving the optimal control policy.

Assuming $|\sigma_t^2/b_t^2|$ is small, i.e., assuming the learning process on b is nearly completed, we may expand γ^*s in Taylor series in (σ_t^2/b_t^2).

This approximation results in the control policy

$$u_{N-t} = -a^{2t-1} x_{N-t} b_{N-t} K_{N-t}/(\lambda^2 + L_{N-t})$$

where K_{N-t} and L_{N-t} are complicated algebraic functions of a, λ, b_{N-t}, and σ_{N-t}.

Computational works on this and other approximation schemes are found in a report by Horowitz (1966). See also Sworder (1966) for discussions of suboptimal control policies of a related but simpler problem where b is assumed to be either b_1 or b_2.

Let us return now to (36) and consider its approximate evaluation.

Since $\exp[-\frac{1}{2}(x-\mu)^2]$ in (36) is very small for $|x-\mu|$ large, one may approximately evaluate I by expanding

$$\frac{x^2}{(ax+b)^2+c^2}$$

about $x = \mu$ and retaining terms up to quadratic in $(x-\mu)$, say. When this is carried out,

$$I \approx \frac{(2\pi)^{1/2}\mu^2}{(a\mu+b)^2+c^2}\left\{1 + \frac{b^2+c^2}{\mu^2[(a\mu+b)^2+c^2]} - 2a\mu\frac{b(a\mu+b)^2+c^2(2a\mu+b)}{\mu^2[(a\mu-b)^2+c^2]^2}\right\}.$$

Similarly,

$$\int_{-\infty}^{\infty}\frac{1}{(ax+b)^2+c^2}\exp\left(-\frac{(x-\mu)^2}{2}\right)dx \approx \frac{(2\pi)^{1/2}}{(a\mu+b)^2+c^2}\left\{1 - \frac{5a^2(a\mu+b)^2+a^2c^2}{[(a\mu+b)^2+c^2]^2}\right\}.$$

After $E(\gamma_N^* \mid x^{N-2})$ is approximately carried out, it will be, in general, a complicated function of u_{N-2}. To find u_{N-2}^*, the following sequential scheme (an approximation in policy space; Bellman [1957]) may be used. First, γ_{N-1} is approximated by a quadratic function of u_{N-2} about some u_{N-2}^0. This u_{N-2}^0 could be the optimal u_{N-2} for the control system with some definite b value.

Minimization of this quadratic form gives u_{N-2}^1 as the optimal u_{N-2}. Then, γ_{N-1} is approximated again by a quadratic function of u_{N-2} about u_{N-2}^1. The optimal u_{N-2} now is denoted by u_{N-1}^2. Generally, γ_{N-1} is approximated by a quadratic form in u_{N-2} about u_{N-2}^t, and u_{N-2}^{t+1} is the minimizing u_{N-2}. Under suitable conditions on γ_{N-1}, $u_{N-2}^t \to u_{N-2}^*$ as $t \to \infty$. See Aoki (1962) for an exposition of a similar successive approximation method.

This sequential determination of the optimal control, coupled with the approximate evaluation of $E(\gamma_{t+1}^* \mid x^{t-1})$, for example, by the Taylor series expansion, can generate u_t^*, $0 \leqslant t \leqslant N-1$, approximately.

7.3 Controlled Markov Processes with Partial Observations

We next present a formulation that subsumes the one in the previous section and many like it.

In some cases $\{(x_t, y_t)\}$ is already a first-order Markov sequence. When this is not the case, there is more than one way, depending on the assumptions about noises and parameters in dynamic and observation equations, of augmenting (x_t, y_t) so that the resulting vector becomes a first-order Markov sequence.

Clearly, if $\{\zeta_t\}$ is Markovian, where ζ_t is some augmented state vector, then we do not destroy the Markov property by adding to ζ_t independent random variables with known distribution functions. Simplicity and ease of computing the conditional densities would indicate a particular choice in any problem. The question of the minimum dimension of augmented state vectors ζ_t to make $\{\zeta_t\}$ Markovian is important theoretically but will not be pursued here.

Generally speaking, the a posteriori probability density functions such as $p(x_t \mid y^t)$ and $p(x_t, \xi_t, \eta_t \mid y^t)$ are sufficient in the sense that the corresponding density functions at time $t+1$ are computable from their known value at time t. We can include the a posteriori probability density function as a part of an augmented state vector to make it Markovian. The dimension of such augmented vectors, generally is infinite. We are primarily interested in finite dimensional augmented state vectors. Section 5 provides additional analysis.

Markov Rule

As an example of a system where $\{(x_t, y_t)\}$ is a first-order Markov sequence, consider a purely stochastic dynamic system of Chapter 3,

$$(40)\qquad \begin{aligned} x_{t+1} &= f_t(x_t, u_t, \xi_t) \\ y_t &= h_t(x_t, \eta_t), \qquad t = 0, 1, \ldots, N-1, \end{aligned}$$

where ξs and ηs are mutually independent and independent in time and have known probability densities and (40) contains no unknown parameters.

Consider a vector

$$\zeta_t = (x_t, \xi_t, \eta_t, y_t, u_t). \tag{41}$$

In (41), y_t is the only component observed by the decision maker. We shall see that under certain assumptions $\{\zeta_t\}$-process is a first-order Markov sequence, where the conditional probability of ζ_{t+1} is such that

$$\mathrm{P}[\zeta_{t+1} \in S \mid \zeta_0 = \mathbf{z}_0, \ldots, \zeta_t = \mathbf{z}_t] = \mathrm{P}[\zeta_{t+1} \in S \mid \zeta_t = \mathbf{z}_t]$$

for all t where S is any measurable set in the Euclidean space with the same dimension as ζ. This is the transition probability of $\{\zeta\}$-process. It is assumed furthermore that the transition probability density $p(\zeta_{t+1} \mid \zeta_t)$ exists so that

$$\mathrm{P}[\zeta_{t+1} \in S \mid \zeta_t = \mathbf{z}_t] = \int_{\zeta_{t+1} \in S} p(\zeta_{t+1} \mid \zeta_t = \mathbf{z}_t)\, d\zeta_{t+1}.$$

Let us compute the conditional probability density $p(\zeta_{t+1} \mid \zeta^t)$ of (41) assuming that u_t depends only on y_t, or at most on y_t and u_{t-1}. Such a control is called a Markov plan or Markov rule as in Blackwell (1965, p. 228). We have seen several examples in previous chapters. Generally speaking a Markov rule implies that y_t is the sufficient statistic for x_t, i.e., $p(x_t \mid y^t) = p(x_t \mid y_t)$ and $p(y_{t+1} \mid y^t, u_t) = p(y_{t+1} \mid y_t, u_t)$. Then γ_{t+1} will be a function of y_t and u_{t-1} rather than y^t and u^t, and u_t^* is obtained as a function of y_t rather than of y^t.

For this class of rules, we can write $p(\zeta_{t+1} \mid \zeta^t)$ as

$$\begin{aligned} &p_{\pi^t}(x_{t+1}, \xi_{t+1}, \eta_{t+1}, y_{t+1}, u_{t+1} \mid x^t, \xi^t, \eta^t, y^t, u^t) \\ &\quad = p(x_{t+1} \mid x_t, \xi_t, u_t) p(\xi_{t+1}) p(\eta_{t+1}) p(y_{t+1} \mid x_{t+1}, \eta_{t+1}) p(u_{t+1} \mid y^{t+1}, u^t) \\ &\quad = \delta(x_{t+1} - f_t) p(\xi_{t+1}) p(\eta_{t+1}) \delta(y_{t+1} - h_{t+1}) \delta(u_{t+1} - \pi_t(y_{t+1}, u_t)) \end{aligned} \tag{42}$$

where the independence assumption of the random noises is used. Thus, we see that

$$p_{\pi^t}(\zeta_{t+1} \mid \zeta^t) = p_{\pi_t}(\zeta_{t+1} \mid \zeta_t)$$

and $p_\pi(\zeta_{t+1} \mid \zeta_t)$ is computable as a function of π_t, or as a function of $u_t = \pi_t(y_t, u_{t-1})$. The vector sequence, $\{\zeta_t\}$, is therefore a first-order Markov sequence where each component of ζ_t can take a continuum of values on the real line. Actually it is not necessary to carry (ξ_t, η_t, u_t) in ζs of (41). $\{(x_t, y_t)\}$-process is still Markovian for Markov plans because

$$\begin{aligned} p_{\pi^t}(x_{t+1}, y_{t+1} \mid x^t, y^t) &= p(x_{t+1} \mid x_t, u_t = \pi_t(y_t)) p(y_{t+1} \mid x_{t+1}) \\ &= p_{\pi_t}(x_{t+1}, y_{t+1} \mid x_t, y_t) \end{aligned} \tag{43}$$

where $p(x_{t+1} \mid x_t, u_t)$ is computed from (40) making use of the known distribution of ξ_t and similarly for $p(y_{t+1} \mid x_{t+1})$. If the Markov rule does not hold, however, then u_t depends on y^t and we must consider

$$\zeta_t = (x_t, \xi_t, \eta_t, y^t) \qquad \text{or} \qquad \zeta_t = (x_t, y^t) \tag{44}$$

instead. Then,

$$\begin{aligned} &p_{\pi^t}(x_{t+1}, \xi_{t+1}, \eta_{t+1}, y^{t+1} \mid x^t, \xi^t, \eta^t, y^t) \\ &\quad = p(x_{t+1} \mid x_t, \xi_t, u_t = \pi_t(y^t)) p(\xi_{t+1}) p(\eta_{t+1}) \\ &\qquad \times p(y_{t+1} \mid x_{t+1}, \eta_{t+1}) \end{aligned}$$

and

$$p_{\pi^t}(\zeta_{t+1} \mid \zeta^t) = p_{\pi_t}(\zeta_{t+1} \mid \zeta_t), \tag{45}$$

and $\{\zeta_t\}$ becomes a first-order Markov sequence. Since the dimension of ζ_t grows with t, this process is not a conventional Markov sequence. One way to avoid the growing state vectors is to use sufficient statistics if they exist.

As another example of constructing a first-order Markov sequence $\{\zeta_t\}$, consider the system of (40) again, this time assuming that the distributions of noises ξ and η contain unknown parameters θ_1 and θ_2, respectively. The random noises are still assumed all independent. We can no longer compute $p(x_{t+1}, y_{t+1} \mid x_t, y_t)$ since $p(x_{t+1} \mid x_t, u_t)$ is a function of θ_1 and $p(y_{t+1} \mid x_{t+1})$ contains θ_2, and both are assumed unknown, i.e., $\{(x_t, y_t)\}$ is no longer Markovian.* Consider instead

$$\zeta_t = (x_t, \xi_t, \eta_t, \theta_{1t}, \theta_{2t}, y_t)$$

where

$$\begin{aligned} \theta_{1t} &= \theta_1 \\ \theta_{2t} &= \theta_2, \qquad t = 0, 1, \ldots, N-1. \end{aligned}$$

Then

$$\begin{aligned} &p_{\pi^t}(x_{t+1}, \xi_{t+1}, \eta_{t+1}, \theta_{1,t+1}, \theta_{2,t+1}, y_{t+1} \mid x^t, \xi^t, \theta_1^t, \theta_2^t, y^t) \\ &\quad = p(x_{t+1} \mid x_t, \xi_t, u_t = \pi_t(y_t)) p(\xi_{t+1} \mid \theta_1) \\ &\qquad \times p(\eta_{t+1} \mid \theta_2) \delta(\theta_{1,t+1} - \theta_1) \delta(\theta_{2,t+1} - \theta_2) \\ &\qquad \times p(y_{t+1} \mid x_{t+1}, \eta t+1) \\ &\quad = \delta(x_{t+1} - f_t) p(\xi_{t+1} \mid \theta_1) p(\eta_{t+1} \mid \theta_2) \\ &\qquad \times \delta(\theta_{1,t+1} - \theta_1) \delta(\theta_{2,t+1} - \theta_2) \delta(y_{t+1} - h_{t+1}) \end{aligned}$$

---

* It is conditionally Markovian in the sense that

$$p(x_{t+1}, y_{t+1} \mid x^t, y^t, \theta_1, \theta_2) = p(x_{t+1}, y_{t+1} \mid x_t, y_t, \theta_1, \theta_2).$$

This fact may be used advantageously in some cases.

which is computable again knowing ζ_t in Markov rules, hence $\{\zeta_t\}$ is Markovian.

Also, as indicated earlier, if we change the independence assumption on ξs and ηs, then $\{(x_t, y_t)\}$ is no longer Markovian even if the noise distributions are assumed known. For example, assume that ξs and ηs are the first-order Markov sequences and $\{\xi\}$-process is independent of $\{\eta\}$-process. Assume that $p(\xi_{t+1} \mid \xi_t)$ and $p(\eta_{t+1} \mid \eta_t)$ are known. Then, in (43), $p(x_{t+1} \mid x_t, u_t)$ is not known since ξ_t is not given; $p(y_{t+1} \mid x_{t+1})$ is unknown because η_{t+1} is not known. The augmented state vector $\boldsymbol{\zeta}_t$, where

$$\boldsymbol{\zeta}_t = (x_t, \xi_t, \eta_t, y_t),$$

still forms a Markov sequence, however, with Markov rules. This is seen by

$$\begin{aligned} p_{\pi^t}&(x_{t+1}, \xi_{t+1}, \eta_{t+1}, y_{t+1} \mid x^t, \xi^t, \eta^t, y^t) \\ &= p(x_{t+1} \mid x_t, \xi_t, u_t) p(\xi_{t+1} \mid \xi_t) \\ &\quad \times p(\eta_{t+1} \mid \eta_t) p(y_{t+1} \mid x_{t+1}, \eta_{t+1}). \end{aligned}$$

Problems with Sufficient Statistics

Now we consider the possibility of replacing y^t by some sufficient statistics. We have seen in previous chapters that u_t is generally a function of y^t and not just of y_t. This dependence of u_t on y^t occurs through $p(\cdot \mid y^t)$ in computing λ_{t+1}. Intuitively speaking, if sufficient statistics, s_t, exist so that $p(x_{t+1} \mid y^t) = p(x_{t+1} \mid s_t)$, where s_t is some function of s_{t-1} and y_t, and possibly of u_{t-1}, then the dependence of u_t on past observation is summarized by s_t. By augmenting the observed portion of the state vector with the addition of s_t, say

$$\boldsymbol{\zeta}_t = (x_t, \xi_t, \eta_t, y_t, s_t),$$

the $\{\zeta\}$-process may become Markovian. In order to make this more precise, consider a situation where the conditional density of x_{t+1}, given x_t, θ, and any control u_t, is known, where θ is a random parameter in a known parameter space Θ. That is, it is assumed that

$$p(x_{t+1} \mid x_t, u_t, \theta), \qquad \theta \in \Theta$$

is given, that the observation noise random variables are independent among themselves and of all other random variables, and that $p(y_t \mid x_t)$ is known completely. θ and x_0 are assumed to be independent. Denote their a priori probability density functions by $p_0(\theta)$ and $p_0(x_0)$, respectively.

Such dependence of the conditional density x_{t+1} on θ may arise either through the unknown system parameter θ in the dynamic equation or through the dynamic noises whose probability distribution function contains θ. This dependence on θ can occur, for example, for a system with the dynamic equation

$$\mathbf{x}_{t+1} = \mathbf{A}(\theta)\mathbf{x}_t + \mathbf{B}(\theta)\mathbf{u}_t + \xi_t$$

where $\mathbf{A}$ and $\mathbf{B}$ are known functions of θ and where the ξ_t are serially independent and have a known probability density, or for a system described by

$$\mathbf{x}_{t+1} = \mathbf{A}\mathbf{x}_t + \mathbf{u}_t + \xi_t$$

where $\mathbf{A}$ is a known constant matrix and ξs are independently and identically distributed with a probability density $p(\xi_t \mid \theta)$, $\theta \in \Theta$.

We know from Chapter 3 that, in order to find an optimal u_t, we need the expression for

(46)
$$\begin{aligned}\lambda_t &= E(W_t \mid \mathbf{y}^{t-1})\\ &= \int \mathrm{W}_t(\mathbf{x}_t, \mathbf{u}_{t-1}) p(\mathbf{x}_t \mid \mathbf{y}^{t-1})\, dx_t\\ &= \int \mathrm{W}_t(\mathbf{x}_t, \mathbf{u}_{t-1}) p(\mathbf{x}_t \mid \mathbf{x}_{t-1}, \mathbf{u}_{t-1}, \theta) p(\theta, \mathbf{x}_{t-1} \mid \mathbf{y}^{t-1})\, d(\mathbf{x}_{t-1}, \mathbf{x}_t, \theta)\end{aligned}$$

where $p(\theta, \mathbf{x}_{t-1} \mid \mathbf{y}^{t-1})$ can be generated recursively as such, or we can write

(47)
$$p(\theta, \mathbf{x}_{t-1} \mid \mathbf{y}^{t-1}) = p(\theta \mid \mathbf{y}^{t-1}) p(\mathbf{x}_{t-1} \mid \theta, \mathbf{y}^{t-1})$$

and obtain $p(\theta \mid \mathbf{y}^{t-1})$ and $p(\mathbf{x}_{t-1} \mid \theta, \mathbf{y}^{t-1})$ by two separate recursion equations.

The sequence of observations $\mathbf{y}_0, \ldots$ is related to the sequence $\mathbf{x}_0, \mathbf{x}_1, \ldots$ through the observation equation. Hence it is assumed that the sequence contains information on θ, i.e., the joint probability density function of y^t and θ is assumed to exist and is not identically equal to zero for almost all realization $\mathbf{y}^t$.

Given any control policy, the a posteriori probability density function of θ given $\mathbf{y}^t$ is computed from $p_0(\theta)$ by Bayes' rule. Define

(48)
$$p_t(\theta) \triangleq p(\theta \mid \mathbf{y}^t) = \frac{p_0(\theta) p(\mathbf{y}^t \mid \theta)}{p(\mathbf{y}^t)}.$$

Suppose now that a set of sufficient statistics $\mathbf{s}_t = \mathbf{s}(\mathbf{y}^t)$ exists for θ such that it satisfies the equation

(49)
$$\mathbf{s}_{t+1} = \psi_1(\mathbf{s}_t, \mathbf{y}_{t+1}, \mathbf{u}_t).$$

When a sufficient statistic exists it is known that the probability density of $\mathbf{y}^t$ given θ for any control policy can be factored as

(50) $$p(\mathbf{y}^t \mid \theta) = \mathrm{f}(\theta, \mathbf{s}_t) g(\mathbf{y}^t).$$

See for example Hogg and Craig (1959). A large class of sufficient statistics satisfy Condition (49). For example, consider a class of probability density functions known as the Koopman–Pitman class. The density function of this class has the form

$$p(y \mid \theta) = \exp\{r(\theta)K(y) + S(y) + q(\theta)\}, \qquad -\infty < y < \infty$$

where $r(\theta)$ is a nontrivial continuous funcction of θ, $S(y)$ and $K'(y) \not\equiv 0$ are continuous in y, and $\mathrm{q}(\theta)$ is some function of θ. The density of a Gaussian random variable, for example, belongs to this class. Then,

$$p(\mathbf{y}^t \mid \theta) = \exp\left[r(\theta)\sum_0^n K(\mathbf{y}_t) + \sum_0^n S(\mathbf{y}_t) + (n+1)\mathrm{q}(\theta)\right]$$

$$= R\left[\sum_0^n K(\mathbf{y}_t)\right]\exp\left[r(\theta)\sum_0^n K(\mathbf{y}_t) + (n+1)\mathrm{q}(\theta)\right]\frac{\exp\left(\sum_0^n S(\mathbf{y}_t)\right)}{\mathrm{R}\left[\sum_0^n K(\mathbf{y}_t)\right]}$$

where R is a function that arises in the one-to-one transformation defined by

$$\mathbf{z}_0 = \sum_0^n K(\mathbf{y}_t)$$

$$\mathbf{z}_t = \mathbf{y}_t, \qquad 1 \leqslant t \leqslant n$$

so that $\Sigma_0^n K(\mathbf{y}_t)$ is seen to be sufficient and is in the form of (49). From (48) and (50), the a posteriori probability density of θ is given by

(51) $$p_t(\theta) = \frac{p_0(\theta)\mathrm{f}(\theta, \mathbf{s}_t)}{\int d\theta\, p_0(\theta)\mathrm{f}(\theta, \mathbf{s}_t)} = p(\theta \mid \mathbf{s}_t).$$

Equation (51) shows that $p_t(\theta)$, the tth a posteriori probability density of θ, depends on $\mathbf{y}^t$ only through $\mathbf{s}(\mathbf{y}^t)$ when s is the sufficient statistic for θ. Then (47) becomes

$$p(\theta, \mathbf{x}_{t-1} \mid \mathbf{y}^{t-1}) = p(\theta \mid \mathbf{s}_{t-1}) p(\mathbf{x}_{t-1} \mid \theta, \mathbf{y}^{t-1}).$$

Therefore, instead of deriving the recursive relation for $p(\theta, \mathbf{x}_t \mid \mathbf{y}^t)$, one can obtain $p(\mathbf{x}_t \mid \theta, \mathbf{y}^t)$ recursively. This recursion equation will be generally easier to manage since θ is now assumed to be known and can be generated in the usual manner. First, write

$$p(\mathbf{x}_t, \mathbf{x}_{t+1}, \mathbf{y}_{t+1} \mid \mathbf{y}^t, \theta) = p(\mathbf{x}_t \mid \mathbf{y}^t, \theta) p(\mathbf{x}_{t+1} \mid \mathbf{x}_t, \mathbf{u}_t, \theta) p(\mathbf{y}_{t+1} \mid \mathbf{x}_{t+1}).$$

Now

$$\int p(\mathbf{x}_t, \mathbf{x}_{t+1}, \mathbf{y}_{t+1} \mid \mathbf{y}^t, \theta)\, d\mathbf{x}_t = p(\mathbf{y}_{t+1} \mid \mathbf{y}^t, \theta) p(\mathbf{x}_{t+1} \mid \mathbf{y}^{t+1}, \theta).$$

Therefore,

$$p(\mathbf{x}_{t+1} \mid \mathbf{y}^{t+1}, \theta) = \int p(\mathbf{x}_t \mid \mathbf{y}^t, \theta) p(\mathbf{x}_{t+1} \mid \mathbf{x}_t, \mathbf{u}_t, \theta) \times \frac{p(\mathbf{y}_{t+1} \mid \mathbf{x}_{t+1})\, d\mathbf{x}_t}{\int [\text{numerator}]\, d\mathbf{x}_{t+1}} \tag{52}$$

with

$$p(\mathbf{x}_0 \mid \mathbf{y}_0, \theta) = \frac{p(\mathbf{x}_0, \mathbf{y}_0 \mid \theta)}{p(\mathbf{y}_0 \mid \theta)} = \frac{p_0(\mathbf{x}_0 \mid \theta) p(\mathbf{y}_0 \mid \mathbf{x}_0, \theta)}{\int p_0(\mathbf{x}_0 \mid \theta) p(\mathbf{y}_0 \mid \mathbf{x}_0, \theta)\, d\mathbf{x}_0} = \frac{p_0(\mathbf{x}_0) p(\mathbf{y}_0 \mid \mathbf{x}_0)}{\int p_0(\mathbf{x}_0) p(\mathbf{y}_0 \mid \mathbf{x}_0)\, d\mathbf{x}_0}.$$

Suppose that (52) is such that another sufficient statistic t_t exists such that $p(\mathbf{x}_t \mid \theta, \mathbf{y}^t) = p(\mathbf{x}_t \mid \theta, t_t)$, where t_{t-1} is a function of $t_t, \mathbf{y}_{t+1}$, and $\mathbf{u}_t$, and that $t_{t+1} = \psi_2(t_t, \mathbf{y}_{t+1}, \mathbf{u}_t)$. Then, (46) shows that λ_t will be a function of $\mathbf{u}_{t-1}, \mathbf{s}_{t-1}$, and t_{t-1}. To show that γ_t also depends at most on $\mathbf{u}_{t-1}, \mathbf{s}_{t-1}, t_{t-1}$, and on $\mathbf{y}_{t-1}$, we must next investigate the functional dependence of $E(\gamma_{t+1}^* \mid \mathbf{y}^{t-1})$. We know from the preceding argument that

$$\gamma_N = \lambda_N(\mathbf{u}_{N-1}, \mathbf{s}_{N-1}, t_{N-1})$$

and therefore

$$\mathbf{u}_{N-1}^* = \boldsymbol{\phi}_{N-1}(\mathbf{s}_{N-1}, t_{N-1}).$$

Then, γ_N^* will be a function of $\mathbf{s}_{N-1}$ and t_{N-1}. In computing γ_{N-1}, we need

$$E(\gamma_N^* \mid \mathbf{y}^{N-2}) = \int \gamma_N^* p(\mathbf{s}_{N-1}, t_{N-1} \mid \mathbf{y}^{N-2})\, d\mathbf{y}_{N-1}$$

where

$$\mathbf{s}_{N-1} = \boldsymbol{\psi}_1(\mathbf{s}_{N-2}, \mathbf{y}_{N-1}, \mathbf{u}_{N-2})$$

and

$$t_{N-1} = \psi_2(t_{N-2}, \mathbf{y}_{N-1}, \mathbf{u}_{N-2})$$

by assumption. Now

$$\begin{aligned} p(\mathbf{y}_{N-1} \mid \mathbf{y}^{N-2}) &= \int p(\mathbf{y}_{N-1} \mid \mathbf{x}_{N-1}) p(\mathbf{x}_{N-1} \mid \mathbf{x}_{N-2}, \mathbf{u}_{N-1}, \theta) \\ &\quad \times p(\theta, \mathbf{x}_{N-2} \mid \mathbf{y}^{N-2})\, d(\mathbf{x}_{N-1}, \mathbf{x}_{N-2}, \theta) \\ &= \int p(\mathbf{y}_{N-1} \mid \mathbf{x}_{N-1}) p(\mathbf{x}_{N-1} \mid \mathbf{x}_{N-2}, \mathbf{u}_{N-2}, \theta) \\ &\quad \times p(\theta \mid \mathbf{s}_{N-2}) p(\mathbf{x}_{N-2} \mid \theta, t_{N-2})\, d(\mathbf{x}_{N-1}, \mathbf{x}_{N-2}, \theta). \end{aligned}$$

Hence $E(\gamma_N^* \mid \mathbf{y}^{N-2})$ will depend at most on $\mathbf{s}_{N-2}, t_{N-2}$, and $\mathbf{u}_{N-2}$. By similar reasonings, $E(\gamma_{t+1}^* \mid \mathbf{y}^{t-1})$ is seen to depend at most on $\mathbf{s}_{t-1}, t_{t-1}$, and $\mathbf{u}_{t-1}$. Thus,

$$\mathbf{u}_{t-1}^* = \pi_{t-1}(\mathbf{s}_{t-1}, t_{t-1}), \qquad 1 \leqslant t \leqslant n.$$

To summarize, $\mathbf{s}_t$ is the sufficient statistic for θ and t_t is the sufficient statistic for $\mathbf{y}^t$. $\mathbf{s}_{t+1}$ and t_{t+1} are computed as known functions of $\mathbf{s}_t, t_t, \mathbf{u}_t$, and $\mathbf{y}_{t+1}$. Now we are ready to show that $\{\zeta_t\}$ is a first-order Markov sequence where

(53) $$\zeta_t = (\mathbf{x}_t, \mathbf{y}_t, t_t, \mathbf{s}_t), \qquad 0 \leqslant t \leqslant N-1.$$

This can be shown by computing

$$\begin{aligned} &p(\mathbf{x}_{t+1}, \mathbf{y}_{t+1}, t_{t+1}, \mathbf{s}_{t+1} \mid \mathbf{x}^t, \mathbf{y}^t, t^t, \mathbf{s}^t) \\ &\quad = p(\mathbf{x}_{t+1} \mid \mathbf{x}^t, \mathbf{y}^t, t^t, \mathbf{s}^t) p(\mathbf{y}_{t+1} \mid \mathbf{x}_{t+1}) \\ &\qquad \times p(t_{t+1} \mid t_t, \mathbf{y}_{t+1}, \mathbf{u}_t) p(\mathbf{s}_{t+1} \mid \mathbf{s}_t, \mathbf{y}_{t+1}, \mathbf{u}_t) \end{aligned}$$

where the assumptions of the dependence of t_{t+1} and $\mathbf{s}_{t+1}$ on $t_t, \mathbf{s}_t, \mathbf{y}_{t+1}$, and $\mathbf{u}_t$ are used. The first factor can be written as

$$p(\mathbf{x}_{t+1} \mid \mathbf{x}^t, \mathbf{y}^t, t^t, \mathbf{s}^t) = \int p(\mathbf{x}_{t+1} \mid \mathbf{x}_t, \mathbf{u}_t, \theta) p(\theta \mid \mathbf{s}_t)\, d\theta$$

since

$$\mathbf{u}_t = \pi_t(t_t, \mathbf{s}_t).$$

Thus

$$p(\zeta_{t+1} \mid \zeta^t) = p(\zeta_{t+1} \mid \zeta_t).$$

The observed portion of the vector ζ_t is $(\mathbf{y}_t, t_t, \mathbf{s}_t)$. It can also be shown that $\{\zeta_t\}$, where $\zeta_t = (\mathbf{x}_t, \theta_t, \mathbf{y}_t, \mathbf{s}_t, t_t)$, $\theta_t = \theta$ is also Markovian.

Optimal Control Policies

Suppose that $\{\zeta_t\}$-process, where ζ_t is derived by augmenting $(\mathbf{x}_t, \mathbf{y}_t)$ appropriately, is a first-order Markov sequence with a known transition probability density function. In the previous section, several ways of constructing such ζs were discussed.

Components of the vector ζ_t can usually be grouped into two classes, one group consisting of components not observed and not available for control signal synthesis, the other group consisting of known (vector) functions of components observed and stored by the controller. Denote them by $\boldsymbol{\mu}_t$ and $\mathbf{v}_t$, respectively,

$$\zeta_t = (\boldsymbol{\mu}_t, \mathbf{v}_t).$$

Therefore, $\boldsymbol{\mu}_t$ contains some function of $\mathbf{x}_t$ among others and $\mathbf{v}_t$ contains some function of $\mathbf{y}^t$.* For example, if $\zeta_t = (\mathbf{x}_t, \xi_t, \boldsymbol{\eta}_t, \mathbf{y}_t)$, then $\boldsymbol{\mu}_t = (\mathbf{x}_t, \xi_t, \boldsymbol{\eta}_t)$ and $\mathbf{v}_t = \mathbf{y}_t$. The available data at time t to the controller is $\mathbf{v}^t$, and $\mathbf{u}_t$ is to be determined by choosing $\boldsymbol{\pi}_t$ where

$$\mathbf{u}_t = \boldsymbol{\pi}_t(\mathbf{u}^{t-1}, \mathbf{v}^t).$$

Last Stage

Let us determine u_{N-1}^*, assuming $u_0^*, \ldots, u_{N-2}^*$ have already been chosen. $E(W_N)$ is minimized by minimizing $E(W_N \mid v^{N-1})$ for every possible v^{N-1}. Now, as a function of u_{N-1},

$$\begin{aligned} E(W_N \mid v^{N-1}) &= \int W_N(x_N, u_{N-1}) p(x_N, u_{N-1} \mid v^{N-1})\, d(x_N, u_{N-1}) \\ &= \int W_N(x_N, u_{N-1}) p(u_{N-1} \mid v^{N-1}) p(x_N \mid u_{N-1}, v^{N-1})\, d(x_N, u_{N-1}) \\ &= \int W_N(x_N, u_{N-1}) p(u_{N-1} \mid v^{N-1}) p(\mu_N \mid \mu_{N-1}, u_{N-1}, v_{N-1}) \\ &\qquad \times p(\mu_{N-1} \mid u_{N-1}, v^{N-1})\, d(\mu_N, \mu_{N-1}, u_{N-1}) \end{aligned}$$

assuming $p(\mu_{N-1} \mid v^{N-1}, u_{N-1})$ is available. The density $p(\mu_N \mid \mu_{N-1}, v_{N-1}, u_{N-1})$ is computed from the known transition density $p(\mu_N, v_N \mid \mu_{N-1}, v_{N-1}, u_{N-1})$.

Following the line of reasoning in Chapter 3, one sees that the nonrandomized control is optimal and is given by

$$p^*(u_{N-1} \mid v^{N-1}) = \delta(u_{N-1} - u_{N-1}^*)$$

where u_{N-1}^* minimizes λ_N,

$$\begin{aligned} \lambda_N(u_{N-1}, v^{N-1}) \triangleq \int & W_N(x_N, u_{N-1}) p(\mu_N \mid \mu_{N-1}, u_{N-1}, v_{N-1}) \\ & \times p(\mu_{N-1} \mid u_{N-1}, v^{N-1})\, d(\mu_N, \mu_{N-1}). \end{aligned}$$

---

* In some cases, it may be convenient to take ζ_t such that both $\boldsymbol{\mu}_t$ and $\mathbf{v}_t$ are functions of observed quantities by the controller.

Define

$$\gamma_N^*(v^{N-1}) = \min_{u_{N-1}} \lambda_N.$$

General Case

Generally u_{t-1}^* is found by minimizing γ_t where

$$\lambda_t \triangleq \int W_t(x_t, u_{t-1}) p(\mu_t \mid \mu_{t-1}, v_{t-1}, u_{t-1}) p(\mu_{t-1} \mid v^{t-1}, u_{t-1})\, d(\mu_t, \mu_{t-1})$$

(54) $$\gamma_t \triangleq \lambda_t + \int \gamma_{t+1}^* p(v_t \mid u_{t-1}, v^{t-1})\, dv_t, \qquad t = 1, \ldots, N$$

$$\gamma_t^* = \min_{u_{t-1}} \gamma_t$$

where $p(\mu_{t-1} \mid u_{t-1}, v^{t-1})$ and $p(v_t \mid u_{t-1}, v^{t-1})$ are assumed known. We see that the optimal control policies are obtainable if $p(\mu_t \mid u_t, v^t)$ and $p(v_{t+1} \mid u_t, v^t)$, $t = 0, \ldots, N-1$, are known.

Therefore, our attention is next turned to computing these conditional probability density functions. Here, it should be pointed out that this approach may be computationally advantageous even when $\{x_t\}$ and $\{y_t\}$ are Markovian by themselves.

Derivation of Conditional Probability Densities

It is not generally true that $\{v_t\}$ itself is Markovian even though $\{\zeta_t\}$ is. We want to compute the conditional probability densities of parts of the components of a multidimensional Markov sequence conditioned on components that are observed. We assume that the a priori probability density of ζ_0,

(55) $$p_0(\zeta_0) = \mathrm{p}_0(\mu_0, v_0)$$

is given. Let us now obtain the recursion equation for $p(\mu_t \mid v^t)$ and $p(v_{t+1} \mid v^t)$. By now, the method of obtaining such a recursion relation should be routine for us. We consider

$$\mathrm{p}(\mu_t, \mu_{t+1}, v_{t+1} \mid v^t).$$

By the chain rule, we can write it as

$$\begin{aligned}\mathrm{p}(\mu_t, \mu_{t+1}, v_{t+1} \mid v^t) &= p(\mu_t \mid v^t) p(\mu_{t+1}, v_{t+1} \mid \mu_t, v^t) \\ &= p(\mu_t \mid v^t) p(\mu_{t+1}, v_{t+1} \mid \mu_t, v_t)\end{aligned}$$

where the last line is obtained from the Markovian property of $\{\zeta_t\}$.

Integrating both sides with respect to μ_t, we have

$$\begin{aligned}p(\mu_{t+1}, v_{t+1} \mid v^t) &= p(v_{t+1} \mid v^t) p(\mu_{t+1} \mid v^{t+1}) \\ &= \int p(\mu_t \mid v^t) p(\mu_{t+1}, v_{t+1} \mid \mu_t, v_t)\, d\mu_t.\end{aligned}$$

Therefore,

$$(56)\qquad p(\mu_{t+1} \mid v^{t+1}) = \frac{\int p(\mu_t \mid v^t) p(\mu_{t+1}, v_{t+1} \mid \mu_t, v_t)\, d\mu_t}{\int [\text{numerator}]\, d\mu_{t+1}}$$

and the denominator gives $p(v_{t+1} \mid v^t)$.

7.4 Examples

Example 6: System with Unknown Random Time Constant. Let us rework Example 4 of Chapter 3 by another method, using the idea of the augmented state vector. We know from our previous investigation of this example that a sufficient statistic exists for θ,

$$(57)\qquad \begin{aligned} p(\theta \mid x^t, u^{t-1}) &= p(\theta \mid s_t, \sigma_{\theta,t}) \\ &= \frac{1}{(2\pi)^{1/2}\sigma_{\theta,t}} \exp\left(-\frac{(\theta - s_t)^2}{2\sigma_{\theta,t}^2} \right) \end{aligned}$$

where we identify s_t and μ_{t-1} and $\sigma_{\theta,t}$ with Σ_{t-1}. With this identification of the sufficient statistics, they are seen to satisfy the recursion equation

$$(58)\qquad \begin{aligned} s_{t+1} &= \frac{\sigma_{\theta,t+1}^2}{\sigma^2} a_j + \frac{\sigma_{\theta,t+1}^2}{\sigma_{\theta,t}^2} s_t \\ \frac{1}{\sigma_{\theta,t+1}^2} &= \frac{1}{\sigma_{\theta,t}^2} + \frac{1}{\sigma^2} \\ s_0 &= \theta_0, \qquad \sigma_{\theta,0}^2 = \sigma_0^2 \\ \sigma_{\theta,t}^2 &\triangleq \sigma_0^2\sigma^2/(\sigma^2 + t\sigma_0^2) \end{aligned}$$

and

$$s_t = \frac{\sigma_0^2 \Sigma_0^{t-1} (x_{t+1} - bu_t/x_t) + \sigma^2\theta_0}{\sigma^2 + t\sigma_0^2}.$$

Note that $(x_{t+1} - bu_t)/x_t = a_t$ in (58), showing that the value the random variable takes at time t is exactly computable because of no measurement errors.

Since s_t summarizes all the information contained in x^t about θ, it is seen that controls u_t depending on s_t and x_t are just as good as controls depending on x^t.

Therefore, we consider the class of nonrandomized control policies such that

$$u_t = \pi_t(x_t, s_t), \qquad 0 \leqslant t \leqslant N-1.$$

Define

$$\zeta_t' = (x_t, s_t).$$

Therefore, the augmented state vector ζ_t obeys the augmented dynamic equation

$$\zeta_{t+1} = \begin{pmatrix} a_t & 0 \\ 0 & \dfrac{\sigma_{\theta,t+1}^2}{\sigma_{\theta,t}^2} \end{pmatrix} \zeta_t + \begin{pmatrix} bu_t & \\ \dfrac{\sigma_{\theta,t+1}^2}{\sigma^2} & a_t \end{pmatrix}. \tag{59}$$

Since u_t depends only on x_t and s_t and since a_t is mutually independent, (59) shows that $\{\zeta_t\}$ is a first-order Markov chain. This is a rather special case where $m_t = x_t$ and $v_t = s_t$, i.e., every component of ζ_t is observed.

Its transition probability density is given by

$$p(x_{t+1}, s_{t+1} \mid x_t, s_t, u_t) = p(x_{t+1} \mid x_t, s_t, u_t) p(s_{t+1} \mid x_{t+1}, x_t, s_t, u_t). \tag{60}$$

The right-hand side of (60) is computable since

$$\begin{aligned} p(x_{t+1} \mid x_t, s_t, u_t) &= \int p(x_{t+1} \mid x_t, s_t, u_t, \theta) p(\theta \mid x_t, s_t, u_t)\, d\theta \\ &= \int p(x_{t+1} \mid x_t, u_t, \theta) p(\theta \mid s_t)\, d\theta \end{aligned}$$

where $p(\theta \mid s_t)$ is given by (57). From (58),

$$p(s_{t+1} \mid x_{t+1}, x_t, s_t, u_t) = \delta\left[s_{t+1} - \left(\frac{\sigma_{\theta,t+1}^2}{\sigma_{\theta,t}^2} s_t + \frac{\sigma_{\theta,t+1}^2}{\sigma^2} \left(\frac{x_{t+1} - bu_t}{x_t} \right)\right)\right].$$

To derive an optimal control policy for the problem, we first compute λ_N by

$$\begin{aligned} \lambda_N &= E(x_N^2 \mid x^{N-1}) \\ &= E(x_N^2 \mid x_{N-1}, s_{N-1}) \\ &= \int x_N^2 p(x_N, s_N \mid x_{N-1}, s_{N-1}, u_{N-1})\, d(x_N, s_N) \\ &= \int x_N^2 p(x_N \mid x_{N-1}, u_{N-1}, \theta) p(\theta \mid s_{N-1})\, d(x_N, \theta). \end{aligned} \tag{61}$$

The probability density, $p(x_N \mid x_{N-1}, u_{N-1}, \theta)$, is Gaussian with mean $\theta x_{N-1} + bu_{N-1}$ and variance $\sigma^2 x_{N-1}^2$. Therefore,

$$\begin{aligned} \lambda_N &= \int [(\theta x_{N-1} + bu_{N-1})^2 + \sigma^2 x_{N-1}^2] p(\theta \mid s_{N-1})\, d\theta \\ &= (s_{N-1} x_{N-1} + bu_{N-1})^2 + x_{N-1}^2 (\sigma^2 + \sigma_{\theta,N-1}^2). \end{aligned}$$

We derive

$$u_{N-1}^* = -s_{N-1}x_{N-1}/b$$

and

$$\gamma_N^* = (\sigma^2 + \sigma_{\theta,N-1}^2)\sigma_{N-1}^2.$$

Since this is a final-value problem,

$$\lambda_t = 0, \qquad t = 1, \ldots, N-1$$

and

$$\gamma_{N-1} = \int \gamma_N^* p(x_{N-1}, s_{N-1} \mid x_{N-2}, s_{N-2}, u_{N-2})\, d(x_{N-1}, s_{N-1})$$

$$= (\sigma^2 + \sigma_{\theta,N-1}^2) \int \sigma_{N-1}^2 p(x_{N-1} \mid x_{N-2}, s_{N-2}, u_{N-2})\, dx_{N-1}. \tag{62}$$

Comparing (62) with (61), we see immediately that, aside from the multiplicative constant factor, minimization of λ_N with respect to u_{N-1} is identical to that of γ_{N-1} with respect to u_{N-2}. Therefore,

$$u_{N-2}^* = -s_{N-2}x_{N-2}/b$$

or, in general

$$u_t^* = -s_t x_t/b, \qquad t = 0, 1, \ldots, N-1$$

where s_t is given by (61). Note that, as $\sigma_0^2/\sigma_2^2 \to 0$, i.e., as our knowledge of the unknown mean θ becomes more precise,

$$s_t \to \theta_0$$

as expected.

Example 7: System with Markovian Gain. Consider a one-dimensional control system described by

$$x_{t+1} = ax_t + b_t u_t, \qquad t = 0, 1, \ldots, N-1$$

where a and x_0 are assumed known. The gain of the system b_t is assumed to form a first-order Markov chain with two possible states,

$$b_t = \begin{cases} \beta_1 \\ \beta_2 \end{cases}$$

with known stationary transition probabilities

$$\mathrm{P}[b_{t+1} = \beta_i \mid b_t = \beta_j] \triangleq p_{ij}, \qquad i,j = 1, 2 \qquad t = 0, 1, \ldots, N-1$$

and $\mathrm{P}[b_0]$ given. The observations are assumed perfect;

$$y_t = x_t, \qquad t = 0, 1, \ldots, N-1.$$

Hence, xs are used throughout, instead of ys. The cost index is taken to be

$$J = \sum_{1}^{N} (x_t^2 + \lambda u_{t-1}^2).$$

Since

$$b_t = \frac{x_{t+1} - a x_t}{u_t} \qquad \text{if } u_t \neq 0,$$

the knowledge of x^t implies that the value of b_{t-1} is known. This fact is used to obtain the probability distribution for b_t that we need when evaluating λ_t.*

The augmented state vector for this example can be taken to be

$$\zeta_t = (b_{t-1}, x_t), \qquad t = 1, 2, \ldots, N.$$

This example problem is special in that all components of ζ_t are known at time t.**

The fact that $\{\zeta_t\}$ is a first-order Markov sequence is verified by computing the conditional density $P(\zeta_{t+1} \mid \zeta^t)$

$$\begin{aligned} p(\zeta_{t+1} \mid \zeta^t) &= p(b_t, x_{t+1} \mid b^{t-1}, x^t) \\ &= p(b_t \mid b_{t-1}) p(x_{t+1} \mid x_t, u_t, b_t) \\ &= p(\zeta_{t+1} \mid \zeta_t) \end{aligned} \tag{63}$$

assuming $u_t = \pi_t(x_t, b_{t-1})$. We shall see shortly that the optimal control variables have the assumed functional dependence.

To derive the optimal control policy, we consider the last control stage first.

Since it is convenient to indicate the dependence of the conditional expectation $E(W_N \mid \zeta_{N-1})$ on the value of b_{N-2}, let us define

$$\begin{aligned} \lambda_{N,k} &\triangleq E(W_N \mid x_{N-1}, b_{N-2} = \beta_k) \\ &= \int (x_N^2 + \lambda u_{N-1}^2) P(x_N, b_{N-1} \mid x_{N-1}, b_{N-2} = \beta_k)\, d(x_N, b_{N-1}) \\ &= \sum_{i=1}^{2} \int (x_N^2 + \lambda u_{N-1}^2) P(x_N \mid x_{N-1}, b_{N-1} = \beta_i, u_{N-1}) p_{ik}\, dx_N \\ &= \sum_{i=1}^{2} [(a x_{N-1} + \beta_i u_{N-1})^2 + \lambda u_{N-1}^2] p_{ik}. \end{aligned} \tag{64}$$

---

* If $u_{t-1} = 0$, then b_{t-1} is unknown. In this unlikely case, we must work with

$$P(b_t \mid b_{t-2}) = \sum_{1}^{2} P(b_t \mid b_{t-1} = \beta_k) P(b_{t-1} = \beta_k \mid b_{t-2}).$$

** ζ_0 and $p(\zeta_0)$ require an obvious special handling because of $P[b_0]$.

Minimizing (64) with respect to u_{N-1}, we obtain the optimal control at time $N-1$

(65) $$u_{N-1,k}^{*} = -\Lambda_{N-1,k}x_{N-1}, \qquad k = 1,2$$

where

(66) $$\Lambda_{N-1,k} \triangleq \frac{\sum_{i-1}^{2} \beta_i p_{ik}}{\lambda + \sum_{i=1}^{2} \beta_i^2 p_{ik}} a.$$

In the denominator (66), note that $\Sigma_{i=1}^{2}\beta_i p_{ik} = E(b_{N-1} \mid b_{N-2} = \beta_k)$. Thus, from (65) and (66), we see that u_{N-1}^{*} depends on the value of b_{N-2}, i.e., $u_{N-1}^{*} = \phi_{N-1}(x_{N-1}, b_{N-2})$ as assumed in connection with (63). Substituting (66) into (64), we obtain

(67) $$\lambda_{N,k}^{*} \triangleq \min_{u_{N-1}} \lambda_{N,k} = C_{N-1,k}x_{N-1}^2$$

where

$$C_{N-1,k} = a^2\{1 - [E(b_{N-1} \mid b_{N-2} = \beta_k]^2/[\lambda + E(b_{N-1}^2 \mid b_{N-2} = \beta_k)]\}.$$

Equation (67) shows that $\lambda_{N,1}$ and $\lambda_{N,2}$ are both quadratic in x_{N-1}. Now we compute the optimal control variable generally. Assume that

(68) $$\begin{aligned} \gamma_{t+1,k}^{*} &= \min_{u_t,\ldots,u_{N-1}} E\left(\sum_{t=t+1}^{N} W_t \mid x_t, b_{t-1} = \beta_k\right) \\ &\triangleq C_{t,k}x_t^2, \qquad k = 1,2, \qquad t = 0,1,\ldots,N-1. \end{aligned}$$

By definition,

$$C_{N,k} = 0, \qquad k = 1,2.$$

Then, from (68),

$$\begin{aligned} \gamma_{t,k}^{*} &= \min_{u_{t-1},\ldots,u_N} E\left(\sum_{s=t}^{N} W_s \mid x_{t+1}, b_{t-2} = \beta_k\right) \\ &= C_{t-1,k}x_{t-1}^2. \end{aligned}$$

On the other hand,

$$\gamma_{t,k}^{*} = \min_{u_{t-1}}\left[\lambda_{t,k} + \sum_{m=1}^{2} E(\gamma_{t+1,m}^{*} \mid x_{t-1}, b_{t-2} = \beta_k)\right]$$

where

$$\lambda_{t,k} = E(x_t^2 + \lambda u_{t-1}^2 \mid x_{t-1}, b_{t-2} = \beta_k)$$

and where

(69) $$\begin{aligned} &\sum_{m=1}^{2} E[\gamma_{t-1,m}^{*} \mid x_{t-1}, b_{t-2} = \beta_k] \\ &= \sum_{m=1}^{2} E(C_{t,m}x_t^2 \mid x_{t-1}, b_{t-2} = \beta_k) \\ &= C_{t,1}(ax_{t-1} + \beta_1 u_{t-1})^2 p_{1k} + C_{t,2}(ax_{t-1} + \beta_2 u_{t-1})^2 p_{2k}. \end{aligned}$$

Therefore, making use of (69),

$$\begin{aligned}\gamma_{t,k} &= \lambda u_{t-1}^2 + (C_{t,1}+1)p_{1k}(ax_{t-1}+\beta_1 u_{t-1})^2 \\ &\quad + (C_{t,2}+1)p_{2k}(ax_{t-1}+\beta_2 u_{t-1})^2, \qquad k = 1,2.\end{aligned} \tag{70}$$

By minimizing (70) with respect to u_{t-1}, we obtain $u_{t-1,k}^*$ to be

$$\mathrm{u}_{t-1,k}^* = -\Lambda_{t-1,k} x_{t-1} \tag{71}$$

where

$$\Lambda_{t-1,k} \triangleq \frac{p_{1k}\beta_1(C_{t,1}+1) + p_{2k}\beta_2(C_{t,2}+1)}{\lambda + p_{1k}\beta_1^2(C_{t,1}+1) + p_{2k}\beta_2^2(C_{t,2}+1)}\, a \tag{72}$$

where $k = 1$ or 2, depending on $b_{t-2} = \beta_1$ or β_2, respectively. Substituting (71) into (70), the recursion equation for $C_{t,k}$ is obtained:

$$\begin{aligned}C_{t-1,k} = a^2 \Bigg\{ & 1 + p_{1k}C_{t,1} + p_{2k}C_{2,k} \\ & - \frac{[p_{1k}\beta_1(C_{t,1}+1) + p_{2k}\beta_2(C_{t,2}+1)]^2}{\lambda + p_{1k}\beta_1^2(C_{t,1}+1) + p_{2k}\beta_2^2(C_{t,2}+1)} \Bigg\}, \qquad k = 1,2.\end{aligned} \tag{73}$$

Equation (73) can be simplified somewhat by writing it in a vector form. Define

$$\mathbf{C}_{t-1} = \begin{pmatrix} C_{t-1,1} \\ C_{t-1,2} \end{pmatrix}. \tag{74}$$

Then

$$\mathbf{C}_{t-1} = a^2 \begin{pmatrix} 1 \\ 1 \end{pmatrix} + a^2 \begin{pmatrix} p_{11} & p_{21} \\ p_{12} & p_{22} \end{pmatrix} \mathbf{C}_t + \mathbf{D}_t, \qquad t = 2,3,\ldots,N \tag{75}$$

where

$$\mathbf{D}_t = \begin{pmatrix} d_{t1}^2 \mid \Delta_{t1} \\ d_{t2}^2 \mid \Delta_{t1} \end{pmatrix}, \tag{76}$$

$$\begin{pmatrix} d_{t,1} \\ d_{t,2} \end{pmatrix} = a \begin{pmatrix} p_{11}\beta_1 & p_{21}\beta_2 \\ p_{12}\beta_1 & p_{22}\beta_2 \end{pmatrix} \left[\mathbf{C}_t + a \begin{pmatrix} 1 \\ 1 \end{pmatrix} \right], \tag{77}$$

and

$$\begin{pmatrix} \Delta_{t1} \\ \Delta_{t2} \end{pmatrix} = \begin{pmatrix} p_{11}\beta_1^2 & p_{21}\beta_2^2 \\ p_{12}\beta_1^2 & p_{22}\beta_2^2 \end{pmatrix} \left[\mathbf{C}_t + \lambda \begin{pmatrix} 1 \\ 1 \end{pmatrix} \right]. \tag{78}$$

Note that the initial vector $\mathbf{C}_0$ must be computed using the a priori probability $\mathrm{P}[b_0 = \beta_i]$, $i = 1, 2$, rather than the transition probabilities. From (72) and (75)–(78),

$$\mathbf{\Lambda}_{t-1} \triangleq \begin{pmatrix} \Delta_{t-1,1} \\ \Delta_{t-1,2} \end{pmatrix} = -\begin{pmatrix} d_{t1} \mid \Delta_{t1} \\ d_{t2} \mid \Delta_{t2} \end{pmatrix}. \tag{79}$$

Since $\mathbf{C}$s and $\mathbf{\Lambda}$s can be precomputed, the only operation that the optimal controller must perform on-line at time t is the determination of b_{t-1} to be either β_1 or β_2 by

$$b_{t-1} = \frac{x_t - x_{t-1}}{u_{t-1}}, \qquad t = 1, 2, \ldots, N-1, \tag{80}$$

i.e., $\mathrm{P}[b_{t-1} = \beta_k \mid x^t, u^{t-1}] = 1$, where β_k is given by (80). Once $b_{t-1} = \beta_k$ is determined, then $u_t^* = -\Lambda_{t,k} x_t$.

Remark: If instead, the system is governed by

$$\begin{aligned} x_{t+1} &= a_t x_t + bu_t \\ y_t &= x_t \end{aligned} \tag{81}$$

where b is now assumed known, and where a_t is such that

$$\begin{aligned} &a_t = \begin{cases} \alpha_1 \\ \alpha_2 \end{cases} \\ &\mathrm{P}[a_{t+1} = a_j \mid a_t = \alpha_k] \triangleq p_{jk} \qquad \text{for all } t = 0, 1, \ldots, \end{aligned} \tag{82}$$

then the optimization problem of the preceding system with respect to the performance index

$$J = \sum_1^N (x_t^2 + \lambda u_t^2)$$

can be similarly carried out.

Next consider the same dynamic equation, the Markov parameter b_t, and the performance index with *noisy* observation

$$y_t = x_t + \eta_t.$$

The previous development has shown that, if the properties of the noise are such that the values of b_t can be determined exactly from y^{t+1} and u^t, namely

$$\mathrm{P}[b_t = \beta_k \mid y^{t+1}, u^t] = 1 \tag{83}$$

for $k = 1$, or 2, then the previously derived optimal control policy is still optimal for this modified noisy observation problem. When (83) is not true, however, then the preceding policy will no longer be optimal. The reader is invited to rework this example by suitably defining an augmented state vector.

Example 8: Discrete-State Stochastic System. Consider a three-stage discrete-state, discrete-time, stochastic control process whose state at time t is denoted by x_t and the observation of x_t at time t is given by y_t. It is given that x_t is either a_1 or a_2 and y_t is either b_1 or b_2. The criterion of control is to minimize the expected value of

$$J = |x_1 - a_1| + |x_2 - a_1| - u_0^2 + u_1^2$$

where

$$x_{t+1} = \mathrm{f}_t(x_t, u_t, \xi_t)$$

$$y_t = \mathrm{h}_t(x_t, \eta_t), \qquad t = 0, 1, 2.$$

ξ_t and η_t are some noise processes. The control u_t is assumed to take on only two values, 0 and m.

The vector $\boldsymbol{\zeta}_t = (x_t, y_t)$ is assumed to be a Markov chain with known stationary transition probabilities, given u_t. The initial probabilities of ζ_0 are also assumed known.

Since this example deals with discrete-state variables, the developments in the main body of the chapter must be modified in an obvious way to deal with probabilities rather than probability densities. Such modifications will be made without further comments.

Four possible states of $\boldsymbol{\zeta}_0$ are labeled as follows:

$$\mathbf{c}_1 = (a_1, b_1),$$

$$\mathbf{c}_2 = (a_2, b_1),$$

$$\mathbf{c}_3 = (a_1, b_2),$$

$$\mathbf{c}_4 = (a_2, b_2).$$

Given

$$\mathrm{P}(\boldsymbol{\zeta}_0 = \mathbf{c}_i) = g_i, \qquad 1 \leqslant i \leqslant 4$$

where

$$g_1 = 0.45,$$

$$g_2 = 0.05,$$

$$g_3 = 0.1,$$

$$g_4 = 0.4,$$

and the stationary transition probabilities

$$g_{ij}(u) = \mathrm{P}(\zeta_{k+1} = \mathbf{c}_i \mid \zeta_k = \mathbf{c}_j, u), \qquad k = 0, 1$$

where

$$p(0) = [g_{ij}(0)] = \begin{pmatrix} 0.4 & 0.05 & 0.5 & 0.05 \\ 0.1 & 0.35 & 0 & 0.55 \\ 0.45 & 0 & 0.45 & 0.1 \\ 0.05 & 0.6 & 0.05 & 0.3 \end{pmatrix}$$

and

$$p(m) = [g_{ij}(m)] = \begin{pmatrix} 0.05 & 0.4 & 0.05 & 0.5 \\ 0.5 & 0 & 0.45 & 0.05 \\ 0.05 & 0.55 & 0 & 0.4 \\ 0.4 & 0.05 & 0.5 & 0.05 \end{pmatrix},$$

it is desired to find the optimal sequence of control (u_0, u_1).

Suppose a particular realization of η_t is such that $y_0 = b_1$ and $y_1 = b_2$ are observed at times 0 and 1. Let us now obtain an optimal u_1. Let

$$\gamma_2^*(y^1, u_0) = \min_{u_1} E(|x_2 - a_1| + u_1^2 \mid y^1, u_0).$$

Define

$$w_0(a_i) = \mathrm{P}(x_0 = a_i \mid y_0).$$

Then, since $y_0 = b_1$,

$$w_0(a_i) = \frac{p(a_i, b_1)}{p(a_1, b_1) + p(a_2, b_1)}$$

$$w_0(a_1) = \frac{0.45}{0.45 + 0.05} = 0.9$$

$$w_0(a_2) = \frac{0.05}{0.45 + 0.05} = 0.1.$$

Let us define

$$w_1(a_i \mid u) = \mathrm{P}(x_1 = a_i \mid y^1, u), \qquad i = 1, 2.$$

The transition probabilities for x are given by

$$\mathrm{P}(x_1 = a_1 \mid x_0 = a_1, u_0) = \mathrm{P}(\zeta_1 = \mathbf{c}_3 \mid \zeta_0 = \mathbf{c}_1, u_0) = g_{31}(u)$$

and so on. Therefore,

$$w_1(a_1 \mid u) = \frac{g_{31}(u)w_0(a_1) + g_{32}(u)w_0(a_2)}{[g_{31}(u) + g_{41}(u)]w_0(a_1) + [g_{32}(u) + g_{42}(u)]w_0(a_2)}$$

and

$$w_1(a_2 \mid u) = \frac{g_{41}(u)w_0(a_1) + g_{42}(u)w_0(a_2)}{[g_{31}(u) + g_{41}(u)]w_0(a_1) + [g_{32}(u) + g_{42}(u)]w_0(a_2)}.$$

Thus,

$$\gamma_2^*(y^1, u_0 = 0) = \min_u \left\{ \sum_{r_1, \zeta_2} (|x_2 - a_1| + u^2) p(\zeta_2 \mid \zeta_1) w_1(x_1 \mid 0) \right\}$$

and

$$\gamma_2^*(y^1, u_0 = m) = \min_u \left\{ \sum_{r_1, \zeta_2} (|x_2 - a_1| + u^2) p(\zeta_2 \mid \zeta_1) w_1(x_1 \mid m) \right\}$$

where the summation over x_1 and ζ_2 ranges over all possible x_1 and ζ_2 states. Since $\zeta_1 = (x_1, b_2)$, defining $d = |a_2 - a_1|$,

$$\gamma_2^*(y', u_0 = 0) = \min \left\{ \begin{array}{l} d[(g_{23}(0) + g_{43}(0))w_1(a_2 \mid 0) \\ \qquad + (g_{24}(0) + g_{44}(0))w_1(a_2 \mid 0)], \\ m^2[(g_{13}(m) + g_{33}(m))w_1(a_1 \mid 0) \\ \qquad + (g_{14}(m) + g_{34}(m))w_1(a_2 \mid 0)] \\ \qquad + (d + m^2)[(g_{23}(m) + g_{43}(m))w_1(a_1 \mid 0) \\ \qquad + (g_{24}(m) + g_{44}(m))w_1(a_2 \mid 0)] \end{array} \right\}.$$

A similar expression for $\gamma_2^*(y^1, u_0 = m)$ is obtained by replacing $w_1(a_i \mid 0)$ with $w_1(a_i \mid m)$. In performing numerical calculations, the optimal u_1, when $u_0 = 0$ and (b_1, b_2) have been observed, is given by

$$u_1(y^1, u_0 = 0) = 0$$

and

$$\begin{aligned} u_1(y_1, u_0 = m) &= 0 \qquad \text{if } d < 3.6m^2 \\ &= m \qquad \text{if } d > 3.6m^2. \end{aligned}$$

Example 9: Convergence of an Adaptive System. Dynamic systems can be converted into Markov processes by suitable definitions of state vectors that are mixtures of "physical" and "informational" states. Consider the next example adapted from Meyn and Caines (1986):

$$\begin{aligned} y_t &= \theta_t y_{t-1} + u_{t-1} + w_t, \\ \theta_t &= \rho\theta_{t-1} + v_t, \qquad |\rho| < 1, \end{aligned}$$

where w_t and v_t are independent and identically distributed mean-zero random variables with finite variances and u_t is a control variable. Assume that y_t is observed without error but θ_t is not. Then $\{y_t\}$ itself is not a Markov process because the knowledge of y_{t-1} and u_{t-1} does not suffice to produce the transition probability for y_t.

Suppose that a certainty equivalent control

$$u_{t-1} = y_t^d - \hat{\theta}_t y_{t-1}$$

is employed where y_t^d is a target value so that the dynamics is

$$y_t = \tilde{\theta}_t y_{t-1} + w_t \tag{84}$$

where $y_t^d = 0$ by a shift of the origin.

Suppose further that $\theta_{t+1} = \rho\theta_t + v_t$, and

$$\hat{\theta}_{t+1} = \rho\hat{\theta}_t + K_t y_t$$

so that the error evolves over time according to

$$\tilde{\theta}_{t+1} = \rho\tilde{\theta}_t - K_t y_t + v_t$$

where

$$\begin{aligned} K_t &= \frac{\rho R_t y_{t-1}}{R_t y_{t-1}^2 + \sigma_w^2} \\ R_{t+1} &= E(\tilde{\theta}_{t+1}^2 \mid \mathscr{F}_t) = \sigma_v^2 + \rho^2 \sigma_w^2 R_t/(R_t y_{t-1}^2 + \sigma_w^2). \end{aligned} \tag{85}$$

Then augmenting y_t by the "informational" state variables $\tilde{\theta}_t$ and R_t, the augmented vector

$$\mathbf{s}_t = \begin{bmatrix} y_t \\ \tilde{\theta}_{t+1} \\ R_{t+1} \end{bmatrix}$$

is a state vector of a Markov process since

$$\mathbf{s}_{t+1} = \Phi(\mathbf{s}_t, u_t, v_t). \tag{86}$$

Note that all elements of $\mathbf{s}_{t+1}$ can be calculated from $\mathbf{s}_t$ and parametrized by the noise vectors u_t and v_t, i.e., (86) specifies a Markov process that determines the learning process of the parameter $\{\theta_t\}$. When the probability distribution functions for u_t and v_t have densities, the transition probability $P(x, A) = \text{Prob}[\mathbf{s}_{t+1} \in A \mid \mathbf{s}_t = x]$ for a Borel set A in R^3 also has the probability density $\text{p}(x, \mathbf{s})$ which is uniformly (in x) integrable with respect to $\mathbf{s}$.

If $E\|\mathbf{s}_t\|^2$ is uniformly bounded for some probability distribution for $\mathbf{s}_0$, then the invariant probability distribution will exist. Eq. (84) implies that

$$E(y_t^2 \mid \mathscr{F}_{t-1}) = R_t y_{t-1}^2 + \sigma_w^2.$$

From (85),

$$R_t = E(\tilde{\theta}_t^2 \mid \mathscr{F}_{t-1}) \geqslant \sigma_v^2.$$

If $\sigma_v^2 > 1$, then $R_t \geqslant 1$ and $E(y_t^2 \mid \mathscr{F}_{t-1})$ becomes unbounded. On the other hand, if $\sigma_v^2 < 1$, then (85) implies that

$$R_{t+1} \leqslant \sigma_v^2 + \rho^2 R_t$$

or

$$R_t \leqslant \sigma_v^2/(1 - \rho^2)$$

for all $t \geqslant 1$, and

$$E(R_t) \leqslant \kappa$$

where $\kappa = \sigma_v^2/(1 - \rho^2)$. If $\kappa < 1$,* then

$$E(y_t^2/\sigma_w^2) \leqslant \kappa E(y_{t-1}^2/\sigma_w^2) + 1$$

or

$$E(y_t^2/\sigma_w^2) \leqslant \frac{1}{1 - \kappa}$$

and

$$E(y_t^2) < \infty$$

for all t. Thus $E\|s_t\|^2$ is uniform bounded in t. By the mean ergodic theorem,

$$1/N\, \Sigma_1^N f(s_t) \to E[f \mid \mathscr{T}] \text{ a.s.}$$

for any $f \in L^1$ where $\mathscr{T}$ is the σ-field of invariant sets, e.g.,

$$1/N\, \Sigma_1^N \|s_t\|^2 \to E(\|s_0\|^2 | \mathscr{T}).$$

---

* Let $\gamma_t = E(y_t^2 | \mathscr{F}_{t-1})$. Then

$$\begin{aligned} \gamma_t^2 &= R_t(y_{t-1}^2 - \gamma_{t-1}) + R_t \gamma_{t-1} + \sigma_w^2 \\ &= R_t(y_{t-1}^2 - \gamma_{t-1}) + \sigma_w^2 + \sigma_v^2 \gamma_{t-1} \\ &\quad + \rho^2 \sigma_w^2 R_{t-1}. \end{aligned}$$

Hence

$$\begin{aligned} E(y_t^2) = E(\gamma_t^2) &\leqslant \sigma_w^2 + \sigma_v^2 E(y_{t-1}^2) + \rho^2 \sigma_w^2 \sigma_v^2/(1 - \rho^2) \\ &\leqslant \frac{\sigma_w^2}{1 - \sigma_v^2}\left(1 + \frac{\rho^2 \sigma_v^2}{1 - \rho^2}\right), \text{ if } \sigma_v^2 < 1. \end{aligned}$$

Exercise 1: Perfect Observation System with Unknown Noise in the Dynamics
Consider a one-dimensional linear control system described by

$$x_{t+1} = ax_t + bu_t + \xi_t, \qquad u_t \in (-\infty, \infty), \qquad 0 \leqslant t \leqslant N-1$$

and assumed to be perfectly measured,

$$y_t = x_t, \qquad 0 \leqslant t \leqslant N-1,$$

where a and b are known constant plant parameters and where ξs are random variables assumed to be independently and identically distributed with the distribution function

$$\mathscr{L}(\xi_t) = \mathrm{N}(\theta, \sigma_1^2)$$

for each θ on the real line, where θ is the unknown random parameter of the distribution function for ξ. The variance σ_1^2 is assumed given.

The a priori distribution function for θ is assumed given by

$$\mathscr{L}_0(\theta) = \mathrm{N}(\mu, \sigma_0^2)$$

where μ and σ_0^2 are assumed known.

The initial state x_0 of the model is assumed to be a random variable independent of θ and assumed to be distributed according to

$$\mathscr{L}(x_0 = \mathrm{N}(\alpha, \sigma_2^2)$$

where α and σ_2 are assumed known. The criterion function is taken to be

$$J = \sum_1^N q_t x_t^2, \qquad q_t > 0, \qquad 1 \leqslant t \leqslant N.$$

Show that

$$u_t^* = -(ax_t + \mu_t)/b$$

$$\gamma_t = \sum_{j=1}^{N-1} \Gamma_j^2, \qquad 0 \leqslant t \leqslant N-1$$

where $\Gamma_t^2 = \operatorname{var}(\xi_t \mid x^t)$ and $\mu_t = E(\xi_t \mid x^t)$. Since the xs are perfectly observed, the knowledge of x^t is equivalent to that of ξ^{t-1} since $\xi_t = x_{t+1} - ax_t - bu_t$. Therefore,

$$p(\xi_t \mid x^t) = p(\xi_t \mid \xi^{i-1}) = \int p(\xi_t \mid \theta) p(\theta \mid \xi^{t-1})\, d\theta,$$

where

$$p(\xi_0 \mid x_0) = \text{const} \exp\left(-\frac{(\xi_0 - \mu)^2}{2(\sigma_0^2 + \sigma_1^2)} \right).$$

Namely,

$$\mu_0 = \mu$$

$$\Gamma_0^2 = \sigma_0^2 + \sigma_1^2.$$

Assume

$$p(\theta \mid \xi^t) = \text{const} \exp\left[-\frac{(\theta - \theta_t)^2}{2\Sigma_t^2} \right].$$

This is certainly true for $t = 0$ with

$$\theta_0 = \frac{\mu/\sigma_0^2 + \xi_0/\sigma_1^2}{1/\sigma_0^2 + 1/\sigma_1^2},$$

and

$$1/\Sigma_0^2 = 1/\sigma_0^2 + 1/\sigma_1^2.$$

Use

$$\begin{aligned} p(\theta, \xi_{t+1} \mid \xi^t) &= p(\theta \mid \xi^t) p(\xi_{t+1} \mid \theta, \xi^t) \\ &= p(\theta \mid \xi^t) p(\xi_{t+1} \mid \theta), \qquad 0 \leqslant t \leqslant N-1 \end{aligned}$$

to derive

$$p(\theta \mid \xi^{t+1}) = \frac{p(\theta \mid \xi^t) p(\xi_{t+1} \mid \theta)}{\int p(\theta \mid \xi^t) p(\xi_{t+1} \mid \theta)\, d\theta}, \qquad 0 \leqslant t \leqslant N-1$$

and show that θ_t and Γ_t are sufficient statistics for θ with

$$\begin{aligned} \theta_{t+1} &= \frac{\theta_t/\Sigma_t^2 + \xi_{t+1}/\sigma_1^2}{1/\Sigma_t^2 + 1/\sigma_1^2} \\ 1/\Sigma_{t+1}^2 &= 1/\Sigma_1^2 + 1/\sigma_1^2 \end{aligned}$$

and that

$$\begin{aligned} \mu_t &= \theta_{t-1} \\ \Gamma_t^2 &= \sigma_1^2 + \Sigma_{t-1}^2, \qquad 0 \leqslant t \leqslant N-1. \end{aligned}$$

Show that γ_N^* and all subsequent γ^*s become constants, that each control stage can be optimized separately, and that

$$\gamma_1^* = \min EJ = \sum_0^{N-1} \Gamma_j^2.$$

Exercise 2: System with Unknown Random Time Constants

Consider a one-dimensional dynamics

$$x_{t+1} = a_t x_t + u_t, \qquad 0 \leqslant t \leqslant N-1, \qquad u_t \in (-\infty, \infty)$$

where as are independently and identically distributed random variables with

$$\mathscr{L}(a_t) = \mathrm{N}(\theta, \sigma^2)$$

for each θ where θ is assumed to have an a priori distribution function

$$\mathscr{L}(\theta) = \mathrm{N}(\theta_0, \sigma_0^2).$$

The system is assumed to be perfectly observed:

$$y_t = x_t, \qquad 0 \leqslant t \leqslant N-1.$$

The criterion function is taken to be

$$J = x_N^2.$$

Define

$$\hat{a}_t = E(a_t \mid a^{t-1}) \qquad 0 \leqslant t \leqslant N-1$$

$$\hat{\sigma}_t^2 = \operatorname{var}(a_t \mid a^{i-1}), \qquad 0 \leqslant t \leqslant N-1.$$

Show that

$$u_t^* = -\hat{a}_t x_t, \qquad 0 \leqslant t \leqslant N-1$$

is the optimal control policy and

$$\min EJ = \gamma_1^* = \left(\prod_{i=0}^{N-1} \hat{\sigma}_t^2\right) x_0^2.$$

Now from the conditional independence assumption on θ,

$$p(a_t \mid a^{t-1}) = \int p(a_t \mid \theta) p(\theta \mid a^{t-1})\, d\theta$$

where by assumption

$$p(a_t \mid \theta) = \text{const} \exp\left(-\frac{(a_t - \theta)^2}{2\sigma^2}\right).$$

To compute $p(\theta \mid a^{t-1})$, use the recursion formula

$$p(\theta \mid a^t) = \frac{p(\theta \mid a^{t-1}) p(a_t \mid \theta, a^{t-1})}{\int p(\theta \mid a^{t-1}) p(a_t \mid \theta, a^{t-1})\, d\theta}, \qquad 1 \leqslant t \leqslant N-1$$

where its initial condition probability density function is given by

$$p(\theta \mid a_0) = \frac{p_0(\theta) p(a_0 \mid \theta)}{\int p_0(\theta) p(a_0 \mid \theta)\, d\theta} = \text{const} \exp\left(-\frac{(\theta - \mu_0)^2}{2\Sigma_0^2}\right),$$

where

$$\mu_0 = \frac{a_0/\sigma^2 + \theta_0/\sigma_0^2}{1/\sigma^2 + 1/\sigma_0^2}$$

and

$$1/\Sigma_0^2 = 1/\sigma^2 + 1/\sigma_0^2.$$

Show that

$$p(\theta \mid a^t) = \text{const}\exp\left(-\frac{(\theta - \mu_t)^2}{2\Sigma_t^2}\right),$$

$$1/\Sigma_t^2 = 1/\Sigma_{t-1}^2 + 1/\sigma^2,$$

and

$$p(a_t \mid s^{t-1}) = \text{const}\exp\left(-\frac{(a_t - u_{t-1})^2}{2(\sigma^2 + \Sigma_{t-1}^2)}\right),$$

i.e.,

$$\hat{a}_t = \mu_{t-1}, \qquad 1 \leqslant t \leqslant N-1,$$

$$\hat{a}_0 = \theta_0,$$

and

$$\hat{\sigma}_1^2 = a^2 + \Sigma_{t+1}^2.$$

Or, more explicitly,

$$1/\Sigma_t^2 = 1/\sigma^2 + (j+1)/\sigma^2$$

$$\hat{a}_{t+1} = \mu_t = \frac{\theta_0/\sigma_0^2 + \Sigma_0^t a_t/\sigma^2}{1/\sigma_0^2 + (j+1)/\sigma^2}$$

$$= \frac{\theta_0/\sigma_0^2 + \Sigma_0^t\left(\dfrac{x_{t+1} - \mu_t}{x_t\sigma^2}\right)}{1/\sigma_0^2 + (t+1)/\sigma^2}.$$

By letting $\sigma_0 \to 0$, we see that $\hat{a}_t$ all reduce to θ_0 and Σ_t to 0. As expected, the optimal control policy reduces to the optimal control policy for the corresponding stochastic case.

Exercise 3

In this exercise, treat a in

$$x_{t+1} = ax_t + bu_t, \qquad u_t \in (-\infty, \infty)$$

$$y_t = x_t + \eta_t$$

$$J = x_N^2$$

as a random variable with known mean and variance and where ηs are assumed to be independent. It is further assumed that a is independent of ηs and that

$$E(a) = \alpha$$

$$\text{var}(a) = \sigma_1^2$$

where α and σ_1 are assumed known. When a is a known constant, then

$$u_{N-1}^* = -a\mu_{N-1}/b$$

is optimal. Show that the rule

$$u_{N-1} = -\alpha_{N-1}\mu_{N-1}/b$$

where the random variable a is replaced by its a posteriori mean value

$$\alpha_{N-1} \triangleq E(a \mid y^{N-1})$$

is not optimal by showing that

$$E(x_N^2 \mid y^{N-1}, u^{N-1}) = (\widehat{ax}_{N-1} + bu_{N-1})^2 + \Sigma_{N-1}^2,$$

and that on the assumption that Σ_{N-1}^2 is independent of u_{N-1}, the optimal control at time $N - 1$ is given by

$$u_{N-1}^* = -\widehat{ax}_{N-1}/b$$

but

$$\widehat{ax}_{N-1} \neq E(x_{N-1} \mid \mathrm{a}, y^{N-1}, u^{N-2})E(\mathrm{a} \mid y^{N-1}, u^{N-1}).$$

The certainty equivalence does not yield optimal rules for this problem.

Exercise 4

Next assume that

$$x_{t+1} = \mathrm{a}x_t + bu_t + \xi_t, \qquad u_t \in (-\infty, \infty),\ x_0 \text{ given,}$$

and

$$y_t = x_t, \qquad 0 \leqslant t \leqslant N - 1$$

where a is a known constant but where b is now assumed to be a random variable, independent of ξs with finite mean and variance. The disturbance ξs are assumed to be independently and identically distributed random variables with

$$E(\xi_t) = 0$$

$$\operatorname{var}(\xi_t) = \Sigma_0^2, \qquad 0 \leqslant t \leqslant N - 1.$$

The certainty equivalence principle replaces the model by

$$x_N = \mathrm{a}x_{N-1} + b_{N-1}u_{N-1}$$

where

$$b_{N-1} \triangleq E(b \mid x^{N-1})$$

and hence derives

$$u_{N-1} = -\mathrm{a}x_{N-1}/b_{N-1}.$$

Show that with this control,

$$E(x_N^2 \mid x^{N-1}, u^{N-1}) = E\left[\left(\mathrm{a}x_{N-1} - \frac{b}{b_{N-1}}\mathrm{a}x_{N-1} + \xi_{N-1}\right)^2 \middle| x^{N-1}\right]$$
$$= \frac{\sigma_{N-1}^2(\mathrm{a}x_{N-1})^2}{b_{N-1}^2} + \Sigma_0^2$$

where

$$\sigma_{N-1}^2 = \mathrm{var}\,(b \mid x^{N-1}),$$

but the control

$$u_{N-1} = -\frac{b_{N-1}}{b_{N-1}^2 + \sigma_{N-1}^2}(\mathrm{a}x_{N-1})$$

produces a smaller cost

$$E(x_N^2 \mid x^{N-1}, u^{N-1}) = E\left[\left(\mathrm{a}x_{N-1} - \frac{bb_{N-1}}{b_{N-1}^2 + \sigma_{N-1}^2}\mathrm{a}x_{N-1} + \xi_{N-1}\right)^2 \middle| x^{N-1}\right]$$
$$= \frac{\sigma_{N-1}^2}{b_{N-1}^2 + \sigma_{N-1}^2}(\mathrm{a}x_{N-1})^2 + \Sigma_0^2.$$

This exercise shows that the optimal control for the deterministic system is not optimal for this model with uncertain b.

7.5 Gihman and Skorohod Formulation*

The information structure of many sequential decision problems contains only parts of the information on the states of the underlying Markovian processes. To put it differently, the vectors observed by the decision makers do not form Markovian processes by themselves but must be augmented by other variables not directly observed by the decision makers to make them Markovian. We call these processes partially observed Markovian processes. Suppose that $\{\zeta_t\}$ is a Markovian process, where $\zeta_t = (\rho_t, s_t)$ of which only s_t is observed, and $\{s_t\}$ is not a Markovian process. We can still find a sequence of nonrandomized controls (Gihman and Skorohod [1979, p. 36]). Control u_t must be at most a function of s^t and u^{t-1}.

This section shows heuristically that to each optimization problem with the Markovian process $\{\zeta_t\}$, we can construct another optimization problem

* This section follows Gihman and Skorohod (1979, Chapter 1).

involving only observable variables. We assume that the underlying state space $(Z, \mathscr{A})$ is representable as $(Z_\rho \times Z_s, \mathscr{A}_\rho \times \mathscr{A}_s)$. In such situations, if a pair of conditional probability distribution defined by

$$P_t^\rho(d\rho_t \mid \rho^{t-1}; s^t; u^{t-1}) \tag{87}$$

and

$$P_t^s(ds_t \mid \rho^{t-1}; s^{t-1}, u^{t-1})$$

are given, then the conditional probability distribution for $\{\zeta_t)$ is then expressible as

$$\begin{aligned} &P_t(A \times B \mid \zeta^{t-1}; u^{t-1}) \\ &\quad = \int_B P_t^\rho(A \mid \rho^{t-1}; s^t; u^{t-1}) \cdot P_t^s(ds_t \mid \rho^{t-1}; s^{t-1}; u^{t-1}). \end{aligned} \tag{88}$$

Conversely, given the conditional probability distribution $\{\zeta_t\}$, $\{s_t\}$ can be obtained by

$$\begin{aligned} P_t^s(B \mid \rho^{t-1}; s^{t-1}; u^{t-1}) &= P(Z_\rho \times B \mid \zeta^{t-1}; u^{t-1}) \\ &= P(Z_\rho \times B \mid \zeta_{t-1}, u_{t-1}). \end{aligned} \tag{89}$$

The probability distribution P_t^ρ is then calculated from (88) as the Radon-Nikodym derivative. (See Rényi [1970, p. 352], for example.)

Admissible nonrandomized controls are sequences of functions $u_t = \psi_t(y^t)$. For any subsets $A_0, A_1, \ldots, A_t \in \mathscr{A}_t$, (87) defines a probability

$$\begin{aligned} &\Pi_t(u^{t-1}, A_0, \ldots A_t) \\ &\quad = \int_{A_0} P_0^s(ds_0) \int_{z_\rho} P_0^\rho ; d\rho_0 \mid s_0) \int_{A_1} P^{s1}(ds_1 \mid \zeta_0, u_0) \\ &\quad\quad \cdots \int_{A_t} P_t^s(ds_t \mid \zeta^{t-1}, u^{t-1}). \end{aligned} \tag{90}$$

Define a probability μ_t on $\mathscr{A}_s^t$ by

$$\mu_t(A_t \mid C, u^{t-1}) = \Pi_t(u^{t-1}, A_0, \ldots, A_t)$$

for

$$C = A_0 \times A_1 \times \ldots \times A_{t-1}$$

where u^{t-1} is treated as a parameter.

Define

$$\mu_t^*(C; u^{t-2}) = \Pi_{t-1}(u^{t-2}, A_0, \ldots, A_{t-1}).$$

Note from the consistency requirement that

$$\mu_t(A_t \mid C, u^{t-1}) \ll \mu_t(Z_s \mid C, u^{t-1}) = \mu_t^*(C; u^{t-1}),$$

i.e., μ_t is absolutely continuous with respect to μ_t^*. Denote the Radon-Nikodym derivative by

$$\frac{d\mu_t}{d\mu_t^*} = \bar{P}_t(A_t \mid s^{t-1}, u^{t-1}) \tag{91}$$

which is $\mathscr{A}_s^t \times \mathscr{B}^t$-measurable. When Z_s is a complete separable metric space and $\mathscr{A}_s$ is a Borel σ-field, then $\bar{P}_t(A_t \mid \cdot, \cdot)$ is a measure on $\mathscr{A}_s$.

Let π be a sequence of nonrandomized control $\{u_t = \psi_t(y^t), t = 0, 1, \ldots\}$, and consider the resulting trajectory of the model defined by (87). Then for any bounded $\mathscr{A}_\rho^{t+1} \times \mathscr{A}_s^{t+1} = \mathscr{B}^{t+1}$-measurable function $\mathrm{f}(\rho^t, s^t, u^t)$ we can calculate

$$\begin{aligned} E_\pi \mathrm{f}(\rho^t, s^t, u^t) = \int \ldots \int &\mathrm{f}(\rho^t, s^t, \phi_0(s_0), \ldots, \phi_t(s^t)) P_0(ds_0) \\ &\times P_0^\rho(d\varrho_0 \mid s_0) \ldots P_t^\rho(ds_t \mid \rho^{t-1}, s^{t-1}, \phi_0(s_0), \ldots, \phi_{t-1}(y^{t-1})) \\ &\times P_t^\rho(d\rho_t \mid \rho^{t-1}, s^t, \phi_0(s_0), \ldots, \phi_{t-1}(s^{t-1})). \end{aligned} \tag{92}$$

Set $\mathrm{f}(\rho^t, s^t, u^t) = \pi_{k=0}^t \chi_{A_k}(s_k)$ where $\chi(\cdot)$ is the indicator function. Then the probability is defined on $\mathscr{A}_s^t$ by

$$\begin{aligned} P_\pi(s_0 \in A_0, \ldots, s_t \in A_t) = \int_{A_0} P_0^s(ds_0) \int_{Z_\rho} P_0^\rho(d\rho_0 \mid s_0) \int_{A_1} P_1^s(ds_1 \mid \rho_0, s_0, \phi_0(s_0)) \\ \ldots \int_{A_t} P_t^s\, P_t^s(ds_t \mid \rho^{t-1}, s^{t-1}, \phi_0(s_0), \ldots, \phi_{t-1}(s^{t-1})), \end{aligned}$$

and

$$P_\pi(s_t \in A_t) = \bar{P}_t(A_t \mid s^{t-1}, \phi_0(s_0), \ldots, \phi_{t-1}(s^{t-1}))$$

follows.

Define $\Gamma C(s^t, \phi_0(s_0), \ldots, \phi_t(s^t))$ as the Radon–Nikodym derivative of the probability (92) with respect to that of (90) where $u_k = \phi_k(s^k)$ is substituted. ΓC thus constructed does not depend on $u_{t+1}, u_{t+2}, \ldots$.

More concretely, $\Gamma C(\underline{s}, \underline{u})$ when $C(\underline{s}, \underline{u}) = \mathrm{f}_t(\rho_t, s^t, u^t)$ is the ratio of

$$P_0^s(ds_0) \int P_0^\rho(d\rho_0 \mid s_0) \ldots \int \mathrm{f}_t(\rho^t, s^t, u^t) P_t^\rho(d\rho_t \mid \rho^{t-1}, s^t, u^t)$$

over the similar expression that results by setting $f_t = 1$. If $\mathrm{f}_t(\rho^t, s^t, u^t)$ is $\mathrm{f}_{t-1}(p^{t-1}, s^{t-1}, u^{t-1})$, then $P_t^\rho(Z_\rho \mid \rho^{t-1}, s^t, u^t) = 1$ in the numerator and $P_t^s(ds_t \mid \rho^{t-1}, s^{t-1}, u^{t-1})$ in the numerator and denominator cancel out. This shows that $\Gamma C(\underline{s}, \underline{u})$ in this case is independent of u_t.

To each partially observed Markov process with cost function $C(\underline{\rho}, \underline{s}, \underline{u})$, we can construct, as just shown, a control process described by a sequence of probabilities $\bar{P}_t(\cdot \mid \cdot)$ given by (91) and the cost function $\Gamma C(\underline{s}, \underline{u})$. An optimal control of the latter is also optimal for the former. This heuristic argument can be made rigorous under certain regularity conditions. Sufficient conditions for this to be equivalent when $(Z_\rho, \mathscr{A}_\rho)$ and $(Z_s, \mathscr{A}_s)$ are complete separable metric spaces and U compact all with Borel σ-fields are

(1) There exists in $(Z_s, \mathscr{A}_s)$ probability $m(ds)$ such that for all t, ρ^{t-1}, $s^{t-1}, u^{t-1}, p_t^s(\cdot \mid \rho^{t-1}, s^{t-1}, u^{t-1})$ is absolutely continuous with respect to m and its density (Radon-Nikodym derivative) is bounded and continuous jointly in the variables and
(2) The transition probability $P_t^\rho(d\rho_t \mid \rho^{t-1}, s^t, u^{t-1})$ has the property that

$$\int g(\rho^t, s^{t-1}, u^{t-1})\, P_t^\rho(d\rho_t \mid \rho^{t-1}, s^t, u^{t-1})$$

is continuous jointly in the variable for any bounded continuous function g. See Gihman and Skorohod (1979, lemma 1.10) for proof.

Appendices

Discrete-Time Markov Processes

A new method for establishing the convergence of adaptive schemes has been proposed by Meyn and Caines (1986), and it is based on the existence of invariant measures for Markov processes. This appendix collects some facts on the existence and uniqueness of measures and some related matters.

Markov Chains

It is useful to discuss their counterparts in a simpler setting of Markov processes with at most countably infinite, i.e., discrete, state space. They are called Markov chains. A Markov chain with a finite state space, i.e., with a finite number of states, is the simplest to deal with. We associate a Markov chain x_t, $t = 1, 2, \ldots$ with a matrix $\mathbf{P}$, called a transition matrix, $\mathbf{P} = (P_{ij})$ where $P_{ij} = \text{Prob}[x_{t+1} = j \mid x_t = i]$, $0 \leqslant i, j \leqslant n$. A Markov process is stationary or time homogeneous if the probability (matrix) is independent of time. Note that $\mathbf{P} \geqslant 0$, and each row sums to 1, $\Sigma_{j=1}^n P_{ij} = 1$, for each i. A Markov chain is said to be regular if $\mathbf{P}^m > 0$ for some positive integer m. Also, Markov chain is regular if it has 1 as a simple eigenvalue and no other eigenvalue has magnitude 1 [Gantmacher (1959, p. 107)].

Theorem *A regular Markov chain has a unique probability vector* $\mathbf{p}' > 0$ *such that* $\mathbf{p}'\mathbf{P} = \mathbf{p}'$. *The vector* $\mathbf{p}$ *is the limit* $\lim_{M\to\infty} \mathbf{e}_i'\mathbf{P}^m$ *for each i where* $\mathbf{e}_i$ *is a* *vector with* 1 *at the ith component as the only nonzero element. The limit* $\lim \mathbf{P}^m = \bar{\mathbf{P}}$ *exists, and each row is equal to* $\mathbf{p}'$.

These facts follow from the Perron-Frobenius theorem [Gantmacher (1959, p. 65)] that the transition matrix $\mathbf{P}$ has the simple eigenvalue 1 as the dominant eigenvalue.

To state conditions for the existence of an invariant probability vector for a Markov chain with denumerably infinite state space, we need some more notations and terminology. Let $P_{jk}(n)$ be the probability of going from state j to state k in n steps. A Markov chain is said to be irreducible if all pairs of states of the chain communicate, i.e., $P_{ij}(n)$ is nonzero for some n and for all pairs i and j. A regular Markov chain is irreducible since all states communicate, but not all irreducible Markov chain are regular, e.g., $P = \left[\begin{smallmatrix} 0 & 1 \\ 1 & 0 \end{smallmatrix}\right]$, which is irreducible but not regular since $\mathbf{P}^m$ contain two zero entries for all $m > 0$. A closed communicating class can be treated as an irreducible Markov chain. If not, state k is said to be nonrecurrent. An irreducible Markov chain is said to be recurrent (nonrecurrent) if every state in the chain is recurrent (nonrecurrent). Let $f_{jk}(n)$ denote the probability that the first passage from state j to state k occurs in exactly n steps, and define

$$f_{jk} = \sum_{1}^{\infty} f_{jk}(n)$$

State that k is nonrecurrent if $f_{kk} < 1$. Now suppose that k is recurrent, and define

$$m_{jk} = \sum_{n=1}^{\infty} n f_{jk}(n)$$

which is the mean recurrent time. Suppose that state k of an irreducible Markov chain is recurrent. Then

$$\lim 1/n \sum_{m=1}^{n} P_{kk}(m) = \pi_k$$

exists. To see this, define the z-transform

$$\begin{aligned} P_{jk}(z) &= \sum_{0}^{\infty} P_{jk}(n) z^n \\ &= \delta_{jk} + \sum_{1}^{\infty} P_{jk}(n) z^n \end{aligned}$$

and

$$F_{jk}(z) = \sum_{1}^{\infty} f_{jk}(n) z^n,$$

and then noting that

(A1)
$$P_{jk}(n) = \sum_{m=1}^{n} f_{jk}(m) P_{kk}(n-m)$$

takes its z-transform, which is

$$P_{kk}(z) = 1 + F_{kk}(z) P_{kk}(z),$$

or solve it for the transform

$$P_{kk}(z) = \frac{1}{[1 - F_{kk}(z)]}.$$

Then $\lim \Sigma_{m=0}^{n} a_m = S$ exists if and only if $\lim_{z \to 1-} A(z) = S$, and more profoundly

Fact **[Hardy (1949, Th. 96)]** Let $\{a_m\}$ be a sequence of nonnegative numbers and define

$$A(z) = \sum_{0}^{\infty} a_m z^m, \qquad |z| < 1.$$

Then $\lim 1/n \, \Sigma_1^n a_m < \infty$ exists if and only if $\lim_{z \to 1-} (1 - z) A(z) = 1$.

By l'Hopital's rule,

$$\lim_{z \to 1-} \frac{1 - z}{1 - F_{kk}(z)} = \lim_{z \to 1-} 1 \bigg/ \frac{dF_{kk}(z)}{dz} = 1 \bigg/ \sum_{1}^{\infty} n f_{kk}(n) = 1/m_{k,k},$$

and Fact establishes that

$$\lim 1/n \sum_{1}^{n} P_{kk}(n) = \frac{1}{m_{kk}}.$$

Denote this by π_k. Then from (A1), $P_{jk}(z) = F_{jk}(z) P_{kk}(z)$, and we obtain

$$\lim 1/n \sum_{1}^{n} P_{jk}(n) = f_{jk} \pi_k.$$

Theorem [Parzen (1962, p. 220)] *An irreducible Markov chain has a unique stationary distribution if and only if* $\pi_k > 0$ *for some* k *where*

$$\lim_{n \to \infty} P_{jk}(n) = f_{jk} \pi_k$$

where $P_{jk}(n)$ is the n-step transition probability from state j to state k.

Example 10. Here is a simple two-state Markov chain

$$(s_{t+1}^0, s_{t+1}^1) = (s_t^0, s_t^1)\mathbf{P}$$

where

$$\mathbf{P} = \begin{bmatrix} q & 1-q \\ 1-p & p \end{bmatrix},$$

where s_t^0 is the state 0 at time t and s_t^1 denote the state 1 at time t. Therefore,

$$(s_{t+n}^0 \quad s_{t+n}^1) = (s_t^0 \quad s_t^1)\mathbf{P}^n, \qquad n \geqslant 1.$$

The matrix $\mathbf{P}$ has eigenvalue $\lambda_1 = 1$ and $\lambda_2 = -1 + p + q$ with the corresponding unnormalized row eigenvector $\mu'_1 = (1-p, 1-q)$ and $\mu'_2 = (1, -1)$, respectively. Unless $p = 1$ and $q = 1$, the chain is regular since $\mathbf{P} > 0$. Actually $\lambda_1 = 1$ is a simple eigenvalue, and $|\lambda_2| \neq 1$, unless $p = q = 1$. Let

$$\mathbf{M} = \begin{pmatrix} \mu'_1 \\ \mu'_2 \end{pmatrix}$$

and

$$\mathbf{V} = \mathbf{M}^{-1} = \frac{1}{1-p+1-q}\begin{bmatrix} 1 & 1-q \\ 1 & -(1-p) \end{bmatrix}.$$

Since $\mathbf{P} = \mathbf{V\Lambda M}$, where $\Lambda = \operatorname{diag}(1, \lambda_2)$,

$$\mathbf{P}^n = \mathbf{V\Lambda}^n\mathbf{M} = \rho\left\{\begin{pmatrix} 1 \\ 1 \end{pmatrix}(v,\ 1) + \lambda_2^n \begin{pmatrix} 1 \\ -v \end{pmatrix}(1,\ -1)\right\}$$

where

$$v = (1-p)/(1-q)$$

and

$$\rho = (1+v)^{-1}.$$

We can directly read off from $\mathbf{P}^n$ that

$$\lim_{n\to\infty} P_{00}(n) = \lim_{n\to\infty} P_{10}(n) = \rho v = 1/m_{00}$$

and

$$\lim_{n\to\infty} P_{01}(n) = \lim_{n\to\infty} P_{11}(n) = \rho = 1/m_{11},$$

since $\sqrt{/}\lambda_2| < 1$ unless $p = q = 1$. Both states are recurrent. The limit of P^n as n goes to infinity is

$$\mathbf{P}^\infty = \rho\begin{bmatrix} 1 \\ 1 \end{bmatrix}(v \quad 1).$$

The stationary probability distribution of the states is given by $\rho(v, 1)$ since

$$\rho(v \quad 1)\mathbf{P}^{\infty} = \rho(v \quad 1).$$

Thus, the unconditional probability that $s_t = 1$ is ρ, and the unconditional probability that $s_t = 0$ is $\rho v = 1 - \rho$ in this stationary probability distribution.

From the expression for $\mathbf{P}^n$, we have the conditional probability distribution. For example,

$$P(s_{t+n} \mid s_t^0) = \rho[(v, 1) + \lambda_2^n(1, -1)]$$

and

$$P(s_{t+n} \mid s_t^1) = \rho[(v, 1) - \lambda_2^n v(1, -1)],$$

and hence

$$E(s_{t+n} \mid s_t^0) = \rho(1 - \lambda_2^n)$$

and

$$E(s_t \mid s_t^1) = \rho(1 + v\lambda_2^n).$$

Jointly they can be stated as $E(s_{t+n} \mid s_t) = \rho + \lambda_2^n(s_t - \rho)$ where $s_t = 0$ or 1.

For Markov processes with nondenumerable states, the condition for the existence of invariant measures is almost the same as the one for Markov processes with denumerable states. A set $F \in \Phi$ is called transient for a Markov process $P(x, B)$ if $\Sigma_1^\infty P^{(n)}(x, F) < \infty$ m a.s. A set F is called weakly transient if $\Sigma_1^\infty P^{(n_i)}(x, F) < \infty$, m a.s. for some subsequence $\{n_i\}$, when $n_i \to \infty$. If there exists a finite invariant measure which is equivalent to the original probability measure, then Ito (1964, Prop. 3) showed that a transient set of a Markov process has measure zero, $m(F) = 0$. If $\Sigma_{k=1}^\infty P^n(x, F) < \infty$ for some $F \in \Phi$, then $P^{(n)}(x, F) \to 0$ and, by the dominated convergence, then

$$\int_\Omega P^{(n)}(x, F)v(dx) \to 0.$$

By the invariance of v, $v(F) = 0$ and hence $m(F) = 0$ because m and v are equivalent (mutually absolutely continuous), and it is similar for weakly transient states.

Ito (1964) established that the condition "every weakly transient set for $P(x, B)$ has m-measure zero" is equivalent to "$m(B) > 0$ implies $\liminf Q_n(B) > 0$." This condition is analogous to the one for denumerable state Markov chains where $Q_n(B) = \int P^{(n)}(x, B)m(dx)$. Ito (1964) also discussed a number of equivalent necessary and sufficient conditions for the existence of invariant measures for Markov processes.

Invariant Measures for Linear Dynamic Systems

The state vector $\mathbf{x}_t$ of a linear dynamic system

$$\mathbf{x}_{t+1} = \mathbf{A}\mathbf{x}_t + \mathbf{B}\mathbf{w}_t,$$

where $\mathbf{A}$ is asymptotically stable and w_t i.i.d. $N(0, \mathbf{\Sigma})$, is a stationary Gaussian vector with mean 0 and variance $\mathbf{\Pi}$, where

$$\mathbf{\Pi} = \mathbf{A\Pi A}' + \mathbf{B\Sigma B}'.$$

We know that an invariant measure for $\{x_t\}$ exists if $\mathbf{A}$ is asymptotically stable and $(\mathbf{A}, \mathbf{B})$ is controllable.

Filtering in Partially Observed Markov Chains

The data-generating process is specified by

$$\mathbf{s}'_{t+1} = \mathbf{s}'_t\mathbf{P}.$$

Note that $\mathbf{s}_t$ is the state vector of a finite state Markov chain.

The problem is to calculate $p(\mathbf{s}^t \mid \mathbf{y}^t)$ recursively. We can calculate $p(\mathbf{s}_t \mid \mathbf{s}_{t-1})$ and $p(\mathbf{y}_t \mid \mathbf{y}^{t-1}, \mathbf{s}^t)$ from the basic model specification as follows. Note first that

$$\begin{aligned} p(\mathbf{y}_t, \mathbf{s}^t \mid \mathbf{y}^{t-1}) &= p(\mathbf{x}^t \mid \mathbf{y}^{t-1})p(\mathbf{y}_t \mid \mathbf{y}^{t-1}, \mathbf{s}^t) \\ &= p(\mathbf{s}_t \mid \mathbf{s}_{t-1})p(\mathbf{s}^{t-1} \mid \mathbf{y}^{t-1})p(\mathbf{y}_t \mid \mathbf{y}^{t-1}, \mathbf{s}^t) \end{aligned}$$

where $p(\mathbf{s}_t \mid \mathbf{s}^{t-1}) = p(\mathbf{s}_t \mid \mathbf{s}_{t-1})$ is used.

On the other hand

$$p(\mathbf{y}_t, \mathbf{s}^t \mid \mathbf{y}^{t-1}) = p(\mathbf{y}_t \mid \mathbf{y}^{t-1})p(\mathbf{s}^t \mid \mathbf{y}^t),$$

and from these two relations, we obtain the recursion

$$p(\mathbf{s}^t \mid \mathbf{y}^t) = p(\mathbf{s}_t \mid \mathbf{s}_{t-1})p(\mathbf{s}^{t-1} \mid \mathbf{y}^{t-1})p(\mathbf{y}_t \mid \mathbf{y}^{t-1}, \mathbf{s}^t)/p(\mathbf{y}_t \mid \mathbf{y}^{t-1})$$

where

$$p(\mathbf{y}_t \mid \mathbf{y}^{t-1}) = \int (\text{numerator})\, d\mathbf{s}^t$$

since

$$p(\mathbf{y}_t \mid \mathbf{y}^{t-1}) = \int P(\mathbf{y}_t, \mathbf{s}^t \mid \mathbf{y}^{t-1})\, d\mathbf{s}^t.$$

In general, we must carry s^t, which grows with t. In some dynamic models, the dependence of $\mathbf{y}_t$ on past history $\mathbf{y}_t = \psi(\mathbf{y}^{t-1}, \boldsymbol{\eta}^t, \mathbf{s}^t)$ simplifies. As an example, consider

$$y_t - \alpha s_t = \phi(\mathbf{y}_t - \alpha s_{t-1}) + (\sigma_0 + \sigma_1 s_t)n_t$$

where $n_t \sim NID(0, \sigma^2)$. This is an AR (1) version considered by Hamilton (1987). Here the transition probability density $p(y_t | s^t, y^{t-1})$ simplifies to $p(y_t | s_t, s_{t-1}, y_{t-1})$.

Thus

$$p(s^1 | y^1) = p(s_1 | s_0)p(s_0 | y_0)p(y_1 | s_0, s_1, y_0)/p(y_1 | y_0)$$

where

$$p(y_1 | y_0) = \int p(s_1 | s_0)p(s_0 | y_0)p(y_1 | s_0, s_1, y_0)\,\mathrm{d}(s_1, s_0),$$

and in general,

$$p(s^t | y^t) = p(s_t | s_{t-1})p(s^{t-1} | y^{t-1})p(y_t | s_t, s_{t-1}, y_{t-1})/p(y_t | y^{t-1})$$

where

$$p(y_t | y^{t-1}) = \int p(s_t | s_{t-1})p(s^{t-1} | y^{t-1})p(y_t | s_t, s_{t-1}, y_{t-1})\,\mathrm{d}(s_{t-1}, s_t).$$

Since

$$p(y_{t+1}, y_t | y^{t-1}, s^{t-1}) = \int p(y_{t+1}, y_t | y^{t-1}, s^t)p(s_t | s_{t-1})\,\mathrm{d}s_t$$

where

$$\begin{aligned} p(y_{t+1}, y_t | y^{t-1}, s^t) &= p(y_t | y^{t-1}, s^t)p(y_{t+1} | y^t, s^t) \\ &= \int p(y_t | y^{t-1}, s^t)p(y_{t+1} | y^t, s^{t+1})p(s_{t+1} | s_t)\,\mathrm{d}s_{t+1}, \end{aligned}$$

the recursion is

$$\begin{aligned} p(y_{t+1}, y^t) &= p(y_{t+1}, y_t | y^{t-1})/p(y_t | y^{t-1}) \\ &= p(y_{t+1}, y_t | y^{t-1}, s^{t-1})p(s^{t-1} | y^{t-1})/p(y_t | y^{t-1}) \\ &= \frac{p(s^{t-1} | y^{t-1})\int p(y_t | y^{t-1}, s^t)p(y_{t+1} | y^t, s^{t+1})p(s_{t+1} | s_t)p(s_t | s_{t-1})\,\mathrm{d}(s_{t+1}, s_t)}{p(y_t | y^{t-1})}. \end{aligned}$$

Regime Shift

An important class of partially observed Markov processes arises in modeling regime shifts in macroeconomic models or data-generating processes in general. A regime shift may be characterized as occasional discrete changes in the dynamic characteristics of data-generating processes. These discrete changes can often be modeled after those in the parameters in the data-generating dynamics. An important example is a conditionally linear state space model

$$\begin{aligned} \mathbf{x}_{t+1}^{\theta} &= \mathbf{A}(\theta_t)\mathbf{s}_t^{\theta} + \mathbf{B}(\theta_t)\mathbf{e}_t \\ \mathbf{y}_t &= \mathbf{C}(\theta_t)\mathbf{s}_t^{\theta} + \mathbf{e}_t \end{aligned}$$

where $\boldsymbol{\theta}_t \in \Theta = \{\theta^1, \ldots, \theta^m)$ and changes in $\boldsymbol{\theta}$ are modeled by an unobserved Markov chain, which is independent of $\mathbf{e}_t$. The vector $\mathbf{y}_t$ alone is not enough to calculate the transition probability distribution from $\mathbf{y}_t$ to $\mathbf{y}_{t+1}$. If, in addition, $\boldsymbol{\theta}_t$ is given, then the model is a Markov process. Note that in $(\mathbf{y}_t, \boldsymbol{\theta}_t)$, only $\mathbf{y}_t$ is observed.

A finite number of data-generating processes (DGP) lie behind the observed time series $\mathbf{y}_t$. Which of the DGP is currently generating y_t is not certain because of independent regime shifts that occur from time to time. Hamilton (1988) has modeled interest rates by an AR process with regime shifts driven by a two-state Markov chain. See Examples 6 and 7 in the previous section for a description of a regular two-state Markov chain.

CHAPTER **8**

Linear Rational Expectations Model*

8.1 Introduction

Rational expectations models are specified by a set of difference and identity equations for endogenous variables, y_t, some of which appear in conditional form $E(y_{t+h-k} \mid \Omega_{t-k})$, written for short as $y_{t+h-k|t-k}$, for some h and $k \geqslant 0$, where Ω_{t-k} is the information set available at $t-k$. (A vector-valued) rational expectation model generally is of the form

$$y_t = \mathrm{F}(y_{t+h-k|t-k}, y_{t-j}, n_t, h, k, j \geqslant 0),$$

where n_t stands for input variables that drive the model. Lagged variables $y_{t-j}, j \geqslant 0$, and expectations of y_t conditional on y_{t-k}, $k > 0$ provide linkage from the *past* to the present y_t, while the conditional expectations with $h-k > 0$ provide linkages from the *future* to the present if $k = 0$, or to the past if $k > 0$.

Aoki and Canzoneri (1979), Wallis (1980), Visco (1981, 1984), and Broze and Szafarz (1984) discussed the rational expectations models containing

* The author benefited from discussions with M. H. Pesaran who graciously provided Chapters 4 and 5 of his forthcoming book (1988). C. Gourieroux and L. Broze gave me several references I did not have. I also thank L. Broze for several private communications on solution methods for rational expectations models.

only terms $y_{t|t-k}$, $k > 0$. As we show later in this chapter, models with unilateral linkages, either to the future only or to the past only, are simpler to solve. Models with links from the future, i.e., conditional expectations $y_{t+h|t}$, $h > 0$, and from the past, i.e.,$y_{t|t-k}$, $k \geqslant 0$ or y_{t-j}, $j > 0$, are discussed in Broze, Gourieroux, and Szafarz (1985a). When bilateral linkages exist, situations are more complex. Control of rational expectations models introduce some subtle difficulties beyond those present in ordinary stochastic models without conditional expectations.

Several different methods have been used by different authors to solve linear models with rational expectations of endogenous variables (LRE models). The method of undetermined coefficients was introduced by Muth (1961) and since then adopted by Lucas (1972), Taylor (1977), Blanchard (1973), and McCallum (1983), for example. The z-transform or lag transform, including the methods based on backward or forward recursion method, was used extensively by Shiller (1978), Sargent (1979), Gourieroux (1982), and Whiteman (1983). More recently the methods based on martingale or martingale differences were proposed by Pesaran (1981), Gourieroux et al. (1982), and Broze et a (1985). We mostly use the martingale difference approach since it is the most general and encompasses other methods.

A prototype of the first-order rational expectation model for a scalar-valued y_t is given as

(1) $$y_t = ay_{t+1|t} + n_t.$$

When n_t is an ARMA process, this model has been thoroughly discussed by Gourieroux et al. (1982), using the transform method. We also treat this model in detail to show that it has the unique solution only when $|a| < 1$. Some samples of LRE models found in the literature are briefly described next, before we actually show how to solve them.

Example 1: Model of Inflation. A well-known example of the first-order model is the so-called Cagan model of inflation (1956), which describes a portfolio equilibrium (Sargent [1979, p. 192]),

$$m_t - p_t = -\alpha(p_{t+1|t} - p_t) + u_t, \qquad \alpha > 0,$$

where m_t is the logarithm of the money supply, p_t the logarithm of the price level, and $p_{t+1|t}$ the logarithm of the price expected to prevail at time $t+1$, given information available at time t. The input u_t is a weakly stationary disturbance process, typically taken to be a white noise process. Sometimes the model is regarded as the aggregate demand equation, and a separate aggregate supply equation is specified. (See McCallum [1983] for example.) The Cagan model of inflation, as used in McCallum (1983), can be restated in the form of (1)

$$p_t = \alpha m_t + \gamma p_{t+1|t} + u_t$$

where

$$0 < \alpha, \qquad \gamma < 1$$

and where u_t is a mean-zero white noise process. McCallum takes $\{m_t\}$ to be an AR(1) process

$$m_t = \mu_0 + \mu_1 m_{t-1} + v_t, \qquad -1 < \mu_1 < 1,$$

where v_t is a white noise process independent of u_t. The information set is taken to be current and past observations on price and money supply are $\Omega_t = \{p^t, m^t\}$. The n_t in (1) is $\alpha m_t + u_t$ here. This model has been much discussed in the literature. (See Black [1974], Burmeister [1980], Flood and Garber [1980], McCallum [1983], Shiller [1978], Saracoglu and Sargent [1978], Sargent and Wallace [1973] and many others.)

Example 2: Second-Order Models. There are many second-order examples. One of the earliest is the one due to Muth (1961), which is a model of an isolated market with inventory speculation:

$$\begin{aligned}
C_t &= -\beta p_t, \\
Y_t &= \gamma p_t^e + u_t, \\
I_t &= \alpha(p_{t+1|t} - p_t), \\
Y_t &= C_t + (I_t - I_{t-1}),
\end{aligned}$$

where p_t is the price of a storable commodity at t, C_t is the demand for current consumption, I_t is the stock of its inventory, Y_t is the output, and p_t^e is $p_{t|t}$. Finally, u_t is the disturbance process. The model reduces to

$$\alpha p_{t+1|t} - (\alpha + \gamma)p_t^e - (\alpha + \beta)p_t + \alpha p_{t-1} = u_t.$$

If we set $p_t^e = p_t$ (called the perfect foresight assumption, which will be further discussed later), this equation is a second-order difference equation in the price.

Another example of the second-order dynamics is provided by the model discussed in Taylor (1977). It has the reduced form

$$p_{t+1|t} = p_{t|t-1} + \delta_1 p_t + \delta_0 + u_t$$

where u_t is a white noise sequence.

A third well-known second-order example is a simplified version of the Sargent-Wallace model (1975),

$$y_t = a_1 + a_2(p_t - p_{t|t-1}) + u_{1t},$$
$$y_t = b_1 - b_2[r_t - (p_{t+1|t-1} - p_{t|t-1})] + u_{2t},$$
$$m = p_t + c_1 y_t - c_2 r_t + u_{3t},$$

where y_t, p_t, and m are the logarithms of output, its price level at time t, and the money supply which is taken to be constant.

All these second-order models can be rewritten as first-order models for suitably define state vectors which are vector-versions of (1). For example, let $\mathbf{y}_t = (p_t, p_{t-1})'$. Then the Muth example can be restated as

$$\mathbf{A}\mathbf{y}_{t+1|t} = \mathbf{B}\mathbf{y}_t + \alpha u_t$$

where $A = [\alpha, -(\alpha+\gamma)]$ and $B = [\alpha+\beta, -\alpha]$.

The Taylor model becomes

$$(0, 1)\mathbf{y}_{t+1} = (\delta_1, 1)\mathbf{y}_t + \delta_0 + u_t$$

where

$$\mathbf{y}_t = (p_t, p_{t|t-1})'.$$

A systematic procedure for putting LRE models in state space form is given later in this chapter.

Example 3: Mixed Expectations. An example of LRE model with "mixed" expectation is

$$y_t = By_{t+1|t-j} + Cx_t + u_t, \qquad B \neq 0$$

where $\{u_t\}$ is a mean-zero and independent white noise sequence and $\{x_t\}$ is an exogenous nonwhite stochastic process independent of us. Evans (1985) considered this model where y is scalar and $C = 0$. In both Evans (1985) and Evans and Honkapohja (1986), the term $y_{t+1|t-1}$ is missing. Gourieroux et al. (1982) also considered a special scalar case (1). Broze, Gourieroux and Szafarz (1985) considered more general equations where scalar $y_{t+h-k|t-k}$ appear where $0 \leqslant h \leqslant H$ and $0 \leqslant h \leqslant K$, with coefficients a_{kh} subject to what they call the nullity constraint. In particular, we show how expectational errors behave differently in the class of models with $y_{t+1|t}$ and those with $y_{t+1|t-j}$, $j \geqslant 1$. In vector-valued models, the coefficient matrices are not expected to cancel out, i.e., the nullity constraint seems unnatural. We later consider a model with a lagged endogenous variable Dy_{t-1} to it since this produces a model not covered in Broze et al. (1985b). One can readily extend this model by including other lagged y terms or expectations of the current value of y formed earlier, such as the terms $y_{t|t-k}$, $k > 0$. These, however, do not contribute to the characteristic feature of rational expectations models with future expectations, i.e., intertemporal linkage that constrains martingale differences associated with expectational errors. A nonzero B introduces intertemporally linked constraints on martingale differences.

Other economic examples are found in Bray (1982), Bray and Kreps (1984), the special issue of rational expectations analysis of *Journal of Economic Dynamics and Control* (1980), and in many other places.

8.2 Expectation Revision Processes

Rational expectation models specify how expectations of endogenous variables are to be revised as expectations on exogenous variables are revised as time passes and more information becomes available. This section introduces expectation revision processes, shows that they are martingale difference processes, and expresses solutions of difference equations involving rational expectations in terms of them.

Aoki and Canzoneri (1979) used innovations associated with expectation revision processes to derive reduced form models from structural models when LRE models contain only expectational variables from the past. This method was later generalized by Visco (1981, 1984). A related procedure was used by Broze and Szafarz (1984). They based their derivation on Doob's martingale decomposition theorem to obtain reduced forms of linear rational expectations models in terms of martingale differences. Broze, Gourieroux, and Szafarz (1985a, b) have also provided studies of solutions of a more general class of rational expectations models by examining the associated martingale differences. In this section, we show how to convert LRE models into those containing unconditional ys and martingale differences. In the next section, we solve models

(2) $$y_t = \alpha y_{t|t-1} + n_t,$$

where n_t is exogenous, and generalize to LRE models with no lagged endogenous variables

(3) $$y_t = \alpha_1 y_{t|t-1} + \dots + \alpha_p y_{t|t-p} + n_t,$$

and to the most general LRE models of this class,

(4) $$y_t = a_1 y_{t-1} + \dots + a_q y_{t-q} + \alpha_1 y_{t|t-1} + \dots + \alpha_p y_{t|t-p} + n_t$$

for scalar y_t.

Later the models are assumed to be vector-valued. Then we discuss (1) in some detail and follow it by discussing

(5) $$y_t = \alpha_1 y_{t+1|t} + \ldots + \alpha_p y_{t+p|t} + n_t$$

and then

(6) $$y_t = a_1 y_{t-1} + \ldots + a_q y_{t-p} + \alpha_1 y_{t+1|t} + \ldots + \alpha_p y_{t+p|t} + n_t,$$

all for scalar y. Sometimes scalar and vector cases are simultaneously presented. We then discuss vector-valued models and finally models with mixed future and past expectations. When only weakly stationary solutions are sought, the Wiener-Hopf equation method can be used as in Saracoglu and Sargent (1978) or White (1983). However, restricting solutions to be weakly stationary is overly restrictive. Weakly stationary solutions by the Wiener-Hopf equations are treated separately in connection with Wiener filters.

Martingale Difference Processes

Any stochastic process can be decomposed into a predictable part and a martingale. Formally this is the content of Doob composition stated in Appendix. Denote the Doob decomposition of $y_{t+j|t}$ by $M_t^j + P_t^j$, $j = 0, 1, 2, \ldots$ where P_t^j is a predictable process, i.e., $P_{t|t-1}^j = P_t^j$ and M_t^j is a martingale, i.e., $M_{t|t-1}^j = M_{t-1}^j$. From the definitions of predictable processes and martingales, we see that

$$\begin{aligned} y_t - y_{t|t-1} &= M_t^0 + P_t^0 - E(M_t^0 + P_t^0 \mid I_{t-1}) \\ &= M_t^0 - M_{t-1}^0. \end{aligned}$$

Define the error associated with $y_{t|t-1}$ by

$$\begin{aligned} e_t^0 &= y_t - y_{t|t-1} \\ &= M_t^0 - M_{t-1}^0, \end{aligned}$$

which is a martingale difference. It satisfies

$$\begin{aligned} E(e_t^0 \mid I_{t-1}) &= M_{t-1}^0 - M_{t-1}^0 \\ &= 0. \end{aligned}$$

The difference of the expectations of y_{t+j} formed at time t and $t-1$ is called the expectation revision process. It is also a martingale difference process because we have

(7) $$y_{t+j|t} - y_{t+j|t-1} = M_t^j - M_t^{j-1}.$$

Define this as e_t^j. Then

$$E(e_t^j \mid \Omega_{t-1}) = 0.$$

By telescoping the difference $y_{t|t-1} - y_{t|t-k}$ as $\Sigma_{j=1}^{k-1}(y_{t w t-j} - y_{t|t-j-1})$, we can express it as a sum of martingale differences

$$y_{t|t-1} - y_{t|t-k} = \sum_{j=1}^{k-1} e_{t-j}^j.$$

The general relation is

$$y_{t+h-k} = y_{t+h-k|t-k} + \sum_{j=0}^{h-1} e_{t+h-k-j}^j, \qquad h, k \leqslant 0.$$

Revisions of expectations are expressible in terms of these martingale difference processes. For example,

$$y_{t+1|t-1} - y_{t+1|t-2} = e_{t-1}^2,$$

and

$$\begin{aligned} y_{t+1} &= y_{t+1} - y_{t+1|t} + y_{t+1|t} - y_{t+1|t-1} \\ &\quad + y_{t+1|t-1} - y_{t+1|t-2} + y_{t+1|t-2} \\ &= e_{t+1}^0 + e_t^1 + e_{t-1}^2 + y_{t+1|t-2} \end{aligned}$$

or

$$y_{t+1|t-2} = y_{t+1} - e_{t+1}^0 - e_t^1 - e_{t-1}^2.$$

We later see how these martingale difference processes are sometimes constrained by the rational expectations models. In solving LRE models with exogenous inputs, we need to relate the sequence $\{e_t^k\}$ to the expectations revisions of the exogenous sequence $\{n_t\}$, since the latter drives the solution sequence. Nonuniqueness of the solutions also arises from the nonuniqueness of this latter revision sequence. Different information sets imply different revision processes. To see this point simply, suppose that the information set at time t is

$$\Omega_t = \{y^t, n^t, \zeta^t\}$$

where subscript t refers to the sequence up to t, e.g., n^t stands for $n_t, n_{t-1}, n_{t-2}, \ldots$ etc. Here $\{\zeta^t\}$ is a sequence independent of all others. One example of such a ζ is a sunspot i.e., a truly exogenous variable. In some cases, ζ^t may stand for economic variables observed by the agents forming the expectations that do not appear in the model under consideration but do appear in other parts of the total (structural) equation system of which it is only a part. Then the information set will in general contain ζ^t. We generally set ζ^t to zero.

The next example illustrates how to convert rational expectation equations into those with martingale difference processes.

Example 4. Given a model

$$\mathbf{y}_t = \mathbf{B}\mathbf{y}_{t|t-2} + \mathbf{n}_t, \tag{8}$$

we can use

$$\mathbf{y}_t - \mathbf{y}_{t|t-2} = \mathbf{y}_t - \mathbf{y}_{t|t-1} + \mathbf{y}_{t|t-1} - \mathbf{y}_{t|t-2} = \mathbf{e}_t^0 + \mathbf{e}_{t-1}^1$$

to substitute $\mathbf{y}_{t|t-2}$ out so we can rewrite (8) as

$$\mathbf{y}_t = \mathbf{B}(\mathbf{y}_t - \mathbf{e}_t^0 - \mathbf{e}_{t-1}^1) + \mathbf{n}_t$$

(9) or

$$(\mathbf{I} - \mathbf{B})\mathbf{y}_t = -\mathbf{B}(\mathbf{e}_t^0 + \mathbf{e}_{t-1}^1) + \mathbf{n}_t.$$

Then we can decompose (9) into $(\mathbf{I} - \mathbf{B})\mathbf{y}_t^p = \mathbf{n}_t$ and $(\mathbf{I} - \mathbf{B})\mathbf{y}_t^h = -\mathbf{B}(\mathbf{e}_t^0 + \mathbf{e}_{t-1}^1)$ and solve them separately. Alternately, we can tie $\mathbf{e}_t^0$ and $\mathbf{e}_{t-1}^1$ directly to the revision processes for $\mathbf{n}_t$. To carry out this step, subtract the conditional expectation E_{t-1} of (8) from it to see that

$$\mathbf{e}_t^0 = \mathbf{n}_t - \mathbf{n}_{t|t-1}. \tag{10}$$

Take the difference of E_{t-1} and E_{t-2} of (8) to obtain

$$\mathbf{e}_{t-1}^1 = \mathbf{n}_{t|t-1} - \mathbf{n}_{t|t-2}. \tag{11}$$

Substitute (10) and (11) into (9) to produce the equation

$$(\mathbf{I} - \mathbf{B})\mathbf{y}_t = -\mathbf{B}(\mathbf{n}_t - \mathbf{n}_{t|t-1}) - \mathbf{B}(\mathbf{n}_{t|t-1} - \mathbf{n}_{t|t-2}) + \mathbf{n}_t, \tag{12}$$

which exhibits that $\mathbf{n}_t$ and the revision processes $\mathbf{n}_t - \mathbf{n}_{t|t-1}$ and $\mathbf{n}_{t|t-1} - \mathbf{n}_{t|t-2}$ act as forcing terms in this equation. Once they are specified, (12) can be solved as an ordinary stochastic difference equation. When $\mathbf{I} - \mathbf{B}$ is nonsingular, $\mathbf{y}_t$ can be solved out to obtain its reduced-form expression. Note that the martingale differences $\mathbf{e}_t^j$, $j \geqslant 2$ do not affect the solution process.

Example 5. Now modify (8) by adding $\mathbf{A}\mathbf{y}_{t-1}$ to the right side,

$$\mathbf{y}_t = \mathbf{B}\mathbf{y}_{t|t-2} + \mathbf{A}\mathbf{y}_{t-1} + \mathbf{n}_t. \tag{13}$$

Instead of (9) we now obtain

$$(\mathbf{I} - \mathbf{B})\mathbf{y}_t = \mathbf{A}\mathbf{y}_{t-1} - \mathbf{B}(\mathbf{e}_t^0 + \mathbf{e}_{t-1}^1) + \mathbf{n}_t. \tag{14}$$

The martingale differences are now intercoupled:

$$\mathbf{e}_t^0 = \Delta\mathbf{n}_t$$

(15) and

$$\mathbf{e}_{t-1}^1 = \mathbf{A}\mathbf{e}_{t-1}^0 + \Delta\mathbf{n}_{t|t-1}$$

where we define

$$\Delta \mathbf{n}_t = \mathbf{n}_t - \mathbf{n}_{t|t-1}$$

$$\Delta \mathbf{n}_{t|t-1} = \mathbf{n}_{t|t-1} - \mathbf{n}_{t|t-2}.$$

Note that $\mathbf{e}_t^1$ is no longer independent of $\mathbf{e}_t^0$.

When $\mathbf{n}_t$ is tied to $\mathbf{y}$s more explicitly, the solutions can be made more explicit as well. As an example, suppose that $\mathbf{n}_t$ is a feedback signal on $\mathbf{y}_{t-1}$ with random disturbances $\mathbf{u}_t$,

$$\mathbf{n}_t = \mathbf{\Phi y}_{t-1} + \mathbf{u}_t.$$

Then

$$\begin{aligned}\Delta \mathbf{n}_t &= \mathbf{n}_t - \mathbf{n}_{t|t-1} \\ &= \mathbf{\Phi y}_{t-1} - \mathbf{\Phi y}_{t-1} + \mathbf{u}_t - \mathbf{u}_{t|t-1} \\ &= \mathbf{u}_t - \mathbf{u}_{t|t-1}.\end{aligned}$$

If $\mathbf{u}_{t|t-1} = 0$, then $\Delta \mathbf{n}_t = \mathbf{u}_t$ and

$$\begin{aligned}\Delta \mathbf{n}_{t|t-1} &= \mathbf{\Phi y}_{t-1} + \mathbf{u}_{t|t-1} - \mathbf{\Phi y}_{t-1|t-2} + \mathbf{u}_{t|t-2} \\ &= \mathbf{\Phi}(\mathbf{y}_{t-1} - \mathbf{y}_{t-1|t-2}) + \mathbf{u}_{t|t-1} - \mathbf{u}_{t|t-2} \\ &= \mathbf{\Phi e}_{t-1}^0 + \mathbf{u}_{t|t-1} - \mathbf{u}_{t|t-2} \\ &= \mathbf{\Phi e}_{t-1}^0\end{aligned}$$

if $\mathbf{u}_{t|t-1} - \mathbf{u}_{t|t-2}$ is zero. The martingale differences are constrained by

$$\mathbf{e}_t^0 = \mathbf{u}_t,$$

and

$$\begin{aligned}\mathbf{e}_t^1 &= \mathbf{A e}_t^0 + \Delta \mathbf{n}_{t+1|t} \\ &= \mathbf{A e}_t^0 + \mathbf{\Phi e}_t^0 = (\mathbf{A} + \mathbf{\Phi})\mathbf{e}_t^0 = (\mathbf{A} + \mathbf{\Phi})\mathbf{u}_t.\end{aligned}$$

Under this assumption on $\mathbf{n}_t$, (14) specializes to

$$(\mathbf{I} - \mathbf{B})\mathbf{y}_t = (\mathbf{A} + \mathbf{\Phi})\mathbf{y}_{t-1} + (\mathbf{I} - \mathbf{B})\mathbf{u}_t - \mathbf{B}(\mathbf{A} + \mathbf{\Phi})\mathbf{u}_t.$$

If $(\mathbf{I} - \mathbf{B})$ is nonsingular, it can be solved out to yield the reduced form expression for $\mathbf{y}_t$.

Perfect Foresight Models

In the literature, the notion of perfect foresight models has only been used loosely. The distinction between $y_{t|t-1}$ and $y_{t|t-2}$ is disregarded because the model is "perfect foresight," e.g., $y_{t|t-1} = y_{t|t-j}$, $j = 1, \ldots$ may be the defining relations. In Broze et al. (1985a), perfect foresight solutions are taken to mean

$$y_{t+h-k|t-k} = y_{t+h-k}$$

for all $0 \leqslant h \leqslant H$ and $0 \leqslant k \leqslant K$ in the original equation. Two different LRE models may correspond to the same perfect foresight model. Different constraints are then implied for martingale difference processes of the original models. For example, $y_{t+1|t-1} = y_{t+1} - e^0_{t+1} - e^1_t$, and, hence, the constraint $e^1_t + e^0_{t+1} = 0$ for all t is imposed, when $y_{t+1|t-1}$ is equated to y_{t+1}. If y_{t+1} comes from $y_{t+1|t}$, then the perfect foresight model implies the condition that $e^0_t = 0$ for all t is imposed.

Thus, a perfect foresight model

$$\mathbf{y}_t = \mathbf{B}\mathbf{y}_{t+1} + \mathbf{C}\mathbf{x}_t + \mathbf{u}_t, \tag{16}$$

where $\mathbf{x}_t$ is some deterministic sequence, may come from

$$\mathbf{y}_t = \mathbf{B}(\mathbf{y}_{t+1} - \mathbf{e}^0_{t+1}) + \mathbf{C}\mathbf{x}_t + \mathbf{u}_t \tag{17}$$

or from

$$\mathbf{y}_t = \mathbf{B}(\mathbf{y}_{t+1} - \mathbf{e}^0_{t+1} - \mathbf{e}^1_t) + \mathbf{C}\mathbf{x}_t + \mathbf{u}_t \tag{18}$$

where $\mathbf{e}^0_t$ and $\mathbf{e}^1_t$ are constrained by (16). Since $\mathbf{e}^0_t = \mathbf{C}(\mathbf{x}_t - \mathbf{x}_{t|t-1})$ in (17) this yields a perfect foresight solution only if the one-step-ahead forecast error of $\mathbf{x}_{t+k}$ is zero for all $k \geqslant 0$ in general. If $\mathbf{B}$ is nilpotent, $\mathbf{B}^2 = 0$ to be specific, then $\mathbf{e}^0_{t+1} = 0$ if and only if

$$\mathbf{C}(\mathbf{x}_{t+1} - \mathbf{x}_{t+1|t}) + \mathbf{BC}(\mathbf{x}_{t+2|t+1} - \mathbf{x}_{t+2|t}) = 0.$$

This is satisfied if $\mathbf{x}_{t+1} = \mathbf{x}_{t+1|t}$ and $\mathbf{x}_{t+2} = \mathbf{x}_{t+2|t}$, i.e., if $\mathbf{x}_{t+1}$ and $\mathbf{x}_{t+2}$ can be perfectly forecast at time t, even though $\mathbf{x}_{t+k}$, $k \geqslant 3$ may not be known exactly.

With (18), the constraint becomes

$$0 = \mathbf{e}^0_{t+1} = \mathbf{C}\Delta\mathbf{x}_{t+1|t}$$

and

$$0 = \mathbf{e}^1_t = \mathbf{C}\Delta\mathbf{x}_{t+1|t} + \mathbf{BC}\Delta\mathbf{x}_{t+2|t}.$$

Thus in the former case, the constraint is

$$\mathbf{C}(\mathbf{x}_t - \mathbf{x}_{t|t-1}) + \mathbf{BC}(\mathbf{x}_{t+1|t} - \mathbf{x}_{t+1|t-1}) = 0, \tag{18a}$$

while the latter imposes two separate constraints

$$\mathbf{C}(\mathbf{x}_t - \mathbf{x}_{t|t-1}) = 0 \quad \text{and} \quad \mathbf{BC}(\mathbf{x}_{t+1|t} - \mathbf{x}_{t+1|t-1}) = 0. \tag{19}$$

Clearly (19) implies (18a). The converse does not necessarily hold. The perfect foresight model solutions thus may coincide with that of (18a), with $u_t \equiv 0$ for all t, but not with that of the latter. To illustrate, suppose $\mathbf{C} = \mathbf{I}$. Then the former requires that $\mathbf{x}_t$ is exactly known as $t - 1$ and the error made in forecasting $\mathbf{x}_t$ two periods ahead, $\mathbf{x}_{t+1|t} - \mathbf{x}_{t+1|t-1}$ lies in the null space of $\mathbf{B}$, while the latter implies a milder restriction, $\mathbf{n}_t^0 + \mathbf{B}\mathbf{n}_t^1 = 0$, where $\mathbf{n}_t^0 = \mathbf{x}_t - \mathbf{x}_{t|t-1}$ and $\mathbf{n}_t^1 = \mathbf{x}_{t+1|t} - \mathbf{x}_{t+1|t-1}$, i.e., any deterministic forecasting rule that satisfies

$$\mathbf{n}_t^0 + \mathbf{B}\mathbf{u}_t^1 = 0$$

causes the solution of the latter and (17) to be the same over two consecutive time periods.

8.3 Models with Expected Current Variables Formed in the Past

Since the solution of a general LRE model is rather complicated, this chapter uses an organizing strategy of progressing from simple to complex models. All models are solved by using the martingale differences; relations with other solution methods are later established. This section discusses models containing expectations of current endogenous variables formed in the past, followed by models containing expectations of future endogenous variables, and lastly a mixture of the two types of expectations. Within each category, scalar models are discussed generally before vector-valued models. In each subcategory, models with no lagged endogenous variables are treated before those with lagged endogenous variables.

Models with No Lagged Endogenous Variables

By rewriting the model

$$y_t = \alpha y_{t|t-1} + n_t$$

as

$$\begin{aligned} y_t &= \alpha y_{t|t-1} + n_t \\ &= \alpha(y_t - e_t^0) + n_t \\ &= \alpha y_t + n_t - \alpha e_t^0, \end{aligned}$$

we can decompose y_t as $y_t^p + y_i^h$ where

$$y_t^p = \alpha y_t^p + n_t$$

and

$$y_t^h = \alpha y_t^h - \alpha e_t^0.$$

The former has the solution, $n_t/(1-\alpha)$ if $\alpha \neq 1$, and the latter is equal to $-\alpha e_t^0/(1-\alpha)$. The general solution is then given by

$$y_t = \frac{1}{1-\alpha} n_t - \frac{\alpha}{1-\alpha} e_t^0.$$

Since $e_t^0 = n_t - n_{t|t-1}$ in this model, it may be written as $n_t - \alpha(1-\alpha)^{-1} n_{t|t-1}$ also. Suppose $\alpha \neq 1$. Once the mechanism for calculating $n_{t|t-1}$ is specified, the solution is uniquely determined because if y_t^1 and y_t^2 are two solutions, then $y_t^1 - y_t^2 = \alpha(y_{t|t-1}^1 - y_{t|t-1}^2)$, from which $y_{t|t-1}^1 = y_{t|t-1}^2$ and hence $y_t^1 = y_t^2$ follows if $\alpha \neq 1$.

Example 6. The solution procedure of a general model containing expectations of current endogenous variables formed at several past time periods is easily inferred from the next example:

$$y_t = \alpha_1 y_{t|t-1} + \alpha_2 y_{t|t-2} + n_t.$$

Rewrite this model first as

$$y_t = \alpha_1(y_t - e_t^0) + \alpha_2(y_t - e_t^0 - e_{t-1}^1) + n_t,$$

and then decompose it into

$$(1-\alpha_1-\alpha_2)y_t^p = n_t \qquad \text{or} \qquad y_t^p = (1-\alpha_1-\alpha_2)^{-1} n_t$$

and

$$(1-\alpha_1-\alpha_2)y_t^h = -(\alpha_1+\alpha_2)e_t^0 - \alpha_2 e_{t-1}^1$$

or

$$y_t^h = \frac{\alpha_1+\alpha_2}{1-\alpha_1-\alpha_2} e_t^0 - \frac{\alpha_2}{1-\alpha_1-\alpha_2} e_{t-1}^1.$$

The martingale differences are related to n_t and its expectations by

$$e_t^0 = n_t$$

and

$$e_{t-1}^1 = n_{t|t-1} - n_{t|t-2}.$$

The general solution is then equal to

$$\begin{aligned} y_t &= (1-\alpha_1-\alpha_2)^{-1}[n_t - (\alpha_1+\alpha_2)e_t^0 - \alpha_2 e_{t-1}^1] \\ &= n_t - (1-\alpha_1-\alpha_2)^{-1}\alpha_2[n_{t|t-1} - n_{t|t-2}] \end{aligned}$$

if $\alpha_1 + \alpha_2 \neq 1$.

Note that no "true" dynamics are involved in these solutions since only timing differences in forming expectations generate dynamics in this example.

Conversion to State Space Representation

In the Example of the previous section, we converted an LRE model with lagged endogenous variables into state space form. Actually any LRE models can be so converted. This section describes how to convert linear rational expectations models into state space form. State space representation unifies and facilitates solution procedures for rational expectation models with different lag structures of endogenous variables. The idea of the procedure is easily conveyed by the next example.

Example 7. Consider

$$y_t = \alpha_1 y_{t-1} + \alpha_2 y_{t-2} + B_1 y_{t+1|t} + B_2 y_{t+2|t} + n_t,$$

where $B_2 \neq 0$.

Substitute out $y_{t+i|t}$, $i = 1, 2$ by

$$y_{t+1|t} = y_{t+1} - e^0_{t+1}$$

and

$$y_{t+2|t} = y_{t+2} - e^0_{t+2} - e^1_{t+1}$$

to convert the model into

$$\begin{aligned} y_t = \alpha_1 y_{t-1} + \alpha_2 y_{t-2} + B_1 y_{t+1} + B_2 y_{t+2} \\ - B_2 e^0_{t+2} - B_2 e^1_{t+1} - B_1 e^0_{t+1} + n_t. \end{aligned}$$

Dividing through by B_2 and rearranging, we have

$$\begin{aligned} y_{t+2} = \phi_1 y_{t+1} + \phi_2 y_t + \phi_3 y_{t-1} + \phi_4 y_{t-2} \\ + e^2_{t+2} + e^1_{t+1} + \theta_1 e^0_{t+1} + \theta_2 n_t \end{aligned} \tag{20}$$

where

$$\phi_1 = -B_1/B_2, \quad \phi_2 = 1/B_2, \quad \phi_3 = -\alpha_1/B_2, \quad \phi_4 = -\alpha_2/B_2,$$
$$\theta_1 = B_1/B_2, \quad \theta_2 = 1/B_2.$$

Now to convert (20) into state space form, introduce state variables by

$$\begin{aligned} z_{1t} &= y_t - e^0_t, \\ z_{2t-1} &= z_{1t} - \phi_1 y_{t-1} - e^1_{t-1} - \theta_1 e^0_{t-1} \\ &= z_{1t} - \phi_1 z_{1t-1} - (\phi_1 + \theta_1) e^0_{t-1} - e^1_{t-1}, \\ z_{3t-1} &= z_{2t} - \phi_2 y_{-1} - \theta_2 n_{t-1} \\ &= z_{2t} - \phi_2 z_{1t-1} - \phi_2 e^0_{t-1} - \theta_2 n_{t-1}, \end{aligned}$$

and

$$z_{4t-1} = z_{3t} - \phi_3 y_{t-1}$$
$$= z_{3t} - \phi_3 z_{1t-1} - \phi_3 e^0_{t-1}.$$

Then from (20) and the preceding definitional relations, we see that

$$z_{4t} = \phi_4 y_{t-1}$$
$$= \phi_4 z_{1t-1} + \phi_4 e^0_{t-1},$$

and we have

$$z_t = \Phi z_{t-1} + (\psi + \theta_1 i_1) e^0_{t-1} + i_1 e^1_{t-1} + \theta_2 i_2 n_{t-1}$$

where

$$z_t = (z_1 \quad z_2 \quad z_3 \quad z_4)'_t$$

(21)
$$\boldsymbol{\Phi} = \begin{bmatrix} \phi_1 & 1 & 0 & 0 \\ \phi_2 & 0 & 1 & 0 \\ \phi_3 & 0 & 0 & 1 \\ \phi_4 & 0 & 0 & 0 \end{bmatrix}$$

and

$$y_t = (1 \quad 0 \quad 0 \quad 0) z_t + e^0_t,$$

where

$$\psi = (\phi_1 \quad \phi_2 \quad \phi_3 \quad \phi_4)',$$
$$i_1' = (1 \quad 0 \quad 0 \quad 0),$$

and

$$i_2' = (0 \quad 1 \quad 0 \quad 0).$$

Now decompose z_t into $z^p_t + z^h_t$ where

(22) $$z^p_t = \Phi z^p_{t-1} + \theta_2 i_2 n_{t-1}$$

and

(23) $$z^h_t = \Phi z^h_{t-1} + (\psi + \theta_1 i_1) e^0_{t-1} + i_1 e^1_{t-1}$$

where e^0_t and e^1_t are two distinct arbitrary martingale differences. Correspondingly, we also decompose y_t into

(24) $$y_t = y^p_t + y^h_t.$$

Since we know how to analyze (22), we record only the solution method for (23). First, note that $|\lambda I - \Phi| = \lambda^4 - \phi_1 \lambda^3 - \phi_2 \lambda^2 - \phi_3 \lambda - \phi_4$, i.e., the eigenvalues of Φ are the same as the roots of

$$1 = \alpha_1 \lambda^{-1} + \alpha_2 \lambda^{-2} + B_1 \lambda + B_2 \lambda^2$$

associated with the denominator of the transfer function representation of the model (20)

$$(1 - B_1 L^{-1} - B_2 L^{-2} - \alpha_1 L - \alpha_2 L^2)$$

where $Ly_t = y_{t-1}$ and $L^{-1} y_t = y_{t+1}$.

From (23) and (24), we obtain

$$y_t^h = \sum_{i=1}^{t-1} h_i e_{t-i}^0 + \sum_{i=1}^{t-1} k_j e_{t-j}^1 + i_i' \Phi^t z_0^h.$$

If all the eigenvalues of Φ are less than one in magnitude, then we have a unique weakly stationary solution

$$y_t^h = \sum_{i=1}^{\infty} h_i e_{t-i}^0 + \sum_{j=1}^{\infty} k_j e_{t-j}^1$$

where

$$h_i = i_1' \Phi^{i-1}(\psi + \theta_1 i_1) = \theta_1 k_i + k_{i+1},$$

and

$$k_i = i_1' \Phi^{i-1} i_1,$$

where we use the relation

$$\Phi i_1 = \Psi,$$

i.e.,

$$i_1' \Phi^{i-1} \Psi = i_l' \Phi^i i_1 = k_{i+1}.$$

A special case is obtained by setting $\alpha_1 = 0$ and $\alpha_2 = 0$. This is a model with no lagged endogenous variables. It is possible then to represent y_t^h in terms of two distinct but arbitrary martingales as

$$y_t^h = \lambda_1^{-t} m_{1t} + \lambda_2^{-t} m_{2t}$$

where λ_1 and λ_2 are the roots of

$$B_2 \lambda^2 + B_1 \lambda = 1.$$

Models with Lagged Endogenous Variables

When lagged endogenous variables are present, additional dynamics are introduced. Consider the next example.

Example 8. This model contains lagged endogenous variables in addition to the expectations formed at different past periods:

$$y_t = a_1 y_{t-1} + a_2 y_{t-2} + \alpha_1 y_{t|t-1} + \alpha_2 y_{t|t-2} + \alpha_3 y_{t|t-3} + n_t.$$

Since

$$y_{t|t-1} = y_t - e_t^0,$$

$$y_{t|t-2} = y_t - e_t^0 - e_{t-1}^1,$$

and

$$y_{t|t-3} = y_t - e_t^0 - e_{t-1}^1 - e_{t-2}^2,$$

substitute them into the preceding to rewrite it as

$$(1 - \alpha_1 - \alpha_2 - \alpha_3) y_t = a_1 y_{t-1} + a_2 y_{t-2} + n_t - (\alpha_1 + \alpha_2 + \alpha_3) e_t^0 - (\alpha_2 + \alpha_3) e_{t-1}^1 - \alpha_3 e_{t-2}^2,$$

and decompose the model into

$$(1 - \alpha_1 - \alpha_2 - \alpha_3) y_t^p = a_1 y_{t-1}^p + a_2 y_{t-2}^p + n_t \tag{25}$$

and

$$(1 - \alpha_1 - \alpha_2 - \alpha_3) y_t^h = a_1 y_{t-1}^h + a_2 y_{t-2}^h - (\alpha_1 + \alpha_2 + \alpha_3) e_t^0 - (\alpha_2 + \alpha_3) e_{t-1}^1 - \alpha_3 e_{t-2}^2. \tag{26}$$

Let (25) be rewritten as

$$y_t^p = \phi_1 y_{t-1}^p + \phi_2 y_{t-2}^p + \theta_0 n_t$$

where ϕ_1, ϕ_2, and θ_0 are defined by (25) and (27). Define state variables

$$z_{1t} = y_t^p - \theta_0 n_t$$

and

$$\begin{aligned} z_{2t-1} &= z_{1t} - \phi_1 y_{t-1}^p \\ &= z_{1t} - \phi_1 (z_{1t-1} + \theta_0 n_{t-1}). \end{aligned}$$

Then advance time by one unit in the preceding to note that

$$\begin{aligned} z_{2t} &= z_{1t+1} - \phi_1 y_t^p \\ &= y_{t+1}^p - \theta_0 n_{t+1} - \phi_1 y_t^p \\ &= \phi_2 y_{t-1}^p \\ &= \phi_2 (z_{1t-1} + \theta_0 n_{t-1}). \end{aligned}$$

We have constructed a state space model for (27) with the transition dynamics

$$\begin{bmatrix} z_1 \\ z_2 \end{bmatrix}_t = \begin{bmatrix} \phi_1 & 1 \\ \phi_2 & 0 \end{bmatrix} \begin{bmatrix} z_1 \\ z_2 \end{bmatrix}_{t-1} + \theta_0 \begin{bmatrix} \phi_1 \\ \phi_2 \end{bmatrix} n_{t-1} \tag{28}$$

and the observation equation

$$y_t^p = (1 \quad 0) z_t + \theta_0 n_t. \tag{27a}$$

Rewrite (26) as

$$y_t^h = \phi_1 y_{t-1}^h + \phi_2 y_{t-2}^h + \mu_0 e_t^0 + \mu_1 e_{t-1}^1 + \mu_2 e_{t-2}^2,$$

where μs are defined by the equality of (26) and (27a). Introduce another set of state variables s_{1t} and s_{2t} analogously to the preceding

$$s_{1t} = y_t^h - \mu_0 e_t^0$$

$$s_{2t-1} = s_{1t} - \phi_1 y_{t-1}^h - \mu_1 e_{t-1}^1,$$

and hence

$$s_{2t} = \phi_2 y_{t-1}^h + \mu_2 e_{t-1}^2,$$

from which we have constructed a state space model for (27a) as

$$\begin{aligned} \begin{bmatrix} s_1 \\ s_2 \end{bmatrix}_t &= \begin{bmatrix} \phi_1 & 1 \\ \phi_2 & 0 \end{bmatrix} \begin{bmatrix} s_1 \\ s_2 \end{bmatrix} + \mu_0 \begin{bmatrix} \phi_1 \\ \phi_2 \end{bmatrix} e_{t-1}^0 \\ &\quad + e_{t-1}^1 + \mu_1 \begin{bmatrix} 1 \\ 0 \end{bmatrix} e_{t+1}^1 + \mu_2 \begin{bmatrix} 0 \\ 1 \end{bmatrix} e_{t-1}^2 \end{aligned} \tag{29}$$

and

$$y_t^h = (1 \quad 0) \mathbf{s}_t + \mu_0 e_t^0.$$

As before, we can relate e_t^0 and e_{t-1}^1 to the expectation revisions of exogenous variables. In this model,

$$\begin{bmatrix} 1 & 0 \\ -a_2 & 1 \end{bmatrix} \begin{bmatrix} e_t^1 \\ e_t^2 \end{bmatrix} = a_1 \begin{bmatrix} 1 \\ 1 \end{bmatrix} e_t^0 + \begin{bmatrix} n_{t+1|t} - n_{t+1|t-1} \\ n_{t+2|t} - n_{t+1|t-1} \end{bmatrix}, \tag{30}$$

where

$$e_t^0 = n_t - n_{t|t-1}.$$

Solve (30) to yield

$$\begin{bmatrix} e_t^1 \\ e_t^2 \end{bmatrix} = a_1 \begin{bmatrix} 1 \\ 1 + a_2 \end{bmatrix} e_t^0 + \begin{bmatrix} 1 & 0 \\ a_2 & 1 \end{bmatrix} \begin{bmatrix} n_{t+1|t} - n_{t+1|t-1} \\ n_{t+2|t} - n_{t+2|t-1} \end{bmatrix}.$$

Both (28) and (29) can be treated as ordinary stochastic difference equations. Solve (28) as

$$z_t = \theta_0[\boldsymbol{\phi} n_{t-1} + \boldsymbol{\Phi}\boldsymbol{\phi} n_{t-2} + \ldots + \boldsymbol{\Phi}^{t-1}\boldsymbol{\phi} n_0] + \boldsymbol{\Phi}^t z_0,$$

where

$$\boldsymbol{\phi}' = (\phi_1 \quad \phi_2) \qquad \text{and} \qquad \boldsymbol{\Phi} = \begin{bmatrix} \phi_1 & 1 \\ \phi_2 & 0 \end{bmatrix},$$

from which we obtain

(31) $$y_t^p = \theta_0(n_t + h_1 n_{t-1} + h_2 n_{t-2} + \ldots + h_{t-1} n_0) + (1 \quad 0)\boldsymbol{\Phi}^t z_0$$

where

$$h_i = (1 \quad 0)\boldsymbol{\Phi}^{i-1}\boldsymbol{\phi}$$

is the dynamic multiplier of the model (28). Note that the characteristic polynomial of the matrix $\boldsymbol{\Phi}$, $\lambda^2 - \phi_1\lambda - \phi_2$, is the same as the roots of the lag transform of (25) or (26): $(1 - \alpha_1 - \alpha_2 - \alpha_3) = a_1 L + a_1 L - a_2 L^2$. Similarly (29) can be solved as

$$\begin{aligned} \mathbf{s}_t = \mu_0[\boldsymbol{\phi} e_{t-1}^0 &+ \boldsymbol{\Phi}\boldsymbol{\phi} e_{t-2} + \ldots + \boldsymbol{\Phi}^{t-1}\boldsymbol{\phi} e_0^0] \\ &+ [\boldsymbol{\psi} e_{t-1}^1 + \boldsymbol{\Phi}\boldsymbol{\psi} e_{t-2}^1 + \ldots + \boldsymbol{\Phi}^{t-1}\boldsymbol{\psi} e_0^1] \\ &+ \boldsymbol{\Phi}^t s_0, \end{aligned}$$

where

$$\boldsymbol{\psi}' = (\mu_1 \quad \mu_2),$$

from which we obtain

(32) $$\begin{aligned} y_t^h &= \mu_0(e_t^0 + h_1 e_{t-1}^0 + \ldots + h_{t-1} e_0^0) + \mu_1(k_1 e_{t-1}^1 + \ldots + k_{t-1} e_0^1) \\ &\mu_2(l_1 e_{t-1}^2 + \ldots + l_{t-1} e_0^2) + (1 \quad 0)\boldsymbol{\Phi}^t \mathbf{s}_0 \end{aligned}$$

where

$$k_t = (1 \quad 0)\boldsymbol{\Phi}^{t-1}\begin{pmatrix} 1 \\ 0 \end{pmatrix},$$

and

$$l_t = (1 \quad 0)\boldsymbol{\Phi}^{t-1}\begin{pmatrix} 0 \\ 1 \end{pmatrix}, \qquad i \geqslant 1,$$

are the dynamic multipliers of (29). The general solution is the sum of (31) and (32), which depends on $\mathbf{z}_0$ and $\mathbf{s}_0$ and on the expectation-revision processes for the exogenous variable, $n_t, n_t - n_{t|t-1}, n_{t|t-1} - n_{t|t-2}$. Note that the stability property of $\boldsymbol{\Phi}$ is determined by ϕ_1 and ϕ_2, i.e., jointly by the coefficient multiplying the endogenous variable, a_1, a_2, and those on the expectational variables α_1, α_2, and α_3.

When the coefficients of LRE models are matrices, no new conceptual problem arises, except for the definition of minimal dimensional state vectors, which is resolved by the method due to Gilbert (1963) when no repeated eigenvalues is present. See Aoki (1987c, p. 57) for its exposition.

We are now ready to discuss a class of rational expectations models represented by

$$\mathbf{y}_t = \mathbf{A}\mathbf{y}_{t-1} + \mathbf{B}_1\mathbf{y}_{t|t-1} + \dots + \mathbf{B}_p\mathbf{y}_{t|t-p} + \mathbf{C}\mathbf{x}_t + \mathbf{u}_t$$

where $\mathbf{y}_t$ is a vector of endogenous variables, $\mathbf{x}_t$ is a vector of exogenous variables, and $\mathbf{u}_t$ is a mean-zero unobserved-error vector, i.e., $\mathbf{n}_t = \mathbf{C}\,\mathbf{x}_t + \mathbf{u}_t$. The rational expectations are defined by

$$\mathbf{y}_{t|t-j} = E(\mathbf{y}_t \mid \Omega_{t-j})$$

where Ω_{t-j} is the information set available at time $t-j$. We use

$$\Omega_t = \{\mathbf{y}^t, \mathbf{x}^{t-1}, \mathbf{u}^{t-1} \dots\},$$

but others are also possible. Discussion in this section is a slight generalization of Aoki and Canzoneri (1979) and Visco (1981, 1984).

Case of Single Lag

Project the equation

(33) $$\mathbf{y}_t = \mathbf{A}\mathbf{y}_{t-1} + \mathbf{B}_1\mathbf{y}_{t|t-1} + \mathbf{C}\mathbf{x}_t + \mathbf{u}_t$$

onto Ω_{t-1} to produce

(34) $$\mathbf{y}_{t|t-1} = \mathbf{A}\mathbf{y}_{t-1} + \mathbf{B}_1\mathbf{y}_{t|t-1} + \mathbf{C}\mathbf{x}_{t|t-1}.$$

Subtract the latter from the former to obtain

(35) $$\mathbf{e}_t^0 = \mathbf{y}_t - \mathbf{y}_{t|t-1} = \mathbf{C}(\mathbf{x}_t - \mathbf{x}_{t|t-1}) + \mathbf{u}_t,$$

and note that $E(\mathbf{e}_t^0 \mid \Omega_{t-1}) = 0$, i.e., $\{\mathbf{e}_t^0\}$ is a martingale different sequence introduced earlier.

From (33) and (35), (33) is restated as

$$\mathbf{y}_t = \mathbf{A}\mathbf{y}_{t-1} + \mathbf{B}_1(\mathbf{y}_t - \mathbf{e}_t) + \mathbf{C}\mathbf{x}_t + \mathbf{u}_t.$$

Equation (35) shows that the martingale difference $\mathbf{e}_t^0$ represents the "surprise" component of the exogenous vector, Equation (34) is solved as

$$\mathbf{y}_t = (\mathbf{I} - \mathbf{B}_1)^{-1}\mathbf{A}\mathbf{y}_{t-1} + (\mathbf{I} - \mathbf{B}_1)^{-1}\mathbf{C}\mathbf{x}_{t|t-1} + \mathbf{e}_t^0,$$

which shows the exogenous vector expected at time $t-1$ and its unexpected component separately. It can be put into a final form

$$\mathbf{y}_t = \mathbf{S}_t\mathbf{y}_0 + \sum_{\tau=1}^{t} \mathbf{S}_{t,\tau}(\mathbf{I}-\mathbf{B}_1)^{-1}\mathbf{C}\mathbf{x}_{\tau|\tau-1} + \sum_{j=1}^{t} \mathbf{S}_{t-j}\mathbf{e}_j^0,$$

where $\mathbf{F} = (\mathbf{I}-\mathbf{B}_1)^{-1}\mathbf{A}$, $\mathbf{G} = (\mathbf{I}-\mathbf{B}_1)^{-1}\mathbf{C}$, $\mathbf{S}_t = \mathbf{F}^t$, and $\mathbf{S}_{t,\tau} = \mathbf{F}^{t-\tau}\mathbf{G}$. We need to specify the exogenous process $\{\mathbf{x}_t\}$ and how $\mathbf{x}_{t|t-1}$ is formed. Note that the solution procedure can be extended to time-varying $\mathbf{A}$, $\mathbf{B}$, and $\mathbf{C}$ matrices.

Case of Multiple Lags

Next, we treat a slightly generalized version

$$\mathbf{y}_t = \sum_{k=1}^{p} \mathbf{B}_k\mathbf{y}_{t|t-k} + \sum_{i=1}^{q} \mathbf{A}_i\mathbf{y}_{t-i} + \mathbf{C}\mathbf{x}_t + \mathbf{u}_t \tag{36}$$

and obtain its reduced form expression by extending the method in Aoki-Canzoneri and Visco. This problem is also considered by Broze et al. (1984). The information set is defined by $\Omega_t = \{\mathbf{y}^t, \mathbf{x}^t\}$. Assume that for every positive t and j

$$E(\mathbf{u}_t) = 0$$

and

$$E(\mathbf{u}_t \mid \Omega_{t-j}) = 0.$$

Subtract the conditional expectation of (36) from itself to obtain

$$\mathbf{e}_t^0 = \mathbf{u}_t + \mathbf{C}(\mathbf{x}_t - \mathbf{x}_{t|t-1}). \tag{37}$$

Use the relation obtained earlier in this chapter on martingale differences,

$$\begin{aligned}\mathbf{y}_t &= (\mathbf{y}_t - \mathbf{y}_{t|t-1}) + (\mathbf{y}_{t|t-1} - \mathbf{y}_{t|t-k}) + \mathbf{y}_{t|t-k}\\ &= \mathbf{e}_t^0 + \sum_{j=1}^{k-1} \mathbf{e}_{t-j}^j + \mathbf{y}_{t|t-k}\\ &= \sum_{j=0}^{k-1} \mathbf{e}_{t-j}^j + \mathbf{y}_{t|t-k},\end{aligned}$$

to rewrite (36) as

$$\mathbf{y}_t = \sum_{k=1}^{p} \mathbf{B}_k\mathbf{y}_t - \sum_{k=1}^{p}\sum_{j=0}^{k-1} \mathbf{B}_k\mathbf{e}_{t-j}^j + \sum_{i=1}^{q} \mathbf{A}_i\mathbf{y}_{t-i} + \mathbf{C}\mathbf{x}_t + \mathbf{u}_t, \tag{38}$$

and, by changing the order of summation, rearrange it as

$$\begin{aligned}\mathbf{V}_p\mathbf{y}_t &= -\sum_{k=1}^{p}\sum_{j=0}^{k-1}\mathbf{B}_k\mathbf{e}_{t-j}^{j} + \sum_{i=1}^{q}\mathbf{A}_i\mathbf{y}_{t-i} + \mathbf{C}\mathbf{x}_t + \mathbf{u}_t, \\ &= -\sum_{j=0}^{p-1}\mathbf{W}_{j+1}\mathbf{e}_{t-j}^{j} + \sum_{i=1}^{q}\mathbf{A}_i\mathbf{y}_{t-i} + \mathbf{C}\mathbf{x}_t + \mathbf{u}_t\end{aligned} \tag{39}$$

where

$$\mathbf{V}_p = \mathbf{I} - \sum_{k=j}^{p}\mathbf{B}_k$$

and

$$\mathbf{W}_j = \sum_{k=j}^{p}\mathbf{B}_k.$$

We next successively relate $\mathbf{e}_{t-j}^{j}$ to $\mathbf{e}_{t-j}^{i}$, $i \leqslant j$.

We assume that $\mathbf{V}_j$ is nonsingular, $j = 1, \ldots p$. (When some $\mathbf{V}_j$ is singular, the solutions are not unique.) Project (36) onto $\mathbf{\Omega}_{t-j}$ and $\mathbf{\Omega}_{t-j-1}$, respectively, to obtain

$$\begin{aligned}\mathbf{y}_{t|t-j} = \sum_{k=1}^{j-1}\mathbf{B}_k\mathbf{y}_{t|t-j} + \sum_{j}^{p}\mathbf{B}_k\mathbf{y}_{t-k} + \sum_{t=1}^{j-1}\mathbf{A}_i\mathbf{y}_{t-i|t-j} \\ + \sum_{i=j}^{q}\mathbf{A}_i\mathbf{y}_{t-j} + E(\mathbf{C}\mathbf{x}_t + \mathbf{u}_t \mid \mathbf{\Omega}_{t-j})\end{aligned}$$

and

$$\begin{aligned}\mathbf{y}_{t|t-j-1} = \sum_{k=1}^{j}\mathbf{B}_k\mathbf{y}_{t|t-j-1} + \sum_{j+1}^{p}\mathbf{B}_k\mathbf{y}_{t-k} + \sum_{t=1}^{j}\mathbf{A}_i\mathbf{y}_{t-i|t-j-1} \\ + \sum_{i=j}^{q}\mathbf{A}_i\mathbf{y}_{t-1} + E(\mathbf{C}\mathbf{x}_t + \mathbf{u}_t \mid \mathbf{\Omega}_{t-j-1}).\end{aligned}$$

Taking the difference,

$$\begin{aligned}\mathbf{y}_{t|t-j} - \mathbf{y}_{t|t-j-1} = {} & \sum_{k=1}^{j}\mathbf{B}_k(\mathbf{y}_{t|t-j} - \mathbf{y}_{t|t-j-1}) \\ & + \sum_{t=1}^{j}\mathbf{A}_i(\mathbf{y}_{t-i|t-j} - \mathbf{y}_{t-i|t-j-1}) \\ & + \mathbf{C}(\mathbf{x}_{t|t-j} - \mathbf{x}_{t|t-j-1})\end{aligned}$$

or recalling that $\mathbf{y}_{t|t-j} - \mathbf{y}_{t|t-j-1}$ is equal to $\mathbf{e}_{t-j}^{j}$ and $\mathbf{y}_{t-i|t-j} - \mathbf{y}_{t-i|t-j-1}$ to $\mathbf{e}_{t-j}^{j-i}$,

$$\mathbf{V}_j\mathbf{e}_{t-j}^{j} = \sum_{t=1}^{j}\mathbf{A}_i\mathbf{e}_{t-j}^{j-i} + \mathbf{C}(\mathbf{x}_{t|t-j} - \mathbf{x}_{t|t-j-i}) \tag{40}$$

where

$$\mathbf{V}_j = \mathbf{I} - \sum_{k=1}^{j} \mathbf{B}_k, \qquad i = 1, \ldots, p.$$

This set of difference equations governs the martingale differences **e**s. These **e**s determine the ways estimates are updated, i.e., **e**s govern the forecast errors.

$\{\mathbf{e}_t^i\}$ is governed by (37) and (39). Equation (39) becomes

$$\mathbf{e}_{t-j}^j = \mathbf{V}_j^{-1} \sum_{i=1}^{j} \mathbf{A}_i \mathbf{e}_{t-j}^{j-i} + \mathbf{V}_j^{-1} \mathbf{C} \zeta_{t,t-j}$$

where

$$\zeta_{t,j} = \mathbf{x}_{t|t-j} - \mathbf{x}_{t|t-j-1}.$$

Equation (39) can be solved sequentially starting with $j = 1$. For example,

$$\mathbf{V}_1 \mathbf{e}_{t-1}^1 = \mathbf{A}_1 \mathbf{e}_{t-1}^0 + \mathbf{C}\zeta_{t,1}$$

and

$$\mathbf{V}_2 \mathbf{e}_{t-2}^2 = \mathbf{A}_1 \mathbf{e}_{t-2}^1 + \mathbf{A}_2 \mathbf{e}_{t-2}^0 + \mathbf{C}\zeta_{t,2}.$$

Since

$$\mathbf{e}_{t-2}^1 = \mathbf{V}_1^{-1} \mathbf{A}_1 \mathbf{e}_{t-2}^0 + \mathbf{V}_1^{-1} \mathbf{C}\zeta_{t-1,1},$$

we obtain

$$\mathbf{V}_2 \mathbf{e}_{t-2}^2 = [\mathbf{A}_2 + \mathbf{A}_1 \mathbf{V}_1^{-1} \mathbf{A}_1] \mathbf{e}_{t-2}^0 + \mathbf{A}_1 \mathbf{V}_1^{-1} \mathbf{C}\zeta_{t-1,1} + \mathbf{C}\zeta_{t,2}.$$

The general formula is

$$\text{(41)} \qquad \mathbf{V}_j \mathbf{e}_{t-j}^j = \mathbf{K}_j \mathbf{e}_{t-j}^0 + \sum_{i=1}^{j} \mathbf{L}_{i,j} \mathbf{C}\zeta_{t-j+1,i}$$

where

$$\text{(42)} \qquad \begin{aligned} \mathbf{K}_0 &= \mathbf{I}, \\ \mathbf{K}_j &= \mathbf{A}_j + \sum_{t=1}^{j-1} \mathbf{A}_{j-i} \mathbf{V}_i^{-1} \mathbf{K}_i, \\ \mathbf{L}_{j,j} &= \mathbf{I}, \\ \mathbf{L}_{i,j} &= \sum_{k=1}^{j-1} \mathbf{A}_k \mathbf{V}_{j-k}^{-1} \mathbf{L}_{i,j-k}, \qquad i < j, \end{aligned}$$

which may be verified inductively. From (36), (41), and (42),

$$\mathbf{V}_p \mathbf{y}_t = \sum_{t=1}^{q} \mathbf{A}_i \mathbf{y}_{t-1} + \mathbf{C}\mathbf{x} + \mathbf{u}_t + \boldsymbol{\phi}_t^P$$

where

$$\text{(43)} \qquad \boldsymbol{\phi}_t^p = -\sum_{j=0}^{p-1} \mathbf{W}_{j+1} \mathbf{V}_j^{-1} \Big(\mathbf{K}_j \mathbf{C}\zeta_{t-j,1} + \mathbf{K}_j \mathbf{u}_{t-j} + \sum_{t=1}^{j} \mathbf{L}_{i,j} \mathbf{C}\zeta_{t-j+1,i}\Big).$$

Equation (43) shows that $\mathbf{V}_p^{-1} \boldsymbol{\phi}_t^p$ represents the stochastic past of (36).

8.4 Models with Future Expectations

Models with No Endogenous Variables

The model discussed by Gourieroux et al. (1982) is the simplest one in this class

$$y_t = \alpha y_{t+1|t} + n_t = \alpha(y_{t+1} - e^0_{t+1}) + n_t,$$

which is decomposed into

(44) $$y^p_t = \alpha y^p_{t+1} + n_t$$

and

(45) $$y^h_t = \alpha y^h_{t+1} - \alpha e^0_{t+1}.$$

Equation (44) is rewritten as $y_{t+1} = \alpha^{-1} y^p_t - \alpha^{-1} n_t$. When $|\alpha| < 1$, (44) can be iterated forward in time to produce

$$y^p_t = n_t + \alpha n_{t+1|t} + \ldots + \alpha^k n_{t+k|t} + \alpha^{k+1} y_{t+k+1|t},$$

which converges to

$$y^p_t = \sum_0^\infty \alpha^i n_{t+i|t}$$

if $\alpha^k y^p_{t+k|t} \to 0$ as $k \to \infty$. If $|\alpha| > 1$ (44) is solved normally (backward in time) to yield

$$y^p_{t+1} = -\alpha^{-1}[n_t + \alpha^{-1} n_{t-1} + \ldots + \alpha^{-k} n_{t-k} - \alpha^{-k} y_{t-k}]$$

and if $\alpha^{-k} y_{t-k} \to 0$ as $k \to \infty$, then we obtain $y^p_t = -\alpha^{-1} \Sigma_{j=0}^{\infty} (\alpha^{-j}) n_{t-j-1}$. From (45) we obtain

(46) $$y^h_t = \frac{1}{\alpha} y^h_{t-1} + e^0_t.$$

When $|\alpha| > 1$, this produces

(47) $$\begin{aligned} y^h_t &= e^0_t + \alpha^{-1} e^0_{t-1} + \ldots + \alpha^{-t} e^0_0 + \alpha^{-t-1} y^h_0 \\ &= \alpha^{-t} m_t + \alpha^{-t-1} y^k_0 \end{aligned}$$

where

$$m_t = \alpha^t e^0_t + \alpha^{t-1} e^0_{t-1} + \ldots + e^0_0.$$

Since $m_{t+1|t} = m_t$, m_t is a martingale, i.e., the reduced form (46) has the final form (47) in terms of y^h_0 and an arbitrary martingale.

Solution Uniqueness

Some LRE models have a multitude of solutions. A simple example is the model discussed by Gourieroux et al. (1982):

$$y_t = \alpha y_{t+1|t} + n_t.$$

Decompose the solution into $y_t = y_t^p + y_t^h$ where

$$y_t^p = \alpha y_{t+1|t}^p + n_t$$

and

$$y_t^h = \alpha y_{t+1|t}^h. \tag{48}$$

Gourieroux et al. pointed out that when $|\alpha| > 1$, if y_t^h satisfies (48), so does $y_t^h + \alpha^{-t} m_t$ where m_t is an arbitrary martingale, since (44) is linear and

$$\alpha E(\alpha^{-t-1} m_{t+1} \mid I_t) = \alpha^{-t} m_t.$$

Thus, at least when $|\alpha| > 1$, there are an infinite amount of solutions for (48).*

Advance t by one in (48) and take the conditional expectation E_t to see that $y_{t+1|t}^h = \alpha y_{t+2|t}^h$, i.e.

$$\begin{aligned} y_t^h &= \alpha^2 y_{t+2|t}^h \\ &= \alpha^k y_{t+k|t}^h, \qquad k \geqslant 1. \end{aligned}$$

Therefore, if y_t^1 and y_t^2 are two solutions of (48), then $y_t^1 - y_t^2 = \alpha^k(y_{t+k|t}^1 - y_{t+k|t}^2)$ will converge to 0 as k goes to infinity if $y_{t+k|t}^i$, $i = 1, 2$ remains bounded with $|\alpha| < 1$. Thus with $|\alpha| < 1$, Eq. (48) has at most one solution.

Equation (48) can be rewritten, dropping subscript h, as

$$y_t = \alpha(y_{t+1} - e_{t+1}^0) \tag{49}$$

where

$$e_t^0 = y_t - y_{t|t-1}$$

is the expectation revision martingale difference.

With $|\alpha| > 1$, rewrite (49) as

$$y_{t+1} = \frac{1}{\alpha} y_t + e_{t+1}^0.$$

---

* With $|\alpha| < 1$, $m_t = 0$. Otherwise $\alpha^{-t} m_t$ is not a martingale since its expectation becomes unbounded as t becomes large.

Iterating this relation backward in time, we obtain

$$y_t = e_t^0 + \frac{1}{\alpha} e_{t-1}^0 + \ldots + \frac{1}{\alpha_t} e_0^0 + \left(\frac{1}{\alpha}\right)^t y_0.$$

Set y_0 to zero and write the preceding as

$$y_t = \alpha^{-t} m_t$$

where

$$m_t = \alpha^t e_t^0 + \alpha^{t-1} e_{t-1}^0 + \ldots + e_0^0.$$

Then

$$m_{t+1} = \alpha^{t+1} e_{t+1}^0 + \alpha^t e_t^0 + \ldots + e_0^0$$

and

$$m_{t+1|t} = m_t,$$

i.e., m_t is a martingale that is an integrated version of the expectation-revision martingale difference. With $|\alpha| < 1$, let y_t^1 and y_t^2 be the solutions of form (47) with $y_0^1 = y_0^2$. Then

$$y_t^1 - y_t^2 = \alpha^{-1}(m_t^1 - m_t^2) = \alpha^{-t} m_t$$

because the difference of two martingales is also a martingale. Then

$$E|\,y_t^1 - y_t^2| = |\alpha|^{-t} E|m_t|.$$

Since

$$\begin{aligned} E|m_t| &= E|m_{t+1|t}| \\ &= E|E(m_{t+1}\,|\,I_t)| \\ &\leqslant E|E(|m_{t+1}\,|\,I_t) \\ &= E|m_{t+1}|, \end{aligned}$$

$E|\,y_t^1 - y_t^2|$ becomes unbounded as $t \to \infty$ unless $m_t = 0$ a.s. By imposing the boundedness of the solution of the homogeneous equation, both y_t^p and y_t^h are unique, so y_t is too.

Example 8. The next model has no lagged endogenous variable:

$$y_t = a(y_{t+2|t} - y_{t+2|t-1}) + by_{t+1|t} + n_t.$$

Substitute e_t^2 into $y_{t+2|t} - y_{t+2|t-1}$ and $y_{t+1} - e_{t+1}^0$ into $y_{t+1|t}$ to rewrite (44) as $y_t = ae_t^2 + b(y_{t+1} - e_{t+1}^0) + n_t$ or, solving for y_{t+1},

$$y_{t+1} = b^{-1} y_t + e_{t+1}^0 - ab^{-1} e_t^2 - b^{-1} n_t.$$

Note that

$$e_t^0 = be_t^1 + ae_t^2 + n_t - n_{t|t-1}$$

and

$$e_t^1 = be_t^2 + n_{t|t-1} - n_{t|t-2}.$$

Solving out e_t^1 from this relation and substituting it back into y_t we obtain the reduced-form solution as

$$y_t = ay_{t-1} + n_t - (n_t - n_{t|t-1}) + e_t^0.$$

Here $\{e_t^0\}$ is an arbitrary martingale difference sequence.

Example 9. We add a lagged endogenous variable in the next example,

$$y = ay_{t-1} + b(y_{t+1|t} - y_{t+1|t-1}) + n_t, \qquad b \neq 0,$$

which is the same as

$$y_t = ay_{t-1} + be_t^1 + n_t.$$

As usual, $\{n_t\}$ is some exogenous process. The martingale difference sequence is generated as follows:

$$e_t^0 = be_t^1 + n_t - n_{t|t-1}.$$

These two equations are solved to obtain

$$e_t^2 = (a + b^2)^{-1}\{e_t^0 - (1 + b)(n_t - n_{t|t-1})\},$$

which is then used in (45) to yield the reduced form solution of (44) as

$$\begin{aligned} y_{t+1} &= b^{-1}y_t - b^{-1}n_t + e_{t+1}^0 - ab^{-1}(a + b^2)^{-1}e_t^0 \\ &\quad + ab^{-1}(a + b^2)^{-1}(1 + b)(n_t - n_{t|t-1}). \end{aligned}$$

Here $\{e_t^0\}$ is an arbitrary martingale difference sequence.

Solving LRE Models with Transform Methods

In economic models, (weakly) stationary solutions are often important because they are related to the notion of stationary relations among economic variables. This section describes the transform method useful in obtaining stationary solutions. Futia (1981), Gourieroux et al. (1983), McCallum (1983), Broze et al. (1985), and Whiteman (1983) all discuss stationary solutions when exogenous processes in LRE models are generated by an elementary process $\{e_t\}$, which is mean-zero, i.i.d., with a finite variance.

If y_t is of the form

$$y_t = \sum_{i=0}^{\infty} \alpha_i e_{t-i}$$

where the coefficients are constrained by $\Sigma \alpha_i^2 < \infty$, then the revision process is proportional to e_t as in

$$y_{t+j|t} - y_{t+j|t-1} = \alpha_j e_t$$

because

$$y_{t+j|t} = \sum_{i \geqslant j} \alpha_i e_{t+j-i}.$$

The transform method seeks solutions that are weakly stationary.

Consider an example dicussed by Gourieroux et al. (1982):

(50) $$y_t = \alpha y_{t+1|t} + n_t.$$

If we assume that $n_t = A(L)e_t$ where $A(L) = \Sigma a_i L^i$, $\Sigma a_i^2 < \infty$, and we look for a weakly stationary solution of the form $y_t = c(L)e_t$, where $c(L) = \Sigma_0^\infty c_i L^i$, $\Sigma c_i^2 < \infty$ is a transfer function from the elementary process to y_t. The transfer function in L should be analytic inside the unit circle in the L-plane or $c(z^{-1})$ is analytic outside the unit circle in the z-plane:

$$y_{t+1|t} = c_1 e_t + c_2 e_{t-1} + \ldots = L^{-1}[c(L) - c_0]e_t.$$

Substituting this into the original equation, we obtain

$$c(L)e_t = \{\alpha L^{-1}[c(L) - c_0] + A(L)\}e_t$$

as an identity, which yields

$$c(L) = [LA(L) - \alpha c_0]/(L - \alpha)$$

where setting $L = \alpha$ in the preceding yields the unique $c_0 = A(\alpha)$, if $A(\alpha)$ exists, e.g., if $|\alpha| < 1$. For this procedure to be valid, one needs to verify that $c(L)e_t$ belongs to l_2. To this end, equate the coefficients of L^{k+1} to see that

$$c_k = \alpha c_{k+1} + a_k, \qquad k = 0, 1, \ldots, i.e.,$$

the coefficients satisfy the equation

$$\mathbf{\Phi c} = \underline{a}$$

where $\mathbf{c} = (c_0, c_1, \ldots)'$ and $\underline{a} = (a_0, a_1, \ldots)'$, which has the doubly infinite matrix $\mathbf{\Phi}$ with one along the main diagonal and $-\alpha$ along the superdiagonal line immediately above. The inverse $\mathbf{\Phi}^{-1}$ exists as a bounded linear operation in l_2 if and only if the Crone's theorem is satisfied. In particular, $|\alpha| < 1$ must hold. Then $c_0 = A(\alpha)$ is also obtained. Having obtained the conditions for y_t to belong to $l_2, y_t = c(L)e_t$ is the solution whenever $A(L)$ exists. To verify, substitute it into (50) to see that

$$\begin{aligned}(L - \alpha)y_{t+1} &= y_t - \alpha y_{t+1} \\ &= A(L)e_t - \alpha c_0 e_{t+1} \\ &= n_t - \alpha c_0 e_{t+1}\end{aligned}$$

where

$$y_{t+1} - y_{t+1|t} = e^0_{t+1} = c_0 e_{t+1},$$

and hence, $\alpha(y_{t+1} - c_0 e_{t+1}) = \alpha y_{t+1|t}$. The coefficient c_0 is not arbitrary but must satisfy $c_0 = A(\alpha)$, which is assumed to exist.

With $|\alpha| > 1$, the coefficient c_0 is arbitrary. There are an infinite number of solutions that may be written as

$$y_t = \left(\frac{c_0}{1 - \alpha^{-1}L} - \frac{\alpha^{-1}LA(L)}{1 - \alpha^{-1}L}\right) e_t.$$

Since

$$\frac{1}{1 - \alpha^{-1}L} = 1 + \frac{\alpha^{-1}L}{1 - \alpha^{-1}L},$$

this solution form can be restated as

$$\begin{aligned}y_t &= c_0 e_t - \alpha^{-1}A(L)e_{t-1} + \alpha^{-1}y_{t-1} \\ &= \alpha^{-1}y_{t-1} - \alpha^{-1}n_{t-1} + c_0(n_t - n_{t|t-1})\end{aligned}$$

where

$$n_t = A(L)e_t.$$

Sometimes a method of undetermined coefficients is used. This method requires one to guess the structure of the solution. It is employed by McCallum (1983, 1985), for example, who posits $y_t = \pi(L)x_{t-1} + \psi(L)e_t$ as a solution for (50). This method is not systematic.

Example 10. Consider a scalar valued equation

$$y_t = ay_{t+2|t} + by_{t+1|t} + cx_t + u_t \tag{51}$$

where $\{u_t\}$ is a mean-zero white noise process.

Taking the conditional expectation $E_{t-1}(\cdot)$ and subtracting it from the equation, we obtain

$$\begin{aligned} e_t^0 &= y_t - y_{t|t-1} \\ &= a(y_{t+2|t} - y_{t+2|t-1}) + b(y_{t+1|t} - y_{t+1|t-1}) + c(x_t - x_{t|t-1}) + u_t \\ &= ae_t^2 + be_t^1 + c(x_t - x_{t|t-1}) + u_t. \end{aligned}$$

The deterministic equation corresponding to (51) is obtained by taking its conditional expectation

$$y_{t|t-1} = ay_{t+2|t-1} + by_{t+1|t-1} + cx_{t|t-1}$$

and replacing conditional variables as in

$$\rho_t = a\rho_{t+2} + b\rho_{t+1} + cw_t.$$

The sequence $\{\rho_t\}$ is specified uniquely in terms of two initial conditions, ρ_0 and ρ_1, say. Since

$$y_t = \rho_t + e_t^0$$

and

$$y_{t+1} = \rho_{t+1} + e_{t+1}^0 + e_t^1,$$

two additional conditions must be used to specify e_t^0 and e_t^1 as a function of t.

These martingale differences are governed by

$$e_t^0 = be_t^1 + ae_t^2 + cx_{t,t} + u_t, \tag{52}$$

where $x_{t,t} = x_t - x_{t|t-1}$. Define the transform

$$\hat{e} = e_t^0 + e_t^1 q + e_t^2 q^2 + \dots$$

on the assumption that $\{e_t^k\}_{k=0}^{\infty}$ is a bounded sequence. From (52), it follows that

$$\hat{e} = b[\hat{e} - e_t^0]q^{-1} + a[\hat{e} - e_t^0 - e_t^1 q]q^{-2} + c\hat{x} + \hat{u} \tag{53}$$

where

$$\hat{x} = x_{t,t} + x_{t+1}q + x_{t+2,t}q^2 + \dots \tag{54}$$

and $\hat{u}$ is defined analogously. Here (53) and (54) are treated as formal series in q. Solve (53) for $\hat{e}$ as

$$\hat{e} = (1 - bq^{-1} - aq^{-2})^{-1}[\hat{u} + c\hat{x} - be_t^0 q^{-1} - a(e_t^0 q^{-2} + e_t^1 q^{-1})]. \tag{55}$$

Write

$$\frac{1}{1-bq^{-1}-aq^{-2}}=\frac{q^2}{\lambda_1\lambda_2(1-q/\lambda_1)(1-q/\lambda_2)}$$

where λ_1 and λ_2 are the roots of

$$q^2-bq-a=(q-\lambda_1)(q-\lambda_2).$$

Use the partial fraction expansion

$$\frac{1}{(1-q/\lambda_1)(1-q/\lambda_2)}=\frac{1}{\lambda_2-\lambda_1}\left(\frac{\lambda_2}{1-q/\lambda_2}-\frac{\lambda_1}{1-q/\lambda_2}\right),$$

and note that the coefficients in the expansion

$$\frac{1}{1-q/\lambda_i}=1+\frac{1}{\lambda_i}q+\frac{1}{\lambda_i^2}q^2+\dots,\qquad i=1,2$$

remain bounded for $|\lambda_i|\geqslant 1$, $i=1,2$. Then (55) shows that because $\lambda_1\lambda_2=-a$ and $\lambda_1+\lambda_2=b$, the coefficients of q^0 and q^1 imply that any e_t^0 and e_t^1 satisfy (55). Then, comparing the coefficients of q^2 (or more directly from (51) and (52)), we obtain

$$e_t^2=(e_t^0-be_t^1-cx_{t,t}-u_t)/a.$$

Here, e_t^0 and e_t^1 serve as initial conditions for (51) and (52):

$$e_t^3=[(b^2+a)e_t^1-be_t^0]/a^2+\frac{c}{a}x_{t+1,t}+\frac{b}{a^2}(cx_{t,t}+u_t)\text{ etc.}$$

We have shown that e_t^0 and e_t^1 serve as initial conditions only if $|\lambda_i|\geqslant 1$, $i=1,2$, i.e., a and b must be such that

(a) if $a>0$, then $a>1-|b|$ and
(b) if $a<0$, then $|a|>|b|-1$.

For example, with $a=0$, $|b|>1$. If $b=0$, then $|a|>1$. When $|\lambda_1|>1$ and $|\lambda_2|<1$,

$$q^2u_t-(ae_t^1+be_t^0)q-ae_t^0$$

must have the factor $(q-\lambda_2)$ for a bounded solution to exist, i.e.,

$$(a+\lambda_2 b)e_t^0=\lambda_2^2u_t-a\lambda_2e_t^1$$

must be satisfied. Then (55) is replaced by

$$(q-\lambda_1)\hat{e}=(q+\lambda_2)u_t-(ae_t^1+be_t^0)+\frac{c\hat{x}q^2}{(q-\lambda_2)}.$$

Since the coefficients of

$$\frac{\hat{x}q^2}{q-\lambda_2}$$

also diverge if $\hat{x}$ does not contain the factor $q-\lambda_2$, a bounded solution exists only if $\hat{x}$ contains the factor $(q-\lambda_2)$ or is identically equal to zero. When $|\lambda_i|<1$, $i=1,2$, then

$$q^2u_t-(ae_t^1+be_t^0)q-ae_t^0$$

must be proportional to q^2-dq-a, i.e.,

$$e_t^0=u_t$$

and

$$e_t^1=0.$$

Example 11.

$$y_t=ay_{t+2|t-1}+by_{t-1}+u_t,\qquad a_1\neq 0,$$

where u_t is a mean-zero stationary process, is a special case of the class of models discussed by Evans and Honkapolya (1986). Replacing the conditional expectation by

$$(56)\qquad y_{t+2}-e_{t+2}^0-e_{t+1}^1-e_t^2,$$

rewrite it as $ay_{t+2}-y_t+by_{t-1}=a(e_{t+2}^0+e_{t+1}^1+e_t^2)+u_t$ to obtain a reduced-form solution $y_{t+2}=a^{-1}y_t-a^{-1}by_{t-1}+u_{t+2}-a^{-1}u_t+e_{t+1}^1+e_t^2$ where e^1 and e^2 are arbitrary martingale differences. The recursion martingale difference is constrained by

$$e_t^0=u_t$$

$$e_t^k=ae_t^{k+2}+be_t^{k-1},\qquad k\geqslant 1.$$

Using the lag transform, assume that $y_t=G(L)u_t$ where $G(L)=c_0+c_1L+c_2L^2+c_3L^3h(L)$. Then (56) implies that

$$G(L)=[a(c_0+c_1L+c_2L^2)-L^2]/(a-L^2b+L^3),$$

which produces a weakly stationary solution provided $a-bL^2+L^3=0$ has three roots inside the unit circle. The coefficients c_0, c_1, and c_2 are determined by the requirements that the three residues are all zero.

Example 12. Solve the model

$$y_t = \alpha y_{t+1|t} + n_t$$

where

$$n_t = \varepsilon_t - \varepsilon_{t-1}.$$

Since the solution is spanned by ε_s, $s \leqslant t$, posit

$$y_t = [\beta + Lh(L)]\varepsilon_t.$$

Then $y_{t+1|t} = h(L)\varepsilon_t$, and the model requires that $h(L)$ satisfies

$$[\beta + Lh(L)]\varepsilon_t = \alpha h(L)\varepsilon_t + \varepsilon_t - \varepsilon_{t-1}$$

for all realization of $\{\varepsilon_t\}$, i.e.,

$$\beta + Lh(L) = \alpha h(L) + 1 - L,$$

from which we see that $\beta = 1 - \alpha$ and $h(L) = -1$, i.e., $y_t = (1 - \alpha - L)\varepsilon_t$ is a solution for all α.

Example 13. Solve the model

$$y_t = \alpha y_{t+1|t} + n_t$$

where

$$n_t + \phi n_{t-1} = \varepsilon_t.$$

Hypothesize a solution from

$$y_t = [\beta + Lh(L)]\varepsilon_t$$

for some polynomial $h(L)$. Then

$$y_{t+1|t} = h(L)\varepsilon_t,$$

an the model determines β and $h(L)$ by the relation

$$\beta + Lh(L) = \alpha h(L) + \frac{1}{1 + \phi L},$$

i.e.,

$$\beta = \frac{1}{1 + \phi\alpha}$$

and

$$h(L) = -\phi/(1 + \phi L)(1 + \phi\alpha),$$

i.e.,

$$y_t = \frac{1}{(1 + \phi\alpha)(1 + \phi L)}\varepsilon_t = \frac{1}{1 + \phi\alpha}n_t.$$

This is indeed a solution, because $n_{t+1|t} = -\phi n_t$ implies that $y_{t+1|t} = -\phi n_t/(1 + \phi\alpha)$ unless $\phi\alpha = -1$. This solution method does not require $|\phi\alpha| < 1$, which is needed in solving the equation successively forward.

Example 14. Suppose now that

$$y_t = \alpha y_{t+1|t} + n_t$$

where

$$n_t = [\theta(L)/\phi(L)]\varepsilon_t.$$

To ensure that the disturbance term has the finite variance $\operatorname{var} n_t < \infty$, we impose that

$$\frac{1}{2\pi j}\oint \frac{\theta(z^{-1})\theta(z)}{\phi(z^{-1})\phi(z)}\frac{dz}{z} < \infty,$$

i.e.,

$$\phi(z^{-1}) = \prod (z^{-1} - \lambda_i), \qquad |\lambda_i| > 1, \text{ all } i.$$

Again, y_t lies in the subspace of εs. We posit then

$$y_t = [\beta + Lh(L)]\varepsilon_t.$$

The conditional expectation is expressible as

$$y_{t+1|t} = h(L)\varepsilon_t.$$

On substituting this back into the original equation, we obtain the relation for $h(L)$ as

$$\beta + Lh(L) = \alpha h(L) + \frac{\theta(L)}{\phi(L)},$$

where

$$\beta = \theta(\alpha)/\phi(\alpha), \qquad \text{if } \phi(\alpha) \neq 0$$

and

$$h(L) = [\theta(L)/\theta(L) - \phi(\alpha) - \phi(\alpha)]/(L - \alpha).$$

This function produces

$$\begin{aligned} y_t &= [\theta(\alpha)/\phi(\alpha) + Lh(L)]\varepsilon_t \\ &= \frac{\theta(\alpha)}{\phi(\alpha)}\left\{\frac{\phi(L)}{\theta(L)} + \frac{L}{L - \alpha}\left[\frac{\phi(\alpha)}{\theta(\alpha)} - \frac{\phi(L)}{\theta(L)}\right]\right\} n_t \end{aligned}$$

as a well-defined solution in terms of $n_t, n_{t-1}, \ldots$, provided $\theta(L)$ has all its zeros outside the unit circle.

Example 15. We solve the same equation

$$y_t = \alpha y_{t+1|t} + n_t \tag{57}$$

using the expectation-revision variables. To be concrete, suppose that

$$n_t + \phi n_{t-1} = \varepsilon_t$$

where ε_t is the basic mean-zero white noise process. For $\{n_t\}$ to exist in l_2, we require that $|\phi| < 1$. This model has been solved by transform method as Example 13. There, the solution obtained by the transform method is $y_t = n_t/(1 + \phi\alpha)$. Since the conditioning variables are ε_s, $s \leqslant t - 1$, in $n_{t|t-1}$, we have $n_{t|t-1} = -\phi n_{t-1}$ and hence $n_t - n_{t|t-1} = \varepsilon_t$. Next consider $n_{t+1|t} - n_{t+1|t-1}$. From $n_{t+1|t} = -\phi n_t$ and $n_{t+1|t-1} = -\phi n_{t|t-1}$, we note that $n_{t+1|t} - n_{t+1|t-1} = -\phi(n_t - n_{t|t-1}) = -\phi\varepsilon_t$.

Suppose that we take this solution. It implies

$$\begin{aligned}\varepsilon_t^0 = y_t - y_{t|t-1} &= (n_t - n_{t|t-1})/(1 + \phi\alpha)\\ &= (n_t + \phi n_{t-1})/(1 + \phi\alpha)\\ &= \varepsilon_t/(1 + \phi\alpha).\end{aligned}$$

Next

$$\begin{aligned}\varepsilon_t^1 &= y_{t+1|t} - y_{t+1|t-1}\\ &= (n_{t+1|t} - n_{t+1|t-1})/(1 + \phi\theta)\\ &= -\phi\varepsilon_t/(1 + \phi\alpha).\end{aligned} \tag{58}$$

The model constrains the revision sequences by

$$\begin{aligned}\varepsilon_t^0 &= \alpha\varepsilon_t^1 + n_t - n_{t|t-1}\\ &= \alpha\varepsilon_t^1 + \varepsilon_t.\end{aligned}$$

This result agrees with that obtained by the transform method. What is the perfect foresight model counterpart of (57)? If we define the perfect foresight model as that which results by setting all the expectation-revision variables to zero, i.e., $\varepsilon_t^j = 0$ for all $j \leqslant 0$ and t, (58) implies that it requires that $\varepsilon_t = 0$ for all t. Then, the model implies that

$$n_t = -\phi n_{t-1} \qquad \text{for all } t.$$

Let $\eta_t = y_{t|t-1}$. Since ε_t^0 is zero, the model is the same as

$$\eta_t = \alpha\eta_{t+1} + n_t.$$

This is the perfect foresight model. Solving the equations forward in time,

$$\eta_t = n_t + \alpha n_{t+1} + \alpha^2 n_{t+2} + \dots$$

if $|\alpha| < 1$.*

From $\varepsilon_t = 0$ for all t, $n_{t+1} = -\phi n_t$ and $n_{t+j} = (-\phi)^j n_t$ in general. Thus

$$\begin{aligned}\eta_t &= [1 - \alpha\phi + (-\alpha\phi)^2 + \dots]n_t \\ &= n_t/(1 + \alpha\phi)\end{aligned}$$

if $|\alpha\phi| < 1$, i.e., $y_t = n_t/(1 + \alpha\phi)$ is also the perfect foresight solution in this model where $n_t = -\phi n_{t-1}$. Once, the solution form is obtained, it may appear that no restriction such as $|\alpha\phi| < 1$ is needed for the perfect foresight solution to exist, provided $1 + \alpha\phi \neq 0$. For (58) to make sense for all j, we need $|\phi| < 1$.

Example 16. Now change the model in Example 13 slightly into

$$\begin{cases} y_t = \alpha y_{t+1|t-1} + n_t, \\ n_t + \phi n_{t-1} = \varepsilon_t. \end{cases}$$

To solve the model by the transform method, try $y_t = (\alpha + \beta L)n_t$. Then $y_{t+1|t-1} = \alpha n_{t+1|t-1} + \beta n_{t|t-1}$, where $n_{t+1|t-1} = \phi^2 n_{t-1}$ and $n_{t|t-1} = -\phi n_{t-1}$. The model then specifies that

$$y_t = \left(1 + \frac{\alpha\phi^2 L}{1 + \alpha\phi}\right) n_t$$

is a solution.

The first member of the revision process is determined by

$$\begin{aligned} e_t^0 &= y_t - y_{t|t-1} \\ &= n_t - n_{t|t-1} \\ &= \varepsilon_t. \end{aligned}$$

---

* If $|\alpha| > 1$, we solve $\eta_{t+1} = (\eta_t - n_t)/\alpha$ as

$$\eta_t = -\frac{1}{\alpha}\left(n_{t-1} + \frac{1}{\alpha}n_{t-2} + \frac{1}{\alpha^2}n_{t-3} + \dots\right) = -\frac{\phi}{1 + \alpha\phi}n_{t-1} = \frac{n_t}{1 + \alpha\phi}$$

if $|\alpha\phi| > 1$.

Similarly,

$$\begin{aligned} e_t^1 &= y_{t+1|t} - y_{t+1|t-1} \\ &= n_{t+1|t} - n_{t+1|t-1} + \frac{\alpha\phi^2}{1+\alpha\phi}(n_t - n_{t|t-1}) \\ &= -\frac{\phi}{1+\alpha\phi}\varepsilon_t \text{ etc.} \end{aligned}$$

Alternatively, the revision sequence is specified by

$$\begin{aligned} e_t^0 &= n_t - n_{t|t-1} \\ &= \varepsilon_t \\ e_t^1 &= \alpha\varepsilon_t^2 + n_{t+1|t} - n_{t+1|t-1} \\ &= \alpha\varepsilon_t^2 - \phi\varepsilon_t \text{ etc.} \end{aligned}$$

8.5 Heterogeneous Information

Heterogeneous information sets in linear rational expectations models are discussed by Futia (1979), Aoki (1983, Section 13.3), and Pesaran (1988, Appendix B). When agents' information sets are not the same, the expectation $E(x_{jt} \mid \Omega_{it})$ of agent j's variable x_{jt} conditional on agent i's information set Ω_{it} is generally not easy to calculate unless some special assumptions are made regarding the common part and agent-specific part of the information set. The assumption that Ω_{it} is spanned by e_{ot} and e_{it} where e_{it} is independent of e_{ot} and $e_{jt}, j \neq i$ is such an assumption enabling us to approximate conditional expectations by orthogonal projections for non-Gaussian random variables.

We consider two examples. In the one, we examine a static model

$$y_{it} = \kappa_i(y_t \mid \Omega_{it}) + u_{it}, \tag{59}$$

where $y_t = \Sigma\, w_i y_{it}$, $w_i \geqslant 0$, $\Sigma\, w_i = 1$. In the other, we have a dynamic model

$$y_{it+1} - \gamma y_{it} = \mu E(y_{it+1} - y_{t+1}) + u_{it}. \tag{60}$$

The first equation is basically the same as the one discussed in Pesaran (1988). The second is basically the same as the one in Aoki (1983).

Example 17: Static Model

To discuss (60), define a linear operator A by

$$A(y_t) = \Sigma_i \lambda_i \pi_{it} y_t / \lambda,$$

where $\lambda_i = \kappa_i w_i$ and $\lambda = \Sigma \lambda_i$, and define

$$u_t = \Sigma w_i u_{it},$$

where π_{it} is the orthogonal projection operation onto the subspace spanned by I_{it}. We assume that λ_i is of the same sign for all i, i.e., $\lambda \cdot \lambda_i > 0$ for all i. Since π_{it} is a projection, we have

$$|A(y_t)| \leqslant |y_t|,$$

i.e., $A(\cdot)$ is a contraction map. Summing (59) with weights ws,

$$\begin{aligned} y_t &= \Sigma w_i y_{it} \\ &= \lambda \Sigma \frac{\lambda_i}{\lambda} \pi_{it} y_t + u_t \\ &= \lambda A(y_t) + u_t. \end{aligned}$$

By repeated application we can express y_t as

(61) $$y_t = u_t + \lambda A(u_t) + \lambda^2 A(u_t) + \ldots$$

provided $\lambda^k A^k(y_t) \to 0$ as k goes to infinity. For example, if $|\lambda| < 1$, this condition is satisfied since $A(\cdot)$ is bounded. To go beyond (61), we need to specify u_{it} and the relationship among I_{it} in more detail. Suppose that I_{it} is spanned by e_{ot} and e_{it} where e_{ot} and e_{jt} are independent mean-zero processes. Assume further that

$$u_{it} = \theta_i[\phi_0 + Lf_0(L)]e_t^0 + [\phi_i + Lf_i(L)]e_{it}.$$

Then with $j \neq i$,

$$\pi_{it} u_{jt} = \theta_j[\phi_0 + Lf_0(L)]e_t^0$$

and

$$\pi_{it} u_{it} = u_{it}.$$

Thus

$$\begin{aligned} \pi_{it} u_t &= w_i u_{it} + \pi_{it} \Sigma_{j \neq 1} w_j u_{jt} \\ &= w_i u_{it} + (\Sigma_{j \neq i} w_j)\theta_j[\phi_0 + Lf_0(L)]e_t^0 \\ &= \theta[\phi_0 + Lf_0(L)]e_t^0 + w_i[\phi_i + Lf_i(L)]e_{it} \end{aligned}$$

where $\theta = \Sigma \theta_i w_i$ and

$$A(u_t) = \theta[(\phi_0 + Lf_0(L)]e_t^0 + \frac{1}{\lambda} \Sigma_i \theta_i w_i \lambda_i [\phi_i + Lf_i(L)]e_{it}.$$

Proceeding analogously

$$A^2(u_t) = \theta\Sigma_i \frac{\lambda_i}{\lambda}\pi_{it}[\phi_0 + Lf_0(L)]e_t^0$$

$$+ \Sigma_i \frac{\lambda_i}{\lambda}\pi_{it}\{w_i\lambda_i[\phi_i + Lf_i(L)]\}e_{it}$$

$$= \theta[\phi_0 + Lf_0(L)]e_t^0 + \Sigma_i \frac{\lambda_i^2 w_i}{\lambda^2}[\phi_i + Lf_i(L)]e_{it}$$

since $e_{0t} \in I_{it}$, but $e_{jt} \perp I_{it}$. In general, we have

$$A^k(u_t) = \theta[\phi_0 + Lf_0(L)]e_t^0 + \Sigma_i \frac{\lambda_i^k w_i}{\lambda^k}[\phi_i + Lf_i(L)]e_{it},$$

and (62) becomes

$$y_t = \frac{\theta}{1-\lambda}[\phi_0 + Lf_0(L)]e_{0t} + \sum_{k=0}^{\infty} \Sigma_i \lambda_i^k w_i[\phi_i + Lf_i(L)]e_{it}.$$

When u_{it} consists of only the first term

$$u_{it} = \theta_i[\phi_0 + Lf_0(L)]e_t^0,$$

then

$$y_t = \frac{\theta}{1-\lambda}[\phi_0 + Lf_0(L)]e_{ot}.$$

Example 18: Dynamic Model. Forming the weighted sum of (60), we derive

(62) $$\begin{aligned} y_{t+1} - \gamma y_t &= \mu\Sigma_i w_i E_{it}(y_{it+1} - y_{t+1}) + \Sigma w_i v_{it} \\ &= \mu\Sigma w_i \pi_{it}\eta_{it+1} + v_t \end{aligned}$$

where $\eta_{it} = y_{it} - y_t$ and $v_t = \Sigma w_i v_{it}$. The difference of (60) and (62) yields

(63) $$\eta_{it+1} - \gamma n_{it} = \mu\pi_{it}\eta_{it+1} - \Sigma_j \mu w_j \pi_{jt}\eta_{jt+1} + v_{it} - v_t.$$

Summing the preceding over i with weights w_u, we note that

$$\Sigma_i w_i(\eta_{it+1} - \gamma\Sigma\eta_{it}) = 0.$$

Suppose that $\Sigma w_i \eta_{i0} = 0$. Then $\Sigma w_i \eta_{it} = 0$ for all $t > 0$. We assume that $\eta_{it} \in I_t$ for all i. Projection of (63) on the subspace spanned by I_t yields

$$\pi_{it}(\eta_{it+1} - \gamma\eta_{it}) = \mu\pi_{it}\eta_{it+1} - \pi_{it}(\Sigma_j \mu w_j \pi_{jt}\eta_{jt+1}) + \pi_{it}(v_{it} - v_t)]$$

where $\pi_{it}\eta_{it} = \eta_{it}$ is used. Summing the preceding,

$$\Sigma w_i \pi_{it}\eta_{it+1} = (1-\mu)^{-1}[\gamma\Sigma w_i \eta_{it} - \mu\Sigma w_i \pi_{it}(\Sigma_j w_j \pi_{jt}\eta_{jt+1}) + \Sigma w_i \pi_{it}(v_{it} - v_t)]$$

$$= -\frac{\mu}{1-\mu}\Sigma_i w_i \pi_{it}(\Sigma_j w_j \pi_{jt}\eta_{jt+1}) + \frac{1}{1-\mu}\Sigma w_i \pi_{it}(v_{it} - v_t).$$

Let $\eta^*_{t+1} = \Sigma w_i \pi_{it}\eta_{it+1}$. Then

$$\eta^*_{t+1} = -\frac{\mu}{1-\mu}A(\eta^*_{t+1}) + \frac{1}{1-\mu}\Sigma w_i \pi_{it}(v_{it} - v_t) \tag{64}$$

where

$$A(\eta) = \Sigma w_i \pi_{it}\eta$$

is a linear operator similar to the one in the previous example. To proceed further from (64), we also make similar assumptions on v_{it},

$$v_{it} = \theta_i[\gamma_0 + \log_v(L)]e_{0t} + [\gamma_i + \log_i(L)]e_{it},$$

where e_{0t} and e_{it} are all independent.

Appendix

Doob Decomposition

Suppose that X_t is measurable with respect to an increasing sequence of σ-algebra $\{\mathscr{F}_t\}$, i.e., $\mathscr{F}_t \subset \mathscr{F}_{t+1}$. Define

$$M_{t+1} = M_t + (X_{t+1} - X_{t+1|t}), \qquad M_0 = X_0$$

and

$$A_{t+1} = A_t - (X_t - X_{t+1|t}), \qquad A_0 = 0.$$

Clearly A_t is predictable, i.e., $A_{t+1|t} = A_t$, and M_t is a martingale since $E(M_{t+1} - M_t \mid \mathscr{F}_t) = 0$. Finally note that $X_0 = M_0 + A_0$. If $X_t = M_t + A_t$, then $M_{t+1} + A_{t+1} = X_t + X_{t+1} - X_t = X_{t+1}$. We have thus expressed X_t as a sum of a predictable process and a martingale, $M_t + A_t$. This is known as the Doob decomposition (Kopp [1984, p. 67]).

CHAPTER 9

Approximations in Sequential Decision Processes

Sequential decision problems do not generally admit closed-form analytic solutions. Even if we restrict our attention to a class of problems with linear equations and quadratic performance indices, we can obtain optimal Bayesian control policies in the analytically closed form only for special cases.

As a rule then, we must solve optimal control problems approximately. There are many ways to obtain approximately optimal decision rules: linearization of nonlinear dynamics, quadratic approximations of non-quadratic cost indices, approximations in policy spaces (Bellman [1957]), substitution of open-loop rules for closed-loop rules, application of linear or nonlinear programming techniques, such as gradient techniques of various kinds, stochastic approximation, and separation of control from estimation with possibly additional approximations being made for control and/or estimation parts, to name only a few possibilities.

One common idea for approximation is to replace or approximate original decision problems with a "large" (unaccountable) state space or a "large" number of decisions by another with suitable subsets such as a finite number of decisions. Whitt (1978) discusses a general procedure for constructing and analyzing approximations of dynamic programming models including Markov decision processes and optimization of discounted present value over an infinite horizon or finite-stage dynamic programs. Flåm (1987) approximates complex sequential decision problems by concentrating probability distributions. A similar idea is presented later in which parameter adaptive sequential decision processes are approximated by two (or more) stochastic decision problems in which the weights in the combination are adjusted as learning progresses. Two types of approximation are possible: in the time dimension and in the state space dimension (unaccountable) infinite state space in R^n is replaced by finite Markov chain ones or countable state infinite-horizon Markov decision problems are approximated by finite state ones. Tauchen (1988) develops one such discrete-state space approximate solution for a class of nonlinear rational expectations models. Flåm (1987) discusses finite-state approximations for countable state problems. In the time domain, infinite-horizon problems are approximated by finite-horizon, possibly plus terminal cost (valuation) problems, or the frequency of decision making is reduced.

9.1 Infinite-Horizon Approximation

By dividing the planning horizon into N segments (of possibly unequal length) and assuming decision variables are held fixed in each segment, the number of decision variables is reduced from ∞ to N. This approximation effectively reduces an infinite-horizon problem into a finite-horizon one. For example, consider an optimal saving plan $V(k) = \max \Sigma_0^\infty u(c_t)\beta^{-t}$, and split the decision problems into the initial $N-1$ period plus one decision for the remainder

$$V(k) = \max_{s_0^{N-1}, s} [\Sigma_0^N u(c_t)\beta^{-t} + \Phi(k_N, s)],$$

where s_0^{N-1} stands for the saving decisions over the initial N periods and the second term $\Phi(k_N, s)$ stands for $\Sigma_{N+1}^\infty u(c_t)\beta^{-1}$ using a constant saving rate s starting from the capital stock k_N, which is the result of s_0^{N-1}. An indication that such an approximation scheme may be reasonable is reported in Aoki (1973) in which numerical experiments were run to show how the error of approximation varies with N.

9.2 Finite-State Approximation

Let $\{\mathbf{x}_t\}$ be a sequence of state vectors, $\mathbf{x} \in X$, with X countably infinite where $P(\mathbf{x}_{t+1} \mid \mathbf{x}_t, \mathbf{u}_t)$ is the transition probability distribution from $\mathbf{x}_t$ to $\mathbf{x}_{t+1}$ under the decision vector $\mathbf{u}_t$. The immediate cost is $c(\mathbf{x}_t, \mathbf{u}_t)$. Gihman and Skorohod (1979) proved that only stationary Markovian and nonrandomized policies need be considered in minimizing

$$v(\mathbf{x}) = \inf_{\pi} \{c[\mathbf{x}, \pi(\mathbf{x})] + \beta \sum_{\mathbf{y}} V(\mathbf{y}) P[\mathbf{y} \mid \mathbf{x}, \pi(\mathbf{x})]\}$$

where $\mathbf{u} \in U$, U is compact, and $P(\cdot \mid \cdot, \cdot)$ is such that $\Sigma \mathbf{y} f(\mathbf{y}) P(\mathbf{y} \mid \mathbf{x}; \mathbf{u})$ depends continuously on $\mathbf{u}$ for any bounded f: $X \rightarrow R$ and $c(\mathbf{x}, \mathbf{u})$ is lower semicontinuous in $\mathbf{u}$ and bounded. Let the initial state x_0 have the distribution $\mu(\cdot)$. Let $X_1 \subset X_2 \subset \ldots \subset$ be such that $\bigcup_1^\infty X_n = X$. Define

$$c_n(\mathbf{x}, \mathbf{u}) = \begin{cases} c(\mathbf{x}, \mathbf{u}) & \text{if } x \in X_n, \\ 0 & \text{otherwise} \end{cases}$$

and

$$P_n(\mathbf{y} \mid \mathbf{x}; \mathbf{u}) = \begin{cases} P(\mathbf{y} \mid \mathbf{x}; \mathbf{u}) & \text{if } \mathbf{x} \in X_n, \\ 1 & \text{concentrated at } \mathbf{x}. \end{cases}$$

Otherwise, the problem with state space X_n, the cost structure $c_n(\mathbf{x}, \mathbf{u})$, and the transition probability $P_n(\mathbf{y} \mid \mathbf{x}; \mathbf{u})$ is an approximation to the Markovian decision problem. Define an operator $A_n(\pi)$ on l_∞ (the Banach space of all bounded functions f: $X \rightarrow R$ under the supreme norm) by

$$A_n(\pi)[v(\mathbf{x})] = c_n[\mathbf{x}, \pi(\mathbf{x})] + \beta \Sigma_y f(y) P_n[y \mid x; \pi(x)].$$

This is a contraction operator on l_∞. If we assume that f is zero outside X_n, then

$$A_n(\pi)[v(\mathbf{x})] = \begin{cases} c[\mathbf{x}, \pi(\mathbf{x})] + \beta \Sigma_{\mathbf{y} \in X_n} f(\mathbf{y}) P[\mathbf{y} \mid \mathbf{x}; \pi(\mathbf{x})]_{\mathbf{x} \in X_n} \\ 0 \quad \text{otherwise.} \end{cases}$$

For a finite-state Markov chain, approximation of AR models have been proposed by Tauchen (1986).

9.3 Approximation by Bilinear Models

Optimization problems with either nonlinear dynamics or nonquadratic performance indices generally do not admit closed-form analytical solutions. A well-known exception is the problem with log–linear criterion functions with a Cobb-Douglas production function used by Long and Plosser (1983), and by Hansen (1985). In general, then, optimization problems require approximation of one kind or another to solve then "approximately." In deterministic optimization problems, iterative solution procedures are sometimes possible: nonlinear dynamics can be successively linearized and nonquadratic performance indices successively approximated by quadratic ones and the convergence can be proven. See, for example, Aoki (1964). For stochastic optimization problems, no such general iterative procedure is available. This section collects some procedures for solving stochastic optimization problems with nonlinear dynamics or nonquadratic cost functions.

Consider a general model specified by a setr of nonlinear stochastic difference equations

$$\mathbf{z}_{t+1} = \mathrm{f}(\mathbf{z}_t, \mathbf{u}_t, \xi_t) \tag{1}$$

where $\mathbf{u}_t$ is a vector of decision variables and ξ_t are random disturbances with mean $\bar{\xi}_t$ and finite conditional variance, say. Equation (1) can be expanded in the Taylor series in at least three ways. In one, we use

$$\begin{aligned} \mathbf{z}_{t+1} &= \mathrm{f}(\bar{\mathbf{z}}_t, \mathbf{u}_t^0, \bar{\xi}_t) + \mathbf{A}_0(t)(\mathbf{z}_t - \mathbf{z}_t^0) \\ &\quad + \mathbf{B}_0(t)(\mathbf{u}_t - \mathbf{u}_t^0) + \mathbf{C}_0(t)(\xi_t - \bar{\xi}_t) + \text{nonlinear terms} \end{aligned} \tag{2}$$

where $\mathbf{z}_{t+1} = \mathrm{f}(\mathbf{z}_t^0, \mathbf{u}_t^0, \bar{\xi}_t)$ where $\mathbf{A}_0(t)$ and $\mathbf{B}_0(t)$ are possibly time-varying but deterministic matrices. In the other we have

$$\begin{aligned} \mathbf{z}_{t+1} &= \mathbf{f}(\mathbf{z}_t^0, \mathbf{u}_t^0, \xi_t) + \mathbf{A}_0(t, \xi_t)(\mathbf{z}_t - \mathbf{z}_t^0) + \mathbf{B}_0(t, \xi_t)(\mathbf{u}_t - \mathbf{u}_t^0) \\ &\quad + \text{nonlinear terms} \end{aligned} \tag{3}$$

where $\mathbf{z}_t^0 = \mathbf{f}(\mathbf{z}_t^0, \mathbf{u}_t^0, \xi_t)$. In (3), the coefficients are random. This may be one way in which random coefficient models arise, as in Aoki (1975). Note that in (2), the reference path $\{\mathbf{z}_t^0\}$ is deterministic, but it is stochastic in (3). This corresponds to the two basically different decompositions of time paths used in the time-series literature in which deterministic or random "trends" are extracted and the remainder fitted by short-run dynamic models. See Aoki (1987c) for this type of decomposition.

A third way to retain the cross-product terms

$$\begin{aligned} \mathbf{z}_{t+1} &= \mathbf{f}(\mathbf{z}_t^0, \mathbf{u}_t^0, \bar{\xi}_t) + \mathbf{A}_0(t)(\mathbf{z}_t - \mathbf{z}_t^0) + \mathbf{B}_0(t)(\mathbf{u}_t - \mathbf{u}_t^0) \\ &\quad + \mathbf{C}_0(t)(\mathbf{z}_t - \mathbf{z}_t^0)(\xi_t - \xi_t^0) + \mathbf{D}_0(t)(\xi_t - \bar{\xi}_t)(\mathbf{u}_t - \mathbf{u}_t^0) \\ &\quad + \mathbf{E}_0(t)(\mathbf{z}_t - \mathbf{z}_t^0)(\mathbf{u}_t - \mathbf{u}_t^0) + \text{higher order terms.} \end{aligned} \tag{4}$$

Equation (4) gives rise to bilinear models. Sometimes, the expansions into the Taylor series need to be made in two steps to interpret the nature of approximations as in Aoki (1974). There, a two-dimensional vector $\mathbf{k}_t = (d_t^p, k_t^g)'$ is governed by

$$\mathbf{k}_{t+1} = \mathrm{g}(\mathbf{k}_t, s_t) + \mathrm{H}(\mathbf{k}_t, s_t)\mathbf{u}_t$$

where

$$g(\mathbf{k}_t, s_t) = \left[\mathbf{k}_t + s_t \begin{pmatrix} 1 \\ 0 \end{pmatrix} f(\mathbf{k}_t) \right] \Big/ (1+\gamma)$$

$$H(\mathbf{k}_t, s_t) = [\mathrm{k}_1(\mathbf{k}_t, s_t), k_2(\mathbf{k}_t)/s_t)]$$

$$\mathbf{k}_1(\mathbf{k}_t, s_t) = s_t \begin{pmatrix} -1 \\ 1 \end{pmatrix} f(\mathbf{k}_t)/(1+\gamma)$$

$$\mathbf{k}_2(\mathbf{k}_t, s_t) = (1 - s_t) \begin{pmatrix} 0 \\ 1 \end{pmatrix} f(\mathbf{k}_t)/(1+\gamma), \qquad 0 < \gamma$$

where s_t is the saving rate by the private sector and $f(k_t)$ is the per capita Cobb-Douglas production function. (This is (4) in Aoki [1974].)

The control variable $\mathbf{u}_t$ is also two-dimensional, composed of tax rates on saving and consumption. Consider the disturbed time path for the capital stock is denoted by $\mathbf{k}_t + \boldsymbol{\xi}_t + \boldsymbol{\eta}_t$ in response to deviation in the saving rates and tax rates to $s_t + \sigma_t$ and $\mathbf{u}_t + \mathbf{v}_t$. Then ξ_t and η_t are generated by

$$\begin{aligned} \xi_{t+1} &= g(\mathbf{k}_t + \xi_t, s_t + \hat{\sigma}_t) - g(\mathbf{k}_t, s_t) \\ &\quad + \{H(\mathbf{k}_{st} + \xi_t, s_{st} + \hat{\sigma}_t) - H(\mathbf{k}_s, s_t)\}\mathbf{u}_t \\ &\quad + H(\mathbf{k}_t + \xi_t, t + \hat{\sigma}_t)\mathbf{u}_t, \qquad \xi_0 = 0 \end{aligned}$$

where

$$\hat{\sigma}_t = \frac{\partial s_t}{\partial u_t} v_t + \frac{\partial s_t}{\partial \mathbf{k}_t} \xi_t$$

and

$$\begin{aligned} \boldsymbol{\eta}_{t+1} &= g(\mathbf{k}_t + \xi_t + \boldsymbol{\eta}_t, s_t + \sigma_t) - g(\mathbf{k}_t + \xi_t, s_t + \hat{\sigma}_t) \\ &\quad + [H(\mathbf{k}_t + \xi_t + \boldsymbol{\eta}_t, s_t + \sigma_t) - H(\mathbf{k}_t + \xi_t, s_t + \hat{\sigma}_t)]\mathbf{u}_t \\ &\quad + [H(\mathbf{k}_t + \xi_t + \boldsymbol{\eta}_t, s_t + \sigma_t) - H(\mathbf{k}_t + \xi_t, s_t + \hat{\sigma}_t)]\mathbf{v}_t. \end{aligned}$$

In the preceding, $\boldsymbol{\xi}_t$ is independent of $\boldsymbol{\eta}_t$. If $\boldsymbol{\eta}_0 = 0$, then $\boldsymbol{\eta}_t = 0$ for all $t > 0$. The path of $\boldsymbol{\eta}_t$ represents the part of the perturbed time path of capital stocks due to an initial redistribution of capital stocks between the two sectors.

Small perturbations in u_t are expected to produce only small changes in $\|\boldsymbol{\xi}_t\|$ and σ_t. Under this assumption, the $\boldsymbol{\eta}_t$-dynamics are approximated by

$$\boldsymbol{\eta}_{t+1} = (\mathbf{A}_t + \mathbf{B}_t^s v_t^s + \mathbf{B}_t^c v_t^c + \mathbf{c}_t \sigma_t)\boldsymbol{\eta}_t$$

where

$$(1+\gamma)\mathbf{A}_t = I + \left[s_t \begin{pmatrix} 1 \\ 0 \end{pmatrix} + s_t \begin{pmatrix} -1 \\ 1 \end{pmatrix} u_t^s + (1 - s_t) \begin{pmatrix} 0 \\ 1 \end{pmatrix} u_t' \right] f'(\mathbf{k}_t + \xi_t)$$

$$(1+\gamma)\mathbf{B}_t^s = \mathbf{k}_1'(\mathbf{k}_t + \xi_t, s_t),$$

$$(1+\gamma)\mathbf{B}_t^c = \mathbf{k}_2'(\mathbf{k}_t + \xi_t, s_t)$$

$$\mathbf{c}_t = \left[\begin{pmatrix} 1 \\ 0 \end{pmatrix} + u_t^s \begin{pmatrix} -1 \\ 1 \end{pmatrix} u_t^c \begin{pmatrix} 0 \\ 1 \end{pmatrix} \right] f(k_{st})/(1+\gamma).$$

This example suggests how bilinear dynamics could arise in economic optimization.

9.4 Quadratic Approximation

Approximations by quadratic expressions of nonquadratic ones are frequently used. Least squares estimations, Newton-Raphson, and related second-order optimization methods, such as Fletcher-Powell or Davidson methods in nonlinear programming, are all based on quadratic approximations of functions being optimized.

In dynamic optimization problems, we need to approximate dynamic transition equations as well. Often linear approximation of dynamics is used in conjunction with quadratic approximations of functions being optimized. This chapter illustrates the procedure with several examples. When no stochastic processes are involved, such an approximation scheme is known to find the optimal solutions iteratively. (See, for example, Aoki [1964] for one such demonstration.) In some optimization problems containing random variables or stochastic processes, replacing them by their mean values sometimes renders the problem tractable. This procedure is known as certainty-equivalence procedure. In general, however, we know that $E\phi(X) \neq \phi(EX)$ where X is some random variable. For this simple reason, after random variables are replaced by their means, some further approximation or iterative procedures must be employed for certainty-equivalence procedure to be useful.

There are problems where certainty-equivalence procedure does not work or approximations by a sequence of linear-quadratic control problems are not appropriate. Often, the dynamics and the objective functions suggest the functional forms of approximations or converting the original problems into stochastic approximation ones. The first example involves no essential dynamics and illustrates the quadratic approximation of objective functions.

Example 1: Optimal Saving Plan. The steady state (distribution) of consumption (per worker) plan is formulated as

$$\max \sum_{t=0}^{\infty} \beta^t u(c_t)$$

subject to

$$s_t + c_t \leqslant y_t$$
$$y_t = r_t f(k_t)$$
$$s_t = k_{t+1}$$

where y_t is the output which depends on the capital stock k_t, which is equal to the saving of the previous period. The random variable r_t is assumed to be $r_t = e^{v_t}$ where v_t is some random variable. We assume that v_t is a martingale, i.e., $E_t v_{t+1} = v_t$. Some alternative assumptions such as serially correlated shocks are examined elsewhere. This type of problem have been examined by Danthine and Donaldson (1981), for example. This simple formulation has no intertemporal dynamics as in an alternative formulation with explicit investment and it allows the simple analysis that follows.

Assume that $f(k) = k^\alpha$, $\alpha < 1$, and $u(c) = \ln c$. The Bellman equation is

$$V(k_t, v_t) = \max \{u(y_t - s_t) + \beta E_t V(s_t, v_{t+1})\}.$$

The solution takes the form

$$V(k, v) = a + b \ln k + cv.$$

Substitute this into the preceding. The first-order condition is

$$\frac{1}{y_t - s_t} = \frac{b\beta}{s_t}$$

from which

(5) $$s_t = \frac{b\beta}{1 + b\beta} y_t \qquad \text{and} \qquad c_t = \frac{y_t}{1 + b\beta}$$

are obtained. The constraints are determined by comparing the coefficients of $\ln k_t$ and v_t on both sides (recall that $v_{t+1|t} = v_t$):

$$a = [\alpha\beta(1 - \alpha\beta)^{-1} \ln \alpha\beta + \ln(1 - \alpha\beta)]/(1 - \beta),$$

$$b = \frac{\alpha}{1 - \alpha\beta},$$

$$c = [(1 - \beta)(1 - \alpha\beta)]^{-1}.$$

Substitute the constant b in (5) to obtain

$$s_t = \alpha\beta y_t \tag{6}$$
$$= \alpha\beta k_t^{\alpha} e^{v_t}.$$

When the value of v_t is not known, (6) is not operational. Instead, one may use

$$\hat{s}_t = \alpha\beta k_t^{\alpha} e^{v_{t-1}} = \alpha\beta f(k_t)\hat{r}_t$$

since $v_{t|t-1} = v_{t-1}$, and we let $\hat{r}_t = e^{v_{t-1}}$. Now approximate $u(f(k)r - s)$ by expanding it about $f(k_t)\hat{r}_t - \hat{s}_t = (1 - \alpha\beta)f(k_t)\hat{r}_t$. Denote this by γ_t for short. Then

$$\ln[f(k_t)r_t - s_t] = \ln[f(k_t)\hat{r}_t - \hat{s}_t + f(k_t)\Delta r - \Delta s] \tag{7}$$

where

$$\Delta r = r_t - \hat{r}_t \qquad \text{and} \qquad \Delta s = s_t - \hat{s}_t.$$

Without knowing v_t exactly, r_t is not known. Proceed on the assumption that an estimate of v_t is available to focus on the aspect of quadratic approximation here. Expand (7) using $\ln(1 + x) \approx x - \frac{1}{2}x^2$ into

$$LHS = \ln[f(k_t)\hat{r}_t - \hat{s}_t] + x - \tfrac{1}{2}x^2$$

where

$$x = [f(k_t)\Delta r - \Delta s]/r_t.$$

Similarly, $\beta E_t V(x_t, v_{t+1})$ is expanded about $\beta V(\hat{s}_t, v_{t-1})$ as

$$E_t[a + b\ln s_t + cv_{t+1} \cong a + b\ln\hat{s}_t$$
$$+ cv_{t-1} + c\Delta v + \frac{b\Delta s}{s_t} - \frac{b\Delta s}{2s_t^2}$$

where $\Delta v = v_t - v_{t-1}$. The quadratic expression

$$\frac{f(k_t)\Delta\gamma - \Delta s}{r_t} - \frac{1}{2}\frac{[f(k_t)\Delta r - \Delta s]^2}{r_t^2} + \beta b\left(\frac{\Delta s}{s_t} - \frac{\Delta s^2}{2s_t^2}\right)$$

is minimized with respect to Δs to yield

$$\Delta s = \frac{f(k_t)\Delta r(\alpha\beta)^2}{(\alpha\beta)^2 + (1 - \alpha\beta)^2}$$

as the next correction term to $\hat{s}_t$, i.e.,

$$s_t = \hat{s}_t + \Delta s$$
$$= \alpha\beta f(k_t)\left\{\hat{r}_t + \frac{\alpha\beta}{(\alpha\beta)^2 + (1 - \alpha\beta)^2}\Delta r\right\}.$$

Reexpanding about this s again and iterating, we see that the process converges to

$$\hat{r}_t + \frac{(\alpha\beta)^2 + (1 - \alpha\beta)^2}{(\alpha\beta)^2 + (1 - \alpha\beta)^2 - \alpha\beta} \Delta r,$$

which is equal to r_t only if $\alpha\beta \ll 1$.

Example 2: Social Planner's Saving Rates

A version of the textbook social planner's problem is to maximize a discounted stream of his utilities

$$\max \sum_0^\infty \beta^t u(c_t) \tag{8}$$

subject to

$$k_{t+1} + c_t \leqslant f(k_t). \tag{9}$$

With a finite horizon, (8) may be replaced by

$$\max \left[\Sigma_0^T \beta^t u(c_t) + \phi(k_{t+1})\right]$$

where $\phi(k_{t+1})$ is introduced to value the remainder $\Sigma_{T+1}^\infty \beta^t u(c_t)$, rather than setting k_{T+1} to zero since a finite planning horizon is usually a convenience rather than a true end of "everything." See Aoki (1973) for this alternative formulation. The inequality constraint (9) is actually binding as the equality constraint, and the first order conditions are

$$u'(c_{t-1}) = \beta u'(c_t) f'(k_t). \tag{10}$$

As in other optimal saving plan solutions, c_t is proportional to $f(k_t)$, i.e., the optimal $c_t = (1 - s_t) f(k_t)$, i.e., where s_t is the optimal saving rate at time t. To be concrete let $u(c) = \ln c$ and $f(k) = k^\alpha$. Then (10) becomes

$$\frac{1}{(1 - s_{t-1})k_{t-1}^\alpha} = \frac{\alpha\beta}{(1 - s_t)k_t^\alpha} k_t^{\alpha-1} = \frac{\alpha\beta}{k_t(1 - s_t)} = \frac{\alpha\beta}{s_{t-1}k_{t-1}^\alpha(1 - s_t)}$$

or the optimal sequence of saving rates are governed by the next Riccati equation

$$s_t = 1 + \alpha\beta - \frac{\alpha\beta}{s_{t-1}}. \tag{11}$$

Let $s_t = u_t/u_{t-1}$. This equation is converted into $u_t - (1 + \alpha\beta)u_{t-1} + \alpha\beta u_{t-2} = 0$. This equation has the characteristic root 1 and $\alpha\beta$, and the solution of (11) has the form

$$s_t = \frac{A + B(\alpha\beta)^t}{A + B(\alpha\beta)^{t-1}}.$$

With the terminal condition $s_T = 0$, s_t is given by

$$s_t = \alpha\beta \frac{1 - (\alpha\beta)^{T-t}}{1 - (\alpha\beta)^{T-t} + 1}, \qquad t = 0, \ldots, T.$$

Suppose $\alpha\beta$ is of the order of .5 or so; then $(.5)^5 = .03$, $(.5)^{10} = .98 \times 10^{-3}$ and $(.5)^{20} = .95 \times 10^{-6}$. Thus, for $T - t$ (time to the end of the planning horizon) greater than 10, $s_t \approx \alpha\beta$ is a good approximation.

Exercise

To add to a small set of nonquadratic optimization problems that admit analytically closed solutions, modify and redo Example 1 of Chapter 1 by allowing the planning horizon to be infinite and by changing $g(c)$ to $(c - h)^\sigma$. Here h stands for "subsistence" level of consumption which depends on past consumptions by $h_{t+1} = -fh_t + bc_t$, where f is a positive constant and less than one in magnitude. Investigate restrictions on the values of the parameters of the problem statements.

Hint: The value function depends on the wealth level and the level of subsistence consumption. Try $R(x - \xi h)^\sigma$ where R and ξ are some positive constants. The optimal expression for $c - h$ is also proportional to $x - \xi h$. See Constantinides (1988) for a related continuous-time problem.

9.5 Approximation with Open-Loop Feedback Control Rules

In the previous section, we discussed the method that approximately synthesizes optimal closed-loop control policies for adaptive systems from optimal closed-loop policies for the corresponding purely stochastic systems.

In this section, we shall discuss a scheme that approximates optimal closed-loop control policies with what is sometimes called optimal open-loop feedback control policies for the same systems (Dreyfus [1964]).

An open-loop control policy specifies the sequence of control decisions to be followed from a given initial point, i.e., all control decisions are given as functions of the initial point and time. An open-loop feedback control policy computes the current and all future control variables u_j,

$i \leqslant j \leqslant N-1$, at time i from the past and current observed state variables of the system y^t, but incorporates feedback in that only u_t is actually used, and the new observation y_{t+1} on the attained state variable x_{t+1} is used to recompute the open-loop control u_{t+1} as the functions of y^{t+1} at time $t+1$.

The discussion will be for systems whose state vectors are exactly observable. This assumption of exact measurements is not essential for the development of this section. The systems with measurement noise can be treated similarly but with added complexities in the derivation.

One starts from the assumption of the dynamic equation given by

$$\mathbf{x}_{t+1} = \mathbf{A}\mathbf{x}_t + \mathbf{B}\mathbf{u}_t + \mathbf{C}\xi_t, \qquad t = 0, 1, \ldots, N-1$$

where $\mathbf{x}_t$ is the state vector, $\mathbf{u}_t$ is the control vector, and ξ_t is the random disturbance vector. The matrices $\mathbf{A}$, $\mathbf{B}$, and $\mathbf{C}$ are assumed unknown, with given a priori joint probability density function $p_0(\mathbf{A}, \mathbf{B}, \mathbf{C})$. The matrix $\mathbf{C}$ is assumed nonsingular.* The assumption of unknown $\mathbf{C}$ amounts to the assumption that the variance of the noise to the system is unknown.

The joint probability density function of $\xi_0, \xi_1, \ldots, \xi_{N-1}$ is assumed to exist and to be known.

It is a straightforward extension of the method of this section to include the case where the joint probability density function is parametrized by an unknown parameter. As before, optimal closed-loop $\mathbf{u}_t$ is to depend only on $\mathbf{y}^t$ and $\mathbf{u}^{t-1}$. The criterion function J is taken to be quadratic,

$$J = \sum_{1}^{N} W_t,$$

where

$$W_t = \mathbf{x}_t'\mathbf{V}_t\mathbf{x}_t + \mathbf{u}_{t-1}'\mathbf{T}_{t-1}\mathbf{u}_{t-1}$$

and where $\mathbf{V}_t$ and $\mathbf{T}_{t-1}$ are positive symmetric matrices, $1 \leqslant t \leqslant N$. The contribution to J from the present and the future at time t is given by

(12) $$J_t \triangleq \sum_{t}^{N} W_s.$$

Equation (12) can be rewritten as

(13) $$J_t = \mathbf{x}_t'\mathbf{V}_t\mathbf{x}_t + \mathbf{u}_{t-1}'\mathbf{T}_{t-1}\mathbf{u}_{t-1} + \mathbf{J}_{t+1}.$$

This difference equation for $\mathbf{J}_t$ is satisfied by a quadratic form in $\mathbf{x}_{t-1}, \mathbf{u}_{t-1}$,

* If $\mathbf{C}$ is singular, it can be shown that a certain number of coefficients can be learned exactly in a finite time.

$\mathbf{u}_t, \ldots, \mathbf{u}_{t-1}$, and $\xi_{t-1}, \xi_t, \ldots, \xi_{N-1}$. Therefore, we write

$$\begin{aligned} J_t = a_t &+ 2 \sum_{i=t-1}^{N-1} \mathbf{b}_i(t)\mathbf{u}_i + \sum_{i=t-1}^{N-1} \sum_{j=t-1}^{N-1} \mathbf{u}_i'\mathbf{K}_{ij}(t)\mathbf{u}_j \\ &+ 2\mathbf{C}_t\mathbf{x}_{t-1} + \mathbf{x}_{t-1}'\mathbf{L}_t\mathbf{x}_{t-1} + 2 \sum_{i=k-1}^{N-1} \mathbf{g}_i(t)\xi_i \\ &+ \sum_{i=k-1}^{N-1} \sum_{j=k-1}^{N-1} \xi_i'\mathbf{N}_{ij}(t)\xi_j + 2 \sum_{j=t-1}^{N-1} \mathbf{u}_j'\mathbf{f}_j(t)\mathbf{x}_{t-1} \\ &+ 2 \sum_{i=t-1}^{N-1} \xi_i'\mathbf{M}_i(t)\mathbf{x}_{t-1} + 2 \sum_{i=t-1}^{N-1} \sum_{j=t-1}^{N-1} \xi_i'\mathbf{O}_{ij}(t)\mathbf{u}_j \end{aligned} \tag{14}$$

where a_t, **b**s, etc., are matrices of appropriate dimensions Substituting Eq. (14) into Eq. (13), we obtain a set of recursion equations for the coefficients of the expansion. They are then solved for all t off-line.

In evaluating

$$\begin{aligned} \gamma_{t+1} &= E\left(\sum_{t+1}^{N} W_s \mid \mathbf{x}^t \right) \\ &= \int J_{t+1} p(\mathbf{A}, \mathbf{B}, \mathbf{C}, \xi_t, \xi_{t+1}, \ldots, \xi_{N-1} \mid \mathbf{x}^t) \\ &\quad \times \mathrm{d}(\mathbf{A}, \mathbf{B}, \mathbf{C}, \xi_t, \ldots, \xi_{N-1}) \\ &= \int J_{t+1} p(\xi_t, \ldots, \xi_{N-1} \mid \mathbf{A}, \mathbf{B}, \mathbf{C}, \mathbf{x}^t) p(\mathbf{A}, \mathbf{B}, \mathbf{C} \mid \mathbf{x}^t) \\ &\quad \times \mathrm{d}(\mathbf{A}, \mathbf{B}, \mathbf{C}, \xi_t, \ldots, \xi_{N-1}), \end{aligned} \tag{15}$$

we replace closed-loop control decisions with open-loop control decisions. For example,

$$\begin{aligned} \overline{\mathbf{b}_s(t+1)\mathbf{u}_s} &\triangleq \int \mathbf{b}(t+1)\mathbf{u}_s p(\mathbf{A}, \mathbf{B}, \mathbf{C} \mid \mathbf{x}^t)\, \mathrm{d}(\mathbf{A}, \mathbf{B}, \mathbf{C}) \\ &\approx \overline{\mathbf{b}_s(t+1)\mathbf{u}_t}, \end{aligned}$$

since $\mathbf{u}_s$ is taken to be a function of x^t only, i.e., open-loop control variable is substituted for closed-loop one. A bar indicates the conditional expectation operation with respect to $\mathbf{A}, \mathbf{B}, \mathbf{C}$.

Similarly,

$$\overline{\mathbf{u}_i'\mathbf{k}_{ij}(t+1)\mathbf{u}_j} \approx \mathbf{u}_i'\overline{\mathbf{k}_{ij}(t+1)}\mathbf{u}_j, \text{ etc.,} \qquad i,j > t.$$

Note that only when $\mathbf{u}_k$ is involved we can write

$$\overline{\mathbf{b}_t(t+1)\mathbf{u}_t} = \overline{\mathbf{b}_t(t+1)}\mathbf{u}_t$$

$$\overline{\mathbf{u}_t'\mathbf{K}_{tt}(t+1)\mathbf{u}_t} = \mathbf{u}_t'\overline{\mathbf{K}_{tt}(t+1)}\mathbf{u}_t$$

and so on.

When $\overline{\mathbf{b}_i(t+1)\mathbf{u}_i}$, $\overline{\mathbf{u}_i'\mathbf{K}_{ij}(t+1)\mathbf{u}_j}$, etc., are equated with $\overline{\mathbf{b}_i(t+1)}\mathbf{u}_i$, $\mathbf{u}_i'\overline{\mathbf{K}_{ij}(t+1)}\mathbf{u}_j$, etc., the control variables $\mathbf{u}_t, \mathbf{u}_{t+1}, \ldots, \mathbf{u}_{N-1}$ are all taken to be functions of $\mathbf{x}^t$ only.

The optimal open-loop control variables $\mathbf{u}_t, \mathbf{u}_{t+1}, \ldots, \mathbf{u}_{N-1}$ which approximate the optimal closed-loop policy is then given by $\mathbf{u}_t, \ldots, \mathbf{u}_{N-1}$, which minimizes $E(J_{t+1} \mid \mathbf{x}^t)$. Hence, by differentiating it with respect to u_j, $j = t, t+1, \ldots, N-1$, we obtain

$$\sum_{i=k}^{N-1} \overline{K_{ji}(t+1)}\mathbf{u}_i^* = -\left(b_j(t+1) + \sum_{i=k}^{N-1} \overline{\mu_i^{k\prime} O_{ij}(t+1)}\right) - \overline{f_j(t+1)}\mathbf{x}_t, \qquad j = t, t+1, \ldots, N-1 \tag{16}$$

where

$$\mu_j^t = \int \xi_j p(\xi_t, \xi_{t+1}, \ldots, \xi_{N-1} \mid \mathbf{A}, \mathbf{B}, \mathbf{C}, \mathbf{x}^t)\, d(\xi_t, \ldots, \xi_{N-1}), \qquad t \leqslant i, \quad j \leqslant N-1,$$

which, when solved, gives $\mathbf{u}_t^*$ among others.

When $\mathbf{u}_t^*$ is applied at time t and the time advances to $t+1$ from t, we have one more observation $\mathbf{x}_{t+1}$. Therefore, rather than using $\mathbf{u}_{t+1}^*, \ldots, u_{N-1}^*$ obtained by solving Eq. (16) at time t, we resolve Eq. (16) after μs are reevaluated conditioned on $\mathbf{x}^{t+1}$ rather than on $\mathbf{x}^t$. In other words, only the immediate control $\mathbf{u}_t^*$ is used from (16).

Thus, at each time instant t, we have a recursive procedure to obtain u_t^*.

This approximation generates an open-loop feedback control policy since a new observation $\mathbf{x}_{t+1}$ is incorporated in computing a new optimal open-loop policy based on the knowledge $\mathbf{x}^{t+1}$. It is easy to see that open-loop policies are much easier to compute than closed-loop policies.

The question of when optimal open-loop feedback policies are good approximations to optimal closed-loop policies must be carefully considered for each individual problem.

The next topic is a method of approximate synthesis of optimal closed-loop control policies for a class of linear adaptive systems from the knowledge of optimal closed-loop control policies for a corresponding class of purely stochastic control systems. Under certain conditions, optimal policies for adaptive systems can even be synthesized exactly, in this manner. Since the amount of computation involved in deriving optimal closed-loop policies for adaptive systems is usually several orders of magnitude larger than that for purely stochastic systems, the saving in computational work could sometimes be very significant.

We examine a problem in which parameter adaptive decision problems admit approximations by problems with known distribution functions. Here, stochastic control programs are regarded as at extreme points in the class of parameter adaptive ones.

9.6 Approximation of Adaptive Systems

The system to be considered is governed by the dynamic equation

$$x_{i+1} = ax_i + u_i + r_i,$$

and observed exactly

$$y_i = x_i, \qquad 0 \leqslant i \leqslant N-1, \qquad u_i \in (-\infty, \infty),$$

where x, a, u, r, and y are all scalar quantities and where r_i is the independently and identically distributed random variable with

(17) $$r_t = \begin{cases} c & \text{with probability } \theta \\ -c & \text{with probability } 1-\theta. \end{cases}$$

We take the usual criterion function

$$J = \sum_{1}^{N} \mathrm{W}_t(x_t, u_{t-1}).$$

When θ in (17) is assumed known, we have a complete-information stochastic control problem. When θ is not known, we have an adaptive control problem.

Here, for the sake of simplicity, we consider the adaptive system to be such that

(18) $$\begin{aligned} \mathrm{P}[\theta = \theta_1] &= z_0 \\ \mathrm{P}[\theta = \theta_2] &= 1 - z_0 \end{aligned}$$

where z_0 is the given a priori probability. The a posteriori probability that θ be equal to θ_1 at time t is denoted by z_t.

We will see that when W_t is quadratic the optimal adaptive control policy can be synthesized, exactly by knowing the optimal stochastic control policies with $\theta = \theta_j$, $j = 1, 2$. These two complete-information problems are the extreme points for a convex class of adaptive control problems. In the next section, we shall see that for more general adaptive problems, approximations to optimal adaptive control policies can be made with similar linear combinations of extreme problems.

For the adaptive problem of this section, the augmented state vector $\{(x_t, z_t)\}$ forms a first-order Markov sequence.

The minimum of the expected cost of control for the adaptive system at time t, γ_t^* is therefore a function of (x_{t-1}, z_{t-1}).

The general recursion equation for γ_t^* is given as

$$\gamma_t^*(x_{t-1}, z_{t-1}) = \min_{u_{t-1}} [\lambda_t + E(\gamma_{t+1}^* \mid x_{t-1}, z_{t-1})] \tag{19}$$

where, from the definition of λ_t,

$$\begin{aligned} \lambda_t &= E(\mathrm{W}_t(x_t, u_{t-1}) \mid x_{t-1}, z_{t-1}) \\ &= z_{t-1} E(W_t \mid x_{t-1}, \theta_1) + (1 - z_{t-1}) E(W_t \mid x_{t-1}, \theta_2) \end{aligned} \tag{20}$$

where

$$\begin{aligned} E(W_t \mid x_{t-1}, \theta_j) = \theta_j \mathrm{W}_t(x_{t-1}^+, u_{t-1}) \\ + (1 - \theta_j) \mathrm{W}_t(x_{t-1}^-, u_{t-1}), \qquad j = 1, 2 \end{aligned} \tag{21}$$

and

$$x_{t-1}^{\pm} \triangleq a x_{t-1} + u_{t-1} \pm c. \tag{22}$$

In (19), we can write

$$\begin{aligned} E(\gamma_{t+1}^* \mid x_{t-1}, z_{t-1}) = z_{t-1} E(\gamma_{t+1}^* \mid x_{t-1}, \theta_1) \\ + (1 - z_{t-1}) E(\gamma_{t+1}^* \mid x_{t-1}, \theta_2) \end{aligned} \tag{23}$$

where

$$\begin{aligned} E(\gamma_{t+1}^* \mid x_{t-1}, z_{t-1}, \theta_j) = \theta_j \gamma_{t+1}^*(x_{t-1}^+, z_{t-1}^+) \\ + (1 - \theta_j) \gamma_{t+1}^*(x_{t-1}^-, z_{t-1}^-), \qquad j = 1, 2, \end{aligned} \tag{24}$$

and where

$$z_{t-1}^+ = \frac{z_{t-1}\theta_1}{z_{t-1}\theta_1 + (1 - z_{t-1})\theta_2}$$

(25) and

$$z_{t-1}^- = \frac{z_{t-1}(1 - \theta_1)}{z_{t-1}(1 - \theta_1) + (1 - z_{t-1})(1 - \theta_2)}$$

are the a posteriori probabilities that $\theta = \theta_1$, given that $r_{t-1} = +c$ and $r_{t-1} = -c$, respectively.

From (20) and (24), we can write (19) as

$$\gamma_t^*(x_{t-1}, z_{t-1}) = \min_{u_{t-1}} [z_{t-1} \langle W_t + \gamma_{t+1}^* \rangle_1 + (1 - z_{t-1}) \langle W_t + \gamma_{t+1}^* \rangle_2] \tag{26}$$

where

$$\langle W_t + \gamma_{t+1}^* \rangle_j \triangleq E(W_t + \gamma_{t+1}^* \mid x_{t-1}, z_{t-1}, \theta_j), \qquad j = 1, 2 \tag{27}$$

for $1 \leqslant t \leqslant N$ and $\gamma_{N+1}^* = 0$. Note that the variable z_{t-1}, which expresses our ignorance about the precise value of θ, appears linearly in the inside of the minimization operation of the recursion equation (26). Note also that the purely stochastic problem with $\theta = \theta_1$ corresponds to the problem with the a priori probability $z_0 = 1$ and the problem with $\theta = \theta_2$ to $z_0 = 0$. The recursion equation for the corresponding stochastic system where $\theta = \theta_1$ therefore is obtained by putting $z_{t-1} = 1$ for $1 \leqslant t \leqslant N$ in (26).

Denoting by $\gamma_{t,j}^*$ the optimal control cost for the stochastic problem with $\theta = \theta_j$, $j = 1, 2$, it satisfies the recursion equation

$$\begin{aligned} \gamma_{t,j}^*(x_{t-1}) &= \min_{u_{t-1}} \langle W_t(x_t, u_{t-1}) + \gamma_{t+1,j}^*(x_t) \rangle_j \\ &= \min_{u_{t-1}} \{\theta_j [\mathrm{W}_t(x_{t-1}^+, u_{t-1}) + \gamma_{t+1,j}^*(x_{t-1}^+)] \\ &\quad + (1 - \theta_j)[\mathrm{W}_t(x_{t-1}^-, u_{t-1}) + \gamma_{t+1,j}^*(x_{t-1}^-)]\}, \qquad j = 1, 2 \end{aligned} \tag{28}$$

where

$$x_{t-1}^{\pm} = a x_{t-1} + u_{t-1} \pm c, \qquad t = 1, 2, \ldots, N.$$

Now suppose $\mathrm{W}_t(x, u)$ is quadratic in x and u:

$$\mathrm{W}_t(x_t, u_{t-1}) = v_t x_t^2 + t_{t-1} u_{t-1}^2.$$

Then the optimal control at the last stage for the stochastic system with $\theta = \theta_j$, denoted as $u_{N-1,j}^*$, is given by

$$u_{N-1,j}^* = -\frac{v_N[a x_{N-1} + c(2\theta_j - 1)]}{v_N + t_{N-1}}, \qquad j = 1, 2. \tag{29}$$

The optimal control at $N - 1$ for the adaptive system, denoted as u_{N-1}^*, is given from (26) by

$$u_{N-1}^* = -\frac{v_N[a x_{N-1} + c(2\hat{\theta}_{N-1} - 1)]}{v_N + t_{N-1}} \tag{30}$$

where

$$\hat{\theta}_{N-1} \triangleq z_{N-1}\theta_1 + (1 - z_{N-1})\theta_2. \tag{31}$$

By comparing (29) with (31), we see that the optimal adaptive control is given by a linear combination of the optimal stochastic controls

(32) $$u_{N-1}^* = z_{N-1} u_{N-1,1}^* + (1 - z_{N-1}) u_{N-1,2}^*$$

where z_{N-1} is the a posteriori probability that $\theta = \theta_1$ at the $(N-1)$th stage, i.e., after $x_0, x_1, \ldots, x_{N-1}$ have been observed. Thus, at least for this example, the last optimal adaptive control is obtained once the last optimal stochastic controls for $\theta = \theta_1$ and θ_2 are known.

We can show from (26) and (28) that the inequality

(33) $$\gamma_t^*(x, z) \geqslant z\gamma_{t,1}^*(x) + (1 - z)\gamma_{t,2}^*(x)$$

holds. Furthermore, W_t need not be quadratic for (33) to be true (Aoki, 1960). More precisely, we shall show later on that

(34) $$\gamma_t^*(x, z) = z\gamma_{t,1}^*(x) + (1 - z)\gamma_{t,2}^*(x) + \Delta\gamma_t(z)$$

where $\Delta\gamma_t(z)$ depends only on z.

Numerical experiments done for some stable final-value control problems show that $\Delta\gamma_t$ is small compared with $\gamma_{t,1}^*$ and $\gamma_{t,2}^*$ for t of order 10. Thus, for the example under consideration, not only the optimal adaptive control is obtainable exactly from the corresponding optimal stochastic controls but also (33) gives a good approximation for $\gamma_t^*(x, z)$ as well. See Aoki (1960) for some computation results.

Derivation of Approximate Relations of Optimal Control Policies for Adaptive and Stochastic Systems

We now generalize the preceding observation and discuss the relation of the optimal adaptive and stochastic control policies for linear systems with quadratic criterion functions and with exact state vector measurements. The following developments are based in part on Fukao (1964). When the state vector measurements are noisy, the problem of approximating adaptive policies becomes much more difficult and requires further analytical and computational investigation.

The system is now assumed given by

(35) $$\mathbf{x}_{t+1} = \mathbf{A}_t \mathbf{x}_t + \mathbf{B}_t \mathbf{u}_t + \mathbf{C}_t \xi_t$$

(36) $$\mathbf{y}_t = \mathbf{x}_t$$

where $\mathbf{x}_t$ is n vector, $\mathbf{u}_t$ is r vector, ξ_t is m-dimensional random vector, $\mathbf{A}$ is $n \times n$ matrix, $\mathbf{B}$ is $n \times r$ matrix, and $\mathbf{C}$ is $n \times m$ matrix.

Let us use R_t as a generic symbol of the random variables. We consider the problems such that, at time $t + 1$, $R_0, \ldots, R_t$ will be known exactly from the known collection of the state vectors $\mathbf{x}_0, \ldots, \mathbf{x}_{t+1}$.

At the end of this Section, we consider two examples, one with $\mathbf{R}_t = \xi_t$ and the other with $\mathbf{R}_t = \mathbf{A}$.

Since the problem to be considered is adaptive, the probability distribution function for R will not be completely known. Let us assume that a parameter θ characterizes the probability distribution function for R and the a priori information on θ is given as

$$\mathrm{P}[\theta = \theta_j] = z_{0,j}, \qquad 1 \leqslant j \leqslant S$$

where

$$\sum_1^S z_{0,j} = 1$$

and where the first subscript 0 of $z_{0,j}$ refers to time 0 and the second subscript j indicates that it is the probability of θ being equal to θ_j. In other words, when θ is specified to be θ_j, the random variables R^t, $t = 0, 1, \ldots,$ are distributed with known probability distribution function $F(R^t \mid \theta_j)$.

We shall use the notation $\langle \cdot \rangle_k$ introduced by (27) to indicate the expectation operation when the distribution function involved has the parameter value θ_k. For example, $\langle \Phi(R) \rangle_k = \int \Phi(R) p(R \mid \theta_k)\, \mathrm{d}R$ when the indicated integral exists.

In keeping with the notation just introduced, we denote the a posteriori probability that $\theta = \theta_j$ at time t by $z_{t,j}$.

By the Bayes rule, when Rs are independent for each θ, the recursion relation for $z_{t,j}$ is given by

$$(37) \qquad z_{t+1,i} = \frac{p(R_t \mid \theta_i) z_{t,i}}{\sum_{i=1}^{S} p(R_t \mid \theta_i) z_{t,i}}, \qquad t = 0, 1, \ldots, \qquad 1 \leqslant i \leqslant S$$

where $p(R_t \mid \theta_i)$ is the probability density function of R_t when θ is θ_i.

Let $\mathbf{z}_t = (z_{t,1}, z_{t,2}, \ldots, z_{t,S})$. Then the augmented state vector $(\mathbf{x}_t, \mathbf{z}_t)$ forms a first-order Markov sequence.

We now state a series of four observations that will serve as a basis of our approximation scheme. The notations used are summarized here. $\gamma_t^*(\mathbf{x}, \mathbf{z})$ is the minimum of $E[\Sigma_{s=t}^N W_s(\mathbf{x}_s, \mathbf{u}_{s-1}) | \mathbf{x}, \mathbf{z}]$ when $\mathbf{x}_{t-1} = \mathbf{x}$ and the probability of the parameter at time t is given by z. $\gamma_{t,k}^*(\mathbf{x}) \equiv \gamma_t^*(\mathbf{x}, \mathbf{z})$ where $\mathbf{z} = (0, \ldots, 0, 1, 0, \ldots, 0)$ where the only nonzero component of z is the kth component, which is one.

Thus, $\gamma_{t,k}^*(\mathbf{x})$ is the minimum of $E[\Sigma_{j=1}^* \mathrm{W}_j(\mathbf{x}_j, \mathbf{u}_{j-1}) | \mathbf{x}]$ when θ is known to be θ_k. Let us call γ_t^* the adaptive control cost and $\gamma_{t,k}^*$ the stochastic control cost.

Observation 1

Assume that if $\gamma_t^*(x, z)$ is separable in x and in the components of z. Then the adaptive control cost is

$$\gamma_t^*(\mathbf{x}, \mathbf{z}_t) = \sum_{k=1}^{S} \mu_k(\mathbf{z}_{t,k})\nu_k(\mathbf{x}).$$

Assume further that $\mu_k(z_{t,k})$ is proportional to $z_{t,k}$. Then

$$\mu_k(1) \neq 0, \qquad \mu_k(0) = 0, \qquad 1 \leqslant k \leqslant S.$$

Then

(38)
$$\gamma_t^*(\mathbf{x}, \mathbf{z}_t) = \sum_{k=1}^{S} z_{t,k}\gamma_t^*(\mathbf{x}).$$

Remember that

(39)
$$\gamma_t^*(\mathbf{x}) \triangleq \gamma_t^*(\mathbf{x}, \mathbf{z}_t = (0, \ldots, 0, 1, 0, \ldots, 0))$$

where only the kth component of z_t is nonzero and is equal to one in (39). Thus, (38) shows that, if the adaptive control cost is a separable function of $\mathbf{x}$ and components of z, then it is a linear combination of the stochastic control cost. This is a useful fact in approximating the adaptive control policy by those of the stochastic control systems.

Proof of Observation 1. *If* $z_{t,S} = 0$, *i.e.*, $\mathbf{z} = (0, 0, \ldots, 0, 1)$, *then*

$$\gamma_t^*(\mathbf{x}, \mathbf{z}) = \gamma_{t,S}^*(\mathbf{x}) = \mu_S(1)\nu_S(\mathbf{x})$$

If $z_{t,k} = 0$, *then*

$$\gamma_t^*(\mathbf{x}, \mathbf{z}) = \gamma_{t,k}^*(\mathbf{x}) = \mu_k(1)\nu_k(\mathbf{x}).$$

Then

$$\gamma_t^*(\mathbf{x}, \mathbf{z}_t) = \sum_{k=1}^{S} \frac{\mu_k(z_{t,k})}{\mu_k(1)} \gamma_{t,k}^*(\mathbf{x})$$

but

$$\mu_k(z_{t,k})/\mu_k(1) = z_{t,k}$$

and we obtain (38).

Observation 2

As one of the components in z_t approaches 1,

(40)
$$\gamma_t^*(\mathbf{x}, \mathbf{z}_t) \to \sum_{k=1}^{S} z_{t,k}\gamma_{t,k}^*(\mathbf{x}).$$

This shows that, if the a priori probability of θ being equal to one of $\theta_1, \ldots, \theta_S$ is close to 1, or if the a posteriori probability z_t is such that most of the probability mass is concentrated as one of S possible values for θ (i.e., when one is fairly sure which value θ is), then the adaptive control cost will approach the form assumed in Observation 1.

Proof of Observation 2

Expand $\gamma_t^*(\mathbf{x}, \mathbf{z})$ *about* $\mathbf{z}_t^* = (0, \ldots, 1(j\text{th}), 0, \ldots, 0)$ *retaining only linear terms in the components of* z *and use Observation* 1.

Observation 3

Suppose

$$\gamma_{t+1}^*(\mathbf{x}, \mathbf{z}_t) = \sum_1^S z_{t,k}\gamma_{t+1,k}^*(\mathbf{x}). \tag{41}$$

Then the recursion equation for the adaptive control cost is given by

$$\gamma_t^*(\mathbf{x}_{t-1}, \mathbf{z}_{t-1}) = \min_{u_{t-1}} \sum_{k=1}^S z_{t-1,k}\langle \mathbf{W}_t(\mathbf{x}_t, \mathbf{u}_{t-1}) + \gamma_{t+1,k}^*(\mathbf{x}_t)\rangle_k \tag{42}$$

where the notation introduced earlier by (27) is used to write the conditional expectation

$$\langle \mathbf{W}_t(\mathbf{x}_t, \mathbf{u}_{t-1}) + \gamma_{t+1,k}^*(\mathbf{x}_t)\rangle_k \triangleq \int [\mathbf{W}_t(\mathbf{x}_t, \mathbf{u}_{t-1}) + \gamma_{t+1,k}^*(\mathbf{x}_t)] \times p(\mathbf{x}_t \mid \mathbf{x}, \mathbf{z}_{t-1}, \theta_k)\, d\mathbf{x}_t$$

where

$$\mathbf{x}_t = \mathbf{A}_{t-1}\mathbf{x} + \mathbf{B}_{t-1}\mathbf{u}_{t-1} + \mathbf{C}_{t-1}\xi_{t-1}$$

and where the random variable R_{t-1} is assumed to have the probability distribution with the parameter $\theta = \theta_k$. Thus, by knowing the stochastic control cost $\gamma_{t+1,k}^*$ for $1 \leqslant k \leqslant S$, the optimal adaptive control variable u_{t-1}^* can be obtained from (42) if γ_{t+1}^* has the assumed form (41). Equation (42) shows that, even if γ_{t+1}^* is linear homogeneous in $\mathbf{z}$, γ_t^* is not necessarily linear homogeneous in z.

Proof of Observation 3

The recursion equation for γ_t^* *is given by*

$$\gamma_t^*(\mathbf{x}, \mathbf{z}_{t-1}) = \min_{u_{t-1}} \sum_{k=1}^S z_{t-1,k}\langle W_t(\mathbf{x}', \mathbf{u}_{t-1}) + \gamma_{t+1}^*(\mathbf{x}', \mathbf{z}_{t-1}')\rangle_k \tag{43}$$

where

$$\mathbf{x}' = \mathbf{A}_{t-1}\mathbf{x} + \mathbf{B}_{t-1}\mathbf{u}_{t-1} + \mathbf{C}_{t-1}\xi_{t-1}$$

and $\mathbf{z}'_{t-1}$ is the a posteriori probability when the a priori probability is given by $\mathbf{z}_{t-1}$. Its components are given by

$$z'_{t-1,j} = \frac{p(R_{t-1} \mid \theta_j) z_{t-1,j}}{\sum_{k=1}^{S} p(R_{t-1} \mid \theta_k) z_{t-1,k}}.$$

By Assumption (41),

$$\gamma^*_{t+1}(\mathbf{x}', \mathbf{z}'_{t-1}) = \sum_{1}^{S} \mathbf{z}'_{t-1,j} \gamma^*_{t+1,k}(\mathbf{x}').$$

Thus, in (43),

$$\begin{aligned}
&\sum_{k=1}^{S} z_{t-1,k} \sum_{j=1}^{S} \langle z'_{t-1,j} \gamma^*_{t+1,j}(\mathbf{x}') \rangle_k \\
&= \sum_{j=1}^{S} \sum_{k=1}^{S} z_{t-1,k} \int \frac{p(R_{t-1} \mid \theta_j) z_{t-1,j} \gamma^*_{t+1,j}(\mathbf{x}')}{\left[\sum_{k=1}^{S} p(R_{t-1} \mid \theta_k) z_{t-1,k}\right]} p(R_{t-1} \mid \theta_k) \, \mathrm{d}R_{t-1} \\
&= \sum_{j=1}^{S} \int z_{t-1,j} \gamma^*_{t+1,j}(\mathbf{x}') p(R_{t-1} \mid \theta_j) \, \mathrm{d}R_t \\
&= \sum_{j=1}^{S} z_{t-1,j} \langle \gamma^*_{t+1,j}(\mathbf{x}') \rangle_j,
\end{aligned}$$

establishing (42).

The optimal adaptive control u^*_{t-1} is therefore obtained from (42).

If we have assumed

$$\gamma^*_{t+1}(\mathbf{x}, \mathbf{z}_t) = \sum_{1}^{S} z_{t,k} \gamma^*_{t+1,k}(\mathbf{x}) + \Delta\gamma_{t+1}(\mathbf{z}_t)$$

instead of (41), *then u^*_{t-1} is still obtained by*

$$\min_{u_{t-1}} \sum_{1}^{S} z_{t,k} \langle W_t + \gamma^*_{t+1,k} \rangle_k$$

since the contribution $\Delta\gamma_{t+1}$ is independent of u_{t-1}.

Thus, in view of our discussion at the start of this section and of Observations 2 and 3, we may define a measure of increase in control due to the imprecise knowledge of θ by

$$\Delta\gamma_i \triangleq \gamma^*_i(\mathbf{x}, \mathbf{z}_{t-1}) - \sum_{k=1}^{S} z_{t-1,k} \gamma^*_{t,k}(\mathbf{x}) \tag{44}$$

when the system is in the state $(x_{t-1} = x, z_{t-1})$ at time $t-1$.

When γ_t^* is expressible as (42), *the right-hand side of* (44) *can be written as*

(44a)
$$\delta\gamma_i \triangleq \min_{u_{t-1}} \sum_{k=1}^{S} z_{t-1,k} \langle W_t + \gamma_{t+1,k}^* \rangle_k - \sum_{k=1}^{S} z_{t-1,k} \min_{u_{t-1}} \langle W_t + \gamma_{t+1,k}^* \rangle_k$$

i.e., $\delta\gamma_t$ is the special case of $\Delta\gamma_t$ where γ_t^* has the form (42) *as the result of the assumed form* (41).

Note that the operations of the minimization and the averaging with respect to z are interchanged in (44). We may therefore say in this case the increase in the cost of adaptive control over the cost of stochastic control is given by the interchange of the summation and the minimization operations.

The relation of the optimal policies for adaptive and purely stochastic systems is established by Observation 4.

Observation 4

If

(45)
$$\gamma_t^*(\mathbf{x}, \mathbf{z}_{t-1}) = \min_{u_{t-1}} \sum_{k=1}^{S} z_{t-1,k} \langle W_t + \gamma_{t+1,k}^* \rangle_k,$$

if $\langle W_t + \gamma_{t+1,k}^* \rangle_k$ is quadratic in u, and if no constraints are imposed on u and x (i.e., state vector and control vectors are not constrained in any way), then the optimal control vector for the adaptive system at time $t - 1$ is given by a linearly weighted sum of the optimal control vector of the corresponding S stochastic problems at time $t - 1$. Under the same set of assumptions, $\delta\gamma_t$ of (44) is, at most, quadratic in the optimal control vector for the stochastic problems.

Proof of Observation 4

Since the dependence of $\langle W_t + \gamma_{t+1,k}^* \rangle_k$ on u is quadratic by assumption, by completing the square, if necessary, and recalling the recursion formula for $\gamma_{t,k}^*$ and that the notation $u_{t-1,k}^*$ is used to denote the optimal control for the purely stochastic problem with $\theta = \theta_k$ at time t, we can write it as

(46)
$$\langle W_t + \gamma_{t-1,k}^* \rangle_k = (\mathbf{u} - \mathbf{u}_{t-1,k}^*)' \boldsymbol{\Phi}_{t,k} (\mathbf{u} - \mathbf{u}_{t-1,k}^*) + \phi_{t,k}$$

where $\boldsymbol{\Phi}_{t,k}' = \boldsymbol{\Phi}_{t,k}$ and where $\boldsymbol{\Phi}_{t,k}$ and $\phi_{t,k}$ generally depend on $\mathbf{x}$. Note that $\mathbf{u}_{t-1,k}^*$ will generally depend on θ_k.

Substituting (46) *into* (45), *the optimal control* u_{t-1}^* *for the adaptive problem is given by performing the minimization*

$$\min_{u_{t-1}} \sum_{k=1}^{S} z_{t-1,k}[(\mathbf{u} - \mathbf{u}_{t-1,k}^*)' \boldsymbol{\Phi}_{t,k} (\mathbf{u} - \mathbf{u}_{t-1,k}^*) + \phi_{t,k}],$$

(47) *and*

$$\mathbf{u}_{t-1}^* = \left(\sum_{k=1}^{S} z_{t-1,k} \boldsymbol{\Phi}_{t,k} \right)^{-1} \left(\sum_{k=1}^{S} z_{t-1,k} \boldsymbol{\Phi}_{t,k} \mathbf{u}_{t-1,k}^* \right),$$

proving Observation 4 when the indicated inverse exists.

Equation (47) shows that the adaptive optimal control policy is obtainable by solving S purely stochastic control problems.

Knowing optimal control vectors for the stochastic problems $u_{t-1,k}^*$, $k = 1, \ldots, S$, we can obtain the difference of the adaptive and the stochastic control cost when the assumptions of Observations 4 are met. From (44a) and (46),

$$\begin{aligned} \delta\gamma_t &= \sum_{k=1}^{S} z_{t-1,k}[\langle W_t + \gamma_{t+1,k}^* \rangle_k]_{u_{t-1}^*} \\ &\quad - \sum_{k=1}^{S} z_{t-1,k}[\langle W_t + \gamma_{t+1,k}^* \rangle_k]_{u_{t-1,k}^*} \\ &= \sum_{k=1}^{S} z_{t-1,k} (\mathbf{u}_{t-1}^* - \mathbf{u}_{t-1,k}^*)' \boldsymbol{\Phi}_{t,k} (\mathbf{u}_{t-1}^* - \mathbf{u}_{t-1,k}^*) \qquad (48) \\ &= \sum_{k=1}^{S} (\mathbf{u}_{t-1,k}^*)' z_{t-1,k} \boldsymbol{\Phi}_{t,k} \mathbf{u}_{t-1,k}^* - \left(\sum_{k=1}^{S} z_{t-1,k} \boldsymbol{\Phi}_{t,k} \mathbf{u}_{t-1,k}^* \right)' \\ &\quad \times \left(\sum_{k=1}^{S} z_{t-1,k} \boldsymbol{\Phi}_{t,k} \right)^{-1} \left(\sum_{k=1}^{S} z_{t-1,k} \boldsymbol{\Phi}_{t,k} \mathbf{u}_{t-1,k}^* \right). \end{aligned}$$

As a special case, if the quadratic part of $\gamma_{t,k}^*$ is independent of θ_k, i.e., if

$$\boldsymbol{\Phi}_{t,k} = \boldsymbol{\Phi}_t, \qquad 1 \leqslant k \leqslant S,$$

then from (47), the adaptive optimal control is related to $\mathbf{u}_{t-1,k}^*$, $1 \leqslant k \leqslant S$, by

$$\mathbf{u}_{t-1}^* = \sum z_{t-1,k} \mathbf{u}_{t-1,k}^*. \qquad (49)$$

Namely the optimal adaptive control is a weighted average of the corresponding stochastic optimal control. For this special case, (48), reduces to

$$\delta\gamma_t = \sum_{k=1}^{S} (\mathbf{u}^*_{t-1,k})' z_{t-1,k} \mathbf{\Phi}_t \mathbf{u}^*_{t-1,k} - \left(\sum_{k=1}^{S} z_{t-1,k} \mathbf{u}^*_{t-1,k}\right)' \mathbf{\Phi}_t \left(\sum_{k=1}^{S} z_{t-1,k} \mathbf{u}^*_{t-1,k}\right). \tag{50}$$

Even in the general case, by defining

$$\bar{\mathbf{\Phi}}_t \triangleq \sum_{k-1}^{S} z_{t-1,k} \mathbf{\Phi}_{t,k} \tag{51}$$

and

$$\bar{z}_{t,k} \triangleq z_{i,k} \bar{\mathbf{\Phi}}_t^{-1} \mathbf{\Phi}_{t,k}, \tag{52}$$

we can express u^*_{t-1} and $\delta\gamma_t$ in forms similar to (49) and (50), respectively. We have, from (49), (51), and (52),

$$\mathbf{u}^*_{t-1} = \sum_{k=1}^{S} \bar{z}_{t-1,k} \mathbf{u}^*_{t-1,k}, \tag{53}$$

and, from (48), (51), and (52),

$$\delta\gamma_t = \sum_{k=1}^{S} (\mathbf{u}^*_{t-1,k})' \bar{\mathbf{\Phi}}_t \bar{z}_{t-1,k} \mathbf{u}^*_{t-1,k} - \left(\sum_{k=1}^{S} \bar{z}_{t-1,k} \mathbf{u}^*_{t-1,k}\right)' \bar{\mathbf{\Phi}}_t \left(\sum_{k=1}^{S} \bar{z}_{t-1,k} \mathbf{u}^*_{t-1,k}\right). \tag{54}$$

The difference in control cost $\delta\gamma_t$ generally depends on x. If $\bar{\mathbf{\Phi}}_k$ is independent of the state vector $\mathbf{x}$ and if the stochastic optimal control vector $\mathbf{u}^*_{t,k}$ can be expressed as a sum of functions of x_t only and of θ_k only, then $\delta\gamma_t$ will be independent of $\mathbf{x}$. To see this, we express, by assumption, $\mathbf{u}^*_{t,k}$ as

$$\mathbf{u}^*_{t,k} = a(\mathbf{x}_t) + b_k. \tag{55)*}$$

Then, substituting (55), into (53) and (54), we obtain

$$\mathbf{u}^*_t = \mathbf{a}(\mathbf{x}_t) + \sum_{k=1}^{S} \bar{z}_{t,k} \mathbf{b}_k \tag{56}$$

and

$$\delta\gamma_{t+1} = \sum_{k=1}^{S} \mathbf{b}'_k \bar{\mathbf{\Phi}}_{t+1} \bar{z}_{t,k} \mathbf{b}_k - \left(\sum_{k=1}^{S} \bar{z}_{t,k} \mathbf{b}_k\right)' \mathbf{\Phi}_{t+1} \left(\sum_{k=1}^{S} \bar{z}_{t,k} \mathbf{b}_k\right). \tag{57}$$

---

* Equation (29) shows that the stochastic system with known θ satisfies (55), at least for $t = N - 1$.

Equation (57) shows that $\delta\gamma_t$ is independent of $\mathbf{x}$, when the stochastic problems are such that Φ_k is independent of $\mathbf{x}$ and when their optimal control are given by (55).

Let us now consider two examples of adaptive control systems with quadratic performance indices and illustrate the usefulness of these observations for the systems described by (35) and (36).

Example 3: Adaptive Systems with Unknown Dynamic Noise

If we identify R_t with ξ_t in the development of the previous section, then we have an adaptive control problem where the probability distribution function for the dynamic disturbance random variable contains unknown parameter θ. Assume that ξ_t are the only random variables in (35), that they are independent in time, and that their common distribution function is given by $F(z \mid \theta)$, where θ is chosen from $\theta_1, \ldots, \theta_S$. Thus,

$$\begin{aligned}\lambda_N(\mathbf{x}_{N-1}, z_{N-1}) &= \int \mathrm{W}_N(\mathbf{x}, \mathbf{u}_{N-1}) p(\mathbf{x} \mid \mathbf{x}_{N-1}, z_{N-1}, \mathbf{u}_{N-1})\, \mathrm{d}x \\ &= \int \mathrm{W}_N(\mathbf{x}, \mathbf{u}_{N-1}) p(\mathbf{x} \mid \mathbf{x}_{N-1}, \xi_{N-1}, \mathbf{u}_{N-1}) \\ &\qquad \times p(\xi_{N-1} \mid z_{N-1})\, \mathrm{d}(\mathbf{x}, \xi_{N-1}).\end{aligned} \tag{58}$$

We take W_N to be

$$\mathrm{W}_N(\mathbf{x}, \mathbf{u}) = (\mathbf{x}, \mathbf{V}_N \mathbf{x}) + (\mathbf{u}, \mathbf{P}_{N-1} \mathbf{u})$$

where

$$\mathbf{x} = \mathbf{A}_{N-1} \mathbf{x}_{N-1} + \mathbf{B}_{N-1} \mathbf{u}_{N-1} + \mathbf{C}_{N-1} \xi_{N-1}$$

and where

$$p(\xi_{N-1} \mid \mathbf{z}_{N-1}) = \sum_{k=1}^{S} p(\xi_{N-1} \mid \theta_k) z_{N-1,k}. \tag{59}$$

Thus

$$\gamma_N^*(\mathbf{x}_{N-1}, z_{N-1}) = \min_{u_{N-1}} \sum_{k=1}^{N} z_{N-1,k} \langle W_N \rangle_k \tag{60}$$

where, dropping the subscript $N-1$ from $\mathbf{A}$, $\mathbf{B}$, and $\mathbf{C}$,

$$\begin{aligned}\langle W_N \rangle_k &= \langle \mathrm{W}_N(\mathbf{A}\mathbf{x}_{N-1} + \mathbf{B}\mathbf{u}_{N-1} + \mathbf{C}\xi_{N-1}, \mathbf{u}_{N-1}) \rangle_k \\ &= \mathrm{W}_N(\mathbf{A}\mathbf{x}_{N-1} + \mathbf{B}\mathbf{u}_{N-1}, \mathbf{u}_{N-1}) \\ &\quad + 2\{\mathbf{C}\langle \xi_{N-1} \rangle_k, \mathbf{V}_N(\mathbf{A}\mathbf{x}_{N-1} + \mathbf{B}\mathbf{u}_{N-1})\} \\ &\quad + \langle (\mathbf{C}\xi_{N-1}, \mathbf{V}_N \mathbf{C}\xi_{N-1}) \rangle_k.\end{aligned}$$

Therefore, $\mathbf{u}^*_{N-1,k}$ is obtained by

$$\min_{u_{N-1}} [\mathrm{W}_N(\mathbf{A}\mathbf{x}_{N-1} + \mathbf{B}\mathbf{u}_{N-1}, \mathbf{u}_{N-1}) + 2(\mathbf{C}\langle\xi_{N-1}\rangle_k, \mathbf{V}_N(\mathbf{A}\mathbf{x}_{N-1} + \mathbf{B}\mathbf{u}_{N-1}))]$$

where $\langle\xi_{N-1}\rangle_k$ is the mean of the distribution function $F(\xi_{N-1}\,|\,\theta_k)$. Thus $\mathbf{\Phi}_N$ defined in (46) is seen to be independent of x and k:

$$\mathbf{\Phi}_{N,k} = \mathbf{\Phi}_N = \mathbf{B}'_{N-1}\mathbf{V}_N\mathbf{B}_{N-1} + \mathbf{P}_{N-1}. \tag{61}$$

From (61),

$$\begin{aligned} \mathbf{u}^*_{N-1,k} &= -(\mathbf{P}_{N-1} + \mathbf{B}'_{N-1}\mathbf{V}_N\mathbf{B}_{N-1})^{-1}\mathbf{B}'_{N-1}\mathbf{V}_{N-1}(\mathbf{A}_{N-1}\mathbf{x}_{N-1} + \mathbf{C}_{N-1}\langle\xi_{N-1}\rangle_k) \\ &= \mathbf{a}_{N-1}\mathbf{x}_{N-1} + \mathbf{b}_{k,N-1} \end{aligned} \tag{62}$$

where

$$\mathbf{a}_{N-1} \triangleq -(\mathbf{P}_{N-1} + \mathbf{B}'_{N-1}\mathbf{V}_N\mathbf{B}_{N-1})^{-1}\mathbf{B}'_{N-1}\mathbf{V}_{N-1}\mathbf{A}_{N-1} \tag{63a}$$

$$\mathbf{b}_{k,N-1} \triangleq -(\mathbf{P}_{N-1} + \mathbf{B}'_{N-1}\mathbf{V}_N\mathbf{B}_{N-1})^{-1}\mathbf{B}'_{N-1}\mathbf{V}_{N-1}\mathbf{C}_{N-1}\langle\xi\rangle_k. \tag{63b}$$

From (62), the corresponding stochastic control cost is given by

$$\begin{aligned} \gamma^*_{N,k} = \{&\mathbf{x}_{N-1}, [\mathbf{A}'\mathbf{V}_N\mathbf{A} - \mathbf{A}'\mathbf{V}_N\mathbf{B}(\mathbf{P}_{N-1} + \mathbf{B}'\mathbf{V}_N\mathbf{B})^{-1}\mathbf{B}'\mathbf{V}_N\mathbf{A}]\mathbf{x}_{N-1}\} \\ &- 2\langle\xi\rangle'_k\mathbf{C}'\mathbf{V}_N\mathbf{B}(\mathbf{P}_{N-1} + \mathbf{B}'\mathbf{V}_N\mathbf{B})^{-1}\mathbf{B}'\mathbf{V}_N\mathbf{A}\mathbf{x}_{N-1} \\ &- \langle\xi\rangle'_k\mathbf{C}'\mathbf{V}_N\mathbf{B}(\mathbf{P}_{N-1} + \mathbf{B}'\mathbf{V}_N\mathbf{B})^{-1}\mathbf{B}'\mathbf{V}_N\mathbf{A}\mathbf{C}\langle\xi\rangle_k \\ &+ \langle\xi_{N-1}, \mathbf{C}'\mathbf{V}_N\mathbf{C}\xi_{N-1}\rangle_k. \end{aligned} \tag{64}$$

The optimal adaptive control $\mathbf{u}^*_{N-1}$ is obtained by

$$\min_{\mathbf{u}_{N-1}} [\mathbf{W}_N(\mathbf{A}\mathbf{x}_{N-1} + \mathbf{B}\mathbf{u}_{N-1}, \mathbf{u}_{N-1}) + 2(\mathbf{C}'\bar{\xi}, \mathbf{V}_N(\mathbf{A}\mathbf{x}_{N-1} + \mathbf{B}\mathbf{u}_{N-1}))]$$

where

$$\bar{\xi} \triangleq \sum_{k=1}^{S} z_{N-1,k}\langle\xi\rangle_k.$$

From (51)–(54),

$$\mathbf{u}^*_{N-1} = \mathbf{a}_{N-1}\mathbf{x}_{N-1} + \sum_{1}^{S} z_{N-1,k}\mathbf{b}_{k,N-1}.$$

Thus

$$\gamma^*_N(\mathbf{x}_{N-1}, z_{N-1}) = \sum_{k=1}^{S} \mathbf{z}_{N-1,k}\gamma^*_N(\mathbf{x}_{N-1}) + \Delta\gamma_N(z_{N-1}) \tag{65}$$

where $\Delta\gamma_N$ is given from (44), (61), (63a), and (63b), and is independent of x_{N-1} since $\bar{\Phi}$ and bs are independent of x.

If $\langle\xi\rangle_k$ is the same for all $k = 1, \ldots, S$ (for example, zero for all k), then $\mathbf{u}_{N-1,k}$ is the same for all k and $\mathbf{u}_{N-1}$,

$$\mathbf{u}_{N-1}^* = \mathbf{u}_{N-1,k}^*,$$

for any k. By employing the argument similar to those in the proof of Observation 3, we can show that u_{t-1}^* is obtained from

(66)
$$\min_{u_{t-1}} \sum_{k=1}^{S} z_{t-1,k} \langle W_t + \gamma_{t+1,k}^* \rangle_k,$$

and see that

(67)
$$\gamma_{t+1}^*(\mathbf{x}_t, \mathbf{z}_t) = \sum_{k=1}^{S} z_{t,k} \gamma_{t+1,k}^*(\mathbf{x}_t) + \Delta\gamma_{t+1}$$

where $\Delta\gamma_{t+1}$ is independent of x and is the solution to the recursion equation

(68)
$$\Delta\gamma_t(z_{t-1}) = \delta\gamma_t(z_{t-1}) + \sum_{k=1}^{S} z_{t-1,k} \langle \Delta\gamma_{t+1}(z_{t-1}') \rangle_k$$
$$\Delta\gamma_{N+1} \equiv 0$$

where $\delta\gamma_t(z_{t-1})$ is defined by (44a).

In the system just discussed, $\Delta\gamma_t$ turned out to be a function of z only. Hence we couldd synthesize u_t^* from $u_{t,k}^*$, $1 \leqslant k \leqslant S$ exactly. If $\Delta\gamma_t$ is a function of x, however, then we can no longer synthesize u_t^* so simply. When the random variable R_t contains A_t, B_t, and/or C_t in addition to or instead of ξ_t, $\Delta\gamma_t$ will, in general, be a function of u, x, and ξ. Such an example is briefly discussed next.

Example 4: Adaptive System with Unknown Dynamic Matrix

Suppose $\mathbf{A}$ of (35) is an unknown constant matrix with S possible values $\mathbf{A}^{(1)}, \ldots, \mathbf{A}^{(S)}$. The random variable ξs are assumed to be independently and identically distributed with $E\xi_t = 0$ and assumed to have finite second moments. Assume $\mathbf{C} = \mathbf{I}$ in (35) for the sake of simplicity. Now,

(69)
$$\gamma_N(\mathbf{x}_{N-1}, \mathbf{z}_{N-1}) = \sum_{k=1}^{S} z_{N-1,k} \langle \mathbf{W}_N \rangle_k$$

where $\langle \cdot \rangle_k$ now stands for the expected value for the system with $\mathbf{A}^{(k)}$ since

$$p(\mathbf{x}_N \mid \mathbf{x}_{N-1}, \mathbf{z}_{N-1}) = \sum z_{N-1,k} p(\xi_{N-1} = \mathbf{x}_N - \mathbf{A}^{(k)}\mathbf{x}_{N-1} - \mathbf{B}_{N-1}\mathbf{u}_{N-1}).$$

One can write

$$\langle W_N \rangle_k = (\mathbf{u}_{N-1} - \mathbf{u}^*_{N-1,k})'\mathbf{\Phi}_{N,k}(\mathbf{u}_{N-1} - \mathbf{u}^*_{N-1,k}) + \phi_{N,k} \tag{70}$$

where

$$\mathbf{\Phi}_{N,k} = \mathbf{\Phi}_N \triangleq (\mathbf{B}'_{N-1}\mathbf{V}_N\mathbf{B}_{N-1} + \mathbf{P}_{N-1}) \tag{71a}$$

$$\begin{aligned}\mathbf{u}^*_{N-1,k} &= -(\mathbf{B}'_{N-1}\mathbf{V}_N\mathbf{B}_{N-1} + \mathbf{P}_{N-1})^{-1}\mathbf{B}'_{N-1}\mathbf{V}_N\mathbf{A}^{(k)}\mathbf{x}_{N-1}\\ &\triangleq \mathbf{a}_{N-1,k}\mathbf{x}_{N-1}\end{aligned} \tag{71b}$$

$$\begin{aligned}\phi_{N,k} \triangleq\ & \mathbf{x}'_{N-1}(\mathbf{A}^{(k)})'(\mathbf{V}_N^{-1} + \mathbf{B}'_{N-1}\mathbf{P}_{N-1}\mathbf{B}_{N-1})^{-1}\mathbf{A}^{(k)}\mathbf{x}_{N-1}\\ &+ E[\xi'_{N-1}(\mathbf{V}_N^{-1} + \mathbf{B}'\mathbf{P}_{N-1}\mathbf{B})^{-1}\xi_{N-1}]\end{aligned} \tag{71c}$$

when the indicates inverses exist. From (51),

$$\bar{\mathbf{\Phi}}_N = \mathbf{\Phi}_N.$$

From (52),

$$\bar{z}_{N-1,k} = z_{N-1,k}.$$

From (53),

$$\mathbf{u}^*_{N-1} = \sum_{k=1}^{S} z_{N-1,k}\mathbf{u}^*_{N-1,k}.$$

From (54),

$$\begin{aligned}\delta\gamma_N &= \sum_{k=1}^{S} (\mathbf{a}_{N-1,k}\mathbf{x}_{N-1})'\mathbf{\Phi}_N z_{N-1,k}(\mathbf{a}_{N-1,k}\mathbf{x}_{N-1})\\ &\quad - \left(\sum_{k=1}^{S} z_{N-1,k}\mathbf{a}_{N-1,k}\mathbf{x}_{N-1}\right)'\mathbf{\Phi}_N\left(\sum_{k=1}^{S} z_{N-1,k}\mathbf{a}_{N-1,k}\mathbf{x}_{N-1}\right)\\ &= \mathbf{x}'_{N-1}\left[\sum_{k=1}^{S} z_{N-1,k}(\mathbf{a}'_{N-1,k}\mathbf{\Phi}_N\mathbf{a}_{N-1,k})\right.\\ &\quad \left. - \left(\sum_{k=1}^{S} z_{N-1,k}\mathbf{a}_{N-1,k}\right)'\mathbf{\Phi}_N\left(\sum_{k=1}^{S} z_{N-1,k}\mathbf{a}_{N-1,k}\right)\right]\mathbf{x}_{N-1}.\end{aligned} \tag{72}$$

Appendices

The Appendices collect some useful facts in analyzing asymptotic behavior of sequences of random variables.

Orders of Magnitude in Probability

In analysis, the notions of order of magnitude, $O(\cdot)$ and $o(\cdot)$, are used extensively. That is, given two sequences of real numbers $\{a_n\}$ and $\{b_n\}$, we write $a_n = o(b_n)$ if $a_n/b_n \to 0$ as n goes to infinity, while if a_n/b_n is bounded uniformly in n, then we write $a_n = O(b_n)$.

This section collects corresponding notions, order in probability, $O_p(\cdot)$ and $o_p(\cdot)$. They are discussed in greater detail in Fuller (1976, Chapter 3) and Brockwell and Davis (1987, Chapter 6), for example. In short, when a sequence of random variables $\{X_n\}$ converges in probability to zero, we write it as $X_n = o_p(1)$. When the sequence $\{X_n\}$ is bounded in probability, we write it as $X_n = O_p(1)$. We write $X_n = o_p(a_n)$ when $X_n/a_n = o_p(1)$ and $X_n = O_p(a_n)$ when $X_n/a_n = O_p(1)$. Some useful relations follow.

Given two sequences of random variables $\{X_n\}$ and $\{Y_n\}$ defined on the same probability space and given that $a_n > 0$, $b_n > 0$ are two sequences of real numbers, then $X_n = O_p(a_n)$ and $Y_n = O_p(b_n)$ implies $X_n Y_n = O_p(a_n b_n)$, $X_n + Y_n = O_p\{\max(a_n, b_n)\}$, and $|X_n|^r = O_p(a_n^r)$ for $r > 0$. These relations valid when $O_p(\cdot)$ everywhere is replaced by $o_p(\cdot)$. If $g(\cdot)$ is a continuous function and if $X_n \xrightarrow{P} X$, then $g(x_n) \xrightarrow{P} g(X)$, and hence we can write $g(X_n) - g(X) = o_p(1)$, when $X_n - X = o_p(1)$.

Taylor Series in Probability

The Taylor series expansions are quite useful in approximate analysis of functions of real variables. Given a continuous function $g(\cdot)$ and a sequence of random variables $\{X_n\}$ that is $X_n = a + o_p(1)$, we know that $g(X_n) = g(a) + O_p(1)$, and if $X_n = a + O_p(1)$, then $g(X_n) = g(a) + O_p(1)$. We can go further if $g(\cdot)$ is assumed to be differentiable by using the probabilistic analogue of the Taylor series expansion. If g has s derivatives at a, and if $X_n = a + O_p(r_n)$ and $0 < r_n \downarrow 0$, then

$$g(X_n) = \sum_{j=o}^{s-1} \frac{g^{(j)}(a)}{j^1} (X_n - a)^j + O_p(r_n^s).$$

Similarly, if $X_n = a + o(r_n)$, $0 < r_n \downarrow 0$, then the same expansion is valid with $o_p(\cdot)$ rather than $O_p(\cdot)$. We use these expansions in approximating optimization problems with nonquadratic costs and nonlinear dynamics by those with quadratic costs and linear dynamic constraints.

References

Ahn, S. on G. C. Reinsel, "Estimation for Partially Nonstationary Multivariate Autogressive Models," Presented at the 1987 NSF-NBER Workshop on Time Series, Raleigh, 1987.

Aoki, M., "Cointegration, Error Correction and Aggregation in Dynamic Models: A Comment," *Oxford Bulletin of Economics and Statistics* **50**, p. 89–95, 1988a.

Aoki, M., "On Alternative State Space Representation of Time Series Models," *J. Econ. Dyn. Control* **12**, Special Issue on Economic Time Series with Random Walk and Other Nonstationary Components, 1988b.

Aoki, M., "State Space Models for Vector-Valued Time Series with Real and Complex Random Walk Components," Invited Talk, Business and Economic Section, American Stat. Assoc. Meeting, New Orleans, 1988c.

Aoki, M., "Studies of Economic Interdependence by State Space Modeling of Time Series: US-Japan Example," *Annales d'Economie et de Statistique*, No. 6/7, p. 225–252, 1987a.

Aoki, M., "Time Series Evidence of Real Business Cycle Interdependence of the U.S.A., West Germany and Japan," Presented at the 1987 Far Eastern Meeting of the Econometric Soc., Tokyo, Oct. 1987b.

Aoki, M., *State Space Modeling of Time Series*, Springer-Verlag, New York, 1987c.

Aoki, M., *Notes on Economic Time Series: System Theoretic Perspectives*, No. 220, Lecture Notes in Economics and Mathematical Systems, Springer-Verlag, Heidelberg, 1983.

Aoki, M., *Dynamic Analysis of Open Economies*, Academic Press, 1981.

Aoki, M., "Dynamics and Control of a System Composed of a Large Number of Similar Subsystems," Chapter 11 in *Dynamic Optimization and Mathematical Economics*, P-T. Liu (ed.), 'Plenum Press, New York, 1980a.

Aoki, M., "Note on Comparative Dynamic Analysis," Econometrica **48**, p. 1319–1325, 1980b.

Aoki, M., "On Fluctuations in Microscopic States of a Large System," *Direction in Decentralized Control Many-Person Optimization and Large-Scale Systems*, S. Mitter, Y. C. Ho (eds.), Plenum Press, New York, 1976.

Aoki, M., "On Some Price Adjustment Schemes," *Ann. Econ. Soc. Measurements* **3**, p. 95–115, 1974.

Aoki, M., "Approximation Scheme for Evaluating Some Terminal Capital Stock," *J. Econ. Theory* **6**, p. 317–319, 1973.

Aoki, M., *Introduction to Optimization Techniques*, Macmillan Co., New York, 1971.

Aoki, M., "On Performance Losses in some Adaptive Control Systems," *I. Trans. ASME Ser. D.F. Basic Eng.* **87**, No. 1, p. 90–94, March 1965.

Aoki, M., "On the Approximation of Trajectories and its Application to Control Systems Optimization Problems," *J. Math. Anal. Appl.* **9**, p. 23–41, 1964.

Aoki, M., "Successive Approximations in Solving Some Control System Optimization Problems, II," *J. Math. Anal. Appl.* **5**, No. 3, p. 418–434, December 1962.

Aoki, M., "Dynamic Programming Approach to the Final Value Control System with a Random Variable Having an Unknown Distribution Function," *IRE Trans. Auto. Control* **5**, No. 4, p. 270–282, 1960.

Aoki, M., and M. Canzoneri, "Reduced Forms of Rational Expectations Models," *Quart. J. Econ.* **94**, p. 59–71, 1979.

Aoki, M., and J. R. Huddle, "Estimation of the State Vector of a Linear Stochastic System with a Constrained Estimator," *IEEE Trans. Aut. Control AC-12*, p. 432–433, 1967.

Aoki, M., and A. Leijonhufvud, "The Stock-flow Analysis of Investment," in M. Kohn and S. C. Tsiang (eds.) *Finance Constraints, Expectations, and Macroeconomics*, Clarendon Press, Oxford, 1988.

Aoki, M., M. & R. M. Staley, "On Input Signal Synthesis in Parameter Identification," *Automatica* **6**, p. 431–440, 1970.

Åström, K. J., *Introduction to Stochastic Control Theory*, Academic Press, New York, 1970.

Åström, K. J., and T. Bohlin, "Numerical Identification of Linear Dynamic Systems from Normal Operating Records," IFAC Symposium Self-Adaptive Systems, Teddington, England, 1965.

Athans, M., R. Ku and S. B. Gershwin, "The Uncertainty Threshold Principle: Some Fundamental Limitations of Optimal Decision Making under Dynamic Uncertainty," *IEEE Trans. Aut. Control AC-22*, p. 491–495, 1977.

Bellman, R., *Adaptive Control Processes: A Guided Tour*, Princeton Univ. Press, 1961.

Bellman, R., H. H. Kagiwada, R. E. Kalaba and R. Sridhar, "Invariant Embedding and Nonlinear Fitting Theory," RAND, R17-4374-PR, 1964.

Benveniste, L. M., and J. A. Scheinkman, "On the Differentiability of the Value Function in Dynamic Models of Economics," *Econometrica* **41**, 727–732, 1979.

Bertram, J. E., and P. E. Sarachik, "On the Stability of Circuits and Randomly Varying Parameters," *IRE Tran. Inf. Theory*, p. 260–270, 1959.

Bertekas, D. P., *Dynamic Programming and Stochastic Control*, Academic Press Inc., New York, 1976.

Bertsekas, D. P., and S. E. Shreve, *Stochastic Optimal Controls: The Discrete Time Case*, Academic Press, New York, 1978.

Beveridge, S., and C. R. Nelson, "A New Approach to Decomposition of Economic Time Series into Permanent and Transitory Components with Particular Attention to Measurement of the Business Cycle," J.M.E. **7**, p. 151–174, 1981.

Bewley, R. A., "The Direct Estimation of the Equilibrium Response in a Linear Dynamic Model," *Economics Letters* **3**, p. 357–361, 1979.

Billingsley, P., *Statistical Inference for Markov Processes*, University of Chicago Press, 1961.

Blackwell, D., "Discounted Dynamic Programming," *Ann. Math. Stat.* **36**, p. 226–235, 1965.

Blume, L. E., M. Bray and D. Easley, "Introduction to the Stability of Rational Expectations Equilibrium," *JET* **26** (2), p. 313–317, 1982.

Bray, M. M., "Learning, Estimation and Stability of Rational Expectations Equilibria," *J. Econ. Theory* **26**, p. 318–339, 1982.

Bray, M. M., and D. M. Kreps, "Rational Learning and Rational Expectation," Chapter 9 of Feiwel (ed.), 1987.

Bray, M. M., and N. E. Savin, "Rational Expectations Equilibria, Learning, and Model Specifications," *Econometrica* **54**, p. 1129–1160, 1986.

Breiman, L., *Probability*, Addison-Wesley Pub. Co., Reading, MA, 1968.

Brock, W. A., and L. J. Mirman, "Optimal Economic Growth and Uncertainty: The Discounted Case," *J. Econ. Theory* **4**, p. 479–513, 1972.

Brockwell, P. J., and P. A. Davis, *Time Series: Theory and Methods*, Springer-Verlag, New York, 1987.

Broze, L., C. Gourieroux and A. Szafarz, "Solutions of Dynamic Linear Rational Expectations Models," *Econometric Theory* **1**, p. 341–368, 1985.

Broze, L., and A. Szafarz, "On Linear Models with Rational Expectations Which Admit a Unique Solution," *Europ. Econ. Rev.* **24**, p. 103–111, 1984.

Bucy, R. S., "Stability and Positive Supermartingales," *Journal Diff. Eq. 1*, p. 1356, 1960.

Bucy, R. S., "Nonlinear Filtering Theory," *IEEE Trans. Auto. Control AC-10*, No. 2, p. 198, 1965.

Buiter, W. H., "The Superiority of Contingent Rules over Fixed Rules in Models with Rational Expectations," *Econ. J.* **91**, p. 647–670, 1981.

Buiter, W. H., and M. Miller, "Monetary Policy and International Competitiveness: The Problems of Adjustment," *Oxford Econ. Papers* **33**, Series 2, p. 143–175, 1981.

Burmeister, E., "On Some Conceptual Issues in Rational Expectations Modeling," *J. Mon. Cred. Bank* **12**, p. 800–816, 1980.

Burt, E. G. C., and L. Rigby, "Constrained Minimum Variance Control for Minimum- and Non-Minimum Phase Processes," Imperial College Tech. Report EE. CON. 82.44., London, U.K., 1982.

Cagan, P., "The Monetary Dynamics of Hyperinflation," in M. Friedman (ed.) *Studies in the Quantity Theory of Money*, Univ. of Chicago Press, 1956.

Caines, P., *Linear Stochastic Systems*, Wiley, New York, 1988.

Campbell, J. Y., and R. J. Shiller, "Interpreting Cointegrated Models," in *J. Econ. Dyn. Control Special Issue* **12**, 1988.

Canon, M. D., C. D. Culum Jr. and E. Polak, *Theory of Optimal Control and Mathematical Programming*, McGraw-Hill, 1970.

Chow, Y. S., "Local Convergence of Martingales and the Law of Large Numbers," *Ann. Math. Stat.* **36**, p. 552–558, 1965.

Chow, Y. S., and T. L. Lai, "Limiting Behavior of Weighted Sums of Independent Random Variables," *Ann. Prob.* **1**, p. 810–824, 1973.

Chow, Y. S., and H. Teicher, *Probability Theory: Independence, Interchangeability, Martingales*, Springer-Verlag, Berlin and New York, 1978.

Christiano, L. J., "Dynamic Properties of Two Approximate Solutions to a Particular Growth Model," unpublished, Dec. 1986.

Christiano, L. J., "On The Accuracy of Linear Quadratic Approximations: An Example," Working Paper 303, Federal Reserve Bank of Minneapolis, March 1986.

Chung, K. L., *A Course in Probability Theory*, Harcourt, Brace & World, Inc., New York, 1968.

Clark, P. K., "The Cyclical Component of U.S. Economic Activity," *Q. J. Econ.* **102**, p. 797–814, 1987.

Cochrane, J. H., and A. M. Sbordone, "Multivariate Estimates of the Permanent Components of GNP and Stock Prices," in *J. Econ. Dyn. Control Special Issue* **12**, 1988.

Constantinides, G. M., "Habit Formation: A Resolution of the Equity Premium Puzzle," mimeo, Univ. Chicago, Oct. 1988.

Corrado, C., and M. Greene, "Reducing Uncertainty in Short-term Projections: Linkage of Monthly and Quarterly Models," No. 207, Special Studies Paper, FRB, 1984.

Cramér, H., *Mathematical Methods of Statistics*, Princeton Univ. Press, Princeton, New Jersey, 8th Printing, 1958.

Danthine, J. P., and J. B. Donaldson, "Certainty Planning in an Uncertain World: A Reconsideration," *Rev. Econ. Stud.* **68**, p. 507–510, 1981.

Davidson, J., "Cointegration in Linear Dynamic Systems," Mimeo, London School of Economics, June, 1987.

DeCanio, S., "Rational Expectations and Learning from Experience," *Quart. J. Econ.* **93**, p. 47–58, 1979.

Dickey, D. A., and W. A. Fuller, "Distribution of Estimators for Autoregressive Time Series with a Unit Root," *JASA LXXIV*, p. 427–431, 1979.

Doob, J. L., *Stochastic Processes*, Wiley, New York, 1953.

Dreyfus, S. E., "Some Types of Optimal Control of Stochastic Systems," *Journal Soc. Ind. Applied Mathematics Ser. A. Control 2 No. 1*, p. 120–134, 1964.

Dynkin, E. B., "Sufficient Statistics and Extremal Points," *Ann. Prob.*, **6**, 705–730, 1978.

Evans, G. W., "Output and Unemployment Dynamics in the United States: 1950–1985," mimeo, Dept. of Economics, Stanford University, Oct. 1986, Revised May 1987.

Evans, G. W., "Stability and Learning from Experience," *Quart. J. Econ.* **94**, p. 47–58, 1983.

Evans, G. W., "Expectational Stability and the Multiple Equilibria Problem in Linear Rational Expectations Models," *Quart. J. Econ.* **99**, p. 1217–1233, 1985.

Evans, G. W., and S. Honkapohja, "A Complete Characterization of ARMA Solutions to Linear Rational Expectations Models," *Rev. Econ. Stud.* **53**, p. 227–239, 1986.

Feld'baum, A. A., *Optimal Control Systems*, Academic Press, New York, 1966.

Ferguson, T., "Mathematical Statistics," Academic Press, New York, 1967.

Flåm, S. D., "Finite-state Approximations for Countable-state Infinite Horizon Discounted Markov Decision Processes," *Modeling, Identification and Control* **8**, p. 117–123, 1987.

Flood, R. P., and P. M. Garber, "Market Fundamentals versus Price-level Bubbles: The First Test," *J. Pol. Econ.* **88**, p. 745–770, 1980.

Foldes, L., "Martingale Conditions for Optimal Saving-Discrete Time," *J. Math. Econ.* **5**, p. 83–96, 1978.

Fourgeaud, C., C. Gourieroux, and J. Pradel, "Learning Procedures and Convergence to Rationality," *Econometrica* **54**, p. 845–868, 1986.

Friedman, B. M., *Economic Stabilization Policy: Methods in Optimization*, North Holland, Amsterdam, 1975.

Fukao, T., "Some Fundamental Properties of Adaptive Control Processes," *J. Bull. Electrotech. Lab.* **28**, No. 1, p. 1–19 (in Japanese) Tokyo, January 1964.

Fuller, W. A., *Introduction to Statistical Time Series*, Wiley, New York, 1976.

Futia, C. A., "Rational Expectations in Stationary Linear Models," *Econometrica* **49**, p. 171–192, 1981.

Futia, C. A., "Stochastic Business Cycles," Bell Tel. Lab. Tech. Report, New Jersey, 1979.

Gallant, A. R., *Nonlinear Statistical Models*, John Wiley, New York, 1987.

Gantmacher, F. R., *Applications of the Theory of Matrices*, Interscience Publishers Inc., New York, 1959.

Gihman, I. I., and A. V. Skorohod, *Controlled Stochastic Processes*, Springer Verlag, 1979.

Gilbert, F. G., "Controllability and Observability in Multivariable Control Systems," *SIAM J. Control* **1**, p. 128–151, 1963.

Golub, G. H., and C. F. Van Loan, *Matrix Computations*, John Hopkins Univ. Press, Baltimore, 1983.

Goodrich, R. L., and P. E. Caines, "Linear Systems Identification from Nonstationary Cross-Sectional Data, *IEEE T-AC-24*, p. 403–411, 1979.

Goodwin, G. C., and K. S. Sin, *Adaptive Filtering Prediction and Control*, Prentice Hall, Englewood Cliffs, 1984.

Gourieroux, C., J. Laffont and A. Monfort, "Rational Expectations in Dynamic Linear Models: Analysis of the Solutions," *Econometrica* **50**, p. 409–425, 1982.

Grandmont, J. M., and G. Laroque, "Stability of Cycles and Expectations," *J. Econ. Theory* **40**, p. 138–151, 1986.

Hahn, W., "Theory and Applications of Lyapunov's Direct Method," Prentice Hall, Englewood Cliffs, NJ, 1983.

Hall, P., and C. C. Heyde, *Martingale Limit Theory and Its Applications*, Academic Press, New York, 1980.

Hamilton, J. D., "A New Approach to the Economic Analysis of Nonstationary Time Series and the Business Cycle," Mimeo, Dept. Econ. Univ. Virginia, July, 1987.

Hannan, E. J., "Rates of Convergence for Time Series Regression," *Advances in Applied Prob.* **10**, p. 740–743, 1978.

Hannan, E. J., *Multiple Time Series*, Wiley, New York, 1970.

Hannan, E. J., and D. S. Poskitt, "Unit Canonocal Correlation Between Future and Past," *Ann. Stat.*, forthcoming 1988.

Hansen, G. D., "Indivisible Labor and the Business Cycle," *J. Mon. Econ.* **16**, p. 309–327, 1985.

Hansen, G. D., and T. J. Sargent, "Straight Time and Overtime in Equilibrium," *Journal of Monetary Economics*, forthcoming 1988.

Hansen, L. P., and T. J. Sargent, "Formulating and Estimating Linear Rational Expectations Models," *J. Econ. Dyn. Control* **2**, p. 7–46, 1980.

Hardy, G. H., *Divergent Series*, Oxford Univ. Press, Oxford, UK, 1949.

Harvey, A. C., "Trends and Cycles in Macroeconomic Time Series," *J. Business Econ. Stat.* **3**, p. 216–226, 1985.

Hautus, M. L. J., "Controllability and Observability Conditions of Linear Autonomous Systems," *Ned. Akad. Wetenschappen, Proc. Ser. A* **72**, p. 443–448, 1969.

Hendry, D. F., and G. E. Mizon, "Serial Correlation as a Convenient Simplification, not a Nuisance," *Economic Journal* **88**, p. 549–563, 1978.

Hinderer, K., *Foundation of Nonstationary Dynamic Programming with Discrete Time Parameter*, Springer-Verlag, Berlin and New York, 1970.

Hogg, R. V., and A. T. Craig, "Introduction to Mathematical Statistics," Macmillan, New York, 1959.

Horowitz, E., "Some Suboptimal Control Policies in Optimal Stochastic Control Systems," M.S. Thesis, Dept. of Engineering, UCLA, Los Angeles, 1966.

Hosoya, Y., "On the Granger Condition for Non-Causality," *Econometrica* **45**, p. 1735–1736, 1977.

Ibragimov, I. A., and Yu V. Linnik, *Independent and Stationary Sequences of Random Variables*, Wolters-Noordhoff Pub., Groningen, The Netherlands, 1971.

Ito, Y., "Invariant Measures for Markov Processes," *Amer. Math. Soc. Trans.* **110**, p. 152–184, 1964.

Journal of Economic Dynamics and Control, Special Issue on Economic Time Series with Random Walk and Other Nonstationary Components **12**, No. 2, ed. by M. Aoki, 1988.

Journal of Economic Dynamics and Control, Special Issue on Rational Expectations Analysis **2**, No. 1, ed. by J. B. Taylor and M. Canzoneri, 1980.

Kailath, T., *Linear Systems*, Prentice Hall, Englewood Cliffs, NJ, 1980.

Kalchbrenner, J. H., and P. A. Tinsley, "On Filtering Auxiliary Information in Short-Run Monetary Policy," in K. Brunner and A. Meltzer (eds.), *Optimal Policies, Control Theory and Technology Exports*, Carnegie-Rochester Conference Series on Public Policy Volume 6, p. 39–91, 1977.

Kats, I. I., and N. N. Krazovskii, "On the Stability of Systems with Random Parameters," *Applied Mathematical Mechanics* **24**, p. 809–823, 1960.

Kennedy, D. P., "On Characterizing Optimal Policies in Infinite-Horizon Stochastic Control," *Siam Journal of Control and Optimization* **18**, p. 576–584, 1980.

Kiefer, J., and J. Wolfowitz, "Stochastic Estimation of the Maximum of a Regression Function," *Ann. Math. Stat.* **23**, p. 462–466, 1952.

King, R., C. Plosser, J. Stock and M. Watson, "Stochastic Trends and Economic Fluctuations," Working Paper, January 1987.

Klein Haneveld, W. K., *Duality in Stochastic Linear and Dynamic Programming*, Springer Verlag, Heidelberg, 1986.

Kloek, T., "Dynamic Adjustment when the Target is Nonstationary," *Int'l. Econ. Rev.* **25**, p. 315–326, 1984.

Kollintzas, T. E., and H. L. A. M. Geerts, "Three Notes on the Formulation of Linear Rational Expectations Models," in O. D. Anderson (ed.), *Time Series Analysis: Theory and Practice* **8**, Elsevier, 1984.

Kollintzas, T., "The Symmetric Linear Rational Expectations Models," *Econometrica* **53**, 963–976, 1985.

Kopp, P. E., *Martingales and Stochastic Integrals*, Cambridge U.P., London, 1984.

Kreps, D. M., "Sequential Decision Problems with Expected Utility Criteria, III: Upper and Lower Transience," *Siam Journal of Control and Optimization* **16**, p. 420–428, 1978.

Kucera, V., "The Discrete Riccati Equation of Optimal Control," *Kybernetika* **8**, p. 430–447, 1972.

Kushner, H. J., "Approximation and Weak Convergence Methods for Random Processes, with Applications to Stochastic Systems Theory," MIT Press, Cambridge, 1984.

Kushner, H. J., "On the Stability of Stochastic Dynamical Systems," *Proc. National Academic Sciences* **53**, p. 8–12, 1965a.

Kushner, H. J., "New Theorems and Examples in the Lyapunov Theory of Stochastic Stability," Proc. Joint Auto. Control Conf., p. 613–619, Rensselaer Polytech. Inst., Troy, New York, 1965b.

Kydland, F. E., and E. C. Prescott, "Rules Rather than Discretions: The Inconsistency of Optimal Plan," *J. Pol. Econ.* **85**, p. 473–491, 1977.

LaSalle, J., and S. Lefschetz, *Stability by Lyapunov's Direct Method*, Academic Press, New York, 1961.

Lai, T. L., "Asymptotically Efficient Adaptive Control in Stochastic Regression Models," *Adv. Appl. Math.* **7**, p. 23–45, 1986.

Lai, T. L., and H. Robbins, "Limit Theorems for Weighted Sums and Stochastic Approximations Processes," *Proc. Nat. Acad. Sci. USA* **75**, p. 1068–1070, 1978.

Lai, T. L., and H. Robbins, "Adaptive Design and Stochastic Approximation," *Ann. Prob.* **2**, 1196–1221, 1979.

Lai, T. L., H. Robbins and C. Z. Wei, "Strong Consistency of Least Squares Estimates in Multiple Regression II," *J. Multivariate Anal.* **9**, p. 343–361, 1979.

Lai, T. L., and C. Z. Wei, "On the Concept of Excitations in Least Squares Identification and Adaptive Control," *Stochastics* **16**, p. 227–254, 1986.

Lai, T. L., and C. Z. Wei, "Asymptotic Properties of Multivariate Weighted Sums with Applications to Stochastic Regressions in Linear Dynamic Systems," in D. R. Krishnaiah (ed.), Multivariate Analysis—VI, North Holland, Amsterdam, 1985.

Lai, T. L., and C. Z. Wei, "Extended Least Squares and Their Applications to Adaptive Control and Prediction in Linear Systems," *IEEE Trans. Aut. Control* **AC-31**, p. 898–906, Oct. 1986.

Lai, T. L., and C. Z. Wei, "Asymptotic Properties of Projections with Applications to Stochastic Regression Problems," *J. Multivariate Anal.* **12**, p. 346–370, 1982.

Lanning, L. H. Jr., and R. H. Battin, *Random Processes in Automatic Control*, McGraw-Hill, New York, 1956.

Levhari, D. and T. N. Srinivasan, "Optimal Savings Under Uncertainty," *Rev. Econ. Stud.* **36**, p. 153–163, 1969.

Lindquist, A., G. Picci and G. Ruckebusch, "On Minimal Splitting Subspaces and Markovian Representation," *Math. Systems Theory* **12**, p. 271–279, 1979.

Ljung, L. T., *System Identification: Theory for the User*, Prentice Hall, Englewood Cliffs, New Jersey, 1987.

Ljung, L. T., "Analysis of Recursive Stochastic Algorithms," *IEEE Trans. Aut. Control* **AC-22**, p. 551–575, 1977.

Ljung, L. T., "On Positive Real Transfer Functions and the Convergence of Some Recursive Schemes," *IEEE Trans. Aut. Control* **AC-27**, p. 539–551, 1977.

Ljung, L. T., Söderstrom & I. Gustavsson, "Counterexamples to General Convergence of a Commonly Used Recursive Identification Method," *IEEE Trans. Auto. Control* **AC-20**, p. 643–652, Oct. 1975.

Long, J. G. Jr., and C. I. Plosser, "Real Business Cycle," *J. Pol. Econ.* **91**, p. 39–69, 1983.

Luenberger, D. G., "Observations for Multivariable Systems," *IEEE Trans. Aut. Control AC-11*, p. 190–197, 1966.

Lukacs, E., *Stochastic Convergence*, Academic Press, 2nd Ed., New York, 1975.

Marcet, A., and T. J. Sargent, "Convergence of Least Squares Learning Mechanisms in Self-Referential Linear Stochastic Models," Mimeographed, Oct. 1987.

Meditch, J. S., *Stochastic Optimal Linear Estimation and Control*, McGraw-Hill, 1969.

Malliaris, A. G., and W. A. Brock, *Stochastic Methods in Economics and Finance*, North-Holland, Amsterdam, New York, Oxford, 1982.

McCallum, B. T., "On Non-Uniqueness in Rational Expectations Models," *J. Mon. Econ.* **11**, p. 139–168, 1983.

Meyn, S. P., and P. E. Caines, "Asymptotic Behavior of Stochastic Systems Possessing Markovian Realizations," Working Paper, Department of Systems Engineering, Australia National University, Canberra, 1988.

Meyn, S. P., and P. E. Caines, "A New Approach to Stochastic Adaptive Control," *IEEE Trans. Aut. Control* **AC-32**, p. 220–226, 1987.

Moore, J. B., "Persistance of Excitation in Extended Least Squares," *IEEE Trans. Aut. Control* **AC-28**, p. 60–78, 1983.

Muth, J. F., "Rational Expectations and the Theory of Price Movements," *Econometrica* **29**, p. 315–335, 1961.

Nelson, C. R., and C. Plosser, "Trends and Random Walks in Macroeconomic Time Series: Some Evidence and Implications," *J.M.E.* **10**, p. 139–162, 1982.

Neveu, J., *Mathematical Foundations of the Calculus of Probability*, Holden-Day, San Francisco, 1965.

Neveu, J., *Discrete-Parameter Martingales*, North-Holland/American Elsevier, New York, 1975.

Newton, G. C. Jr., Gould, L. A., and J. F. Kaiser, *Analytical Design of Linear Feedback Control*, Wiley, New York, 1957.

Nickell, S., "Error Correction, Partial Adjustment and All That: An Expository Note," *Oxf. Bull. Econ. Stat.* **47**, p. 119–129, 1985.

Nishimura, T., "On the A-priori Information in Sequential Estimation Problem," *IEEE Aut. Control* **AC-12**, p. 197–204, 1967.

Nummelin, E., *General Irreducible Markov Chains and Nonnegative Operators*, Cambridge UP, Cambridge, 1984.

Osborn, D. R., "A Note on Error-Correction Mechanisms and Steady-State Error," *Econ. J.* **96**, p. 208–211, 1986.

Pagan, A., "Time Series Behavior and Dynamic Specification," *Oxford Bulletin Econ. Stat.* **47**, p. 199–212, 1985.

Palm, F. C., and T. E. Nijman, "Missing Observations in the Dynamic Regression Model," *Econometrica* **52**, p. 1415–1435, 1984.

Patterson, K. D., and J. Ryding, "Dynamic Time Series Models with Growth Effects constrained to Zero," *Econ. J.* **94**, p. 137–43.

Pauls, B. D., "Improving the Forecast Accuracy of Provisional Data: An Application of the Kalman Filter to Retail Sales Estimates," *FRB International Finance Discussion Paper*, No. 318, Washington, DC, 1987.

Parzen, E., *Stochastic Processes*, Holden-Day, Inc., San Francisco, 1962.

Pavon, M., "Stochastic Realization and Invariant Directions of the Matrix Riccati Equation," *SIAM J. Control and Optimization* **18**, p. 155–180, 1980.

Perron, P., "Trends and Random Walks in Macroeconomic Time Series: Further Evidence from a New Approach," Chaier 8650, Dept. of Economics, Univ. of Montreal, Oct. 1986.

Pesaran, M. H., *Limits to Rationality*, Oxford Univ. Press, Oxford, U.K., 1988.

Pesaran, M. H., "Identification of Rational Expectations Models," *J. Econometrics* **16**, p. 375–398, 1981.

Phillips, P. C. B., "Time Series Regression with a Unit Root," *Econometrica* **55**, p. 277–301.

Pliska, S. R., "Duality Theory for Some Stochastic Control Models," *Stochastic Differential Systems*, Kohlman, M., and K. N. Christopeit (eds.), Springer-Verlag, pp. 329–337, 1982.

Renyi, A., *Foundations of Probability*, Holden-Day, San Francisco, 1970.

Rappaport, D. and L. M. Silverman, "Structures and Stability of Discrete-time Optimal Systems," *IEEE Trans. Aut. Control AC* **16**, pp. 227–233, 1971.

Robbins, H., and S. Monro, "A Stochastic Approximation Method," *Ann. Math. Stat.* **22**, p. 400–407, 1951.

Rockafellar, R. T., and R. J-B. Wets, "Nonanticipativity and L^1-Martingales in Stochastic Optimization Problems," *Math. Program. Study* **6**, p. 170–187, 1976.

Rosenbrock, H. H., "The Formulation of Optimal Control, with an Application to Large Systems," *Automatica* **1**, p. 263–288, Dec. 1963.

Saaty, T. L., *Modern Nonlinear Equations*, McGraw Hill, 1967.

Sage, A., *Optimum Systems Control*, Prentice Hall, Englewood Cliffs, NJ, 1968.

Salmon, M., "Error Correction Mechanism," *Econ. J.* **92**, p. 615–29, 1982.

Salmon, M., "Error Correction Models, Cointegration and the Internal Mode Principle," in *J. Econ. Dyn. Control Special Issue* **12**, 1988.

Saracoglu, R. & Sargent, J., "Seasonality and Portfolio Balance Under Rational Expectations," *JME* **4**, p. 435–458, 1978.

Sargent, T. J., "Equilibrium Investment under Uncertainty, Measurement Errors, and the Investment Accelerator," Hoover Institution, Stanford University, March 1987.

Sargent, T. J., (ed.), *Energy and Foresight*, Resources for the Future, Washington, D.C., 1984.

Sargent, T. J., *Macroeconomic Theory*, Academic Press, New York, 1979.

Schweppe, F. C., "Evaluation of Likelihood Function for Gaussian Signals," *IEEE Trans. Inf. Theory* **IT-11**, p. 61–70, 1965.

Sims, C. A., "Money, Income and Causality," *AER* **62**, p. 540–552, 1972.

Solo, V., "Strong Consistency of Least Squares Estimators in Regression with Correlated Disturbances," *Ann. Stat.* **9**, p. 689–693, 1981.

Solo, V., "Topics in Advanced Time Series Analysis," in G. del Dino and R. Rebolledo (eds.), Lectures in Probability and Statistics; Lecture Notes in Mathematics, No. 1215, Springer Verlag, 1986.

Solo, V., "The Convergence of MAL," *IEEE Trans. Aut. Control* **AC-24**, p. 958–962, 1979.

Stout, W. G., *Almost Sure Convergence*, Academic Press, New York, 1974.

Striebel, C., *Optimal Control of Discrete Time Stochastic Systems*, Springer Verlag, Berlin and New York, 1975.

Sworder, D. D., "Minimax Control of Discrete Time Stochastic Systems," *SIAM Ser. A Con Control* **2**, p. 433–449, 1964.

Sworder, D. D., "A study of the Relationship Between Identification and Optimization in Adaptive Control Problems," *J. Franklin Inst.* **281**, p. 198–213, 1966.

Sworder, D. D., and M. Aoki, "On the Control System Equivalents of some Decision Theoretic Theorems," *J. Math. Anal. Appl.* **10**, p. 424–438, 1965.

Tauchen, G., "Finite State Markov-Chain Approximations to Univariate Vector Autoregressions," *Economics Letters* **20**, p. 177–181, 1986.

Taylor, J., "Conditions for Unique Solutions in Stochastic Macroeconomic Models with Rational Expectations," *Econometrica* **45**, p. 1377–1385, 1977.

Toda, M., and R. V. Patel, "Bounds on Estimation Errors of Discrete-Time Filters Under Modeling Uncertainty," *IEEE Trans. Aut. Control* **AC-25**, p. 1115–1121, Dec. 1980.

Townsend, R., "Optimal Contracts and Competitive Markets with Costly State Verification," *JET* **21**, 417–25.

Varga, R. S., *Matrix Iterative Analysis*, Prentice Hall, Englewood Cliffs, NJ, 1962.

Vaughn, D. R., "A Nonrecursive Algebraic Solution for the Discrete Riccati Equation," *IEEE Trans. Aut. Control* **AC-15**, p. 597–599, 1970.

Visco, I., "On Linear Models with Rational Expectations, an Addendum," *EER* **24**, p. 113–115, 1984.

Visco, I., "On the Derivation of Reduced Forms of Rational Expectations Models," *EER* **16**, p. 355–65, 1981.

Wallis, K. F., "Econometrica Implications of the Rational Expectations Hypotheses," *Econometrica* **48**, p. 45–72, 1980.

Watson, M. W., "Recursive Solution Methods for Dynamic Linear Rational Expectations Models," Northwestern University, Nov. 1987.

Whiteman, C. H., *Linear Rational Expectations Models: A User's Guide*, Univ. of Minnesota Press, Minneapolis, 1983.

Whittle, P., "Risk-Sensitive Linear/Quadratic/Gaussian Control," *Adv. Appl. Prob.* **13**, p. 764–777, 1981.

Yoshida, K., and S. Kakutani, "Operator-Theoretic Treatment of Markov's Process and Mean Ergodic Theorem," *Ann. Math.* **42**, p. 188–228, 1941.

Young, P. C., "Recursive Estimation and Time Series Analysis," Springer-Verlag, New York, 1984.

Zellner, A., *An Introduction to Bayesian Inference in Econometrics*, J. Wiley, New York, 1971.

Subject Index

D

E

M

N

O

W

Z

ECONOMIC THEORY, ECONOMETRICS, AND MATHEMATICAL ECONOMICS

Edited by Karl Shell, *Cornell University*

Recent titles

Robert, J. Barro, editor, *Money, Expectations, and Business Cycles: Essays in Macroeconomics*

Ryuzo Sato, *Theory of Technical Change and Economic Invariance: Application of Lie Groups*

Iosif A. Krass and Shawkat M. Hammoudeh, *The Theory of Positional Games: With Applications in Economics*

Giorgio Szego, editor, *New Quantitative Techniques for Economic Analysis*

John M. Letiche, editor, *International Economic Policies and Their Theoretical Foundation: A Source Book*

Murray C. Kemp, editor, *Production Sets*

Andreu Mas-Coleil, editor, *Noncooperative Approaches to the Theory of Perfect Competition*

Jean-Pascal Benassy, *The Economics of Market Disequilibrium*

Tatsuro Ichiishi, *Game Theory for Economic Analysis*

David P. Baron, *The Export-Import Bank: An Economic Analysis*

Real P. Lavergne, *The Political Economy of U.S. Tariffs: An Empirical Analysis*

Halbert White, *Asymptotic Theory for Econometricians*

Thomas G. Cowing and Daniel L. McFadden, *Macroeconomic Modeling and Policy Analysis: Studies in Residential Energy Demand*

Svend Hylleberg, *Seasonality in Regression*

Jean-Pascal Benassy, *Macroeconomics: An Introduction to the Non-Walrasian Approach*

C.W.J. Granger and Paul Newbold, *Forecasting Economic Time Series, Second Edition*

Marc Nerlove, Assaf Razin, and Efraim Sadka, *Household and Economy: Welfare Economics of Endogenous Fertility*

Jean-Michel Grandmont, editor, *Nonlinear Economic Dynamics*

Thomas Sargent, *Macroeconomic Theory, Second Edition*

Yves Balasko, *Foundations of the Theory of General Equilibrium*

Jean-Michel Grandmont, editor, *Temporary Equilibrium: Selected Readings*

J. Darrell Duffie, *Security Markets: Stochastic Systems*

Ross M. Starr, *General Equilibrium Models of Monetary Economics*

Masanao Aoki, *Optimization of Stochastic Systems: Topics in Discrete-Time Dynamics, Second Edition*